ABATEMENT OF ENVIRONMENTAL POLLUTANTS

ABATEMENT OF ENVIRONMENTAL POLLUTANTS

TRENDS AND STRATEGIES

Edited by

PARDEEP SINGH
AJAY KUMAR
ANWESHA BORTHAKUR

ELSEVIER

Elsevier
Radarweg 29, PO Box 211, 1000 AE Amsterdam, Netherlands
The Boulevard, Langford Lane, Kidlington, Oxford OX5 1GB, United Kingdom
50 Hampshire Street, 5th Floor, Cambridge, MA 02139, United States

Notices
Knowledge and best practice in this field are constantly changing. As new research and experience broaden our understanding, changes in research methods, professional practices, or medical treatment may become necessary.

Practitioners and researchers must always rely on their own experience and knowledge in evaluating and using any information, methods, compounds, or experiments described herein. In using such information or methods they should be mindful of their own safety and the safety of others, including parties for whom they have a professional responsibility.

To the fullest extent of the law, neither the Publisher nor the authors, contributors, or editors, assume any liability for any injury and/or damage to persons or property as a matter of products liability, negligence or otherwise, or from any use or operation of any methods, products, instructions, or ideas contained in the material herein.

Library of Congress Cataloging-in-Publication Data
A catalog record for this book is available from the Library of Congress

British Library Cataloguing-in-Publication Data
A catalogue record for this book is available from the British Library

ISBN: 978-0-12-818095-2

For information on all Elsevier publications visit our
website at https://www.elsevier.com/books-and-journals

Publisher: Candice Janco
Acquisition Editor: Marisa LaFleur
Editorial Project Manager: Michael Lutz
Production Project Manager: Vignesh Tamilselvvan
Cover Designer: Miles Hitchen

Typeset by TNQ Technologies

Contents

Contributors   ix

1. Bioremediation: a sustainable approach for management of environmental contaminants

Pardeep Singh, Vipin Kumar Singh, Rishikesh Singh, Anwesha Borthakur, Sughosh Madhav, Arif Ahamad, Ajay Kumar, Dan Bahadur Pal, Dhanesh Tiwary, and P.K. Mishra

1. Introduction   1
2. Application of bioremediation for environmental pollutants cleanup   3
3. Conclusion   17
References   17

2. Pollution status and biodegradation of organophosphate pesticides in the environment

Mohd Ashraf Dar, Garima Kaushik, and Juan Francisco Villareal Chiu

1. Introduction   25
2. Organophosphates and other pesticides   27
3. Effect of pesticides   29
4. Toxicological mechanism of organophosphates   33
5. Status of organophosphate pesticide pollution   34
6. Degradation of organophosphate pesticides   46
7. Conclusion   59
References   59

3. Recent trends in the detection and degradation of organic pollutants

Preetismita Borah, Manish Kumar, and Pooja Devi

1. Introduction   67
2. Persistent organic pollutants: health effects and environmental chemistry   68
3. Method of POPs analysis (soil and water)   70
4. Methods for POPs degradation   73
5. Conclusions   76
Acknowledgments   76
References   76

4. Phytoremediation of organic pollutants: current status and future directions

Sachchidanand Tripathi, Vipin Kumar Singh, Pratap Srivastava, Rishikesh Singh, Rajkumari Sanayaima Devi, Arun Kumar, and Rahul Bhadouria

1. Introduction   81
2. The process of phytoremediation   82
3. Physiological and biochemical aspects of phytoremediation   83
4. Strategies of phytoremediation of organic pollutants   85
5. Role of enzymes   87
6. Role of plant-associated microflora   88
7. Fate and transport of organic contaminants in phytoremediation   93
8. Genetically engineered organisms for phytoremediation   94
9. Research and development in phytoremediation   94

10. Advantages and limitations of
    phytoremediation 95
11. Emerging challenges to
    phytoremediation 96
12. Conclusion 97
Acknowledgments 98
References 98
Further reading 105

5. Bioremediation of dyes from textile and dye manufacturing industry effluent

Razia Khan, Vipul Patel, and Zeenat Khan

1. Introduction 107
2. Importance of characterization of dye-containing
    wastewater 108
3. Factors affecting biological removal of
    textile dyes 109
4. Microorganisms and mechanism involved in
    dye bioremediation process 110
5. Application of enzymes as biocatalyst in
    dye bioremediation 112
6. Advancements in bioreactor systems for
    dye remediation 114
7. Treatment of dye-containing industrial
    effluents using genetically modified
    microorganisms or enzymes 115
8. Current status of bioreactor application in
    CETPs of industrial areas for dye removal 115
9. Microbial fuel cell: a novel system for the
    remediation of colored wastewater 116
10. Potential of constructed wetlands
    for the treatment of dye-contaminated
    effluents 118
11. Conclusion and suggestions 120
References 121

6. Mycoremediation of polycyclic aromatic hydrocarbons

Shalini Gupta and Bhawana Pathak

1. Introduction 127
2. Mycoremediation: intact potential 130
3. Major enzymes 135
4. Biosurfactant production by fungi and its
    application in bioremediation 138
5. Factors affecting growth of fungi 139

6. Conclusion and future perspective 141
References 141
Further reading 149

7. Plant growth–promoting rhizobacteria and their functional role in salinity stress management

Akanksha Gupta, Sandeep Kumar Singh, Manoj Kumar Singh,
Vipin Kumar Singh, Arpan Modi, Prashant Kumar Singh, and
Ajay Kumar

1. Introduction 151
2. Plant growth–promoting rhizobacteria 153
3. Plant growth–promoting rhizobacteria in
    salinity stress 153
4. PGPR and ACC deaminase activity 156
5. Conclusion 156
References 157
Further reading 160

8. Plant growth–promoting bacteria and their role in environmental management

Divya Singh, Sandeep Kumar Singh, Vipin Kumar Singh,
Akanksha Gupta, Mohd Aamir, and Ajay Kumar

1. Introduction 161
2. Plant growth–promoting bacteria 162
3. Xenobiotic compounds and their
    classification 163
4. Effect of xenobiotics on the health of human
    beings 164
5. Effects of xenobiotics on the plant growth 165
6. Future prospective 170
Acknowledgments 171
References 171
Further reading 175

9. Fungi as potential candidates for bioremediation

Rajesh Kumar Singh, Ruchita Tripathi, Amit Ranjan, and
Akhileshwar Kumar Srivastava

1. Introduction 177
2. Fungal bioremediation 182
3. Fungi in bioremediation 184
4. Technology advancement 187
References 188

10. Cyanobacteria: potential and role for environmental remediation

Priyanka, Cash Kumar, Antra Chatterjee, Wang Wenjing, Deepanker Yadav, and Prashant Kumar Singh

1. Introduction 193
2. Conclusions and future perspectives 198
Acknowledgments 199
References 199
Further reading 202

11. An effective approach for the degradation of phenolic waste: phenols and cresols

Tripti Singh, A.K. Bhatiya, P.K. Mishra, and Neha Srivastava

1. Introduction 203
2. Treatment technologies for phenolic compound removal 205
3. Factors influencing bioremediation of phenolic waste 216
4. Limitations of biodegradation 219
5. Photocatalytic degradation 221
6. Factors affecting photocatalytic degradation of TiO_2 222
Acknowledgments 228
References 228

12. Environmental fate of organic pollutants and effect on human health

Manita Thakur and Deepak Pathania

1. Introduction 245
2. Types of persistent organic pollutants 247
3. Conclusion 257
References 257
Further reading 262

13. Rhizospheric remediation of organic pollutants from the soil; a green and sustainable technology for soil clean up

Akanksha Gupta, Amit Kumar Patel, Deepak Gupta, Gurudatta Singh, and Virendra Kumar Mishra

1. Introduction 263
2. Organic contaminants in soil and their sources 264
3. Fate of organic pollutants in soil 267
4. Rhizoremediation: a conventional approach 270

5. Factors affecting rhizoremediation 276
6. Rhizoremediation potential, challenges, and future perspectives 278
Acknowledgments 278
References 279
Further reading 286

14. The role of scanning probe microscopy in bacteria investigations and bioremediation

Igor V. Yaminsky and Assel I. Akhmetova

Summary 287
1. Introduction 288
2. Bacterial biofilms 288
3. Scanning probe microscopy is a necessary tool in bioremediation investigations 289
4. Bacterial electromechanical biosensor 291
5. Scanning ion-conductance microscopy 293
6. Nanolithography 296
7. Scanning probe microscopy measurements of bacteria—manual 298
8. Methods 300
9. Conclusion 308
Abbreviations 309
Acknowledgments 309
References 309

15. Research progress of biodegradable materials in reducing environmental pollution

Kangming Tian and Muhammad Bilal

1. Introduction 313
2. Biodegradable materials used for environmental protection 314
3. Conclusion 328
Appendix A: List of abbreviations 328
References 329
Further reading 330

16. Genetically engineered bacteria for the degradation of dye and other organic compounds

Arvind Kumar, Ajay Kumar, Rishikesh Singh, Raghwendra Singh, Shilpi Pandey, Archana Rai, Vipin Kumar Singh, and Bhadouria Rahul

1. Introduction 331

2. Constructing genetically engineered microorganisms 333
3. Detection of genetically engineered microbes 333
4. Need of genetically engineered microbes 334
5. Dye degradation by engineered microbes 334
6. Organic contaminants degradation by genetically engineered microorganisms 336
7. Agent orange degradation by genetically engineered microorganisms 337
8. Organophosphate and carbamate degradation by genetically engineered microorganisms 337
9. Polychlorinated biphenyls degradation by genetically engineered microorganisms 338
10. Degradation of polycyclic aromatic hydrocarbons 338

11. Degradation of herbicide 339
12. Genetically modified endophytic bacteria and phytoremediation 339
13. Approaches to minimize the risks of genetically engineered microbes 340
14. Challenges associated with the use of genetically engineered microorganism in bioremediation applications 340
15. Factors influencing genetically engineered microorganisms 344
16. Regulation of genetically engineered microorganisms 345
17. Future perspective 345
18. Conclusion 346
References 346
Further reading 350

Index 351

Contributors

Mohd Aamir Center of Advanced Study in Botany, Institute of Science, Banaras Hindu University, Varanasi, India

Arif Ahamad School of Environmental Sciences (SES), Jawaharlal Nehru University, New Delhi, India

Assel I. Akhmetova Advanced Technologies Center, Moscow, Russian Federation

Rahul Bhadouria Department of Botany, Institute of Science, Banaras Hindu University, Varanasi, India; Department of Botany, University of Delhi, Delhi, India

A.K. Bhatiya Department of Biotechnology, GLA University, Mathura, India

Muhammad Bilal College of Biotechnology, Tianjin University of Science and Technology, Tianjin, China

Preetismita Borah CSIR-Central Scientific Instruments Organisation, Chandigarh, India

Anwesha Borthakur Leuven International and European Studies (LINES), Katholieke Universiteit Leuven, Belgium

Antra Chatterjee Molecular Biology Section, Centre for Advanced Study in Botany, Department of Botany, Banaras Hindu University, Varanasi, India

Mohd Ashraf Dar Department of Environmental Science, School of Earth Sciences, Central University of Rajasthan, Ajmer, Rajasthan, India

Pooja Devi CSIR-Central Scientific Instruments Organisation, Chandigarh, India

Rajkumari Sanayaima Devi Deen Dayal Upadhyaya College (University of Delhi), New Delhi, India

Akanksha Gupta Institute of Environment & Sustainable Development, Banaras Hindu University, Varanasi, India

Shalini Gupta School of Environment and Sustainable Development, Central University of Gujarat, Gandhinagar, India

Deepak Gupta Institute of Environment & Sustainable Development, Banaras Hindu University, Varanasi, India

Garima Kaushik Department of Environmental Science, School of Earth Sciences, Central University of Rajasthan, Ajmer, Rajasthan, India

Razia Khan Department of Microbiology, Girish Raval College of Science, Gujarat University, Gandhinagar, India

Zeenat Khan Environmental Genomics and Proteomics Lab, BRD School of Biosciences, Satellite Campus, Sardar Patel University, Vallabh Vidyanagar, India

Arvind Kumar State Key Laboratory of Cotton Biology, Key Laboratory of Plant Stress Biology, School of Life Science, Henan University, Kaifeng, Henan, PR China

Ajay Kumar Agriculture Research Organization (ARO), Volcani Center, Rishon LeZion, Israel

Cash Kumar Cytogenetics Laboratory, Department of Zoology, Institute of Science, Banaras Hindu University, Varanasi, India

Manish Kumar CSIR-Central Scientific Instruments Organisation, Chandigarh, India

Arun Kumar Bihar Agricultural University, Sabour, Bhagalpur, India

Sughosh Madhav School of Environmental Sciences (SES), Jawaharlal Nehru University, New Delhi, India

P.K. Mishra Department of Chemical Engineering and Technology, Indian Institute of Technology (Banaras Hindu University), Varanasi, India

Virendra Kumar Mishra Institute of Environment & Sustainable Development, Banaras Hindu University, Varanasi, India

Arpan Modi Agriculture Research Organization, Ministry of Agriculture and Rural Development Volcani Centre, Rishon LeZion, Israel

Dan Bahadur Pal Department of Chemical Engineering, Birla Institute of Technology, Mesra, Ranchi, India

Shilpi Pandey Department of Botany, Institute of Science, Banaras Hindu University, Varanasi, India

Amit Kumar Patel Institute of Environment & Sustainable Development, Banaras Hindu University, Varanasi, India

Vipul Patel Environment Management Group, Center for Environment Education, Ahmedabad, India

Bhawana Pathak School of Environment and Sustainable Development, Central University of Gujarat, Gandhinagar, India

Deepak Pathania Department of Environmental Sciences, Central University of Jammu, District Samba, India

Priyanka Cytogenetics Laboratory, Department of Zoology, Institute of Science, Banaras Hindu University, Varanasi, India

Bhadouria Rahul Department of Botany, University of Delhi, New Delhi, India

Archana Rai Department of Molecular and Cellular Biology, Sam Higginbotom Institute of Agriculture, Technology and Sciences (SHIATS), Allahabad, India

Amit Ranjan Department of Kayachikitsa Institute of Medical Sciences, Banaras Hindu University, Varanasi, India

Divya Singh Center of Advanced Study in Botany, Institute of Science, Banaras Hindu University, Varanasi, India

Pardeep Singh Department of Environmental Science, PGDAV College, University of Delhi, New Delhi, India

Rajesh Kumar Singh Department of Dravyaguna, Institute of Medical Sciences, Banaras Hindu University, Varanasi, India

Tripti Singh Department of Biotechnology, GLA University, Mathura, India; Department of Chemical Engineering and Technology, Indian Institute of Technology (Banaras Hindu University), Varanasi, India

Sandeep Kumar Singh Center of Advanced Study in Botany, Institute of Science, Banaras Hindu University, Varanasi, India

Vipin Kumar Singh Center of Advanced Study in Botany, Institute of Science, Banaras Hindu University, Varanasi, India

Manoj Kumar Singh Department of Chemistry, Indian Institute of Technology Delhi, Hauzkhas, India

Rishikesh Singh Institute of Environment & Sustainable Development, Banaras Hindu University, Varanasi, India

Gurudatta Singh Institute of Environment & Sustainable Development, Banaras Hindu University, Varanasi, India

Raghwendra Singh Crop Production Division, ICAR-Indian Institute of Vegetable Research, Varanasi, India

Prashant Kumar Singh Agriculture Research Organization, Ministry of Agriculture and Rural Development Volcani Centre, Rishon LeZion, Israel

Pratap Srivastava Department of Botany, Institute of Science, Banaras Hindu University, Varanasi, India; Shyama Prasad Mukherjee Government Degree College, Phaphamau, Prayagraj, India

Akhileshwar Kumar Srivastava The National Institute for Biotechnology in the Negev, Ben-Gurion University of the Negev, Beer-Sheva, Israel

Neha Srivastava Department of Chemical Engineering and Technology, Indian Institute of Technology (Banaras Hindu University), Varanasi, India

Manita Thakur Department of Chemistry, Maharishi Markandeshwar University, Solan, India

Kangming Tian Department of Biological Chemical Engineering, College of Chemical Engineering and Materials Science, Tianjin University of Science and Technology, Tianjin, China

Dhanesh Tiwary Department of Chemistry, Indian Institute of Technology (IIT-BHU), Varanasi, India

Sachchidanand Tripathi Deen Dayal Upadhyaya College (University of Delhi), New Delhi, India

Ruchita Tripathi Department of Dravyaguna, Institute of Medical Sciences, Banaras Hindu University, Varanasi, India

Juan Francisco Villareal Chiu Universidad Autónoma de Nuevo León, Facultad de Ciencias Químicas, Laboratorio de Biotecnología. Av. Universidad S/N Ciudad Universitaria, San Nicolás de los Garza, Nuevo León, Mexico

Wang Wenjing State Key Laboratory of Cotton Biology, Henan Key Laboratory of Plant Stress Biology, School of Life Science, Henan University, Kaifeng, Henan, China

Deepanker Yadav Department of Vegetable and Fruit Science, Institute of Plant Science, Agriculture Research Organization (ARO), The Volcani Center, Rishon LeZion, Israel

Igor V. Yaminsky Lomonosov Moscow State University, Moscow, Russian Federation

Bioremediation: a sustainable approach for management of environmental contaminants

Pardeep Singh[1], Vipin Kumar Singh[2], Rishikesh Singh[3], Anwesha Borthakur[4], Sughosh Madhav[5], Arif Ahamad[5], Ajay Kumar[6], Dan Bahadur Pal[7], Dhanesh Tiwary[8], P.K. Mishra[9]

[1]Department of Environmental Science, PGDAV College, University of Delhi, New Delhi, India; [2]Center of Advanced Study in Botany, Institute of Science, Banaras Hindu University, Varanasi, India; [3]Institute of Environment & Sustainable Development, Banaras Hindu University, Varanasi, India; [4]Leuven International and European Studies (LINES), Katholieke Universiteit Leuven, Belgium; [5]School of Environmental Sciences (SES), Jawaharlal Nehru University, New Delhi, India; [6]Agriculture Research Organization (ARO), Volcani Center, Rishon LeZion, Israel; [7]Department of Chemical Engineering, Birla Institute of Technology, Mesra, Ranchi, India; [8]Department of Chemistry, Indian Institute of Technology (IIT-BHU), Varanasi, India; [9]Department of Chemical Engineering and Technology, Indian Institute of Technology (Banaras Hindu University), Varanasi, India

1. Introduction

Polluted soil resulting from industrial or agricultural processes poses serious health hazards to humans and animals and thus can have damaging consequences on the ecosystems by making land inappropriate for cultivation and other fiscal purposes. Various industries such as carpet, textile, and petrochemical production create intensive problems in the natural environment by disposing toxic wastes on one hand and generating a huge quantity of waste, oily sludge, and petroleum waste enriched soil on the other hand, which constitutes a major confront for hazardous waste management (Farhadian et al., 2008).

Abatement of Environmental Pollutants
https://doi.org/10.1016/B978-0-12-818095-2.00001-1

1

Apart from this, oil shipping is also one of the key causes of environmental contamination where the land and water gets polluted because of the oil spill, ship breakage, and seepage of oil pipelines.

The economically viable and environmentally feasible management of industrial sludge is a major concern worldwide. For disposal of industrial sludge, globally adopted technologies comprise of landfilling, high-temperature drying, sludge spreading on land surface, lime added stabilization, burning, and composting. Because of excessive expenditure on sludge management, the majority of textile industries in India generally released their wastewater effluents in the farming fields, open dumps, fallow land, and ineffectively controlled sanitary landfills and alongside the railway tracks. It further contaminates ground-water causing serious human health hazards. Meanwhile, low availability of landfill area, rigorous national wastes discarding policy, and local people awareness have caused landfilling and land spreading highly costly and impractical. The sludge management practices in most of the developing countries are not well developed. Currently, several factories and municipalities are working on environment-friendly and low-cost sludge treatment practices. Thus, it is imperative to mitigate toxic environmental contaminants for sustainable development (Kümmerer et al., 2019).

A number of physical and chemical methods are currently being employed at large scale for municipal wastewater management chiefly based on sewage treatment plants (STPs). In addition to building expenses, upholding troubles in treatment plants has raised the query of sustainability. Furthermore, surplus sewage sludge formed by these treatment plants has posed more severe confines on release during the previous few decades (Vigueros and Camperos, 2002). Several developing countries cannot meet the expense of construction of STPs, necessitating the development of some eco-friendly and economically feasible machinery for in situ wastewater management. Under such critical conditions, a few eco-friendly techniques can resolve the limitations linked with secure and economically efficient wastewater management machinery.

Bioremediation is an emerging and innovative technology because of its economic feasibility, enhanced competence, and natural environment friendliness. The technology uses various eco-friendly microbial processes to handle the ever-rising environmental pollution problem. In such approaches, microbes adjust themselves against noxious wastes and environmentally adapted microbial strains grow naturally, which subsequently convert a wide variety of toxic chemicals into nontoxic forms. The microbial degradation of xenobiotics is based on enzyme activities. It further includes rhizoremediation, phytoremediation (McCutcheon and Jørgensen, 2008; Chang et al., 2009), and vermicomposting depending on the biological activities involved. Phytoremediation is based on plant-assisted extenuation of pollutant concentrations at the contaminated sites, whereas rhizoremediation includes the elimination of specific pollutants from impure sites by the mutual interface of plant roots and appropriate microbial flora (Rajkumar et al., 2012). Because bioremediation appears to be a promising substitute to conventional cleanup machinery, extensive research is being carried out in this field. Vermicomposting and vermifiltration are natural waste management procedures relying on the utilization of worms to change organic wastes to form soil-enriching compounds (Vettori et al., 2012). Domestic wastewater and industrial sludge management can be accomplished through these processes in a sustainable way (Benitez et al., 2002). In addition, a substantial decrease in pathogens has been observed up to the end-products level. Therefore, it can be securely used for land application. Biological methods have been reviewed and acknowledged for remediation of environmental pollutant. In this chapter, we have tried to focus on the application of biological methods which were used for effective waste treatment. Several different strategies of bioremediation have also been discussed in later sections.

2. Application of bioremediation for environmental pollutants cleanup

In previous decades, rapid industrialization, urbanization, and indiscriminate resource utilization by ever-increasing human population have increased the contamination of atmosphere, land surfaces, and ground and surface waters. The wide-scale degradation of natural resources constitutes a major threat to public health around the globe. Majority of contaminants affecting soil and water system are heavy metals, pesticides, petroleum hydrocarbons, and large amount of toxic industrial effluents. These xenobiotics of anthropogenic origin are recalcitrant in nature.

In the current scenario, restoration of degraded land, water, and soil system is only possible by sustainable and eco-friendly processes. Among the various recent processes being used for the abatement of environmental pollution, bioremediation is recognized as an emerging methodology for the restoration of polluted environments. However, its ground level applicability is restricted because of different climatic factors. Various microbes degrade recalcitrant pollutants under aerobic or anaerobic conditions through complete mineralization or cometabolism by using pollutants as their carbon sources. Bacteria and fungi have been reported as favorable and potential candidates for both in situ and ex situ degradation of organic pollutants present at contaminated sites. Furthermore, the microbes can be genetically engineered for efficient degradation of environmental contaminants. Nevertheless, extensive political and ethical concern restricts the wide-scale applicability of genetically engineered organisms. The current biotechnological progress such as the use of proficient microbial consortia, indigenous microbes, application of specific enzymes, biosurfactant, and rhizoremediation are the new prospects in bioremediation technology. A schematic representation of various methods and techniques applied for bioremediation of different inorganic and organic contaminants has been illustrated in Fig. 1.1.

2.1 Bioremediation strategy for hydrocarbon contaminated water and soil

Bioremediation of hydrocarbon polluted soils and groundwater using bacteria has gained immense consideration recently. Bacteria can degrade a large number of toxic hydrocarbons under both aerobic and anaerobic circumstances. Benzene, toluene, ethyl benzene, and xylene (BTEX) compounds are typical examples of hydrocarbons and are carcinogenic (de Graaff et al., 2011) and neurotoxic in nature. Moreover, Environmental Protection Agency (EPA) classified these hydrocarbons as priority pollutants requiring strict regulation. When organic contaminants such as BTEX are released into the environment (Jin et al., 2013), the function and structure of the microbial communities are generally affected. Although significant researches on biodegradation of BTEX components by bacteria (Li et al., 2012) have been reported, however, most of the studies have concentrated on the degradation of only one or two components by bacterial isolate (Table 1.1). Furthermore, it is recognized that the

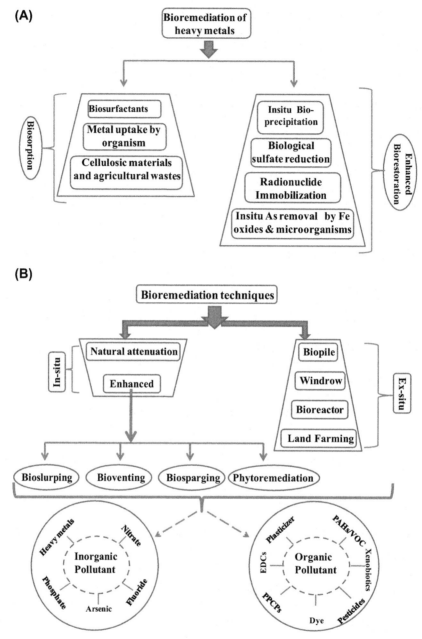

FIGURE 1.1   Different strategies (A) and methods (B) of bioremediation of various inorganic and organic contaminants.

TABLE 1.1　Microorganism involved in organic pollutant degradation.

Organic pollutants	Microbes	Degradation condition	Efficiency	References
Polychlorinated biphenyl	*Acinetobacter* and *Acidovorax*	Aerobic and anaerobic	−	Wang et al. (2016b)
Alkane and PAHs	*Alcanivorax, Marinobacter, Thalassospira, Alteromonas,* and *Oleibacter*	−	−	Catania et al. (2015)
Diesel degradation	*Pseudomonas* sp., *Bacillus subtilis*	−	−	Zhang et al. (2014)
N-alkanes and PAHs	*Pseudomonas* sp. WJ6		92.46%	Xia et al. (2014)
Crude oil	*Colwellia, Cycloclasticus*	−	−	Wang et al. (2016a)
Phenolic compounds	Microbial fuel cells	Bio-electro-chemical-systems	−	Hedbavna et al. (2016)
Crude oil	*Anabaena oryzae, Chlorella kessleri,* and its consortium	Mixotrophic conditions	−	Hamouda et al. (2016)
PAHs and Naphthenic acids	*Fusibacter, Alkaliphilus, Desulfobacterium, Variovorax, Thauera, Hydrogenophaga*	Anaerobic biodegradation	−	Folwell et al. (2016)
Pyrene	*Mycobacterium gilvum*	Immobilized on peanut shell powder	−	Deng et al. (2016)

biodegradation efficiency of one compound in a mixture can be influenced by additional components. The side effect generated through the deprivation of one particular compound can also influence the deprivation of another compound. Therefore, it is imperative to use an integrated bioremediation approach by using the consortium of microbes for degradation of petrochemicals containing hydrocarbons.

2.2 Bioremediation of heavy metal contaminated water

Bioremediation of heavy metal enriched soils and groundwater illustrates an immense perspective for upcoming improvement because of its environmental compatibility and probable expenditure efficiency (Baceva et al., 2014). It relies on microbial actions to diminish, mobilize, or immobilize noxious heavy metals through biosorption, biovolatilization, precipitation, surface complexation, and oxidation−reduction processes (Paul et al., 2014; Teixeira et al., 2014). Microorganism-directed oxidation−reduction reactions involving organic carbon, iron, manganese, and sulfur are the basic mechanisms influencing heavy metals mobility. Under natural environmental conditions, there exists a complicated interface among heavy metal pollutants and interacting microbes. These microorganisms have evolved specific resistance mechanism that permits their existence (Luo et al., 2014) under heavy metal enriched conditions. These microbes are efficient to alleviate the concentration of metal contaminants in their vicinity. Microorganisms are observed to proficiently eliminate dissolved and suspended metals, particularly from a medium having very low concentrations through bioaccumulation, surface complexation, and biosorption; therefore, methods relying on microbial process offer a substitute to the conservative practices of metal remediation (Dundar et al., 2014). Microbes reported for management of heavy metals are tabulated in Table 1.2.

2.3 Bioremediation of dye contaminated water

Dyes are used for the permanent coloring of fibers and other consumer products including foods, cosmetics, pharmaceuticals, papers, etc. The annual production of these dyes is more than 7×10^7 tons, out of which 30,000−150,000 tons are discharged into water (Anjaneya et al., 2011). In textile processing industries, a broad range of structurally varied dyes are utilized, and therefore, effluents from these industries are tremendously erratic in composition. In general, wastewater generated from dyeing industries contains up to 50% dye of the originally used concentration, together with several other chemical components, dispersing agents, fixatives, heavy metals, and inorganic salts. Generally, dyes are visibly detectable at very low quantity ($1 \, \text{mg} \, \text{L}^{-1}$), which may cause water pollution and changes to normal functioning of aquatic ecosystem by reducing dissolved oxygen contents, due to reduced photosynthetic activity of submerged plants caused by very weak penetration of light into water imposed by dyes. Hence, the toxicity exerted by textile effluents due to synergistic actions of physical appearance and chemical constituents is raising an environmental concern for their minimization to permissible limits. It is identified that 90% of reactive dyes get their way into treatment plants that may remain unaffected and released as such into rivers (Abadulla et al., 2000). Several dyes presently being utilized are not susceptible to degradation or elimination with physicochemical methods. Most importantly, often the degradation by-products of physicochemical methods are even more noxious than the parent dye used. Color can be eliminated from wastewater by chemical and physical processes including adsorption, precipitation, flocculation, and oxidation followed by filtration and electrochemical methods. These techniques are relatively expensive and have operational troubles (Kapdan et al., 2000). Numerous troubles, particularly the high dose requirement of chemicals and extensive power consumption, limit their realistic purpose. Consequently, the biodegradation of synthetic dyes has gained rising magnetism because of its natural ecological resemblance, lesser treatment costs, and elevated efficiency.

Currently, several scientific groups across the world are involved in manipulating the bacterial genetic constitution to improve their dye degradation ability to accelerate the bioremediation process. Different microbial consortium can degrade numerous azo-dyes aerobically and anaerobically. Actually, bacterial consortia are supposed to be more beneficial for complete dye degradation. In mixed culture, the toxic intermediates generated because of the activity of one bacterium are degraded by other bacteria of consortium (Forgacs et al., 2004; Jain et al., 2012). Thus, the degradation of dye will depend on origin and chemical nature of dyes and microorganism used in the consortium.

TABLE 1.2 List of biological agents responsible for the remediation of heavy metals.

Heavy metals	Biological agent	Processes	Efficiency	References
Hexavalent chromium	*Geotrichum* sp. and *Bacillus* sp.	Bioleaching system	94.8%	Qu et al. (2018)
Arsenic	*Shewanella* sp.	Transport of arsenic, coupled with microbe-mediated biogeochemical processes. As(V) was reduced to As(III)	—	Lim et al. (2007)
Arsenic	*Eichhornia crassipes* and *Lemna minor*	Bioaccumulation	—	Alvarado et al. (2008)
Arsenic	*Shewanella putrefaciens* 200R and *Shewanella* sp. ANA	Microbial leaching	—	Weisener et al. (2011)
Mercury	Functional genes are merA and merB	Enzyme degradation	—	Dash and Das (2012)
Mercury	*Bacillus cereus*	Bioaccumulating	—	Sinha et al. (2012)
Arsenic	*Pseudomonas, Acinetobacter, Klebsiella,* and *Comamonas*	—	—	Das et al. (2014)
Mercury	Filamentous fungi	Biosorption	97.50% and 98.73%	Kurniati et al. (2014)
Arsenic	*Brevibacillus* sp. KUMAs2	—	—	Mallick et al. (2014)
Inorganic mercury	*Bacillus cereus* BW-03(pPW-05)	Volatilization and biosorption	96.4%	Dash and Das (2015)
Arsenic	*Bacillus* sp. and *Aneurinibacillus aneurinilyticus*	—	51.45%–51.99% for As(III) and 50.37%–53.29% for As(V)	Dey et al. (2016)
Lead	*Rhodobacter sphaeroides*	—	—	Li et al. (2016)
Arsenic	*Bacillus aryabhattai* (NBRI014)	—	—	Singh et al. (2016)
Hexavalent chromium	*Sporosarcina saromensis* M52	Bioreduction	—	Zhao et al. (2016)
Arsenic	*Pseudomonas stutzeri* TS44	—	—	Akhter et al. (2017)
Chromium complex	*Aspergillus tamarii*	In batch and continuous bioreactors	—	Ghosh et al. (2017)
Lead	*Pararhodobacter* sp.	Calcium carbonate precipitation technique	—	Mwandira et al. (2017)

2.3.1 Bioremediation approaches used for dye degradation

Dye decolorization is started by anaerobic reduction reaction performed by azo-reductases or azo-bonds breakage under aerobic or anaerobic condition leading to the formation of aromatic amines because of physiological and biochemical activities of the mixed bacterial community (Sponza and Isik, 2004). A list of microorganisms involved in dye degradation is presented in Table 1.3.

A detailed description on various processes for dye degradation has been presented below:

2.3.1.1 Aerobic treatment

Reports on bacterial degradation of azo-dyes are very limited; however, some microorganisms have shown their capacity of dye reduction. *Pseudomonas aeruginosa* has been demonstrated to degrade commercially exploited textile and tannery dye Navitan Fast Blue SSR in a medium amended with glucose as a carbon source under aerobic condition (Garg and Thripathi, 2017). Kalyani et al. (2009) have also observed that some bacterial strains have capability to degrade dye under the aerobic environments.

TABLE 1.3 List of potential organisms responsible for degradation of dyes.

Dye	Organism	Experimental condition	Efficiency	Reference
Direct Red-31 dye	*Chlorella pyrenoidosa* strain NCIM 2738	Batch experiments/ continuous cyclic bioreactor treatment	96% Decolorization with 40 mg L^{-1} dye at pH 3	Sinha et al. (2016)
Reactive green dye (RGD)	Fungal strain VITAF-1	—	97.9% Decolorization of RGD within 48 h at $30 \pm 2°C$	Sinha and Osborne (2016)
Crystal violet	*Trichoderma asperellum*	—		Shanmugam et al. (2017)
Industrial textile effluent	*Chaetomium globosum* IMA1 KJ472923	Stirred and static batch cultures	Dye removal adsorption (8%) and degradation (92%)	Manai et al. (2016)
Hair dye	*Enterobacter cloacae* DDB I	—	—	Maiti et al. (2017)
Azure B dye	*Bacillus* sp. MZS10		93.55% (0.04 g L^{-1}) within 14 h	Li et al. (2014)
Resonance stabilized and heteropolyaromatic dyes	Bacterial consortium-AVS, consisting of *Pseudomonas desmolyticum* NCIM 2112, *Kocuria rosea* MTCC 1532, and *Micrococcus glutamicus* NCIM 2168	—	—	Kumar et al. (2012)

2.3.1.2 Anaerobic treatment

Under anaerobic condition, azo-dye reduction is achieved by breakage of the azo-bonds. Under the anaerobic situation, cleaving of dyes occurs, which produces toxic aromatic amines by bacterial metabolism (Bhatt et al., 2005).

2.3.1.3 Anoxic treatment

Anoxic degradation of different dyes by facultative anaerobic and mixed aerobic microorganisms is reported in different studies (Kapdan and Alparslan, 2005). Although several microorganisms are capable of growing under aerobic condition, however, the dye is degraded only under anoxic environments. Several pure bacterial cultures including those of *Pseudomonas luteola*, *Aeromonas hydrophila*, *Bacillus subtilis*, and *Proteus mirabilis* are known to anoxically degrade azo-dyes (Sandhya et al., 2005).

2.3.1.4 Sequential degradation of dyes

It has been suggested that aromatic amines produced after anaerobic degradation of azo-dyes can be despoiled subsequently under aerobic environment. The applicability of this approach was firstly proven for Mordant Yellow, a sulfonated azo-dye. After aeration, complete mineralization of amine by microbial activities is observed.

2.4 Vermi-biofiltration of wastewater

Vermi-biofiltration is a natural waste management method relying on worms to alter organic wastes to form soil enriching compounds (Romero et al., 2006). Domestic wastewater and industrial sludge management can be lodged through these methods in a sustainable way (Solis-Mejia et al., 2012). A substantial decrease in pathogens has been observed in the final product to a level that can be securely useful to land (Najar and Khan, 2013). The practice can be used at small scale for organic waste handling or municipal waste management (Karmakar et al., 2012). Vermi-biofiltration is a method that acclimatizes conventional vermicomposting method into an inert wastewater handling procedure by means of epigeic earthworms (Garg et al., 2009; Gupta and Garg, 2009). According to Komarowski (2001), vermi-biofiltration scheme uses suspended solids that are placed on an upper portion of a filter and developed by earthworms and feed to soil microorganisms harnessed in vermifilter. The solubilized and non-soluble organic and inorganic suspensions are removed by adsorption and stabilization process through multifaceted degradation reactions that occurs in soil occupied by aerobic microorganisms and earthworm. In general, acclimatized vermibed earthworms collect numerous organic contaminants in the vicinity of surrounding soil system (Sangwan et al., 2008) through passive absorption by body wall and also by intestinal uptake during the course of soil passage via gut (Mahmood et al., 2013).

2.5 Bioremediation of pesticide contamination

The annual global utilization of pesticides is approximately 2 million tonnes, of which 24% is used by the United States, 45% in Europe, and rest 25% by remaining parts of the world. India is reported to consume about 3.8% of the total pesticide produced worldwide.

The pesticide consumption for agricultural applications is 0.5, 6.6, and 12.0 kg ha^{-1} for India, Korea, and Japan, respectively (Gupta, 2004). Earlier, the pesticide application, as anticipated, was quite helpful in reducing crop yield loss happening because of insect pest attack and thus opened the way for enhancement in crop productivity. However, extensive application of chemically synthesized pesticides has given rise to contamination of the natural environment and also caused several chronic impacts on human society (Bhanti and Taneja, 2007). The stable nature, long-term existence in the natural environment, biological magnification, and accumulation at various trophic levels due to their lack of selectivity, and organochlorines have been reported to be responsible for pest resistance development and hazardous effect on nontarget organisms (Carson, 1962). The possible threats to all living organisms arising from the indiscriminate application of these agricultural chemicals have now emerged as one of the most potent environmental or health problems particularly in Third World countries. Pesticides are mainly criticized for their availability in drinking water, vegetables, mammalian blood, human food, milk products, fat samples, and other food commodities. There is no contradiction regarding the human health hazards associated with the application and chronic exposure to pesticides. Hence, the presence of pesticide residue in different environmental samples via contaminated food chain is a direct indication of acute or chronic exposure and average body risk to persistent pesticides.

2.6 Removal of pharmaceutical and personal care products by biological degradation processes

Microorganisms are known to degrade the environmental pollutants by using the contaminants for their vital physiological and biochemical processes, and under certain conditions, different microbes can coordinate each other to degrade the pollutants. The following subsections represent the role of pure and mixed culture in biodegradation of pharmaceutical and personal care products (PPCPs).

2.6.1 Pure cultures

Several experimental investigations have demonstrated that pure cultures of numerous algae, bacteria, and fungi obtained from different samples including activated sludge, wastewater, or sediment can be utilized to treat the commonly detected pollutants such as iopromide (Liu et al., 2013), carbamazepine (Popa et al., 2014), ibuprofen (Almeida et al., 2013), sulfamethoxazole (Jiang et al., 2012), diclofenac (Hata et al., 2010), paracetamol (Dang et al., 2013), and triclosan (Zhao et al., 2013). The pure cultures of different organisms efficient in degradation of various PPCPs have been presented in Table 1.4. A few pure cultures obtained from activated sludge showed efficiency in removal of a number of PPCPs. For example, *Achromobacter denitrificans* is able to mineralize sulfamethoxazole and other sulfonamides (Reis et al., 2018). Apart from this, several pure cultures can consume specialized PPCPs as sole source of C and energy, though with changed degradation pathways (Dang et al., 2013; Almeida et al., 2013). For example, *Delftia tsuruhatensis*, *P. aeruginosa*, and *Stenotrophomonas* can mineralize paracetamol. The study concluded that *D. tsuruhatensis* and *P. aeruginosa* contributed very less in paracetamol removal, while biosorption by *Stenotrophomonas* contributed significantly in removal of paracetamol. The differences could be due to involvement of different enzymes participating in degradation. Sometimes pure cultures

TABLE 1.4 List of microbes responsible for degradation of pharmaceuticals and pesticides.

Pharmaceuticals/ pesticides	Microbes	Degradation condition	Degradation efficiency	Reference
Atrazine	*Citricoccus*	Alkaline environment	—	Yang et al. (2018)
Tebuconazole	*Serratia marcescens* strain B1	—	70.42%, 96.46% in 30 days	Wang et al. (2018b)
Lincomycin	*Clostridium* sp. strain LCM-B		62.03%	Wang et al. (2018a)
Chlorimuron-ethyl	*Enterobacter ludwigii* sp. CE-1	—	83.1%—83.9%	Pan et al. (2018b)
Ciprofloxacin	*Thermus* sp.	—	—	Pan et al. (2018a)
Levofloxacin	*Chlorella vulgaris*	—	—	Xiong et al. (2017)
Triclosan	*Geitlerinema* sp. and *Chlorella* sp.	—	82.10%—92.83% (3.99 mg L^{-1}) of triclosan in 10 days	Tastan et al. (2017)
Organochlorine pesticide (OCP)	*Streptomyces* consortium	—	99.8%	Fuentes et al. (2017)
Pharmaceutically active compounds	White-rot fungi	Enzymes immobilization, fungal reactors	—	Naghdi et al. (2018)
Mixture of pesticides	*Streptomyces* consortium	—	—	Fuentes et al. (2017)
Atrazine and 2,4-dichlorophenoxyacetic acid	*Pseudomonas* and *Rhodococcus*	Aerobic biodegradation	—	Carboneras et al. (2017)
Acenaphthene and naphthalene	*Pseudomonas mendocina*	—	98.8% and 98.6%	Barman et al. (2017)
Pesticides	Actinobacteria	—	—	Alvarez et al. (2017)
Triazolopyrimidine sulfonamide	*Aspergillus*	—	—	Sondhia et al. (2016)
OCP	Native bacteria	—	—	Kopytko (2016)
Atenolol, gemfibrozil, and ciprofloxacin	—	Activated sludge process, submerged attached biofilter, and membrane bioreactor	—	Arya et al. (2016)
Penicillin G	*Klebsiella pneumoniae* Z1	—	—	Wang et al. (2015)
Pharmaceuticals	Ammonia-oxidizing bacteria	—	80%	Peng et al. (2015)

(Continued)

TABLE 1.4 List of microbes responsible for degradation of pharmaceuticals and pesticides.—cont'd

Pharmaceuticals/ pesticides	Microbes	Degradation condition	Degradation efficiency	Reference
Triclosan	*Aspergillus versicolor*	–	–	Ertit Tastan and Dönmez (2015)
Organophosphorus pesticides	*Stenotrophomonas* sp. G1	–	–	Deng et al. (2015)
Caffeine, sulfamethoxazole, ranitidine, carbamazepine, and ibuprofen	–	Growth reactor	–	Vasiliadou et al. (2013)
Endosulfan	*Alcaligenes faecalis* JBW4	–	87.5% (α-endosulfan) and 83.9% (β-endosulfan) in 5 days	Kong et al. (2013)
Organophosphorus pesticides	*Serratia marcescens*	–	–	Cycoń et al. (2013)
Pharmaceutically active compounds	Active biomass in activated sludge	–	–	Majewsky et al. (2011)
Endosulfan	*Aspergillus niger*	–	–	Bhalerao and Puranik (2007)

are unable to utilize few PPCPs as carbon and energy source because of their substrate specificity. Under such conditions, other substrate needs to be amended in medium to fulfill the requirement of carbon and energy source so as to maintain the vital metabolic activities at optimal rate. For instance, the stable nature of carbamazepine often does not allow the efficient degradation by microbes. But, mixed consortium consisting of an unidentified basidiomycete member (Santoso et al., 2011) and *Streptomyces* MIUG (Popa et al., 2014) was noticed to breakdown the carbamazepine in medium supplemented with glucose. Along with carbamazepine, iopromide was also susceptible to degradation after additional substrates were amended in the medium. Liu et al. (2013) demonstrated that *Pseudomonas* sp. I-24 bears the ability to degrade iopromide with starch as primary supplement. Diclofenac has been reported to display high resistance to biological degeneration in activated sludge system. However, Hata et al. (2010) showed that white-rot fungi is capable to fully remove diclofenac to the safety limits prescribed for the organisms without any added supplement. The microbial enzyme activation has been suggested as another important factor leading to PPCPs degradation. The biological breakdown of PPCPs relies on whether microbes are able to synthesize the essential enzymes responsible for degradation. For example, triclosan is known to stimulate *Nitrosomonas europaea* to synthesize ammonia monooxygenase to facilitate fast degradation of triclosan (Roh et al., 2009). However, few PPCPs including ciprofloxacin, tetracycline, and trimethoprim do not readily stimulate microbes to synthesize the specific enzyme, thus, leading to inefficient biodegradation. Most importantly, there is no report on contribution of any pure culture isolated so far in decomposition of ciprofloxacin, tetracycline, and trimethoprim. Therefore, to facilitate the effective biodegradation of hazardous recalcitrant PPCPs, the primary requirement is to stimulate the microorganisms with the ability to produce the specific degradative enzyme in a given environmental condition.

2.6.2 Mixed cultures

Mixed cultures are more efficient in biological degradation of the PPCPs as compared to pure cultures because under some environmental conditions it is very hard to isolate the pure culture. Limited study has been performed to analyze the efficiency of mixed culture for PPCPs removal (Khunjar et al., 2011). Mixed culture has the efficiency to remove PPCPs. Khunjar et al. (2011) described the role played by mixed culture of heterotrophic and ammonia-oxidizing bacteria for enhanced elimination of 17a-ethinylestradiol. Actually, the commonly employed biological treatment methods rely on synergistic actions of mixed cultures to eliminate PPCPs. However, sometimes activated sludge demonstrates low PPCPs elimination rate. Therefore, experiments have been carried out to enhance the PPCPs degradation through activated sludge process. High rate removal of PPCPs through metabolic activities of mixed culture present in activated sludge has been described by Zhao et al. (2013). Interestingly, mixed culture has been reported to facilitate very fast decomposition of mixed PPCPs as compared to degradation of a single PPCP (Vasiliadou et al., 2013). This can be explained by the fact that few members of mixed PPCPs can serve as C and energy source for mixed culture, thus facilitating further degradation of other PPCPs. Hence, mixed cultures have been proposed as a potential alternative for accelerated degradation of PPCPs.

2.6.3 Activated sludge process

Activated sludge technique has been largely employed for biological treatment in traditionally used wastewater treatment plants. The PPCPs degradation in biological treatment has been attributed to the combinatorial action of volatilization, surface binding, and microbial decomposition. However, the role of volatilization and adsorption processes in degradation of PPCPs is minimal (Li et al., 2015). Generally, volatilization happens simultaneously with aeration. The adsorption of PPCPs during biological treatment is significantly affected by physicochemical attributes of compounds to be degraded. Furthermore, changes in environmental variables including pH, oxygen content, temperature, the composition of the microbial community, and nutrient status can also largely influence the overall efficiency of activated sludge system applied for PPCPs elimination. Biodegradation is suggested as the major phenomenon responsible for PPCPs elimination in activated sludge process. Therefore, to improve the degradation of PPCPs, an effective strategy must be adapted to increase the decomposition. Nevertheless, microbial decomposition is not always efficient in the elimination of environmental contaminants due to the low abundance of microbes responsible for degradation. These impediments can be resolved by prior acclimatization to contaminants, bioaugmentation (Wang et al., 2004) and biostimulation. Through adaptation and bioaugmentation, the abundance of the pure or mixed culture of microorganisms effective in contaminant degradation can be improved in biological treatment methodologies. Plosz et al. (2012) have designed a model for xenobiotic trace chemicals and utilized it to forecast and monitor the environmental factors affecting the elimination of carbamazepine and diclofenac in activated sludge system.

2.7 Vermicomposting of solid wastes

Disposal of industrial solids is becoming a solemn crisis. The indecent and arbitrary discarding of industrial solids is posing an immense challenge to India and other rising countries. They cause odor difficulty and are the probable cause of surface and groundwater contamination. The sludge resulting from diverse industrial actions and wastewater handling plants is managed through unsuitable modes such as landfilling and incineration (Hashemimajd et al., 2006). The inadequate landfill areas, more severe national waste discarding rules, and local awareness have complicated the landfilling even more costly and unfeasible. Taking into consideration all the troubles of waste management, vermicomposting is one of the sustainable modes to degrade the solid and human wastes (Lalander et al., 2013). Furthermore, vermicomposting also alters the waste into compost, which is additionally used as plant nutrients (Gomez-Brandon et al., 2013). Vermicomposting is, therefore, suggested for the management of a broad variety of organic wastes and the production of organic matter rich in soil amendments (Lleo et al., 2013).

2.8 Genetically engineered microorganism—based bioremediation

Presently, several scientific groups across the world are involved in manipulating the bacterial genetic constitution for removal of man-made pollutants. Genetically engineered organisms offer the possibility of degradation of a range of pollutants (Dronik, 1999). They have exhibited bioremediation potential in the management of polluted soil and groundwater system as well as activated sludge environments along with better degradation capabilities for a number of chemical and petrochemicals pollutants (Sayler and Ripp, 2000). Recently, various microorganisms and enzymes have been engineered for biodegradation of polycyclic aromatic hydrocarbons, polychlorinated biphenyls, and persistent organic pollutants (Ang et al., 2005). However, environmental concerns and regulatory constraints limit their large-scale in situ applications (Pandey et al., 2005). Although several experiments dealing with bioremediation have been carried out under in vitro conditions, it has always been too complicated to investigate the fate, behavior, and removal rate of contaminants in the natural ecosystem because of the involvement of varied environmental factors including biosafety (Singh et al., 2011).

2.9 Factors affecting bioremediation with emphasis on petrochemical and other organic pollutants

Bioremediation of petrochemical hydrocarbon contaminated soil and water is a complex process due to their noxious and hydrophobic behavior and multiphasic nature, diversity of microbial community, and the existing environmental attributes. The efficiency of microbial uptake and metabolism of toxic hydrocarbons can be modified by various environmental attributes. Majority of the experimental reports on the degradation of toxic petrochemical wastes have identified different limiting factors (of physical, chemical, and biological nature) affecting the degradation processes, for example, contaminant load, the presence of electron donors and acceptors, dissolved oxygen content, temperature, pH, nutrient status, and microbial adaptability (Mohan et al., 2006). Singh et al. (2017) recently reviewed the bioremediation of petrochemical wastes from soil and water and identified some of the key limiting factors that are described in the following subsections.

2.10 Concentration of pollutant

The substrate interactions have a potential effect over the degradation rates of individual contaminants (Wang et al., 2007). The assessment of substrate interaction at varied concentrations has been considered as an important aspect because of its potential role in bacterial sensitivity and metabolism. The synergistic interactions between different components of a contaminant can improve the decomposition rates and catabolic enzyme activities. For example, in a batch culture study, the growth rate of *Pseudomonas putida* was found to be decreased at high substrate concentrations (Abuhamed et al., 2004). The BTEX compounds (at particular concentrations) showed inhibitory effect over microbiological processes due to their complex interactions (Mathur and Majumder, 2010).

2.11 Nutrients availability

Nutrients (such as C, N, P, K, and Ca) are the basic requirements for the microbial growth and activities, and therefore, their availability has important role in regulating the degradation rates of different contaminants. Moreover, the relative nutrient availability also has a major regulatory impact over contaminant degradation. For example, at a contaminated site having higher levels of organic carbon in the contaminants, the microbial activity is reported to be significantly higher resulting into the rapid depletion of bioavailable nutrients such as nitrogen, phosphorus, potassium, and iron (Cooney, 1984) at early phase and later the degradation decreases because of limitation of these key elements (Hamme et al., 2003). On contrary, excess availability of N, P, and K has also been reported to exert negative impact on degradation of aromatic hydrocarbons (Carmichael and Pfaender, 1997).

2.12 Microbial adaptation (acclimatization)

Microbial adaptation has potential influence on the degradation of petroleum wastes. The acclimatization of microbial communities to aromatic organic compounds enhances their degradation efficiency (Babaarslan et al., 2003). For example, *Alcaligenes xylosoxidans* Y234 degraded benzene and toluene more efficiently under adapted conditions as compared to nonadapted ones (Yeom et al. 1997). However, for single component and multicomponent-based contaminant systems, the nonacclimatized microbes exhibit fastest biodegradation rates.

2.13 Bioavailability

The accessibility of organic pollutants to microorganisms is one of the major factors determining the biodegradation rate (Guthrie and Pfaender, 1998). The bioavailability of hydrocarbons is a function of their physical characteristics and chemical constitution (Foght et al., 1996). Some bacterially derived substances may act as additives which increases the overall degradation of petroleum hydrocarbons (Lee et al., 2001). For example, biosurfactants, a rich carbon source, released from various microbes (e.g., bacteria and fungi) have potential contribution in uptake and mineralization of petroleum hydrocarbons (Leahy and Colwell, 1990).

2.14 Effect of environmental conditions

Various environmental conditions such as temperature, salinity, pH, and oxygen availability have potential control over the bioremediation of petrochemical wastes. An inverse relationship exists between salinity and solubility of petroleum hydrocarbon because increase in salinity enhances the sorption of aromatic hydrocarbon. It has been reported for pyrene in various types of sediments and the possible reason for such increase in degradation was because of salting out effects observed in both solution and solid phases. Furthermore, in some cases, hypersalinity leads to reduced microbial growth and thus hindered metabolic activities together with the growth of unidentified halophytic archaean during the biodegradation. The role of some other environmental variables is described below:

2.14.1 Temperature

Temperature shows a critical control in microbial metabolism and hydrocarbon degradation under both in situ and ex situ conditions (Margesin and Schinner, 2001). A decrease in microbial growth and multiplication is observed at low temperature, which resulted into the slow rate of petrochemical degradation (Gibb et al., 2001). On the other hand, various studies have shown that increase in temperature increases the solubility of hydrocarbons in the medium and thus lead to easy availability of petrochemical hydrocarbons to microorganisms. The rate of degradation of petrochemical wastes is generally high in the temperature range of 30–40°C in soil and 15–20°C in the aqueous or marine environments (Mueller et al., 1989). However, a few thermophilic microorganisms (e.g., *Bacillus thermoleovorans*) have been reported to efficiently transform phenanthrene, naphthalene, and anthracene even at higher temperature ranges.

2.14.2 pH

The variations in pH conditions largely determined the microbial degradation of petrochemical wastes in soil or aqueous medium. Various studies support the favorable mineralization of petroleum hydrocarbons near to neutral pH conditions. A slight variation in pH may exert a significant impact on overall biological degradation processes in an aquatic system. Some fungi and acidophilic microbes have been reported to grow and have biodegradation potential under highly acidic environments (Stapleton et al., 1998).

2.14.3 Oxygen availability

The oxygen availability acts as a deciding factor whether a given environmental condition is aerobic or anaerobic. The decomposition of petrochemical hydrocarbons has been reported mostly under aerobic and in some cases under anaerobic environments. Biodegradation under anaerobic environments mostly occurs in aquifers and submerged marine sediments (Coates et al., 1996) with insignificant degradation rate and was primarily restricted to halogenated aromatics (Boyd and Shelton, 1984). For aerobic degradation of petrochemicals, oxygen plays an essential role for the activity of mono- and dioxygenase enzymes during aromatic ring oxidation. In anaerobic environment, substituted electron acceptors such as ferrous iron, nitrate, and sulfate are essentially required to facilitate the aromatic compound oxidation. However, reduction of electron acceptors such as ferric iron, nitrate, and sulfate under anaerobic conditions leads to liberation of sufficient quantity of phosphorus and ferrous iron, which further contaminated the natural ecosystem. Moreover, release of greenhouse gases (CH_4, NO_2, etc.) and increase in pH have also been observed under anaerobic degradation of petrochemical hydrocarbons (Bamforth and Singleton, 2005). Thus, oxygen availability plays a key role in the bioremediation of aromatic compounds.

3. Conclusion

Bioremediation is an emerging technology which can be simultaneously used with other physical and chemical treatment methods for complete management of diverse group of environmental pollutants. It seems as a sustainable approach for the environmental pollution management, and hence, there is a need for more research in this area. Efforts need to be made to generate a synergistic interaction between the environmental impact on fate and behavior of environmental contaminants and assortment and performance of the most suitable bioremediation technique and another relevant technique that can sustain the effective and successful operation and monitoring of a bioremediation process. Vermi-biofiltration and genetic engineering technology can be promoted and adapted at larger scales for sustainable waste recycling, polluted soil treatment, management of solid wastes, etc., with certain optimized conditions. The continuous efforts of research and development will direct future regulations, dealing with bioremediation targets, contaminant availability, and their potential threat on natural ecosystem and human health. Moreover, the availability and biodegradation of contaminants in any natural or man-made system and degree of threat to human health caused by various environmental pollutants would be more easily forecasted by using multidisciplinary technologies.

References

Abadulla, E., Tzanov, T., Costa, S., Robra, K.H., Cavaco-Paulo, A., Gübitz, G.M., 2000. Decolorization and detoxification of textile dyes with a laccase from *Trametes hirsuta*. Applied and Environmental Microbiology 66, 3357–3362.

Abuhamed, T., Bayraktar, E., Mehmetoğlu, T., Mehmetoğlu, Ü., 2004. Kinetics model for growth of *Pseudomonas putida* F1 during benzene, toluene and phenol biodegradation. Process Biochemistry 39, 983–988.

Akhter, M., Tasleem, M., Mumtaz Alam, M., Ali, S., 2017. In silico approach for bioremediation of arsenic by structure prediction and docking studies of arsenite oxidase from *Pseudomonas stutzeri* TS44. International Biodeterioration and Biodegradation 122, 82–91.

Almeida, B., Kjeldal, H., Lolas, I., Knudsen, A.D., Carvalho, G., Nielsen, K., et al., 2013. Quantitative proteomic analysis of ibuprofen-degrading *Patulibacter* sp. strain I11. Biodegradation 24, 615–630.

Alvarado, S., Guédez, M., Lué-Merú, M.P., Nelson, G., Alvaro, A., Jesús, A.C., Gyula, Z., 2008. Arsenic removal from waters by bioremediation with the aquatic plants Water Hyacinth (*Eichhornia crassipes*) and Lesser Duckweed (Lemna minor). Bioresource Technology 99, 8436–8440.

Alvarez, A., Saez, J.M., Davila Costa, J.S., Colin, V.L., Fuentes, M.S., Cuozzo, S.A., Benimeli, C.S., Polti, M.A., Amoroso, M.J., 2017. Actinobacteria: current research and perspectives for bioremediation of pesticides and heavy metals. Chemosphere 166, 41–62.

Ang, E.L., Zhao, H., Obbard, J.P., 2005. Recent advances in the bioremediation of persistent organic pollutants via biomolecular engineering. Enzyme and Microbial Technology 37, 487–496.

Anjaneya, O., Souche, S.Y., Santoshkumar, M., Karegoudar, T.B., 2011. Decolorization of sulfonated azo dye Metanil Yellow by newly isolated bacterial strains: *Bacillus* sp. strain AK1 and *Lysinibacillus* sp. strain AK2. Journal of Hazardous Materials 190, 351–358.

Arya, V., Philip, L., Murty Bhallamudi, S., 2016. Performance of suspended and attached growth bioreactors for the removal of cationic and anionic pharmaceuticals. Chemical Engineering Journal 284, 1295–1307.

Babaarslan, C., Abuhamed, T., Mehmetoglu, U., Tekeli, A., Mehmetoglu, T., 2003. Biodegradation of BTEX compounds by a mixed culture obtained from petroleum formation water. Energy Sources 25, 733–774.

Baceva, K., Stafilov, T., Matevski, V., 2014. Bioaccumulation of heavy metals by endemic Viola species from the soil in the vicinity of the As-Sb-Tl mine allchar' Republic of Macedonia. International Journal of Phytoremediation 16, 347–365.

Bamforth, S.M., Singleton, I., 2005. Bioremediation of polycyclic aromatic hydrocarbons: current knowledge and future directions. Journal of Chemical Technology and Biotechnology 80 (7), 723–736.

Bhanti, M., Taneja, A., 2007. Contamination of vegetables of different seasons with organophosphorous pesticides and related health risk assessment in northern India. Chemosphere 69, 63–68.

Bhatt, N., Patel, K.C., Keharia, H., Madamwar, D., 2005. Decolorization of diazo-dye Reactive Blue 172 by *Pseudomonas aeruginosa* NBAR12. Journal of Basic Microbiology: An International Journal on Biochemistry, Physiology, Genetics, Morphology, and Ecology of Microorganisms 45, 407–418.

Barman, S.R., Banerjee, P., Mukhopadhayay, A., Das, P., 2017. Biodegradation of acenapthene and naphthalene by *Pseudomonas mendocina*: process optimization, and toxicity evaluation. Journal of Environmental Chemical Engineering 5 (5), 4803–4812.

Benitez, E., Sainz, H., Melgar, R., Nogales, R., 2002. Vermicomposting of a lignocellulosic waste from olive oil industry: a pilot scale study. Waste Management and Research 20, 134–142.

Bhalerao, T.S., Puranik, P.R., 2007. Biodegradation of organochlorine pesticide, endosulfan, by a fungal soil isolate, *Aspergillus niger*. International Biodeterioration and Biodegradation 59 (4), 315–321.

Boyd, S.A., Shelton, D.R., 1984. Anaerobic biodegradation of chlorophenols in fresh and acclimated sludge. Applied and Environmental Microbiology 47 (2), 272–277.

Carboneras, B., Villaseñor, J., Fernandez-Morales, F.J., 2017. Modelling aerobic biodegradation of atrazine and 2,4-dichlorophenoxy acetic acid by mixed-cultures. Bioresource Technology 243, 1044–1050.

Carmichael, L.M., Pfaender, F.K., 1997. The effect of inorganic and organic supplements on the microbial degradation of phenanthrene and pyrene in soils. Biodegradation 8 (1), 1–13.

Carson, R., 1962. Silent Spring. Honghton Mifflin Co., Boston, p. 368.

Catania, V., Santisi, S., Signa, G., Vizzini, S., Mazzola, A., Cappello, S., Yakimov, M.M., Quatrini, P., 2015. Intrinsic bioremediation potential of a chronically polluted marine coastal area. Marine Pollution Bulletin 99 (1–2), 138–149.

Chang, J.-S., Yoon, I.-H., Kim, K.-W., 2009. Heavy metal and arsenic accumulating fern species as potential ecological indicators in As-contaminated abandoned mines. Ecological Indicators 9, 1275–1279.

Coates, J.D., Anderson, R.T., Lovley, D.R., 1996. Oxidation of polycyclic aromatic hydrocarbons under sulfate-reducing conditions. Applied and Environmental Microbiology 62 (3), 1099–1101.

Cooney, J.J., 1984. The fate of petroleum polltants in freshwater ecosystems. In: Atlas, R.M. (Ed.), Petroleum Microbiology. MacMillan Publishing Co, New York, pp. 399–434.

Cycoń, M., Żmijowska, A., Wójcik, M., Piotrowska-Seget, Z., 2013. Biodegradation and bioremediation potential of diazinon-degrading *Serratia marcescens* to remove other organophosphorus pesticides from soils. Journal of Environmental Management 117, 7–16.

Dang, D., Wang, D., Zhang, C., Zhou, W., Zhou, Q., Wu, H., 2013. Comparison of oral paracetamol versus ibuprofen in premature infants with patent ductus arteriosus: a randomized controlled trial. PLoS One 8, e77888.

Das, S., Jean, J.-S., Kar, S., Chou, M.-L., Chen, C.-Y., 2014. Screening of plant growth-promoting traits in arsenic-resistant bacteria isolated from agricultural soil and their potential implication for arsenic bioremediation. Journal of Hazardous Materials 272, 112–120.

Dash, H.R., Das, S., 2012. Bioremediation of mercury and the importance of bacterial mer genes. International Biodeterioration and Biodegradation 75, 207–213.

Dash, H.R., Das, S., 2015. Bioremediation of inorganic mercury through volatilization and biosorption by transgenic *Bacillus cereus* BW-03(pPW-05). International Biodeterioration and Biodegradation 103, 179–185.

de Graaff, M., Bijmans, M.F.M., Abbas, B., Euverink, G.-J.W., Muyzer, G., Janssen, A.J.H., 2011. Biological treatment of refinery spent caustics under halo-alkaline conditions. Bioresource Technology 102, 7257—7264.

Deng, F., Liao, C., Yang, C., Guo, C., Dang, Z., 2016. Enhanced biodegradation of pyrene by immobilized bacteria on modified biomass materials. International Biodeterioration and Biodegradation 110, 46—52.

Deng, S., Chen, Y., Wang, D., Shi, T., Wu, X., Ma, X., Li, X., Hua, R., Tang, X., Li, Q.X., 2015. Rapid biodegradation of organophosphorus pesticides by *Stenotrophomonas* sp. G1. Journal of Hazardous Materials 297, 17—24.

Dey, U., Chatterjee, S., Mondal, N.K., 2016. Isolation and characterization of arsenic-resistant bacteria and possible application in bioremediation. Biotechnology Reports 10, 1—7.

Dronık, J., 1999. Genetically modified organisms (GMO) in bioremediation and legislation. International Biodeterioration and Biodegradation 44, 3—6.

Dundar, E., Sonmez, G.D., Unver, T., 2014. Isolation, Molecular Characterization and Functional Analysis of OeMT2, an Olive Metallothionein with a Bioremediation Potential. Molecular genetics and genomics: MGG.

Ertit Tastan, B., Dönmez, G., 2015. Biodegradation of pesticide triclosan by *A. versicolor* in simulated wastewater and semi-synthetic media. Pesticide Biochemistry and Physiology 118, 33—37.

Farhadian, M., Vachelard, C., Duchez, D., Larroche, C., 2008. In situ bioremediation of monoaromatic pollutants in groundwater: a review. Bioresource Technology 99, 5296—5308.

Foght, J.M., Westlake, D.W.S., Johnson, W.M., Ridgway, H.F., 1996. Environmental gasoline-utilizing isolates and clinical isolates of *Pseudomonas aeruginosaare* taxonomically indistinguishable by chemotaxonomic and molecular techniques. Microbiology 142 (9), 2333—2340.

Folwell, B.D., McGenity, T.J., Price, A., Johnson, R.J., Whitby, C., 2016. Exploring the capacity for anaerobic biodegradation of polycyclic aromatic hydrocarbons and naphthenic acids by microbes from oil-sands-process-affected waters. International Biodeterioration and Biodegradation 108, 214—221.

Forgacs, E., Cserhati, T., Oros, G., 2004. Removal of synthetic dyes from wastewaters: a review. Environment International 30, 953—971.

Fuentes, M.S., Raimondo, E.E., Amoroso, M.J., Benimeli, C.S., 2017. Removal of a mixture of pesticides by a *Streptomyces* consortium: influence of different soil systems. Chemosphere 173, 359—367.

Garg, S.K., Tripathi, M., 2017. Microbial strategies for discoloration and detoxification of azo dyes from textile effluents. Research Journal of Microbiology 12, 1—19.

Garg, V.K., Gupta, R., Kaushik, P., 2009. Vermicomposting of solid textile mill sludge spiked with cow dung and horse dung: a pilot-scale study. International Journal of Environment and Pollution 38, 385—396.

Ghosh, A., Dastidar, M.G., Sreekrishnan, T.R., 2017. Bioremediation of chromium complex dyes and treatment of sludge generated during the process. International Biodeterioration and Biodegradation 119, 448—460.

Gibb, A., Chu, A., Wong, R.C.K., Goodman, R.H., 2001. Bioremediation kinetics of crude oil at 5°C. Journal of Environmental Engineering 127 (9), 818—824.

Gomez-Brandon, M., Lores, M., Dominguez, J., 2013. Changes in chemical and microbiological properties of rabbit manure in a continuous-feeding vermicomposting system. Bioresource Technology 128, 310—316.

Gupta, P.K., 2004. Pesticide exposure—Indian scene. Toxicology 198, 83—90.

Gupta, R., Garg, V.K., 2009. Vermiremediation and nutrient recovery of non-recyclable paper waste employing *Eisenia fetida*. Journal of Hazardous Materials 162, 430—439.

Guthrie, E.A., Pfaender, F.K., 1998. Reduced pyrene bioavailability in microbially active soils. Environmental Science Technology 32 (4), 501—508.

Hamme, J.D., van, Singh, A., Ward, O.P., 2003. Recent advances in petroleum microbiology. Microbiology and Molecular Biology Reviews 67 (4), 503—549.

Hamouda, R.A.E.F., Sorour, N.M., Yeheia, D.S., 2016. Biodegradation of crude oil by *Anabaena oryzae, Chlorella kessleri* and its consortium under mixotrophic conditions. International Biodeterioration and Biodegradation 112, 128—134.

Hashemimajd, K., Kalbas, M., Golchin, A., Knicker, H., Shariatmadari, H., Rezaei-Nejad, Y., 2006. Use of vermicomposts produced from various solid wastes as potting media. European Journal of Horticultural Science 71, 21—29.

Hata, T., Kawai, S., Okamura, H., Nishida, T., 2010. Removal of diclofenac and mefenamic acid by the white rot fungus *Phanerochaete sordida* YK-624 and identification of their metabolites after fungal transformation. Biodegradation 21, 681—689.

Hedbavna, P., Rolfe, S.A., Huang, W.E., Thornton, S.F., 2016. Biodegradation of phenolic compounds and their metabolites in contaminated groundwater using microbial fuel cells. Bioresource Technology 200, 426—434.

Jain, A., Singh, S., Kumar Sarma, B., Bahadur Singh, H., 2012. Microbial consortium—mediated reprogramming of defence network in pea to enhance tolerance against *Sclerotinia sclerotiorum*. Journal of Applied Microbiology 112, 537—550.

Jiang, B., Cui, D., Li, A., Gai, Z., Ma, F., Yang, J., Ren, N., 2012. Genome sequence of a cold-adaptable sulfamethoxazole-degrading bacterium, *Pseudomonas psychrophila* HA-4. Journal of Bacteriology 194, 5721, 5721.

Jin, H.M., Choi, E.J., Jeon, C.O., 2013. Isolation of a BTEX-degrading bacterium, *Janibacter* sp. SB2, from a sea-tidal flat and optimization of biodegradation conditions. Bioresource Technology 145, 57—64.

Kalyani, D.C., Telke, A.A., Dhanve, R.S., Jadhav, J.P., 2009. Ecofriendly biodegradation and detoxification of Reactive Red 2 textile dye by newly isolated *Pseudomonas* sp. SUK1. Journal of Hazardous Materials 163, 735—742.

Kapdan, I.K., Alparslan, S., 2005. Application of anaerobic—aerobic sequential treatment system to real textile wastewater for color and COD removal. Enzyme and Microbial Technology 36, 273—279.

Kapdan, I.K., Kargi, F., McMullan, G., Marchant, R., 2000. Decolorization of textile dyestuffs by a mixed bacterial consortium. Biotechnology Letters 22, 1179—1181.

Karmakar, S., Brahmachari, K., Gangopadhyay, A., Choudhury, S.R., 2012. Recycling of different available organic wastes through vermicomposting. European Journal of Medicinal Chemistry 9, 801—806.

Khunjar, W.O., Mackintosh, S.A., Skotnicka-Pitak, J., Baik, S., Aga, D.S., Love, N.G., 2011. Elucidating the relative roles of ammonia oxidizing and heterotrophic bacteria during the biotransformation of 17 alpha-ethinylestradiol and trimethoprim. Environmental Science and Technology 45, 3605—3612.

Komarowski, S., July 2001. Vermiculture for Sewage and Water Treatment Sludges. Water, Publication of Australian Water and Wastewater Association, pp. 39—43.

Kong, L., Zhu, S., Zhu, L., Xie, H., Su, K., Yan, T., Wang, J., Wang, J., Wang, F., Sun, F., 2013. Biodegradation of organochlorine pesticide endosulfan by bacterial strain *Alcaligenes faecalis* JBW4. Journal of Environmental Sciences 25 (11), 2257—2264.

Kopytko, M., 2016. Comparative study of biodegradation of aged organochlorine pesticides in soil through native bacteria. New Biotechnology 33 (Suppl.), S141.

Kümmerer, K., Dionysiou, D.D., Olsson, O., Fatta-Kassinos, D., 2019. Reducing aquatic micropollutants—increasing the focus on input prevention and integrated emission management. The Science of the Total Environment 652, 836—850.

Kumar, M.A., Kumar, V.V., Premkumar, M.P., Baskaralingam, P., Thiruvengadaravi, K.V., Dhanasekaran, A., Sivanesan, S., 2012. Chemometric formulation of bacterial consortium-AVS for improved decolorization of resonance-stabilized and heteropolyaromatic dyes. Bioresource Technology 123, 344—351.

Kurniati, E., Arfarita, N., Imai, T., Higuchi, T., Kanno, A., Yamamoto, K., Sekine, M., 2014. Potential bioremediation of mercury-contaminated substrate using filamentous fungi isolated from forest soil. Journal of Environmental Sciences 26 (6), 1223—1231.

Lalander, C.H., Hill, G.B., Vinneras, B., 2013. Hygienic quality of faeces treated in urine diverting vermicomposting toilets. Waste Management 33, 2204—2210.

Leahy, J.G., Colwell, R.R., 1990. Microbial degradation of hydrocarbons in the environment. Microbiological Reviews 54 (3), 305—315.

Lee, P.H., Ong, S.K., Golchin, J., Nelson, G.S., 2001. Use of solvents to enhance PAH biodegradation of coal tar. Water Research 35 (16), 3941—3949.

Li, H., Zhang, Q., Wang, X.-L., Ma, X.-Y., Lin, K.-F., Liu, Y.-D., Gu, J.-D., Lu, S.-G., Shi, L., Lu, Q., Shen, T.-T., 2012. Biodegradation of benzene homologues in contaminated sediment of the East China Sea. Bioresource Technology 124, 129—136.

Li, H., Zhang, R., Tang, L., Zhang, J., Mao, Z., 2014. 538601. Evaluation of *Bacillus* sp. MZS10 for decolorizing Azure B dye and its decolorization mechanism. Journal of Environmental Sciences 26, 1125—1134.

Li, W., Shi, Y., Gao, L., Liu, J., Cai, Y., 2015. Occurrence, fate and risk assessment of parabens and their chlorinated derivatives in an advanced wastewater treatment plant. Journal of Hazardous Materials 300, 29—38.

Li, X., Peng, W., Jia, Y., Lu, L., Fan, W., 2016. Bioremediation of lead contaminated soil with *Rhodobacter sphaeroides*. Chemosphere 156, 228—235.

Lim, M.-S., Yeo, I.W., Prabhakar Clement, T., Roh, Y., Lee, K.-K., 2007. Mathematical model for predicting microbial reduction and transport of arsenic in groundwater systems. Water Research 41 (10), 2079—2088.

Liu, Y.C., Young, L.S., Lin, S.Y., Hameed, A., Hsu, Y.H., Lai, W.A., Shen, F.T., Young, C.C., 2013. *Pseudomonas guguanensis* sp. nov., a gammaproteobacterium isolated from a hot spring. International Journal of Systematic and Evolutionary Microbiology 63, 4591–4598.

Lleo, T., Albacete, E., Barrena, R., Font, X., Artola, A., Sanchez, A., 2013. Home and vermicomposting as sustainable options for biowaste management. Journal of Cleaner Production 47, 70–76.

Luo, X., Zeng, X.C., He, Z., Lu, X., Yuan, J., Shi, J., Liu, M., Pan, Y., Wang, Y.X., 2014. Isolation and characterization of a radiation-resistant bacterium from Taklamakan Desert showing potent ability to accumulate Lead (II) and considerable potential for bioremediation of radioactive wastes. Ecotoxicology 23 (10), 1915–1921.

Mahmood, Q., Pervez, A., Zeb, B.S., Zaffar, H., Yaqoob, H., Waseem, M., Zahidullah, Afsheen, S., 2013. Natural treatment systems as sustainable ecotechnologies for the developing countries. BioMed Research International 19. https://doi.org/10.1155/2013/796373.

Maiti, S., Sinha, S.S., Singh, M., 2017. Microbial decolorization and detoxification of emerging environmental pollutant: cosmetic hair dyes. Journal of Hazardous Materials 338, 356–363.

Majewsky, M., Gallé, T., Yargeau, V., Fischer, K., 2011. Active heterotrophic biomass and sludge retention time (SRT) as determining factors for biodegradation kinetics of pharmaceuticals in activated sludge. Bioresource Technology 102 (16), 7415–7421.

Mallick, I., Hossain, S.T., Sinha, S., Mukherjee, S.K., 2014. *Brevibacillus* sp. KUMAs2, a bacterial isolate for possible bioremediation of arsenic in rhizosphere. Ecotoxicology and Environmental Safety 107, 236–244.

Manai, I., Miladi, B., El Mselmi, A., Smaali, I., Hassen, A.B., Hamdi, M., Bouallagui, H., 2016. Industrial textile effluent decolourization in stirred and static batch cultures of a new fungal strain *Chaetomium globosum* IMA1 KJ472923. Journal of Environmental Management 170, 8–14.

Margesin, R., Schinner, F., 2001. Biodegradation and bioremediation of hydrocarbons in extreme environments. Applied Microbiology and Biotechnology 56 (5), 650–663.

Mathur, A.K., Majumder, C.B., 2010. Kinetics modelling of the biodegradation of benzene, toluene and phenol as single substrate and mixed substrate by using *Pseudomonas putida*. Chemical and Biochemical Engineering Quarterly 24 (1), 101–109.

McCutcheon, S.C., Jørgensen, S.E., 2008. Phytoremediation. In: Fath, S.E.J.D. (Ed.), Encyclopedia of Ecology. Academic Press, Oxford, pp. 2751–2766.

Mohan, S.V., Kisa, T., Ohkuma, T., Kanaly, R.A., Shimizu, Y., 2006. Bioremediation technologies for treatment of PAH-contaminated soil and strategies to enhance process efficiency. Reviews in Environmental Science and Biotechnology 5 (4), 347–374.

Mueller, J.G., Chapman, P.J., Pritchard, P.H., 1989. Creosote-contaminated sites. Their potential for bioremediation. Environmental Science and Technology 23 (10), 1197–1201.

Mwandira, W., Nakashima, K., Kawasaki, S., 2017. Bioremediation of lead-contaminated mine waste by *Pararhodobacter* sp. based on the microbially induced calcium carbonate precipitation technique and its effects on strength of coarse and fine grained sand. Ecological Engineering 109 (Part A), 57–64.

Naghdi, M., Taheran, M., Brar, S.K., Kermanshahi-pour, A., Verma, M., Surampalli, R.Y., 2018. Removal of pharmaceutical compounds in water and wastewater using fungal oxidoreductase enzymes. Environmental Pollution 234, 190–213.

Najar, I.A., Khan, A.B., 2013. Management of fresh water weeds (macrophytes) by vermicomposting using *Eisenia fetida*. Environmental Science and Pollution Research 20, 6406–6417.

Pan, L.-J., Li, J., Li, C.-X., Tang, X.-D., Yu, G.-W., Wang, Y., 2018a. Study of ciprofloxacin biodegradation by a *Thermus* sp. isolated from pharmaceutical sludge. Journal of Hazardous Materials 343, 59–67.

Pan, X., Wang, S., Shi, N., Fang, H., Yu, Y., 2018b. Biodegradation and detoxification of chlorimuron-ethyl by *Enterobacter ludwigii* sp. CE-1. Ecotoxicology and Environmental Safety 150, 34–39.

Pandey, G., Paul, D., Jain, R.K., 2005. Conceptualizing "suicidal genetically engineered microorganisms" for bioremediation applications. Biochemical and Biophysical Research Communications 327, 637–639.

Paul, D., Poddar, S., Sar, P., 2014. Characterization of arsenite-oxidizing bacteria isolated from arsenic-contaminated groundwater of West Bengal. Journal of Environmental Science and Health. Part A, Toxic/hazardous Substances and Environmental Engineering 49, 1481–1492.

Peng, L., Chen, X., Xu, Y., Liu, Y., Gao, S.-H., Ni, B.-J., 2015. Biodegradation of pharmaceuticals in membrane aerated biofilm reactor for autotrophic nitrogen removal: a model-based evaluation. Journal of Membrane Science 494, 39–47.

Plósz, B.G., Langford, K.H., Thomas, K.V., 2012. An activated sludge modeling framework for xenobiotic trace chemicals (ASM-X): assessment of diclofenac and carbamazepine. Biotechnology and Bioengineering 109, 2757–2769.

Popa, C., Favier, L., Dinica, R., Semrany, S., Djelal, H., Amrane, A., Bahrim, G., 2014. Potential of newly isolated wild *Streptomyces* strains as agents for the biodegradation of a recalcitrant pharmaceutical, carbamazepine. Environmental Technology 35, 3082–3091.

Qu, M., Chen, J., Huang, Q., Chen, J., Xu, Y., Luo, J., Wang, K., Gao, W., Zheng, Y., 2018. Bioremediation of hexavalent chromium contaminated soil by a bioleaching system with weak magnetic fields. International Biodeterioration and Biodegradation 128, 41–47.

Rajkumar, M., Sandhya, S., Prasad, M.N.V., Freitas, H., 2012. Perspectives of plant-associated microbes in heavy metal phytoremediation. Biotechnology Advances 30, 1562–1574.

Reis, A.C., Čvančarová, M., Liu, Y., Lenz, M., Hettich, T., Kolvenbach, B.A., et al., 2018. Biodegradation of sulfamethoxazole by a bacterial consortium of *Achromobacter* denitrificans PR1 and *Leucobacter* sp. GP. Applied Microbiology and Biotechnology 102, 10299–10314.

Roh, H., Subramanya, N., Zhao, F., Yu, C.P., Sandt, J., Chu, K.H., 2009. Biodegradation potential of wastewater micropollutants by ammonia-oxidizing bacteria. Chemosphere 77, 1084–1089.

Romero, E., Salido, A., Cifuentes, C., Fernandez, J.D., Nogales, R., 2006. Effect of vermicomposting process on pesticide sorption capability using agro-industrial wastes. International Journal of Environmental Analytical Chemistry 86, 289–297.

Sandhya, S., Padmavathy, S., Swaminathan, K., Subrahmanyam, Y.V., Kaul, S.N., 2005. Microaerophilic–aerobic sequential batch reactor for treatment of azo dyes containing simulated wastewater. Process Biochemistry 40, 885–890.

Sangwan, P., Kaushik, C.P., Garg, V.K., 2008. Vermiconversion of industrial sludge for recycling the nutrients. Bioresource Technology 99, 8699–8704.

Santoso, H., Gunawan, T., Jatmiko, R.H., Darmosarkoro, W., Minasny, B., 2011. Mapping and identifying basal stem rot disease in oil palms in North Sumatra with Quick Bird imagery. Precision Agriculture 12, 233–248.

Shanmugam, S., Ulaganathan, P., Sivasubramanian, S., Esakkimuthu, S., Krishnaswamy, S., Subramaniam, S., 2017. *Trichoderma asperellum* laccase mediated crystal violet degradation–optimization of experimental conditions and characterization. Journal of Environmental Chemical Engineering 5, 222–231.

Sayler, G.S., Ripp, S., 2000. Field applications of genetically engineered microorganisms for bioremediation processes. Current Opinion in Biotechnology 11, 286–289.

Singh, J.S., Abhilash, P.C., Singh, H.B., Singh, R.P., Singh, D.P., 2011. Genetically engineered bacteria: an emerging tool for environmental remediation and future research perspectives. Gene 480, 1–9.

Singh, N., Gupta, S., Marwa, N., Pandey, V., Verma, P.C., Rathaur, S., Singh, N., 2016. Arsenic mediated modifications in *Bacillus aryabhattai* and their biotechnological applications for arsenic bioremediation. Chemosphere 164, 524–534.

Singh, P., Jain, R., Srivastava, N., Borthakur, A., Pal, D.B., Singh, R., Madhav, S., Srivastava, P., Tiwary, D., Mishra, P.K., 2017. Current and emerging trends in bioremediation of petrochemical waste: a review. Critical Reviews in Environmental Science and Technology 47 (3), 155e201.

Sinha, A., Osborne, W.J., 2016. Biodegradation of reactive green dye (RGD) by indigenous fungal strain VITAF-1. International Biodeterioration and Biodegradation 114, 176–183.

Sinha, A., Pant, K.K., Khare, S.K., 2012. Studies on mercury bioremediation by alginate immobilized mercury tolerant *Bacillus cereus* cells. International Biodeterioration and Biodegradation 71, 1–8.

Sinha, S., Singh, R., Chaurasia, A.K., Nigam, S., 2016. Self-sustainable *Chlorella pyrenoidosa* strain NCIM 2738 based photobioreactor for removal of Direct Red-31 dye along with other industrial pollutants to improve the water-quality. Journal of Hazardous Materials 306, 386–394.

Sponza, D.T., Isık, M., 2004. Decolorization and inhibition kinetic of Direct Black 38 azo dye with granulated anaerobic sludge. Enzyme and Microbial Technology 34, 147–158.

Solis-Mejia, L., Islas-Espinoza, M., Esteller, M.V., 2012. Vermicomposting of sewage sludge: earthworm population and agronomic advantages. Compost Science and Utilization 20, 11–17.

Sondhia, S., Rajput, S., Varma, R.K., Kumar, A., 2016. Biodegradation of the herbicide penoxsulam (triazolopyrimidine sulphonamide) by fungal strains of *Aspergillus* in soil. Applied Soil Ecology 105, 196–206.

Stapleton, R.D., Savage, D.C., Sayler, G.S., Stacey, G., 1998. Biodegradation of aromatic hydrocarbons in an extremely acidic environment. Applied and Environmental Microbiology 64 (11), 4180–4184.

Tastan, B.E., Tekinay, T., Çelik, H.S., Özdemir, C., Cakir, D.N., 2017. Toxicity assessment of pesticide triclosan by aquatic organisms and degradation studies. Regulatory Toxicology and Pharmacology 91, 208−215.

Teixeira, S., Vieira, M.N., Espinha Marques, J., Pereira, R., 2014. Bioremediation of an iron-rich mine effluent by Lemna minor. International Journal of Phytoremediation 16, 1228−1240.

Vasiliadou, I.A., Molina, R., Martínez, F., Melero, J.A., 2013. Biological removal of pharmaceutical and personal care products by a mixed microbial culture: sorption, desorption and biodegradation. Biochemical Engineering Journal 81, 108−119.

Vettori, L., Missaglia, A., Felici, C., Russo, A., Tamantini, I., Carrozza, G.P.C., Cinelli, F., Toffanin, A., 2012. Erratum to "Bioremedation and phytoremedation: synergism in lead extraction from contamined soils" [J. Biotechnol. 150S (2010) S509−S510]. Journal of Biotechnology 160, 268.

Vigueros, L.C., Camperos, E.R., 2002. Vermicomposting of sewage sludge: a new technology for Mexico. Water Science and Technology 46, 153−158.

Wang, L.Q., Falany, C.N., James, M.O., 2004. Triclosan as a substrate and inhibitor of 3 -phosphoadenosine 5-phosphosulfate-sulfotransferase and UDP-glucuronosyl transferase in human liver fractions. Drug Metabolism and Disposal 32, 1162−1169.

Wang, J., Sandoval, K., Ding, Y., Stoeckel, D., Minard-Smith, A., Andersen, G., Dubinsky, E.A., Atlas, R., Gardinali, P., 2016a. Biodegradation of dispersed Macondo crude oil by indigenous Gulf of Mexico microbial communities. The Science of the Total Environment 557−558, 453−468.

Wang, L., Barrington, S., Kim, J.W., 2007. Biodegradation of pentyl amine and aniline from petrochemical wastewater. Journal of Environmental and Management 83 (2), 191−197.

Wang, M., Cai, C., Zhang, B., Liu, H., 2018a. Characterization and mechanism analysis of lincomycin biodegradation with Clostridium sp. strain LCM-B isolated from lincomycin mycelial residue (LMR). Chemosphere 193, 611−617.

Wang, P., Liu, H., Fu, H., Cheng, X., Wang, B., Cheng, Q., Zhang, J., Zou, P., 2015. Characterization and mechanism analysis of penicillin G biodegradation with Klebsiella pneumoniae Z1 isolated from waste penicillin bacterial residue. Journal of Industrial and Engineering Chemistry 27, 50−58.

Wang, S., Wang, X., Zhang, C., Li, F., Guo, G., 2016b. Bioremediation of oil sludge contaminated soil by landfarming with added cotton stalks. International Biodeterioration and Biodegradation 106, 150−156.

Wang, X., Hou, X., Liang, S., Lu, Z., Hou, Z., Zhao, X., Sun, F., Zhang, H., 2018b. Biodegradation of fungicide Tebuconazole by Serratia marcescens strain B1 and its application in bioremediation of contaminated soil. International Biodeterioration and Biodegradation 127, 185−191.

Weisener, C.G., Guthrie, J.W., Smeaton, C.M., Paktunc, D., Fryer, B.J., 2011. The effect of Ca−Fe−As coatings on microbial leaching of metals in arsenic bearing mine waste. Journal of Geochemical Exploration 110 (1), 23−30.

Xia, W., Du, Z., Cui, Q., Dong, H., Wang, F., He, P., Tang, Y., 2014. Biosurfactant produced by novel Pseudomonas sp. WJ6 with biodegradation of n-alkanes and polycyclic aromatic hydrocarbons. Journal of Hazardous Materials 276, 489−498.

Xiong, J.-Q., Kurade, M.B., Jeon, B.-H., 2017. Biodegradation of levofloxacin by an acclimated freshwater microalga, Chlorella vulgaris. Chemical Engineering Journal 313, 1251−1257.

Yang, X., Wei, H., Zhu, C., Geng, B., 2018. Biodegradation of atrazine by the novel Citricoccus sp. strain TT3. Ecotoxicology and Environmental Safety 147, 144−150.

Yeom, S.H., Kim, S.H., Yoo, Y.J., Yoo, I.S., 1997. Microbial adaptation in the degradation of phenol by Alcaligenes xylosoxidans Y234. Korean Journal of Chemical Engineering 14 (1), 37−40.

Zhang, X., Liu, X., Wang, Q., Chen, X., Li, H., Wei, J., Xu, G., 2014. Diesel degradation potential of endophytic bacteria isolated from Scirpus triqueter. International Biodeterioration and Biodegradation 87, 99−105.

Zhao, R., Wang, B., Cai, Q.T., Li, X.X., Liu, M., Hu, D., Guo, D.B., Wang, J., Fan, C., 2016. Bioremediation of hexavalent chromium pollution by Sporosarcina saromensis M52 isolated from offshore sediments in Xiamen, China. Biomedical and Environmental Sciences 29 (2), 127−136.

Zhao, J.L., Zhang, Q.Q., Chen, F., Wang, L., Ying, G.G., Liu, Y.S., et al., 2013. Evaluation of triclosan and triclocarban at river basin scale using monitoring and modeling tools: implications for controlling of urban domestic sewage discharge. Water Research 47, 395−405.

Pollution status and biodegradation of organophosphate pesticides in the environment

Mohd Ashraf Dar[1], *Garima Kaushik*[1],
Juan Francisco Villareal Chiu[2]

[1]Department of Environmental Science, School of Earth Sciences, Central University of Rajasthan, Ajmer, Rajasthan, India; [2]Universidad Autónoma de Nuevo León, Facultad de Ciencias Químicas, Laboratorio de Biotecnología. Av. Universidad S/N Ciudad Universitaria, San Nicolás de los Garza, Nuevo León, Mexico

1. Introduction

Pesticide is a composite term that covers a wide range of chemical compounds that are utilized to counteract, kill, or control various insect pests such as insecticides (insects), molluscicides (mollusks), fungicides (fungi), herbicides (weeds), rodenticides (rodents), nematocides (nematodes), and plant growth promoters (Aktar et al., 2009). EPA has defined pesticides as any substance or combination of substances proposed for preventing, destroying, repelling, or mitigating the pests, or as plant regulators, desiccant, defoliant, or nitrogen stabilizer. Pesticides consist of two main ingredients: active and inert. Active ingredient performs the main function like control of pests, while inert components, such as edible oils, herbs, spices, cellulose, etc., are added with the active component to make pesticides and perform a vital role in effectiveness and performance of pesticides (EPA 2017a). The recurrent application of pesticides is implemented in modern agriculture at an extensive rate to fulfill the increasing demand of yield. Pesticides are applied annually in millions of tons throughout the world, covering the market of billions of dollars (EPA 2017b). Pesticide consumption rate in diverse countries depends on their agricultural area and type of yield. Top 10 pesticide-consuming (kg/Year) countries in the world are Italy (63,305,000), Turkey (60,792,400), Colombia (48,618,470), India (40,379,240), Japan (36,557,000), Bolivia (31,566,760), Ecuador (31,203,100), Germany (27,585,490), Romania

25

(26,506,740), and Chile (18,032,000). The top 10 countries with higher agricultural lands in square kilometers (Km^2) globally are India (1,797,590), Ecuador (749,770), Colombia (425,030), Ukraine (412,670), Turkey (390,120), Bolivia (369,650), Peru (214,700), United Kingdom (172,240), Germany (167,000), and Chile (157,430) (Verma et al., 2014). Pesticides have played a significant role in Green Revolution by counteracting the pest attack, which would otherwise reduce the quantity and quality of agricultural production (Wilson and Tisdell, 2001), and played an essential role in fulfilling the requirements of tremendously increasing population. However, Green Revolution has led to various problems such as soil fertility loss, acidification of soil, nitrate leaching, resistance of species toward pesticides, and loss of biological diversity (Tilman et al., 2002; Verma et al., 2013). Pesticides also assisted in improving the nutritional value and safety of food (Damalas and Eleftherohorinos, 2011) and save up to 40% of crop damages because of the attack of pests; however, their misuse or overuse leads to environmental contamination (water, soil, air) with the pesticide residues and results in direct and indirect hazards for both humans and the environment (Richardson, 1998; Damalas, 2009). Two million tonnes of pesticides are consumed annually worldwide, out of which 45% is shared by Europe, 25% by the United States, and 25% rest of the world. The share of pesticide consumption all over the world consists of 47.5% of herbicides, 29.5% of insecticide, 17.5% of fungicide, and others account only 5.5% (De et al., 2014). Dichlorodiphenyltrichloroethane (DDT) was the first pesticide produced in 1874. However, 20,000 chemicals were registered by the US EPA as pesticides in 1998 (Garcia et al., 2012). The extensive and injudicious utilization of these chemicals leads to severe environmental problems, as less than 0.1% of the applied pesticide reaches the target and the rest (99.9%) remains in the environment, resulting in detrimental impacts on human health, plants, animals, and also contamination of soil, water, and air environments (Pimentel, 1995). The utilization of pesticides periodically worsens the situation, and repetitions for longer period lead their accumulation in various environments because of their direct relation, risking the entire ecosystem by their manifold toxicity (Javaid et al., 2016). The persistence of these chemicals in the environment is so often that their residues remain in soil and sediments for longer periods after their application and finally find their entrance into water (surface and groundwater) and food chain (Eevers et al., 2017). However, various remediation technologies are available for decontaminating the polluted sites. Conventional technologies include physicochemical methods such as incineration, burning, landfilling, composting, and chemical modification (Kempa, 1997) But because of their ex situ nature, they are costly and time-consuming because the contaminated matrix has to dug-up and transported to the treatment facility and are invasive, resulted in the destruction of ecosystems. Therefore, from several years interest was growing for the development of in situ technologies to remediate contaminated sites because of their eco-friendly, low-cost, low-maintenance and renewable nature (Chaudhry et al., 2005). Remediation technology uses plants and microorganisms (bacteria, fungi etc.) to convert/degrade the toxic chemicals into less, or nontoxic constituents that have been increasingly researched. US EPA has defined bioremediation as a treatment technology which uses biological activity (plants and microorganisms) to break down the pollutant to decrease its concentration and toxicity (Mcguinness and Dowling, 2009). Therefore, the objective of this study is to provide an overview of reports on organophosphate pesticide pollution and application of indigenous microorganisms for their biodegradation and detoxification.

2. Organophosphates and other pesticides

Table 2.1 displays different types of pesticides and their use. However, classification of pesticides can be done in various ways, but they are mostly classified by their chemical composition, which keeps pesticide groups in a uniform and systematic way to create a connection between activity, structure, toxicity, and degradation mechanisms, among others (Laura et al., 2013). According to chemical composition, pesticides are mainly grouped into four categories, namely organochlorine, carbamates, pyrethroids, and organophosphate. Organochlorine pesticides (e.g., DDT, endosulfan, aldrin) are composed of chlorine, carbon, hydrogen, and occasionally oxygen atoms. Because of nonpolar and lipophilic (lipid soluble) nature of these pesticides, they get accumulated in the fatty tissue of animals and are transported through the food chain; they are toxic to a diverse group of animals and insects by interrupting their nervous system, resulting in paroxysms, paralysis, and finally death, and have longer environmental persistence. However, carbamate pesticides (e.g., carbaryl, carbofuran, aminocarb) are carbamic acid derivatives and their chemical structure is based on plant alkaloid *Physostigma venenosum*. They have higher toxicity toward vertebrates and have relatively low environmental persistence. Pyrethroid pesticides (e.g., cypermethrin, permethrin) are chemical compounds similar to synthetic pyrethrins. They have neurotoxicity and are extremely toxic toward fish and other insects but have lower toxicity toward mammals and avifauna and less environmental persistence than other pesticides. However, organophosphate pesticides (e.g., chlorpyrifos, parathion, diazinon) are composed of a central phosphorus atom in the molecule. They are stable and have less toxicity as compared with organochlorine pesticides and can be aliphatic, cyclic, and heterocyclic. They are solvable both in water and organic solvents. They infiltrate into groundwater and have less persistence as compared with chlorinated hydrocarbons and have higher toxicity toward vertebrates and invertebrates because of the inhibition of cholinesterase enzyme resulting in impulse failure, paralysis, and finally death (Ortiz-Hernández and Sánchez-Salinas, 2010; Castrejón-Godínez, Sánchez-Salinas and Ortiz-Hernández, 2014; Yadav and Devi, 2017). Pesticides can be systematic or nonsystematic. Systematic pesticides are absorbed by the animal or plants and can penetrate effectively into the tissues of plant to kill particular pests, whereas nonsystematic (contact) pesticides kill pests when they come in contact with them, and it does not necessarily enter into the tissues of plant (Yadav and Devi, 2017). However, according to toxicity, WHO has classified pesticides as extremely hazardous (class Ia), highly hazardous (class Ib), moderately hazardous (class II), and slightly hazardous (class III) (WHO, 2009).

Organophosphates, comprising of phosphorus, carbon, and oxygen (P−O−C) bonds, are mainly used in controlling pests because of their degradable organic nature and less persistence, as compared with chlorinated and carbamate compounds (Yang et al., 2005). OPs are amides, esters, or thiol derivatives of phosphoric acid. R_1 and R_2 are alkyl or aryl groups, and their linkage with phosphorus through various atoms leads to the formation of various compounds. Phosphate is formed when linked through oxygen, although bonding through sulfur leads to the formation of phosphorothiolate or S-substituted phosphorothioate, the compound in which phosphorus is linked to sulfur via a double bond is termed as phosphorothioates; however, when the carbon atom is bonded with phosphorus through an NH

TABLE 2.1 Pesticide types and their use.

S.No.	Type of pesticide	Purpose (used for)
01	Algicides	Kill algae in lakes, canals, swimming pools, water tanks, and other sites.
02	Antifoulants	Kill or repel organisms that attach to underwater surfaces, such as barnacles that cling to boat bottoms.
03	Fungicides	Kill fungi (including blights, mildews, molds, and rusts).
04	Fumigants	Produce gas or vapor intended to destroy pests in buildings or soil.
05	Herbicides	Kill weeds and other plants that grow where they are not wanted.
06	Insecticides	Kill insects and other arthropods.
07	Miticides (acaricides)	Kill mites that feed on plants and animals.
08	Microbial pesticides	Microorganisms that kill, inhibit, or outcompete pests, including insects or other microorganisms.
09	Molluscicides	Kill snails and slugs.
10	Nematicides	Kill nematodes (microscopic, worm-like organisms that feed on plant roots).
11	Ovicides	Kill eggs of insects and mites.
12	Pheromones	Biochemicals used to disrupt the mating behavior of insects.
13	Repellents	Repel pests, including insects (such as mosquitoes) and birds.
14	Rodenticides	Control mice and other rodents.
15	Defoliants	Cause leaves or other foliage to drop from a plant, usually to facilitate harvest.
16	Desiccants	Promote drying of living tissues, such as unwanted plant tops.
17	Insect growth regulators	Disrupt the molting, maturity from pupal stage to adult or other life processes of insects.
18	Plant growth regulators	Substances (excluding fertilizers or other plant nutrients) that alter the expected growth, flowering, or reproduction rate of plants.
19	Antimicrobials	Kill microorganisms such as bacteria and viruses.
20	Attractants	Lure pests to a trap or bait, for example, attract an insect or rodent into a trap.
21	Biocides	Kill microorganisms.
22	Biopesticides	Derived from natural materials such as animals, plants, bacteria, and certain minerals and are used to control pests.
23	Disinfectants and sanitizers	Kill or inactivate disease-producing microorganisms on inanimate objects.
24	Plant-incorporated protectants	Substances that plants produce from genetic material that has been added to them and provides self-protection against pests.

Adapted from type of pesticide ingredients EPA. What Is a Pesticide? January 2017a. http://www.epa.gov; EPA. Pesticides Industry, Sales and Usage. January 2017b, http://www.epa.gov.

group, the resulted compound is termed as phosphoramidates. X can be aliphatic, aromatic, or heterocyclic substituted or branched groups, bonded to phosphorus through an $-O-$ or $-S-$ to make it more liable and is referred as leaving the group (Vale and Lotti, 2015). General formula, biodegradation pathway, and chemical structures of some organophosphates are illustrated in Fig. 2.1. Organophosphate pesticide (OPP) accounts for about 38% of total consumed pesticides globally (Singh and Walker, 2006). The degradable nature of OPP under environmental conditions made them an important substitute to the persistent pesticides such as DDT, aldrin, dieldrin, etc. Rapid degradation of OPP through hydrolysis on exposure to air, soil, and sunlight has been reported by various studies and has been experimentally demonstrated in mustard fields in Bikaner, Rajasthan (Dhas and Srivastava, 2010). OPP is extensively applied in agriculture, horticulture, veterinary medicine, domestic purposes, and also for the control of disease vectors. Some OPs such as malathion are used to treat head lice, scabies, and crab lice in humans. OP nerve agents are also utilized in terrorist attacks and as warfare agents (Vale and Lotti, 2015. Besides their faster degradation, their residual concentrations have been detected in water, soil, food, human fluids, etc., and are relatively water-soluble, highly toxic, and absorbed through all routes such as inhalation, ingestion, and dermal absorption (Maurya and Malik, 2016).

3. Effect of pesticides

In spite of the beneficial outcomes of pesticide application in agriculture and public health sectors, their indiscriminate usage also leads to harmful impacts on environment and public health and holds a unique place among various contaminants of environment because of their higher biological activity and toxicity. Pesticides are potentially dangerous toward humans, animals, other living creations, and the environment if not used properly (Yadav and Devi, 2017). It has been estimated that poisoning of pesticides results in about 220,000 deaths and 3 million poisoning cases annually. Poisoning of these compounds is about 13-fold more than the poisoning in developed countries (Gunnell and Eddleston, 2003; Chandra et al., 2015), out of which at least 50% of intoxicated and 75% of deaths are triggering in workers who are involved in agricultural activities, and the remaining are being poisoned because of the consumption of contaminated food (Yadav and Devi, 2017). The people working in agricultural fields and inhabitants of agricultural areas are mostly poisoned from drift exposure, and fumigations of soil were a significant threat (Lee et al., 2011).

3.1 Effects on human health

Bioaccumulation process begins, when the runoff from pesticide-contaminated agricultural land areas, during rainfall, storms or through other process reaches into the waterbodies such as streams, rivers, and finally into the oceans. These pesticides are ingested by fishes through gills or scales from the water column, get sequestered into their organs and fat tissues, ultimately get accumulated into the food chain, and finally reach the human body (Maurya and Malik, 2016).

Application of pesticides on agricultural products, mainly fruits and vegetables, gets discharged into the soil, leached into groundwater and finally reaches the drinking water, and also gets drifted leading to air pollution. The harmful effect of pesticides on human

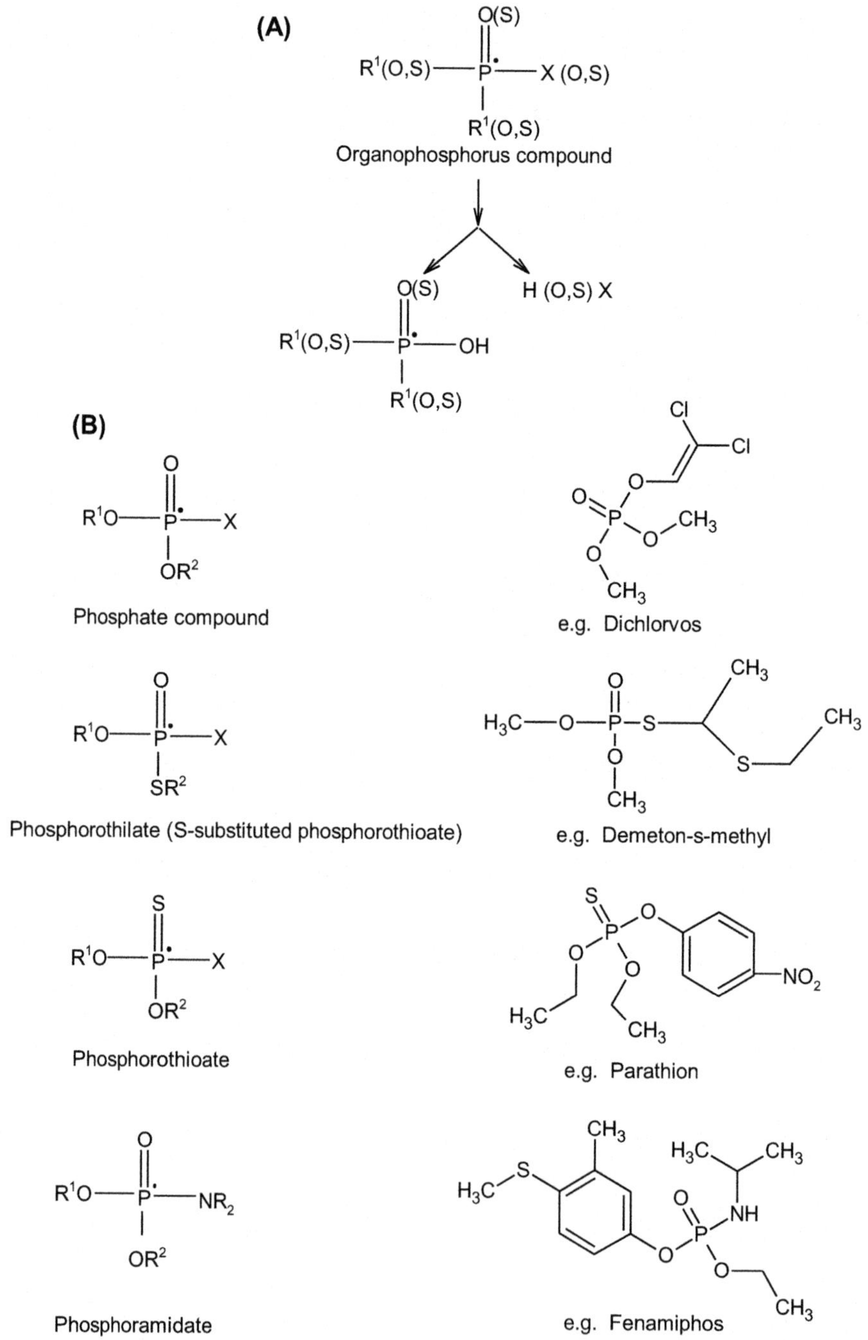

FIGURE 2.1 (A) General formula of organophosphates and their major biodegradation pathway. (B) Chemical structures of some organophosphates. *(A) Adopted from Kumar et al., 2018.*

health is also growing because of their toxicity and environmental persistence and their capability to find their way into the food chain. Pesticides may find their entrance into the human body by direct chemical contact (dermal route), through the ingestion of food especially contaminated fruits, vegetables, and water or through inhalation of pesticide dust, mist or fumes, and polluted air (Sacramento, 2008). The modes of pesticide exposure and their metabolic routes are illustrated in Fig. 2.2. The degree of detrimental impact of these chemicals on human health is determined by their toxicity, length, and magnitude (Lorenz, 2009). Chemical toxicity depends on the nature of toxicant, exposure routes, dosage, and organisms and can be acute or chronic. Acute toxicity is the development of harmful effects in a short period after exposure, whereas chronic toxicity defines the adverse impacts resulting from long period of exposure. The pesticide toxicity is generally expressed as LD_{50} (lethal dose) or LC_{50} (lethal concentration). LD_{50} is the amount of chemical, which leads to the death of 50% of the pest population in a single dose, whereas LC_{50} refers to the chemical concentration in air, water, or surrounding the experimental animals which kills 50% of test population (Yadav and Devi, 2017).

3.1.1 Acute effect

Acute effects in humans are generally caused through pesticide exposure during their application and intentional or unintentional poisoning (Lee et al., 2011; Dawson et al., 2010). The symptoms of acute pesticide poisoning include skin rashes, headaches, body

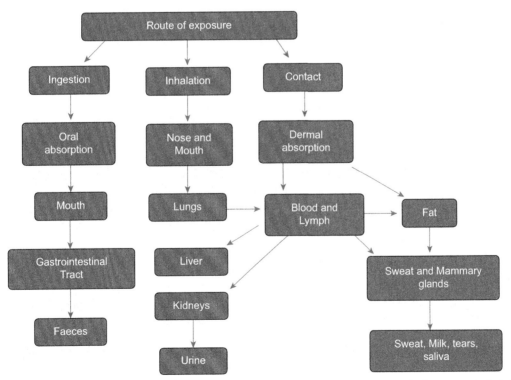

FIGURE 2.2 Pesticide exposure modes and their metabolic routes through different body organs until their excretion. *Modified from Sharma and Goyal, 2014.*

aches, nausea, poor concentration, cramps, impaired vision, panic attacks, dizziness, and in severe cases coma and death (PAN, 2012). It has been reported that annually 3 million cases of acute poisoning take place globally, out of which 2 million are suicide attempts and the remaining are due to occupational or accidental cases of poisoning (Singh and Mandal, 2013).

3.1.2 Chronic effect

Constant exposure to pesticides for an extended period, i.e., several years to decades, leads to chronic diseases in humans. Chronic effects of pesticides include congenital disabilities, benign or malignant tumors, genetic changes, fetus toxicity, nerve disorders, blood disorders, endocrine disruption, and reproduction effects. Symptoms of poisoning are not noticed immediately but appeared in later stage (PAN 2012). Chronic diseases are not easily diagnosed, and recently numerous studies have established an association between pesticide exposure and occurrence of chronic diseases in humans affecting reproductive, nervous, renal, cardiovascular, and respiratory systems (Mostafalou and Abdollahi, 2012).

3.2 Environmental impact

The widespread pesticide application and their subsequent disposal by farmers, industries, and others releases various potential pesticide into the environment. The effects of pesticides are broad even after its application on a small area, as it gets absorbed in soil or drifted into the air or dissolves in waterbodies and ultimately reaches a bigger range of area. Pesticides may have different fates on their release into the environment. The pesticides applied in agricultural fields get drifted in the air and finally end up in soil, sediment, water, etc. However, the pesticide, which is directly applied to soil, gets washed away and reaches the adjacent surface waterbodies via runoff or gets percolated to lower layers of soil and finally to groundwater (Harrison, 1990).

3.3 Impact on nontarget organisms

The impact of pesticides on nontarget creatures has gained researchers attention for decades universally, as less than 0.1% of applied pesticides reach the target (Yadav et al., 2016). It has been reported that pesticide application adversely affects the nontarget arthropods, animals, and green plants (Ware, 1980); additionally, the natural enemies of insects such as parasites and predators are harshly affected because of their vulnerability to these chemicals (Vickerman, 1988). The natural enemies play a crucial role in controlling and regulating the pest population. However, their destruction by the application of pesticides exacerbates pest attacks and results in the additional spraying of chemicals to control these target pests. Pesticides affect the beneficial soil invertebrates such as nematodes, mites, springtails, earthworms, spiders, microarthropods, insects, and other microorganisms which constitute the food web in soil, decompose organic substances such as manure, leaves, residues of plants, etc., and are also crucial for maintaining soil structure, transformation, and mineralization of organic compounds, hence leading to a detrimental impact on some links in the food web (Maurya and Malik, 2016).

3.4 Effects on the microbial diversity of soil

The indiscriminate and frequent employment of pesticides leads their accumulation in soil, which affects the properties of soil and its microflora and may also undergo different types of degradation, transport, and adsorption/desorption procedures (Hussain et al., 2009). The degraded pesticides or their metabolites alter the microbial diversity of soil, biochemical, and enzymatic activity of indigenous microorganisms (Hussain et al., 2009; Munoz-Leoz et al., 2011), which may ultimately affect the critical process in soil such as nitrogen fixation, nitrification, and ammonification through the activation or deactivation of specific micro-organisms of soil or their enzymes (Hussain et al., 2009; Munoz-Leoz et al., 2011). These chemicals interact with a microbial diversity of soil and their metabolic processes (Singh and Walker, 2006) and may result in the alteration of the physiological and biochemical activity of microorganisms. Several studies had also demonstrated the detrimental impact on microbial biomass and respiration of soil because of these pesticides (Pampulha and Oliveira, 2006; Zhou et al., 2006) and usually reduction in soil respiration results in the reduction of microbial biomass (Klose and Ajwa, 2004).

3.5 Pesticide resistance

IRAC has defined pesticide resistance as the inherited variation in the sensitivity of pests, produced because of repeated failure of a pesticide to attain the expected level of control, when applied for a particular pest as per the recommended label (IRAC, 2010). The indiscriminate utilization of pesticides increases the resistant pest population by providing them particular benefit in the presence of a pesticide, which would otherwise be very rare and continue to multiply and then develop into the dominant share of the population over generations. As the number of resistant individuals increases, the effectiveness of pesticide remains no longer and results in the development of pesticide resistance (Maurya and Malik, 2016). It has been reported that indiscriminate pesticide usage has led to the development of resistance toward pesticides in different targeted pest species worldwide (Tabashnik et al., 2009). Even control of pests has become very challenging in some cases such as essential crop pests, livestock parasites, urban pests, and vectors of various diseases because of their pesticide resistance development to higher extant (Van Leeuwen et al., 2010; Gondhalekar et al., 2011). However, pesticide resistance development has been influenced by various factors such as genetic, biological, and operational elements (Georghiou and Taylor, 1977). One of the frequently used methods for detecting resistance is insecticide bioassays using whole insects (Gondhalekar et al., 2013). However, various new approaches from the last two decades, employing biochemical methods, techniques at the molecular level, and insecticide bioassay combinations, have been established for detecting pesticide resistance (Zhou et al., 2002; Scharf et al., 1999).

4. Toxicological mechanism of organophosphates

The acute exposure to OPP primarily affects parasympathetic, sympathetic, and central nervous system. Acetylcholine (Ach) is a neurotransmitter for transmission of nerve impulse in brain, skeletal muscles, and other parts, where nerve impulses occur. OPP prevents the breakdown of Ach (a neurotransmitter) by inhibiting acetylcholinesterase enzyme (AChE), which is responsible for its hydrolysis (Ecobichon and Joy, 1993) and which leads to accumulation of Ach in autonomic ganglia, neuromuscular junctions, and in the central nervous system leading to overstimulation of nerves and subduing of neurotransmission to organs (Paudyal, 2008). OPP binds to AChE very avidly and shares a chemical structure of close similarity. The schematic representation of the possible reactions of inhibition, reactivation, and aging of AChE by OPP is illustrated in Fig. 2.3. The rate and degree of acetylcholinesterase inhibition depends on the structure of OPP and nature of their metabolites (intermediates). The interaction of OPP with AChE leads to the formation of AChE-OP complex, with the occurrence of two reactions. The first reaction involves the spontaneous reactivation of enzyme at slow speed, much slower than the inhibition of enzyme and needs hours to days to take place. The rate of this regenerative process depends exclusively on the type of OPP. Dimethyl and diethyl compounds have spontaneous reactivation half-life of 0.7 and 31 h, respectively. Generally, the complex of AChE-dimethyl reactivates simultaneously within less than 1 day, while AChE-diethyl complex takes several days and in such conditions, the newly activated enzyme can be reinhabited significantly. The reactivation process can be accelerated by the addition of nucleophilic reagents such as oximes, releasing more active enzymes. These agents thus act as an antidote in the poisoning of OP (Eddleston et al., 2002). In the second reaction, with the passage of time, AChE-OP complex loses one alkyl group making it no longer responsive toward reactivating agents. This following time-dependent process is referred to as aging, and its rate depends on many factors such as temperature, pH, and OP compound. Aging half-life of dimethyl is 3.7 h, while diethyl has 33 h (Worek and Diepold, 1999; Worek et al., 2014). Slower the spontaneous reactivation, more significant amount of inactive AChE will be offered for aging. Oximes reduce the amount of inactive AChE that is available for aging by catalyzing the regeneration of active AChE from the complex. Thus, aging will take place more rapidly with dimethyl OPs; hypothetically oximes are useful before 12 h in dimethyl poisoning and many days in case of diethyl OP intoxication days (Worek and Diepold, 1999; Worek et al., 2014). The Ach buildup at the motor nerves results in weakness, muscle cramps, fasciculation, fatigue, and muscular weakness of respiratory muscles. Accumulation at autonomic ganglia increases the heartbeat and blood pressure, pallor, and hypoglycemia. However, visual disorders, chest tightness, and wheezing due to bronchoconstriction and increased bronchial secretions and lacrimation, sweating, peristalsis, salivation, and urination are caused as a result of Ach accumulation at muscarinic receptors, whereas accumulation of Ach in central nervous system results in anxiety, headache, convulsions, confusion, ataxia, depression of respiration and circulation, slurred speech, tremor, and generalized weakness (Eddleston et al., 2002; Sherman, 1995).

5. Status of organophosphate pesticide pollution

The food demand till 2020 has been projected, and it shows that, from 1995 to 2020 years, the requirement of food grains is expected to be doubled, and vegetables 2.5 times and fruits 5 times. Therefore, pesticide consumption is expected to be increased by at least 2–3 times in

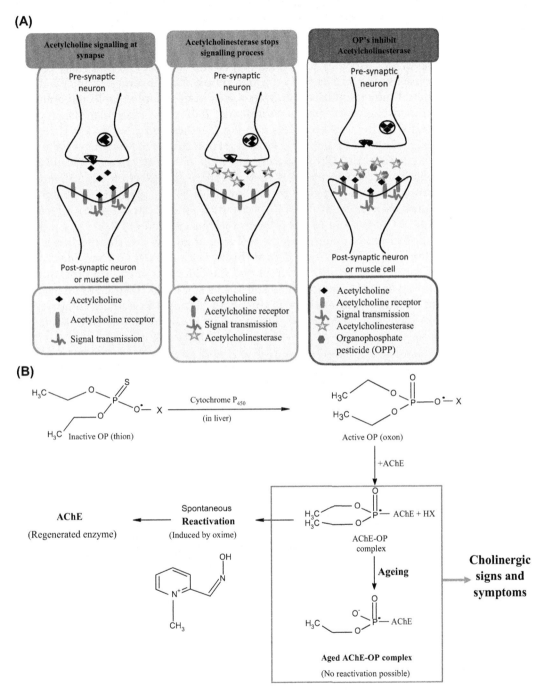

FIGURE 2.3 (A) Reactivation and aging of acetylcholinesterase enzyme by organophosphate pesticide. (B) Schematic representation of the possible reactions of inhibition. *(A) Source: http://depts.washington.edu/opchild/acute. html. (B) Adopted from Kazemi et al., 2012.*

the upcoming years (Sarnaik, 2004). The contamination of OPP has been reported worldwide in agricultural soil, water, sediment, fruits, and also in human fluids, etc., through residual analysis of these pesticides (Table 2.2). It is well-documented that soil acts as a potential source for the transport of pesticides to contaminate water (surface/ground), air, plants, and food and finally reaches the human body through runoff and subsurface drainage, interflow, leaching, transfer of mineral nutrients and pesticides from soil into plants and animals, which constitute the food chain of humans (Abrahams, 2002). The molecular structure of some of the OPPs used worldwide is exemplified in Fig. 2.4. According to SACONH report, the residues of ethion, malathion, and phorate have been detected in the sediments, sampled from Tarnadmund, Nedugula, and Bison Swamp wetlands of Nilgiris district, Tamil Nadu (SACONH report). Similarly, soil samples of paddy-wheat, paddy-cotton, and sugarcane fields of Hisar, Haryana, were reported to be contaminated with chlorpyrifos, malathion, and quinalphos residues (Kumari et al., 2008). Chlorpyrifos residues in the soil samples obtained from nut fields of China were found to be in the concentration of $0.00-77.2$ $\mu g k g^{-1}$ (Han et al., 2017). Another study determines the residues of ethion and chlorpyrifos in the soil samples, taken from tea fields of West Bengal (Bishnu et al., 2009). In a similar type of study carried out by Jacob et al. (2014), they reported the detection of chlorpyrifos, quinalphos, and ethion residues in the cardamom field soil samples of Idukki district, Kerala, above their MRL (maximum residual limit) values. In this series, three OPPs namely chlorpyrifos, diazinon, and ethion have also been traced in soil, sediment, and sludge samples taken from Turia river, Spain, in the range of $0.18-70.3$ ngg^{-1} (Masiá et al., 2015).

Water pollution caused by OPP is also a threat for the deterioration of different environments worldwide. The riverine water pollution and depletion of its resources put the lives of numerous people in danger (Chimwanza et al., 2006). Moreover, it has also been reported that water resources including surface, ground, and drinking water are contaminated with pesticides worldwide (Table 2.2). According to a report, chlorpyrifos-ethyl concentration was detected in the range of $0.0002-0.004$ mgL^{-1} in water samples taken from certain lakes of Bijapur, Karnataka (Pujeri et al., 2011). Similarly, Ahad et al. (2000) have also reported the presence of OPP namely dichlorvos, dimethoate, methyl parathion, fenitrothion, and chlorpyrifos in water samples taken from Mardan division of Pakistan in the range of $0.0-0.45$ $\mu g L^{-1}$. The concentration of ethion pesticide has been detected in a small amount in waterbodies adjacent to tea fields of some West Bengal regions (Bishnu et al., 2009). The detection percentage of methyl parathion and monocrotophos in groundwater, sampled from open wells around intensive cotton-growing districts of Pakistan, was found in the range of $0\%-5.4\%$ and $24.3\%-35.1\%$ in the months of October and July, respectively (Ilyas et al., 2004). Groundwater sampled from tube wells near paddy, sugarcane, and cotton fields in Hisar, Haryana, was testified to be contaminated with chlorpyrifos residues above the regulatory limits (Kumari et al., 2008). Ma et al., (2009) reported the presence of four OPPs, i.e., dichlorvos, methyl parathion, malathion, and parathion in underground water samples taken from Qingdao, China, by the utilization of SPE-GC-MS (solid-phase extraction-gas chromatography-mass spectrometry) technique. In the same study, different pesticides of organophosphate category, such as chlorpyrifos, dichlorvos, ethion, parathion methyl, profenofos, and phorate, have been detected in surface and groundwater samples of Vidarbha region, Maharashtra, and it has been observed that the pesticide levels were more in surface water as compared with groundwater (Lari et al., 2014). Another survey carried out by Dehghani et al. (2012) reported the contamination of

TABLE 2.2 Different samples reported to be contaminated with the residues of organophosphates.

Type of sample	Study area	Detected organophosphate	Technique used	References
Sediment from wetlands	Nilgiris district	Phorate, malathion, and ethion	GC	SACONH report
Soil from paddy-wheat, paddy-cotton, and sugarcane fields	Hisar, Haryana	Chlorpyrifos, malathion, and quinalphos	GC-ECD, GC-NPD	Kumari et al. (2008)
Soil from nut fields	Nut producing belt, China	Chlorpyrifos	GC-ECD, GC-FPD	Han et al. (2017)
Soils from tea fields	West Bengal	Ethion and chlorpyrifos	GC-NPD	Bishnu et al. (2009)
Soil from cardamom field	Idukki, Kerala	Chlorpyrifos, quinalphos, and ethion	GC-ECD, GC-FPD, GC-MS	Jacob et al. (2014)
Soil, sediment, and sludge samples from Túria River	Spain	Chlorpyrifos, diazinon, ethion	LC-MS/MS	Masiá et al. (2015)
Lake water	Bijapur, Karnataka	Chlorpyrifos ethyl	GC-NPD, GC-MS/MS	Pujeri et al. (2011)
Groundwater samples from wells, tube wells, and hand pumps	Mardan, Pakistan	Dichlorvos, dimethoate, methyl parathion, fenitrothion, chlorpyrifos	GC-ECD	Ahad et al. (2000)
Water samples adjacent to the tea garden	West Bengal	Ethion	GLC-NPD	Bishnu et al. (2009)
Water samples from open well of cotton-growing districts	Pakistan	Methyl parathion, monocrotophos	GC-ECD, GC-NPD	Tariq et al. (2004)
Groundwater from tube wells	Hisar, Haryana	Chlorpyrifos	GC-ECD, GC-NPD	Kumari et al. (2008)
Underground water samples	Qingdao, China	Dichlorvos, methyl parathion, malathion, parathion	GC-MS	Ma et al. (2009)
Surface and groundwater	Vidarbha region of Maharashtra	Dichlorvos, ethion, parathion methyl, phorate, chlorpyrifos, and profenofos	GC-ECD, GC-MS	Lari et al. (2014)
Water sample	Zayandehroud spring and Saadabad river, Iran	Chlorpyrifos, diazinon	HP-TLC	Dehghani et al. (2012)
Tea	Across India	Monocrotophos and ethion	Not mentioned	Greenpeace India report (2014)
Tea samples from growers	Balkans region, Spain	Chlorpyrifos	GC-MS/MS	Beneta et al. (2018)

(Continued)

TABLE 2.2 Different samples reported to be contaminated with the residues of organophosphates.—cont'd

Type of sample	Study area	Detected organophosphate	Technique used	References
Tea leaves of *Camellia sinensis* obtained from tea factories	Tamil Nadu	Ethion and quinalphos	GC-NPD	Kottiappan et al. (2013)
Tea samples collected from tea factories	Tamil Nadu	Ethion and quinalphos	GC-ECD, GC-NPD	Seenivasan and Muraleedharan (2011)
Different types of tea samples from retail commercial outlets	USA	Chlorpyrifos, triazophos, dicrotophos	GC-MS, LC-MS	Hayward et al. (2015)
Made tea, fresh tea leaves	Hill and Dooars regions of West Bengal	Ethion and chlorpyrifos	GLC-NPD	Bishnu et al. (2009)
Rice samples from paddy fields	Korea	Chlorpyrifos, fenthion	GC-MS-SIM	Nguyen et al. (2008)
Maize and soy	Piedmont region, Italy	Chlorpyrifos	GC-MS-SIM	Marchis et al. (2012)
Seasonal vegetables	Northern India	Methyl parathion, chlorpyrifos, and malathion	GC-ECD	Bhanti and Taneja (2007)
Vegetable samples from random selling points	Karachi, Pakistan	Chlorpyrifos, dimethoate, fenitrothion, methamidophos, methyl parathion, profenophos, monocrotophos	HPLC, GC-FID	Praveen et al. (2005)
Winter vegetable	Hisar, Haryana	Dimethoate, malathion, fenitrothion, monocrotophos, phosphamidon, quinalphos, and chlorpyrifos	GC-ECD, GC-NPD	Kumari et al. (2003)
Farm gate seasonal vegetables	Hisar, Haryana	Monocrotophos, quinalphos, and chlorpyrifos	GLC-ECD, GC-NPD	Kumari et al. (2004)
Farm gate seasonal vegetables	Ramanagara district of Karnataka	Acephate, chlorpyrifos, dichlorvos, monocrotophos, phorate, and profenofos	GLC-ECD, GLC-FTD	Gowda and Somasekhar (2012)
Samples from open fields and greenhouse	Beijing, China	Acephate	GC-FPD, GC-MS/MS	Chuanjiang, 2010 (residual analysis)
Leafy, root, modified stem, and fruity vegetables from local market	Lucknow City	Anilophos, dichlorvos, dimethoate, diazinon, and malathion	GC-ECD, GC-NPD	Srivastava et al. (2011)
Seasonal vegetables (cauliflower and capsicum)	Nanded, Maharashtra	Chlorpyrifos and monocrotophos	GC-MS	Chandra et al. (2014)

TABLE 2.2 Different samples reported to be contaminated with the residues of organophosphates.—cont'd

Type of sample	Study area	Detected organophosphate	Technique used	References
Farm vegetables	In and around Delhi	Malathion, quinalphos, chlorpyrifos	GC-ECD	Mukharjee (2003)
Tomato and cucumber samples	Almaty, Kazakhstan	Chlorpyrifos, dimethoate, chlorpyrifos ethyl, triazophos	GC-ECD, GC-NPD	Lozowicka et al. (2015)
Farm gate and market vegetables	Jaipur	Monocrotophos, quinalphos, dimethoate, chlorpyrifos	GC-NPD	Singh and Gupta (2002)
Vegetable from street outlets	Hyderabad AP	Chlorpyrifos, triazophos, acephate, fenitrothion, diazinon, monocrotophos, quinalphos, etc.	LC-MS/MS	Sinha et al. (2012)
Cabbage and radish samples from local markets and supermarkets	Central area, Korea	Diazinon	GC-MS-SIM	Nguyen et al. (2008)
Farm gate samples of cauliflower	Punjab	Acephate, chlorpyrifos, profenophos, quinalphos, fenamiphos	GC-ECD, GC-FTD, GC-MS	Mandal and Singh (2010)
Fruits from local market and vegetables from farmers' fields	Kuwait	Malathion, profenofos, monocrotophos, pirimiphos methyl, diazinon, chlorpyrifos-methyl	GC-MS. LC-MS/MS	Jallow et al., 2017
Mango from farmers field	Multan, Pakistan	Monocrotophos, methyl parathion	GC-ECD	Hussain et al. (2002)
Bayberry samples from orchards and retail markets	Zhejiang province, China	Chlorpyrifos	GC-FPD, LC-MS/MS	Yang et al. (2017)
Fruits such as banana, guava, orange, grapes, etc.	Andhra Pradesh	Dimethoate, chlorpyrifos, profenophos, quinalphos, and malathion	GC-MS	Harinathareddy et al. (2014)
Apple and citrus fruit samples from different selling points	Karachi, Pakistan	Chlorpyrifos, methamidophos, methyl parathion, monocrotophos, profenophos, quinalphos	GC-FID, HPLC	Praveen et al. (2004)
Butter and ghee	Cotton growing belt of Haryana	Chlorpyrifos	GC-NPD	Kumari et al. (2005)
Honey from bee keepers	Himachal Pradesh	Malathion, dimethoate, and quinalphos	GC-NPD	Choudhary and Sharma (2008)
Different commercial fruit juices purchased from supermarkets	Madrid, Spain	Diazinon, chlorpyrifos, ethion	GC-NPD, GC-MS	Albero et al., (2003)

(Continued)

TABLE 2.2   Different samples reported to be contaminated with the residues of organophosphates.—cont'd

Type of sample	Study area	Detected organophosphate	Technique used	References
Soft drinks	Gujarat and Maharashtra	Chlorpyrifos and malathion	GC-NPD	CSE report (2006)
Human blood	Punjab	Monocrotophos, chlorpyrifos, malathion, and phosphamidon	GC-NPD	CSE report (2005)
Urban residents and farmers urine samples	Near Shandong province, China	Chlorpyrifos	UPLC-MS/MS	L Wang et al. (2016)
Blood samples of chick, goat, fish (Rita rita), and man	Utter Pradesh, India	Chlorpyrifos	GLC-ECD	Singh et al. (2008)
Tissue samples of different fish species	Chad river, North eastern, Nigeria	Dichlorvos, diazinon, chlorpyrifos, fenitrothion	GC/MS-FD	Akan et al. (2014)
Tissue samples of fishes *Channa striata* and *Catla catla*	Kolleru Lake, Andhra Pradesh	Malathion and chlorpyrifos	GC-ECD	Amaraneni and Pillala (2001)
Breast milk	Bhopal, Madhya Pradesh	Malathion, chlorpyrifos, and methyl parathion	GC-ECD	Sanghi et al. (2003)
Bovine milk	Allahabad	Methyl parathion	GC-ECD	Srivastava et al. (2008)

water sampled from Zayandeh Roud spring and Saadabad river of Iran by chlorpyrifos and diazinon pesticides, and their residual concentrations were found as 11.79 ppb and 22.43 ppm, respectively. To put these records in perspective and protect the aquatic life, UK environmental agency had established the environmental quality standards for annual average exposure of different pesticides in fresh and marine water (EA, 1996).

The extensive pesticide utilization has not threatened our environment alone; contamination of agricultural products including tea, vegetables, fruits, and sugars by OPP residues has also been demonstrated (Table 2.2). According to Greenpeace India report, (2014), a likely mutagenic and neurotoxic pesticide, monocrotophos, has been detected in 27 tea samples, taken across India, within the concentration range of 0.026—0.270 mgkg^{-1} of different brands manufactured by different companies including Golden Tips, Tata, Kho Cha, Hindustan Unilever, Goodricke, Wagh Bakri, and Royal Girnar. Monocrotophos is not permitted for application on tea and is classified by the WHO as a highly hazardous pesticide (class Ib). However, the detection of another pesticide ethion in 22 tea samples is mostly because of its direct application on tea (Greenpeace India report, 2014). Beneta et al. (2018) have reported the detection of chlorpyrifos residues in tea samples, obtained from growers in Balkan region, Spain, at the concentration of 303 μgkg^{-1}. Similarly, tea leaves of *Camellia sinensis* species obtained from tea factories in southern India have been found to be contaminated with the deposits of ethion and quinalphos pesticides (Kottiappan et al., 2013). In a similar study on tea samples obtained from tea factories across different districts of Tamil Nadu, India, it

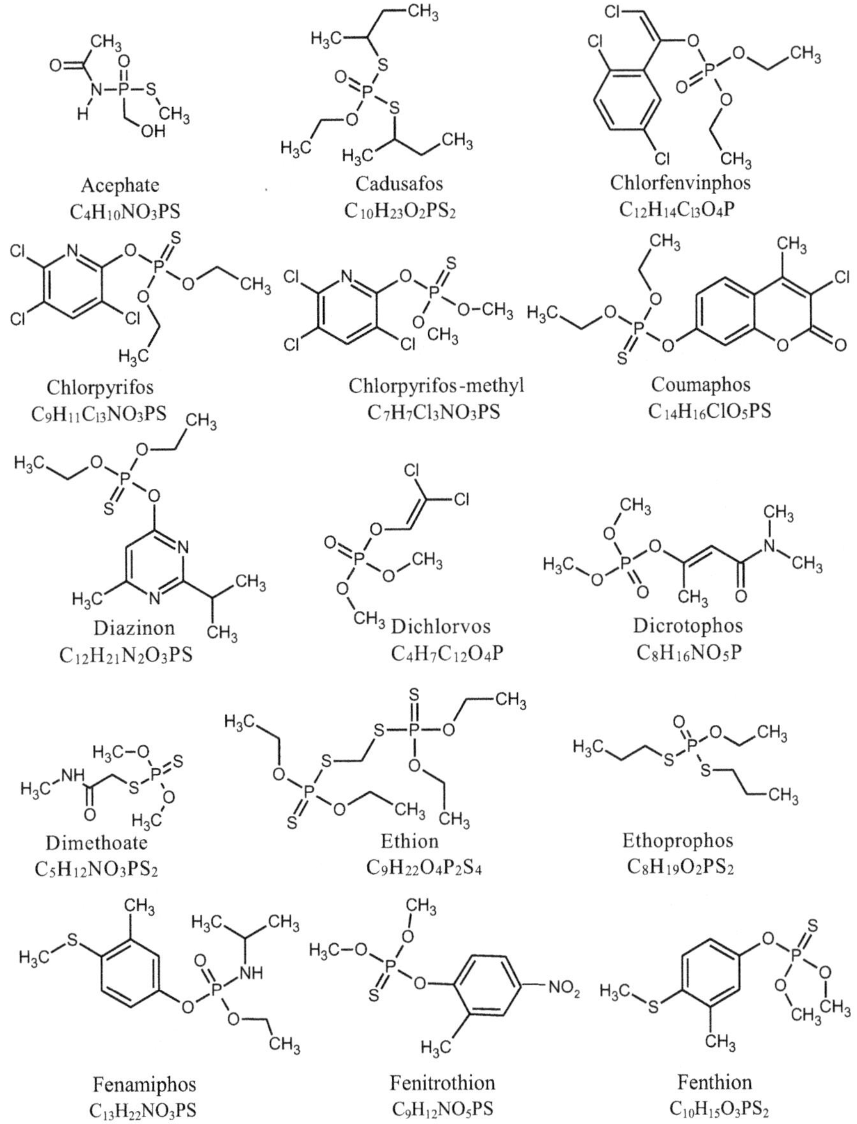

FIGURE 2.4 Molecular structure of organophosphate pesticides used globally.

has been reported that out of 912 samples, only 0.5% samples were contaminated with ethion and quinalphos residues below their MRL values along with other pesticide categories (Seenivasan and Muraleedharan, 2011). Similarly, Hayward et al. (2015) have also testified the presence of three OPPs namely chlorpyrifos, triazophos, and dicrotophos in black, green, white, and oolong tea samples, obtained from retail commercial outlets, USA. Leaves of fresh tea and made tea sampled from the regions of Hill and Dooars of West Bengal were also reported to be contaminated with ethion and chlorpyrifos. However, in this study, pesticides of

Isofenphos
C$_{15}$H$_{24}$NO$_4$PS

Malathion
C$_{10}$H$_{19}$O$_6$PS$_2$

Monocrotophos
C$_7$H$_{14}$NO$_5$P

Parathion
C$_{10}$H$_{14}$NO$_5$PS

Parathion-methyl
C$_8$H$_{10}$NO$_5$PS

Phorate
C$_7$H$_{17}$O$_2$PS$_3$

Phosphamidon
C$_{10}$H$_{19}$ClNO$_5$P

Profenofos
C$_{11}$H$_{15}$BrClO$_3$PS

Quinalphos
C$_{12}$H$_{15}$N$_2$O$_3$PS

Terbufos
C$_9$H$_{21}$O$_2$PS$_3$

Tetrachlorvinphos
C$_{10}$H$_9$Cl$_4$O$_4$P

Triazophos
C$_{12}$H$_{16}$N$_3$O$_3$PS

FIGURE 2.4 cont'd

organophosphate category were found higher in fresh tea leaves as compared with made tea leaves (Bishnu et al., 2009). Similarly, residues of OPP had also been found in food grains. According to a report, residues of chlorpyrifos and fenthion had been detected in rice paddy samples obtained from rice paddy fields of Korea, during the first stage of late growing season. Chlorpyrifos was detected in two samples in the range of 0.10–2.2 mgkg^{-1}, whereas fenthion was found in one sample at the concentration of 0.54 mgkg^{-1} (Dong et al., 2008). In a similar pesticide residual analysis study, carried out on maize and soy samples obtained from northern Italy were also found contaminated with Chlorpyrifos pesticide. In this study, residues of chlorpyrifos were traced in the range of 0.0074—12.4 mgkg^{-1} and 0.010—0.206 mgkg^{-1} in maize and soy, respectively (Marchis et al., 2012).

Vegetables are plant parts and are consumed by people worldwide as one of the essential components in their diet. Because of their easy pest infestation, OPPs are extensively used worldwide for the protection of vegetable cultivations against these pests. The residual contamination of vegetables and fruits with OPP has indicated their extensive and unregulated application in agriculture. Different OPPs, such as methyl parathion, chlorpyrifos, and malathion, have been found in vegetables of distinct seasons in northern India. Their residual concentrations were found to be below the established tolerance limits, but their constant consumption can result in accumulation of these pesticides in the receptors body and may prove deadly in the long term (Bhanti and Taneja, 2007). Similarly, Parveen et al. (2005) have reported the presence of six OPPs namely chlorpyrifos, dimethoate, fenitrothion, methamidophos, methyl parathion, profenophos, and monocrotophos in vegetable samples obtained from various selling points of Karachi, Pakistan. In this study, it was also reported that out of 206 samples of 27 different vegetables, 35% of samples were contaminated with OPP and root/tuberous vegetables were detected to be the most contaminated, whereas MRL violation was found greater in leafy vegetables. Residual concentrations of different OPPs, such as dimethoate, phosphamidon, malathion, quinalphos, fenitrothion, chlorpyrifos, and monocrotophos, were detected to be the highest followed by carbamates, pyrethroids, and organochlorines above their respective MRL levels in the edible portions of winter vegetable samples obtained from wholesale market at Hisar, Haryana, by GC-ECD and GC-NPD techniques (Kumari et al., 2003). In a similar study carried out by Kumari et al. (2004), over 84 seasonal farm gate vegetable samples, including brinjal, knol khol, cauliflower, pea, okra, cabbage, cucumber, etc., residues of chlorpyrifos, quinalphos, and monocrotophos, were detected above the MRL values in some samples. Residues of acephate, dichlorvos, monocrotophos, profenofos, phorate, and chlorpyrifos were detected in farm gate seasonal vegetables such as cauliflower, capsicum, okra, and tomato by GC-ECD and FPD technique; samples were obtained from vegetable growing area of Ramanagara district, Karnataka, India, above their respective MRL levels (Gowda and Somasekhar, 2012). Residues of moderately hazardous (class II) pesticide acephate were detected in the pakchoi (*Brassica campestris* L.) samples by gas chromatography (GC) tandem mass spectrometry (MS)/MS after the application of acephate on an open field and greenhouse pakchoi in Tongzhou district, Beijing, China. Acephate concentration was found higher in greenhouse samples as compared with open field samples as >90% of acephate under field conditions gets degraded into its metabolite (methamidophos) while in greenhouse conditions only >50% of pesticide gets degraded (Chuanjiang et al., 2010). Similarly, vegetables such as leafy, modified stem, root and fruity vegetables (bitter gourd, French bean, onion, jackfruit, capsicum, spinach, potato, carrot, radish, cucumber, beetroot, cauliflower, cabbage, tomato etc.) obtained from a local market of Lucknow, India, were analyzed for residual concentration of pesticides by GC-ECD and GC-NPD, and it was reported that these samples were contaminated with different pesticides especially OP category, i.e., dichlorvos, dimethoate, anilophos, diazinon, and malathion, above their respective maximum residual limit (Srivastava et al., 2011), whereas chlorpyrifos and monocrotophos pesticides were detected below their respective maximum residual limit in some seasonal vegetables (cauliflower and capsicum) collected from a local market of Nanded, Maharashtra, India (Chandra et al., 2014). Similarly, a study carried out by Mukherjee, (2003) reported the residual contamination of three OPPs: malathion, quinalphos, and chlorpyrifos in vegetable samples including cabbage, cauliflower, chili, eggplant, tomato, mustard, onion, and okra, collected in and around Delhi, India, above their prescribed limit of tolerance. As per a report, vegetables including tomato and cucumber obtained from Almaty region of Kazakhstan were found to be contaminated with chlorpyrifos, dimethoate, and triazophos above their respective MRL, whereas chlorpyrifos was detected below its MRL established by the European Union and Custom Union regulation for tomatoes and cucumbers (Lozowicka et al., 2015). In this series, study over different vegetables, viz. tomato, cabbage, brinjal, chili, cauliflower, cucumber, bitter gourd, onion, bottle gourd, and okra, obtained from agricultural fields and vegetable markets near Jaipur, Rajasthan, India, exposed their contamination with OPP residues, viz. monocrotophos, quinalphos, dimethoate, and chlorpyrifos from below to above their residual limits (Singh and Gupta, 2002). In a similar kind of study over vegetables (tomato, eggplant, cabbage, cauliflower, ladyfinger) obtained from farmers' market and local street outlets of Hyderabad, India, contamination of these vegetables with different OPPs by applying LC-MS/MS technique was revealed (Sinha et al., 2012). The residual analysis study over cabbage and radish samples, collected from local and supermarkets in central Korea, revealed that Korean cabbage is contaminated with moderately toxic diazinon pesticide residues while no residues were found in radish samples (Dong et al., 2008). In a similar survey carried out by Mandal and Singh (2010), they reported the contamination of cauliflower (*Brassica oleracea*), obtained from intensive vegetable growing areas of Punjab, India, with different OPPs below their respective MRL. Vegetables and fruits which are commonly consumed in Kuwait such as bell pepper, eggplant, tomato, cucumber, zucchini, carrot, potato, and cabbage and fruits (strawberry, watermelon, apple, and grapes) were found, contaminated with different OPPs including malathion, profenofos, monocrotophos, pirimiphos-methyl, diazinon, and chlorpyrifos-methyl. The samples of these vegetables and fruits were collected from farmers' fields, and local markets of Kuwait and their contamination with the above-mentioned pesticides were found, below to above their MRL (Jallow et al., 2017). However, not only vegetables but also the good source of vitamins and minerals, i.e., fruits, were found to be contaminated with different organophosphates globally (Table 2.2). According to an article, OPPs such as monocrotophos and methyl parathion have been detected in mango (*Mangifera indica*) samples obtained from a farmer's field in Multan division of Pakistan. The residual concentrations of these pesticides were found, below their respective MRLs set by the FAO/WHO Codex Alimentarius Commission (Hussain et al., 2002). Similarly, the study over an economically important Chinese fruit, bayberry (*Myrica rubra*), collected from retail markets and orchards in Zhejiang province, China, revealed that this tasty and healthy fruit is contaminated with the residues of moderately toxic OPP chlorpyrifos below its MRL (Yang et al., 2017). In a similar survey carried out by Harinathareddy et al. (2014), the detection of various OPPs in different fruit samples collected from farmer's field of Andhra Pradesh, India, was reported. The presence of organophosphates such as chlorpyrifos, methamidophos, methyl parathion, monocrotophos, profenophos and quinalphos has also been reported in apple and citrus fruit (orange, grapefruit, Kino, and lemon) samples, obtained from diverse selling points of Karachi, Pakistan, by applying HPLC (high-performance liquid chromatography) detection technique and was found above their respective MRLs in some samples (Parveen et al., 2004).

The presence of OPP residues is not limited only to fruits, vegetables, and other agricultural products, but they have also shown their presence in specific animal tissues and food products including butter, ghee, honey, juices, soft drinks, etc. (Table 2.2). Butter and ghee samples were obtained randomly from rural and urban zones of three cotton-growing belts of Haryana, India, to known the contamination status of ghee and butter from a safety point of view toward consumers and were found to be contaminated with chlorpyrifos above its respective maximum residual limit (Kumari et al., 2005). Honey, being a natural product synthesized by honey bees, is thought to be free from any external material, but the samples of honey obtained from commercial beekeepers of Himachal Pradesh, India, were found to be contaminated with three organophosphates including malathion, dimethoate, and quinalphos residues (Choudhary and Sharma, 2008). According to a study, samples of different commercial juices including apple, grape, peach, pineapple, and orange juices, procured from a supermarket in Madrid, Spain, were testified to be contaminated with the residues of moderately toxic diazinon, chlorpyrifos, and ethion pesticides (Albero et al., 2003). In a similar investigation, carried out over soft drinks (Coco-Cola and Limca), mass-produced in Gujarat and Maharashtra were also demonstrated to be contaminated with chlorpyrifos and malathion pesticides. Chlorpyrifos was reported in 100% of analyzed samples, within the range of 0.17–20.43 ppb, i.e., 200 times the BIS (Bureau of Indian Standards) limit of 0.1 ppb for individual pesticides, whereas malathion was found in 39% samples within the range of 0.00–3.11 ppb (CSE report 2006).

Organophosphates are less persistent as compared with other pesticide classes and have less bioaccumulation potential, but their residues have been found worldwide in human blood, urine, fish tissues, breast milk, bovine milk, etc. (Table 2.2). There are various reports regarding the presence of OPP in human and animal parts and their (pesticides) cumulative exposure comes from water, food, air, soil, dust, etc. The presence of organophosphates in blood samples indicates that these pesticides persist in the body for more extended period. As per a report, chlorpyrifos, monocrotophos, malathion, and phosphamidon were detected in blood samples of human beings, collected from different villages of Punjab, India (CSE report 2005). In another study, residues of chlorpyrifos were detected by UPLC-MS/MS (ultrahigh-performance liquid chromatography system-mass spectrometer) technique in urine samples collected from adult farmers and urban residents in Shandong Province, China (Wang et al., 2016). Similarly, residues of chlorpyrifos were found in fish, chick, goat, and man because of their transport to other ecosystems. During this study, blood samples of fish (*Rita rita* Ham.) were collected from Gomti River, blood samples of chick (*Gallus gallus)* and goat (*Capra hircus* Linn.) from the local market, and donated blood samples of human (*Homo sapiens)* in Utter Pradesh, India. All these blood samples were confirmed to be contaminated with the residues of chlorpyrifos pesticide. However, the maximum level was found in fish and minimal in man in the order of fish ˃ chick > goat > human (Singh et al., 2008). In a similar study, organophosphates such as dichlorvos, diazinon, chlorpyrifos, and fenitrothion were detected in the tissues of four fish species namely *Clarias gariepinus, Hetrotis niloticus, Oreochromis niloticus and Tilapia zilli,* collected from Lake Chad in northeastern Nigeria. In this study, maximum pesticide concentration was found in the liver, and the lowest concentration was recorded in the flesh, with chlorpyrifos most abundant ranging from

1.92 to 3.21 μgg^{-1}, whereas diazinon was recorded in lowest concentration ranging from 1.13 to 1.77 μgg^{-1} and accumulation of pesticides in the tissues was in the order of liver > gills > stomach > flesh (Akan et al., 2014). Similarly, residues of malathion and chlorpyrifos had been found as well in two fish species (*Channa striata* and *Catla catla*), collected from largest natural freshwater lake (Kolleru) of Andhra Pradesh, India. Residual concentration of pesticides was detected higher in *C. striata* than *C. catla*, and this variation might be because of their food habits and lipid content, and pesticide concentration was traced to be higher in the liver followed by gills and the muscles (Amaraneni and Pillala, 2001). Residues of organophosphates such as malathion, chlorpyrifos, and methyl parathion were also detected in the breast milk samples of women (age 19–45 years), belonging to the lower socioeconomic class of Madhya Pradesh, India (Sanghi et al., 2003). OPP contaminates the herbaceous vegetation (livestock feed), and it was reported that cows feeding on pesticide-contaminated fodder produce milk with higher pesticide concentration than those feeding on uncontaminated feed (Fagnani et al., 2011). According to a report, bovine milk samples acquired from rural and urban dairies of Allahabad region, India, were found to be contaminated with residues of extremely hazardous (class Ia) pesticide, methyl parathion. The load of methyl parathion was higher in rural dairy samples, which might be due to large-scale agricultural activities and unsafe handling in those areas (Srivastava et al., 2008).

6. Degradation of organophosphate pesticides

The damage caused by pesticide residues to environmental health forced people to think about their safe elimination in an eco-friendly and economically efficient way and has led to the development of treatment techniques having safe and economically efficient methodology than conventional treatment methods, along with the avoidance of additional environmental damage. Biological methods have been used for the treatment of pesticide-contaminated wastes and sites (Araya and Lakhi, 2004). Several studies reported that degradation of xenobiotic substances by microorganisms is significant. The biological methods can be applied for the treatment of compounds whose chemical structure is rarely or inexistent in the environment because these compounds have been synthesized artificially (Ortiz-Hernández, and Sánchez-Salinas, 2001). The additional pesticide application from several ways to prevent and control pests, the biodegradation of these chemical compounds offers a chief, eco-friendly and effective solution for their disposal or agricultural soil treatment, water contamination or polluted ecosystems. The usage of pesticides is well-documented in agronomy, nonagriculture, and public well-being program. Among different pesticide classes, the world market is dominated by herbicides, whereas Indian markets are leaded by insecticides (Adityachaudhury et al., 1997).

Utilization of pesticides for increasing the agricultural yield has turned out to be indispensable. Owing to the biodegradable nature of organophosphates, they are extensively utilized in broad areas and have triggered inevitable pollution of the environment (Singh et al., 2014). Organophosphates are, however, easily degraded with comparatively low persistence in the environment but are readily soluble in water, which makes them highly vulnerable to human consumption, resulting in serious health threats. Therefore, degradation of OPP through biological or physiochemical methods is investigated intensively. However, degradation or removal of organophosphate contaminants is preferably done by microorganisms under laboratory settings, as its degradation rate is faster (one order of magnitude) than chemical hydrolysis, which is in turn roughly 10 times faster than photolysis (Ragnarsdottir, 2000). Microorganisms mostly degrade organophosphates by using them as carbon, nitrogen, or phosphorus sources. Environmental fate and degradation of pesticides in the environment is illustrated in Fig. 2.5. The first microorganism with organophosphate-degrading capability was isolated and identified as *Flavobacterium* sp. in 1973 (Singh and Walker, 2006), followed by the isolation and identification of diverse microbes (fungi, algae, bacteria) with OPP-degrading ability (Table 2.3).

According to a report, the *Exiguobacterium* sp. (BCH4) and *Rhodococcus* sp. (BCH2) isolated from pesticide-contaminated agricultural soil were found to degrade 75.85% of acephate after 6 days of incubation and also help in reducing the toxicity of acephate on earthworms (Phugare et al., 2012). *Pseudomonas aeruginosa* (strain Is 6) isolated from agricultural soil samples of Tamil Nadu, India, was reported to completely mineralize 100% of acephate (50 mgL^{-1}) within 96 h and was identified by HPLC and ESI-MS (electron spray ionization-mass spectrometry) techniques (Ramu and Seetharaman, 2014). Similarly, another

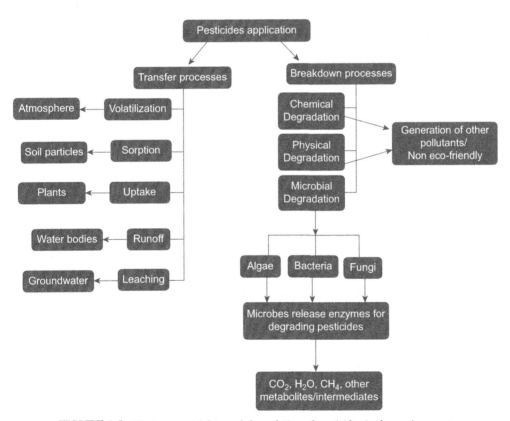

FIGURE 2.5 Environmental fate and degradation of pesticides in the environment.

TABLE 2.3 Bacterial/fungal strains isolated globally capable of degrading organophosphate pesticides.

Organophosphate pesticide	Isolated strains of Bacteria/Fungi	Isolation matrix (location)	Technique used	References
Acephate	*Exiguobacterium* sp., *Rhodococcus* sp.	Agricultural soil (Maharashtra, India)	HPLC, FTIR, and GC-MS	Phugare et al. (2012)
	Pseudomonas aerugtnosa	Agricultural soil (Tamil Nadu, India)	HPLC, ESI-MS	Ramu and Seetharaman (2014)
Cadusafos	*Pseudomonas putida*	Farm soil (Saudi Arabia)	GLC	Abo-Amer (2012)
	Sphingomonas sp., *Flavobacterium* sp.	Potato field soil (Greece)	GC-NPD	Karpouzas et al. (2005)
Chlorfenvinphos	*Arthrobacter* sp., *Mycobacterium* sp.	Petroleum-contaminated soil (Hilo, Hawaii, USA)	GC-MS	Seo et al. (2007)
	Penicillium citrinum,[a] *A. fumigatus*[a] *Aspergillus terreus*[a] *Trichoderma harzianum*[a]	Untreated surface water of Tagus river (Lisbon, Portugal)	HPLC	Oliveira et al. (2015)
Chlorpyrifos	*Stenotrophomonas* sp.	Industrial sludge (Jiangsu Province, China)	GC-FPD, HPLC	Deng et al. (2015)
	Cupriavidus sp.	Industrial sludge (Jiangsu, China)	HPLC	Lu et al. (2013)
	Serratia marcescens	Agricultural soil (Poland)	GC-ECD	Cycon et al. (2013)
	Sphingobacterium sp.	Soil from paddy field (Tamil Nadu, India)	HPLC, GC-MS	Abraham and Silambarasan (2013)
	P. putida, Klebsiella sp., *Pseudomonas stutzeri, Pseudomonas aeruginosa*	Soil from paddy field (Tamil Nadu, India)	LC-MS	Sasikala et al. (2012)
	Streptomyces chattanoogensis, Streptomyces olivochromogenes	Soil from blueberry field (Southern Chile)	HPLC, GC-NPD	Briceno et al. (2012)
	Pseudomonas stuzeri, Enterobacter aerogenes, Pseudomonas pseudoalcaligenes, Pseudomonas maltophilia, Pseudomonas vesicularis	Agricultural soil (Cairo and Giza, Egypt)	Not mentioned	Awad et al. (2011)
	Pseudomonas spp., *Agrobacterium* spp., *Bacillus* spp.	Agricultural farm soil (Varanasi, India)	HPLC	Maya et al. (2011)

TABLE 2.3 Bacterial/fungal strains isolated globally capable of degrading organophosphate pesticides.—cont'd

Organophosphate pesticide	Isolated strains of Bacteria/Fungi	Isolation matrix (location)	Technique used	References
	Pseudomonas aeruginosa, Pseudomonas nitroreducens, P. putida	Effluent storage ponds and moist soil (Iran)	HPLC	Latifi et al. (2012)
	Bacillus sp., *Pseudomonas* sp.	Soil from groundnut fields (Andhra Pradesh, India)	HPLC, GC-FID	Madhuri and Rangaswamy (2009)
	Leuconostoc mesenteroides, L. brevis, L. plantarum, Lactobacillus sakei	Kimchi during fermentation (Korea)	TLC, HPLC	Cho et al. (2009)
	Sphingomonas sp., *Stenotrophomonas* sp., *Bacillus* sp., *Brevundimonas* sp., *Pseudomonas* sp.	Soil and industrial water (Jiangsu, China)	HPLC	Li et al. (2008)
	Pseudomonas sp.	Wastewater irrigated agricultural soil (Uttar Pradesh, India)	HPLC, TLC	Bhagobaty and Malik (2008)
	Pseudomonas fluorescence, Brucella melitensis, Bacillus subtilis, Bacillus cereus, Klebsiella sp., *S. marcescens, P. aeruginosa*	Field soil (Punjab, India)	GC-ECD	Vidya Lakshmi et al. (2008)
	Serratia sp., *Trichosporon* sp.[a]	Activated sludge (Shandong, China)	GC-FPD, GC-MS, HPLC	Xu et al. (2007)
	Sphingomonas sp.	Industrial effluent (Nantong, China)	HPLC	Li et al. (2007)
Chlorpyrifos-methyl	*Burkholderia cepacia*	Soil from paddy field (Korea)	GC, GC-MS	Kim and Ahn (2009)
Coumaphos	*Leuconostoc mesenteroides, L. brevis, L. plantarum, L. sakei*	Kimchi during fermentation (Korea)	TLC, HPLC	Cho et al. (2009)
	Serratia marcescens	Agricultural soil (Saudi Arabia)	GC	Abo-Amer (2011)
Dichlorvos	*Proteus vulgaris, Acinetobacter* sp., *Serratia* sp., *Vibrio* sp.,	Agricultural farm soil (Nigeria)	Syringes	Agarry et al. (2013)
	Bacillus sp., *Pseudomonas* sp.	Soil from groundnut field (Andhra Pradesh, India)	HPLC, GC-FID	Madhuri and Rangaswamy (2009)
	Trichoderma atroviride[a]	Not mentioned	Not mentioned	Tang et al. (2009)

(Continued)

TABLE 2.3 Bacterial/fungal strains isolated globally capable of degrading organophosphate pesticides.—cont'd

Organophosphate pesticide	Isolated strains of Bacteria/Fungi	Isolation matrix (location)	Technique used	References
	Bacillus sp.	Soil from grape wine yard (Maharashtra, India)	UV-visible spectrophotometer, GC-MS	Pawar and Mali (2014)
	Ochrobactrum sp.	Sludge from wastewater (Jining, China)	GC-FID	Zhang et al. (2006a)
Diazinon	Stenotrophomonas sp.	Industrial sludge (China)	GC-FPD, HPLC	Deng et al. (2015)
	Serratia marcescens	Agricultural soil (Saudi Arabia)	GC	Abo-Amer (2011)
	Lactobacillus brevis	Not mentioned	GC-FPD	Zhang et al. (2014)
	Leuconostoc mesenteroides, L. brevis, L. plantarum, L. sakei	Kimchi during fermentation (Korea)	TLC, HPLC	Cho et al. (2009)
	Serratia liquefaciens, S. marcescens, Pseudomonas sp.	Agricultural soil (Poland)	GC-TSD	Cycon et al. (2009)
	Arthrobacter sp., Mycobacterium sp.	Petroleum-contaminated soil (Hilo, Hawaii, USA)	GC-MS, GC-FID	Seo et al. (2007)
Dimethoate	Paracoccus sp.	Activated sludge (china)	GC-MS, MS/MS	Li et al. (2010)
	Raoultella sp.	Industrial soil sample (China)	TLC	Liang et al. (2009)
	Aspergillus niger[a]	Sewage and soil from cotton fields (China)	GC	Liu et al. (2001)
Ethoprophos	Sphingomonas sp., Flavobacterium sp.	Potato field soil (Greece)	GC-NPD	Karpouzas et al. (2005)
	P. putida	Soil (Northern Greece)	GLC	Karpouzas et al. (2000)
Fenamiphos	P. putida, Acinetobacter rhizosphaerae	Soil farm banana field (Eastern Crete, Greece)	HPLC	Chanika et al. (2011)
	Microbacterium esteraromaticum	Turf green soil (South Australia)	HPLC	Caceres et al. (2009)
	Brevibacterium sp.	Soil (Adelaide Hills, Australia)	HPLC, GC-MS	Megharaj et al. (2003)

TABLE 2.3 Bacterial/fungal strains isolated globally capable of degrading organophosphate pesticides.—cont'd

Organophosphate pesticide	Isolated strains of Bacteria/Fungi	Isolation matrix (location)	Technique used	References
Fenitrothion	*Serratia marcescens*	Agricultural soil (Poland)	GC-ECD	Cyco'n et al. (2013)
	Burkholderia sp.	Wastewater sludge (China)	GC-FPD	Zhang et al. (2006b)
	Burkholderia spp., *Pseudomonas* spp., *Sphingomonas* spp., *Cupriavidus* sp., *Corynebacterium* sp., *Arthrobacter* sp.	Soil from agricultural field and golf course (Korea)	HPLC	Kim et al. (2009)
	Bartonella spp., *Rhizobium* sp., *Burkholderia* spp., *Cupriavidus* sp., *P. putida*	Soil (Japan)	HPLC	Tago et al. (2006)
Fenthion	*Bacillus safensis*	Pesticide contaminate soil (Hasahisa, Sudan)	GC, HPLC, GC-MS	Abdelbagi et al. (2018)
Parathion	*Stenotrophomonas* sp.	Industrial sludge (China)	GC-FPD, HPLC	Deng et al. (2015)
	Serratia marcescens	Agricultural soil (Poland)	GC-ECD	Cycon et al. (2013)
	Leuconostoc mesenteroides, *L. brevis*, *L. plantarum*, *L. sakei*	Kimchi during fermentation (Korea)	TLC, HPLC	Cho et al. (2009)
Phorate	*Ralstonia eutropha*, *Pseudomonas aeruginosa*, *Enterobacter cloacae*	Agricultural soil (Maharashtra, India)	GC-ECD	Rani and Juwarkar (2012)
	Bacillus sp., *Pseudomonas* sp.	Soil from groundnut fields (Andhra Pradesh, India)	HPLC, GC-FID	Madhuri and Rangaswamy (2009)
	Rhizobium sp., *Pseudomonas* sp., *Proteus* sp.	Agricultural soil (Aligarh, India)	HPLC	Bano and Musarrat (2003)
Tetrachlorvinphos	*Stenotrophomonas maltophilia*, *P. vulgaris*, *Vibrio metschnikovii*, *Serratia ficaria*, *Serratia* sp., *Yersinia enterocolitica*	Cornfield soil (Mexico)	GC-MS	Ortiz-Hernández and S'anchez-Salinas, 2010
Triazophos	*Stenotrophomonas* sp.	Industrial sludge (China)	GC-FPD, HPLC	Deng et al. (2015)

(Continued)

TABLE 2.3   Bacterial/fungal strains isolated globally capable of degrading organophosphate pesticides.—cont'd

Organophosphate pesticide	Isolated strains of Bacteria/Fungi	Isolation matrix (location)	Technique used	References
	Bacillus sp.	Soil samples from wastewater treatment plant (China)	HPLC	Tang and You (2012)
	Diaphorobacter sp.	Triazophos contaminated soil (Jiangsu, China)	HPLC, MS/MS	Yang et al. (2011)
Malathion	Acinetobacter johnsonii	Soil (suburbs of Beijing, China)	GC-NPD	Shan et al. (2009)
	Bacillus sp., Pseudomonas sp.	Agricultural soil (Assam, India)	Not mentioned	Baishya and Sharma (2014)
	Pseudomonas sp.	Agricultural soil (Pakistan)	HPLC	Jilani (2013)
	Pseudomonas sp., P. putida, Micrococcus lylae, Pseudomonas aureofaciens, Acetobacter liquefaciens	Soil from agricultural field (Cairo, Egypt)	HPLC, GC-ECD	Goda et al. (2010)
	Enterobacter aerogenes, Bacillus thuringiensis	Agricultural wastewater (Egypt)	HPLC, GC-MS	Mohamed et al. (2010)
Isofenphos	Arthrobacter sp.	Cornfield soil (USA)	GLC, TLC	Racke and Coats (1988)
	Arthrobacter sp.	Turf green soil (Japan)	GC-MS, HPLC	Ohshiro et al. (1997)
Monocrotophos	Arthrobacter atrocyaneus, Bacillus megaterium	Vegetable farm soil (Maharashtra, India)	TLC, GC-FID	Bhadbhade et al. (2002)
	Paracoccus sp.	Wastewater sludge (China)	HPLC, GC-NPD	Jia et al. (2006)
	A. fumigatus[a]	Not mentioned	Not mentioned	Pandey et al. (2014)
Parathion methyl	Stenotrophomonas sp.	Industrial sludge (China)	GC-FPD, HPLC	Deng et al. (2015)
	Agrobacterium sp.	Activated sludge (Shandong, China)	GC, GC-FPD	Wang et al. (2012)
	Bacillus sp., Pseudomonas sp.	Soil from groundnut fields (Andhra Pradesh, India)	HPLC, GC-FID9	Madhuri and Rangaswamy (2009)

TABLE 2.3 Bacterial/fungal strains isolated globally capable of degrading organophosphate pesticides.—cont'd

Organophosphate pesticide	Isolated strains of Bacteria/Fungi	Isolation matrix (location)	Technique used	References
	Leuconostoc mesenteroides, L. brevis, L. plantarum, L. sakei	Kimchi during fermentation (Korea)	TLC, HPLC	Cho et al. (2009)
	Pseudomonas pseudoalcaligenes, Micrococcus luteus, Bacillus sp., *Exiguobacterium aurantiacum*	Natural lake water (Antequera, Spain)	GC-MS	Lopez et al. (2005)
	Plesiomonas sp.	Not mentioned	Not mentioned	Zhongli et al. (2001)
	Flavobacterium sp.	Agricultural soil (Mexico)	GC	Ortiz-Hernandez et al. (2001)
Ethion	*Pseudomonas* sp. *Azospirillum* sp.	Ethion contaminated soil (Australia)	GC-MS	Foster et al. (2004)
Profenofos	*Stenotrophomonas* sp.	Industrial sludge (China)	GC-FPD, HPLC	Deng et al. (2015)
	Pseudomonas aeruginosa	Profenophos contaminate soil (Hanchuan, China)	GC-MS, GC-ECD	Malghani et al. (2009)
	Bacillus subtilis	Grapevines or grape rhizosphere (Maharashtra, India)	GC-MS, LC-MS/MS	Salunkhe et al. (2013)
Quinalphos	*Ochrobactrum* sp.	Pesticide contaminated soil sample (Karnataka, India)	HPLC, GC-MS	Talwar et al. (2014)
	Bacillus sp. *Pseudomonas* sp.	Agricultural soil samples (Punjab, India)	HPLC	Dhanjal et al. (2014)
	Pseudomonas strains (14)	Soil samples of grape wine yards (Maharashtra, India)	GC-MS	K.R.Pawar and G.V.Mali (2014)

pesticide cadusafos was found to be completely degraded by *Pseudomonas putida*, (PC1) isolated from agricultural soil in Saudi Arabia, within 6 days at the concentration of 20 mgL^{-1} and 40 mgL^{-1} (Abo-amer, 2012). *Sphingomonas* sp. and *Flavobacterium* sp. isolated from commercial potato grown fields in northern Greece were equally efficient in 100% degradation of cadusafos within 48 h (Karpouzas et al., 2005). Seo et al. (2007) isolated bacterial isolates namely *Arthrobacter* sp. (P1-1) and *Mycobacterium* sp. (JS19b1) from petroleum contaminated soil of an oil gasification unit in Hawaii, USA, and reported that these strains are capable of degrading chlorfenvinphos pesticide. In a similar work, four fungal strains namely *Penicillium citrinum, Aspergillus fumigatus, Aspergillus terreus, and Trichoderma harzianum,* isolated from water samples of Tagus River, Portugal, were found to be resistant and capable of degrading chlorfenvinphos in the aquatic environment (Oliveira et al., 2015).

Stenotrophomonas sp. (G1), isolated from sludge sample of chlorpyrifos manufacturing unit, Jiangsu Province, China, has been found to degrade 63% of chlorpyrifos within 24 h at the initial concentration of 50 mgL^{-1}. This strain was also reported to degrade 100% of methyl parathion, methyl paraoxon, diazinon, and phoxim, 95% of parathion, 38% of profenofos, and 34% of triazophos in 24 h at the same concentration (Deng et al., 2015). A similar kind of study reported 100% degradation of chlorpyrifos in liquid culture within 6 h at the initial concentration of 100 mgL^{-1}, by *Cupriavidus* sp. (DT1), isolated from industrial sludge, Jiangsu Province, China (Lu et al., 2013). *Serratia marcescens* isolated from diazinon-contaminated soil, Poland; degrades 45.3%, 61.4%, and 68.9% of chlorpyrifos; 61.4%, 79.7%, and 81% of fenitrothion, and 72.5%, 64.2%, and 63.6% of parathion in sandy, sandy loam, and silty soil, respectively, after 42 days of experiment at the initial concentration of 100 mgkg^{-1} (Cycoń et al., 2013). Similarly, *Sphingobacterium* sp. JAS3 strain, isolated from paddy field soil samples of Tamil Nadu, India, has been found more efficient in chlorpyrifos degradation and degrades 100% of chlorpyrifos within 5 days in liquid media at the initial concentration of 300 mgL^{-1} and could also tolerate 400 mgL^{-1} of chlorpyrifos (Abraham and Silambarasan, 2013). In a similar kind of study, four bacterial strains viz. *P. putida, Klebsiella* sp., *Pseudomonas stutzeri,* and *P. aeruginosa,* isolated from paddy field soil of Tamil Nadu, India, were reported to degrade 65.87% of chlorpyrifos in the soil when used as a consortium. Degradation studies were carried out at neutral pH and 37°C temperature with chlorpyrifos concentration OF 500 mgkg^{-1} (Sasikala et al., 2012). Bacterial strains such as *Streptomyces chattanoogensis* (AC5), *Streptomyces olivochromogene* (AC7) isolated from soil samples of blueberry farms of southern Chile and *Pseudomonas* sp., *Enterobacter* sp. isolated from chlorpyrifos-contaminated agricultural soil samples of Giza and Cairo, Egypt, were found to be able to degrade chlorpyrifos by utilizing it as a sole carbon and phosphorus source (Briceño et al., 2012; Nabil et al., 2011). The bacterial strain *P. aeruginosa* (IRLM 1) isolated from effluent and moist soil samples of pesticide producing factories in Iran degrades 100% of chlorpyrifos within 8–9 days at chlorpyrifos concentration of 140 mgL^{-1} (Latifi et al., 2012). Similarly, different species of *Pseudomonas,* including *P. putida, P. stutzeri, P. aeruginosa, Pseudomonas nitroreducens,* and *Pseudomonas fluorescence,* isolated from agricultural soil samples and contaminated effluents from different regions have confirmed biodegradation of chlorpyrifos at different concentrations and time periods (Bhagobaty and Malik, 2010; Vidya Lakshmi, Kumar and Khanna, 2008; Maya et al., 2011; Latifi et al., 2012; Sasikala et al., 2012). Other bacterial strains such as *Leuconostoc mesenteroides* (WCP907), *L. brevis* (WCP902), *L. plantarum* (WCP931), and *Lactobacillus sakei* (WCP904) were isolated by Cho et al., 2009 from kimchi during fermentation at initial chlorpyrifos concentration of 200 mgL^{-1}. These strains utilize chlorpyrifos and other pesticides such as coumaphos, diazinon, parathion, and methyl parathion as a sole carbon and phosphorus sources. Similarly, *Bacillus* sp. and *Pseudomonas* sp. isolated from groundnut field soil, Andhra Pradesh, India, were found to degrade 75% of chlorpyrifos and phorate and 50% of dichlorvos, methyl parathion, and methomyl within a week (Madhuri and Rangaswamy, 2009). In a similar study, four bacterial species viz. *Sphingomonas* sp. (Dsp-2), *Stenotrophomonas* sp. (Dsp-4), *Bacillus* sp. (Dsp-6), *Brevundimonas* sp. (Dsp-7), and *Pseudomonas* sp. (Dsp-1, 3, 5), isolated from industrial water and agricultural soil, were found to be able of degrading chlorpyrifos by utilizing it as a sole carbon source. The degradation rate varies from 37 $mgL^{-1}d^{-1}$ to 100 $mgL^{-1}d^{-1}$ among these seven strains. However, *Sphingomonas* sp. Dsp-2 strain was most efficient and degrades 100 mgL^{-1} of chlorpyrifos completely within 24 h in liquid media, whereas in soil *Pseudomonas* sp. (Dsp-1) showed the highest degradation of chlorpyrifos (Li et al., 2008). Vidya Lakshmi et al. (2008) reported 75%−87% chlorpyrifos degradation within 20 days, *by Brucella melitensis, P. fluorescence, Bacillus subtilis, Serratia marcescens, Bacillus cereus, Klebsiella* sp., and *P. aeruginosa,* isolated from chlorpyrifos contaminated soil samples, Punjab, India. A bacterial strain (*Serratia* sp.) and a fungal strain (*Trichosporon* sp.) isolated from activated sludge sample, in Shandong, China, were found to mineralize chlorpyrifos completely. *Serratia* sp. (TCR strain) transforms/degrades 100% of chlorpyrifos (50 mgL^{-1}) within 4 days into TCP (3,5,6-trichloro-2-pyridinol), which is completely (100%) mineralized by *Trichosporon* sp. (TCF strain) within 5 days at the concentration of 50 mgL^{-1} (Xu et al., 2007), whereas *Sphingomonas* sp. Dsp-2 strain, isolated from chlorpyrifos-contaminated water samples, in Nantong, China, degrades 100% of chlorpyrifos (100 mgL^{-1}) within 24 h in cell culture, 90% in soil within 7 days, and could also utilize other pesticides such as parathion, fenitrothion, methyl parathion, and profenofos (Li et al., 2007). Chlorpyrifos-methyl, a slightly toxic, nonsystematic pesticide-degrading bacteria *Burkholderia cepacia* (KR 100), isolated from Korean rice paddy soil was found to be capable of hydrolyzing chlorpyrifos-methyl (300 μgmL^{-1}) completely into TCP within 132 h, which in turn (TCP, 100 μgmL^{-1}) was completely disappeared within 144 h. This study also reported that KK-100 strain could also degrade chlorpyrifos, dimethoate, fenitrothion, malathion, and monocrotophos at 300 μgmL^{-1}, but diazinon, dicrotophos, parathion, and parathion methyl at 100 μgmL^{-1} (Kim and Ahn, 2009).

Dichlorvos is a moderately hazardous pesticide, with mammalian oral LD_{50} of 56−108 $mgkg^{-1}$ and dermal LD_{50} of 75−210 $mgkg^{-1}$ (WHO, 2009). Degradation of dichlorvos has been studied by various researchers. From an agricultural farm of Nigeria, four bacterial strains namely *Proteus vulgaris, Acinetobacter* sp., *Serratia* sp., and *Vibrio* sp. were isolated and found to be capable of degrading dichlorvos; however, *Proteus vulgaris* showed highest degradation rate (70%), and *Vibro* sp. showed lowest degradation rate (45%) and degradation capacity of isolated strains was in the order of *Proteus vulgaris* > *Acinetobacter* sp. > *Serratia* sp. > *Vibrio* sp (Agarry et al., 2013). Tang et al., (2009) employed REMI (restriction enzyme-mediated integration) method to construct transformants of *Trichoderma atroviride* (T23) for the dichlorvos degradation. Out of 247 transformants, 76% showed higher degradation ability as compared to *T. atroviride* strain, and degradation rate of transformants ranged from 81% to 96% as compared to 72% of the parent strain (T23). Similarly, *Bacillus* sp. isolated from different field soil samples of India were found, capable of degrading dichlorvos at different concentrations (Madhuri and Rangaswamy, 2009; Pawar and Mali, 2014). However, a bacterial species was isolated successfully from activated sludge of Jiangsu Province, China, and identified as *Ochrobactrum* sp. (DVD-1). This bacterial strain utilizes dichlorvos as sole carbon source at pH 7 and temperature 30°C and was also reported to completely degrade dichlorvos in soil within 24 h at the concentration of 100 mgL^{-1} or 500 mgL^{-1} when inoculated with 0.5% or 1% (v/v) (Zhang et al., 2006a, b).

Diazinon is moderately toxic organophosphate pesticide with mammalian oral LD_{50} of 26−300 $mgkg^{-1}$ and dermal LD_{50} of 379 $mgkg^{-1}$ [1] (Kumar et al., 2018). Diazinon has a half-life of 40 days in soil. However, it has 138 days of hydrolytic half-life (Kegley et al., 2014). According to a study, *Serratia marcescens* (D1101) isolated from agricultural soil samples of Taif Province, Saudi Arabia, was capable of degrading 100% of Diazinon (50 mgL^{-1}) within 11 days in liquid media and 14 days in sterilized soil at the concentration of 100 mgkg^{-1}. The rate of degradation by D1101 strain was efficient at 25°C–30°C of temperature and 7–8 pHs. The D1101 strain was also able to degrade other pesticides such as chlorpyrifos (91%), coumaphos (89%), parathion (85%), and isazofos (87%) when provided as sole carbon and phosphorus source (Abo-Amer, 2011). Similarly, lactic acid bacteria (LAC) including *Lactobacillus brevis, Lactobacillus plantarum,* and *Lactobacillus sakei* were found to be very efficient in diazinon degradation when provided as sole carbon and phosphorus source (Zhang et al., 2014; Cho et al., 2009). Another study carried out by Cycon et al., (2009) isolated three bacterial strains, viz. *Serratia liquefaciens (DDS-1), Serratia marcescens (DDS-2),* and *Pseudomonas* sp. (DDS-3) from agricultural soil samples of southern Poland and were found to be capable of degrading 80%, 89%, and 84% of Diazinon, respectively, after 14 days of incubation at the initial concentration of 50 mgL^{-1}. However, their consortium degrades diazinon more efficiently, i.e., 92% with the same concentration and time as compared to diazinon degradation by single isolate (Cycon et al., 2009). Similarly *Arthrobacter* sp. and *Mycobacterium* sp. isolated from petroleum contaminated soil of Hilo, Hawaii, were found to degrade diazinon by utilizing them as a growth substrate (Seo et al., 2007).

Dimethoate is a moderately toxic, possibly carcinogenic organophosphate pesticide with mammalian oral LD$_{50}$ of 235 mgkg^{-1} and dermal LD$_{50}$ of >400 mgkg^{-1} (WHO, 2009). According to a study, *Paracoccus* sp. (Lgjj-3), isolated from activated sludge of a dimethoate manufacturing wastewater treatment pool in Dafeng, China, was found to be capable of degrading 100 mgL^{-1} of dimethoate within 6 h to nondetectable level by utilizing it as a sole carbon source. LGjj-3 strain could efficiently degrade dimethoate at different temperatures ranging from 25°C to 40°C and pH values ranging from 6 to 9; however, 35°C and 7.0 was the optimum temperature and pH, respectively, for dimethoate degradation (Li et al., 2010). Similarly, Liang et al. (2009) isolated *Raoultella* sp. (X1) from dimethoate contaminated soil samples, China, and found to be capable of removing 75% of dimethoate cometabolically. Similarly, a dimethoate-degrading enzyme, isolated from a fungus, *Aspergillus niger* (ZHY256), was found to be capable of degrading approximately 87% of dimethoate (Liu et al., 2001).

Karpouzas et al. (2005) isolated bacterial species which were capable of degrading another pesticide ethoprophos. These strains viz. *Sphingomonas* sp. and *Flavobacterium* sp. were isolated from soil samples of commercial potato fields of Northern Greece and were found capable of degrading ethoprophos efficiently. *Sphingomonas* sp. degraded ethoprophos and cadusafos completely within 8 and 4 days, respectively, while as *Flavobacterium* sp. degraded ethoprophos and cadusafos within 13 and 8 days, respectively, in liquid culture. In a similar kind of study; *P. putida* (epI and epII), isolated from soil samples of Northern Greece, was found to be capable of degrading ethoprophos completely in both fumigated and nonfumigated samples within 5 and 4 days, respectively, after inoculation with high inoculum mass, but removed from fumigating soil only at low inoculum density. These isolates are also capable of degrading other pesticides such as cadusafos, isofenphos, fenamiphos, and isazofos, but only at a slow rate (Karpouzas et al., 2000).

Fenamiphos is highly hazardous (Ib Class) OPP with mammalian oral LD_{50} of 10 $mgkg^{-1}$ and dermal LD_{50} of $^>200$ $mgkg^{-1}$ (WHO, 2009) and is also acetylcholinesterase inhibitor. Bacterial species such as *P. putida* and *Acinetobacter rhizosphaerae* isolated from banana plantation field soil were found to be capable of degrading fenamiphos within 4 days in minimal salt media (MSM) and in MSM supplemented with nitrogen source (MSMN) (Chanika et al., 2011). Similarly, *Microbacterium esteraromaticum* and *Brevibacterium* sp. isolated from different field soil samples of Australia were found to hydrolyze fenamiphos and its toxic oxidation products (fenamiphos sulfoxide and fenamiphos sulfone) efficiently to less toxic phenols (Cáceres et al., 2009; Megharaj et al., 2003).

Fenitrothion is a moderately toxic OPP (WHO, 2009), with a mammalian oral LD50 of approximately 500 to1 416 $mgkg^{-1}$ and mammalian dermal LD50 of 1416 $mgkg^{-1}$ (Kumar et al., 2018). Fenitrothion has less natural persistence than chlorpyrifos with 2.7 d of half-life in soil and about 183 d of hydrolysis half-life (Kegley et al., 2014). Species of *Burkholderia*, *Pseudomonas*, *Arthrobacter*, *Sphingomonas*, *Corynebacterium*, and *Cupriavidus* isolated from different environmental matrices around Asia have been found to be very efficient in fenitrothion degradation (Tago et al., 2006; Zhang et al., 2006b; Kim et al., 2009). Fenthion-degrading bacterium, *Bacillus safensis*, isolated from pesticide-contaminated soil of Hasahisa, Sudan, was found to reduce the concentration of fenthion from 400 mgL^{-1} to 275 mgL^{-1} after the incubation of 30 days; however, the strain was also capable of degrading temphos and reduce its concentration from 400mgL^{-1} to 89.3 mgL^{-1} within 30 days (Abdelbagi et al., 2018).

Similarly *Pseudomonas* sp, *Ralstonia eutropha*, *Enterobacter cloacae*, *Bacillus* sp., *Rhizobium* sp., and *Proteus* sp. isolated from different soil samples of India were found to be efficient in degrading phorate (Rani and Juwarkar, 2012; Madhuri and Rangaswamy, 2009; Bano and Musarrat, 2003), which is an extremely hazardous compound (WHO, 2009) with mammalian oral LD50 of 2—4 $mgkg^{-1}$ and mammalian dermal LD50 of 20—30 $mgkg^{-1}$ (Kumar et al., 2018). A bacterial consortium composed of six strains namely *Stenotrophomonas malthophilia*, *Yersinia enterocolitica*, *Serratia ficaria*, *Vibrio metschinkouii*, *Serratia* spp., and *Proteus vulgaris*, isolated from cornfield soil samples of Mexico, were found to be capable of degrading Tetrachlorvinphos pesticide (Ortiz-Hernández and Sánchez-Salinas, 2010). Similarly, a novel bacterial isolate TAP-1 (*Bacillus* sp.), isolated from sewage sludge samples of pesticide manufacturing unit in Fijian, China, degrades a highly toxic pesticide triazophos, efficiently through co-metabolism. TAP-1 could degrade 98.5% of triazophos in the medium within 5 days at the concentration of 100 mgL^{-1}, and optimal pH and temperature for degradation study were 6.5—8 and 32 degrees, respectively (Tang and You, 2012). In a similar study, *Diaphorobacter* sp. (TPD-1) isolated from triazophos contaminated soil in Jiangsu, China, utilized triazophos and its metabolite (1-phenyl-3-hydroxy-1,2,4-triazole) as a sole carbon source and could completely degrade both of them to a nondetectable level within 24 and 56 h, respectively (Yang et al., 2011). *Acinetobacter johnsonii* (MA19), isolated from malathion polluted soil of China, could degrade malathion co-metabolically as it was unable to consume malathion as the sole source of carbon and energy but could degrade it in the presence of another carbon source. Malathion was completely (100%) degraded in the presence of sodium acetate or sodium succinate in 84 h (Shan et al., 2009). Similarly species of *Pseudomonas, Bacillus, Micrococcus, Acetobacter* and *Enterobacter* isolated from different matrices across diverse countries and were found to be very efficient in malathion degradation (Baishya and Sharma, 2014; Jilani, 2013; Goda et al., 2010; Mohamed et al., 2010). In the same way, *Arthrobacter* sp. isolated from cornfield and turf green soil samples in USA and China and its Hydrolases enzyme were found to degrade isofenphos pesticide efficiently (Racke and Coats, 1988; Ohshiro et al., 1997). Monocrotophos, a highly toxic compound was reported to be degraded by *Arthrobacter atrocyaneus* (MCM B-425) and *Bacillus megaterium* (MCM B-423) isolated from monocrotophos contaminated field soil in Maharashtra, India. MCM B-425 and MCM B-423 degrade 93% and 83% of Monocrotophos in liquid media, respectively, within 8 days at the concentration of 1000 mgL^{-1} under shaking conditions at 30°C temperature. Phosphatase and esterase enzymes were reported to be involved in biodegradation of OPP and were detected in both the organisms (Bhadbhade et al., 2002). *Paracoccus* sp. (M-1) isolated from wastewater sludge of a chemical factory, China; degraded 79.92% of pure Monocrotophos (300 mgL^{-1}) under aerobic conditions within 6 h. The optimal conditions for degradation were 300 mgL^{-1} Monocrotophos concentration, 30 °C temperature and 7.5 pH (Jia et al., 2006). Similarly fungal strain, *A. fumigatus* was found to degrade Monocrotophos till 1% concentration and its growth was observed to be increased in presence of Tween 80 (Pandey2014).

Parathion methyl, classified as an extremely hazardous pesticide (class Ia) by WHO was found to be completely degraded by *Agrobacterium* sp. (Yw12) by consuming it as a sole source of carbon, phosphorus, and energy. Under optimal conditions, Yw12 strain degrades 50 mgL^{-1} of parathion methyl within 2 h and completely mineralizes it within 6 h. Moreover, this strain could also degrade other pesticides such as chlorpyrifos, phoxim, carbofuran, deltamethrin, methamidophos, and atrazine when provided as the sole sources of carbon and energy (Wang et al., 2012). Similarly, species of *Bacillus, Pseudomonas,* and *Plesiomonas,* isolated from soil and water samples of different countries were found to be efficient in degrading parathion methyl (Madhuri and Rangaswamy, 2009; Lopez et al., 2005). Similarly *Plesiomonas* sp. and *Flavobacterium* sp. were found to hydrolyze parathion methyl to Dimethyl phosphorothinate and p-nitrophenol (Zhongli et al., 2001; Ortiz-Hernandez et al., 2001). In rural Australia, Ethion is a major environmental contaminant with moderate to higher persistence in the soil. Its biodegradation potential decreases due to its hydrophobicity and strong adsorption toward organic matter and soil particles. Therefore, because of its higher toxicity and persistence, its remediation from the contaminated environments should be done with priority (Foster et al., 2004). In this regard mesophilic bacteria viz *Pseudomonas* (WAI-21) and *Azospirillum species* (WAI-19) isolated from Ethion contaminated soil were found to degrade Ethion rapidly within 6−7 h of incubation of approximately 42 and 30 mgL^{-1}h^{-1}, respectively, followed by a slower rate of 3−4 mgL^{-1}h$^-$1 and finally attained 70% and 58% degradation of ethion, respectively (Foster et al., 2004). Similarly, another pesticide profenophos was found to be degraded by *P. aeruginosa*, isolated from profenophos contaminated soil, Hanchuan, China up to 86.81% within 48 h in liquid media (Malghani et al., 2009). In another study, four strains of *Bacillus subtilis* namely DR-39, CS-126, TL-171 and TS-204 isolated form grapevines or grape rhizosphere were reported to degrade 90% (CS-126, TL-171 and TS-204) or 79% (DR-39) of profenophos even in the presence of other carbon sources (Salunkhe et al., 2013). Similarly, species of *Ochrobacterium, Bacillus,* and *Pseudomonas* have been isolated from different soil samples across India and were found to be efficient in the degradation of a moderately toxic quinalphos (Talwar et al., 2014; Dhanjal et al., 2014; Pawar and Mali, 2014).

7. Conclusion

The production of pesticides is increasing, and their consumption has become inevitable because of urbanization and tremendously growing world population. Pesticides are usually classified by the target pest, chemical composition, pesticide characteristics, mode of action, and entry. Among several pesticide classes, organophosphates are the most widely used category because of their degradable and efficient nature. However, their widespread, nonregulated, and inappropriate use has led to significant threats for all ecosystems and the living beings. Organophosphates have less environmental persistence, but their residues have been detected globally by many researchers in water, soil, vegetables, fruits, and other food products. Besides this, their residues were also detected in human fluids such as blood, breast milk, and urine, which indicate their persistence in the human body in good quantity. They also involve the destruction of beneficial nontarget organisms and loss of biodiversity. However, a diverse group of microorganism has been isolated and identified all over the world which has the capability to degrade or eliminate organophosphates from different environments by utilizing them as carbon, phosphorus, or energy sources. However, further research is required to interpret the successful laboratory trail experiments into vigorous field application, and additional degradation approaches are still needed to reduce their accumulation probabilities, and related health problem will be minimized.

References

Abdelbagi, A.O., et al., 2018. Biodegradation of fenthion and temphos in liquid media by *Bacillus safensis* isolated from pesticides polluted soil in the Sudan. African Journal of Biotechnology 17 (12), 396−404.

Abo-Amer, A.E., 2012. Characterization of a strain of *Pseudomonas putida* isolated from agricultural soil that degrades cadusafos (an organophosphorus pesticide). World Journal of Microbiology and Biotechnology 28 (3), 805−814.

Abo-Amer, A., 2011. Biodegradation of diazinon by *Serratia marcescens* DI101 and its use in bioremediation of contaminated environment. Journal of Microbiology and Biotechnology 21 (1), 71−80.

Abraham, J., Silambarasan, S., 2013. Biodegradation of chlorpyrifos and its hydrolyzing metabolite 3,5,6-trichloro-2-pyridinol by *Sphingobacterium* sp. JAS3'. Process Biochemistry 48 (10), 1559−1564.

Abrahams, P.W., 2002. Soils: their implications to human health. The Science of the Total Environment 291 (1−3), 1−32.

Adityachaudhury, N., Banerjee, H., Kole, R.K., 1997. An apprisal of pesticides use in Indian agriculture with special reference to their consumption in West Bengal. Science and Culture 63 (9−10), 223−228.

Agarry, S.E., Aremu, M.O., Jimoda, L.A., 2013. Biodegradation of Dichlorovos (organophosphate pesticide) in soil by bacterial isolates. Biodegradation 3 (8), 12−17.

Ahad, K., et al., 2000. Determination of insecticide residues in groundwater of Mardan Division, NWFP, Pakistan: a case study. Water SA-Pretoria 26 (3), 409−412.

Akan, J.C., Abdulrahman, F.I., Chellube, Z.M., 2014. Organochlorine and organophosphorus pesticide residues in fish samples from Lake Chad, Baga, North Eastern Nigeria. International Journal of Innovation Management and Technology 5 (2), 87−92.

Aktar, W., Sengupta, D., Chowdhury, A., 2009. Impact of pesticides use in agriculture: their benefits and hazards. Interdisciplinary Toxicology 2 (1), 1−12.

Albero, B., Sánchez-Brunete, C., Tadeo, J.L., 2003. Determination of organophosphorus pesticides in fruit juices by matrix solid-phase dispersion and gas chromatography. Journal of Agricultural and Food Chemistry 51 (24), 6915−6921.

Amaraneni, S.R., Pillala, R.R., 2001. Concentrations of pesticide residues in tissues of fish from Kolleru Lake in India. Environmental Toxicology 16 (6), 550−556.

Araya, M., Lakhi, A., 2004. Response to consecutive nematicide application using the same product in Musa AAA cv. Grande Naine originated from in vitro propagative material and cultivated on a virgin soil. Nematologia Brasileira 28 (1), 55–61.

Awad, N.S., et al., 2011. Isolation, characterization and fingerprinting of some chlorpyrifos- degrading bacterial strains isolated from Egyptian pesticides-polluted soils. African Journal of Microbiology Research 5 (18), 2855–2862.

Baishya, K., Sharma, H.P., 2014. Isolation and characterization of organophosphorus pesticide degrading bacterial isolates. Achives of Applied Science Research 6 (5), 144–149.

Bano, N., Musarrat, J., 2003. Isolation and characterization of phorate degrading soil bacteria of environmental and agronomic significance. Letters in Applied Microbiology 36 (6), 349–353.

Beneta, A., et al., 2018. Multiresidue GC-MS/MS pesticide analysis for evaluation of tea and herbal infusion safety. International Journal of Environmental Analytical Chemistry 1–18.

Bhadbhade, B.J., Sarnaik, S.S., Kanekar, P.P., 2002. Biomineralization of an organophosphorus pesticide, Mono-crotophos, by soil bacteria. Journal of Applied Microbiology 93 (2), pp224–234.

Bhagobaty, R.K., Malik, A., 2010. Utilization of chlorpyrifos as a sole source of carbon by bacteria isolated from wastewater irrigated agricultural soils in an industrial ares of western Uttar Pradesh, India. Research Journal of Microbiology 3 (5), 293–307.

Bhanti, M., Taneja, A., 2007. Contamination of vegetables of different seasons with organophosphorous pesticides and related health risk assessment in northern India. Chemosphere 69 (1), 63–68.

Bishnu, A., et al., 2009. Pesticide residue level in tea ecosystems of Hill and Dooars regions of West Bengal, India. Environmental Monitoring and Assessment 149 (1–4), 457–464.

Briceño, G., et al., 2012. Chlorpyrifos biodegradation and 3,5,6-trichloro-2-pyridinol production by actinobacteria isolated from soil. International Biodeterioration and Biodegradation 73, 1–7.

Cáceres, T.P., et al., 2009. Bioresource Technology Hydrolysis of fenamiphos and its toxic oxidation products by Microbacterium sp. In: Pure Culture and Groundwater. Bioresource Technology, vol. 100. Elsevier Ltd, pp. 2732–2736 (10).

Castrejón Godínez, M.L., Sánchez-Salinas, E., Ortiz Hernández, M.L., 2014. Plaguicidas: Generalidades, usos e impactos sobre el ambiente y la salud. Los plaguicidas en México. Aspectos generales, toxicológicos y Ambientales, pp. 11–35.

Chandra, I., et al., 2015. Current status of persistent organic pesticides residues in air , water , and soil , and their possible effect on neighboring countries : a comprehensive review of India. The Science of the Total Environment 511, 123–137.

Chandra, S., et al., 2014. Effect of washing on residues of Chlorpyrifos and Monocrotophos in vegetables. International Journal of Advance Research 2 (12), 744–750.

Chanika, E., et al., 2011. Bioresource technology isolation of soil bacteria able to hydrolyze both organophosphate and carbamate pesticides. Bioresource Technology 102 (3), 3184–3192.

Chaudhry, Q., et al., 2005. Utilising the synergy between plants and rhizosphere microorganisms to enhance breakdown of organic pollutants in the environment. Environmental Science and Pollution Research 12 (1), 34–48.

Chimwanza, B., et al., 2006. The impact of farming on river banks on water quality of the rivers. International Journal of Environmental Science and Technology 2 (4), 353–358.

Cho, K.M., et al., 2009. Biodegradation of chlorpyrifos by lactic acid bacteria during kimchi fermentation. Journal of Agricultural and Food Chemistry 57 (5), 1882–1889.

Choudhary, A., Sharma, D.C., 2008. Pesticide residues in honey samples from Himachal Pradesh (India). Bulletin of Environmental Contamination and Toxicology 80 (5), 417–422.

Chuanjiang, T., et al., 2010. Residue analysis of acephate and its metabolite methamidophos in open field and greenhouse pakchoi (Brassica campestris L.) by gas chromatography – tandem mass spectrometry. Environmental Monitoring and Assessment 165 (1–4), 685–692.

CSE Report, March 2005. Analysis of Pesticide Residues in Blood Samples from Villages of Punjab. CSE/PML/PR-21/2005. http://indiaenvironmentportal.org.in/files/Punjab-blood-report.pdf.

CSE Report, August 2006. Analysis of Pesticide Residues in Soft Drinks. http://www.indiaenvironmentportal.org.in/files/labreport2006.pdf.

Cycoń, M., et al., 2013. Biodegradation and bioremediation potential of diazinon-degrading Serratia marcescens to remove other organophosphorus pesticides from soils. Journal of Environmental Management 117, 7–16.

Cycon, M., Wojcik, M., Piotrowska-Seget, Z., 2009. Biodegradation of the organophosphorus insecticide diazinon by Serratia sp. and Pseudomonas sp. and their use in bioremediation of contaminated soil. Chemosphere 76 (4), 494–501.

Damalas, C.A., Eleftherohorinos, I.G., 2011. Pesticide exposure, safety issues, and risk assessment indicators. International Journal of Environmental Research and Public Health 8 (5), 1402–1419.

Damalas, C.A., 2009. Understanding benefits and risks of pesticide use. Scientific Research and Essays 4 (10), 945–949.

Dawson, A.H., et al., 2010. Acute human lethal toxicity of agricultural pesticides: a prospective cohort study. PLoS Medicine 7 (10).

Dehghani, R., et al., 2012. Detrmination of organophosphorus pesticides (diazinon and chlorpyrifos) in water resources in Barzok, Kashan. Zahedan Journal of Research in Medical Sciences 14 (10), 66–72.

Deng, S., et al., 2015. Rapid biodegradation of organophosphorus pesticides by Stenotrophomonas sp: G1. Journal of Hazardous Materials 297, 17–24.

De, A., et al., 2014. Worldwide pesticide use. In: Targeted Delivery of Pesticides Using Biodegradable Polymeric Nanoparticles. Springer, New Delhi, pp. 5–6.

Dhanjal, N.I.K., Kaur, P., Sud, D., Cameotra, S.S., 2014. Persistence and biodegradation of quinalphos using soil microbes. Water Environment Research 86 (5), 457–461.

Dhas, S., Srivastava, M., 2010. An assessment of phosphamidon residue on mustard crop in an agricultural field in Bikaner, Rajasthan (India). European Journal of Applied Science 2, 55–57.

Dong, T.D., et al., 2008. A multiresidue method for the determination of 107 pesticides in cabbage and radish using QuEChERS sample preparation method and gas chromatography mass spectrometry. Food Chemistry 110 (1), 207–213.

Ecobichon, D.J., Joy, R.M., 1993. Pesticides and Neurological Diseases. CRC Press.

Eddleston, M., et al., 2002. Oximes in acute organophosphorus pesticide poisoning: a systematic review of clinical trials. International Journal of Medicine 95 (5), 275–283.

Eevers, N., et al., 2017. Bio- and phytoremediation of environments: a Review. Advances in Botanical Research 83, 277–318.

Environment Agency, 1996. Pesticides in the Aquatic Environment. Environment Agency, Wallingford.

EPA, January 2017a. What Is a Pesticide? http://www.epa.gov.

EPA, January 2017b. Pesticides Industry, Sales and Usage. http://www.epa.gov.

Fagnani, R., et al., 2011. Organophosphorus and carbamates residues in milk and feedstuff supplied to dairy cattle. Pesquisa Veterinária Brasileira 31 (7), 598–602.

Foster, L.J.R., Kwan, B.H., Vancov, T., 2004. Microbial degradation of the organophosphate pesticide, Ethion. FEMS microbiology letters 240 (1), 49–53.

Garcia, F.P., et al., 2012. Pesticides: classification , uses and toxicity. Measures of exposure and genotoxic risks. Journal of Research in Environmental Science and Toxicology 1 (11), 279–293.

Georghiou, G.P., Taylor, C.E., 1977. Genetic and biological influences in the evolution of insecticide resistance. Journal of Economic Entomology 70 (3), 319–323.

Goda, S.K., et al., 2010. Screening for and isolation and identification of malathion-degrading bacteria: cloning and sequencing a gene that potentially encodes the malathion-degrading enzyme, carboxylestrase in soil bacteria. Biodegradation 21 (6), 903–913.

Gondhalekar, A.D., et al., 2013. Implementation of an indoxacarb susceptibility monitoring program using field-collected German cockroach isolates from the United States. Journal of Economic Entomology 106 (2), 945–953.

Gondhalekar, A.D., Song, C., Scharf, M.E., 2011. Development of strategies for monitoring indoxacarb and gel bait susceptibility in the German cockroach (Blattodea: Blattellidae). Pest Management Science 67 (3), 262–270.

Gowda, S.R., Somashekar, R., 2012. Monitoring of pesticide residues in farmgate samples of vegetables in Karnataka, India. International Journal of Science and Nature 3 (3), 563–570.

Greenpeace India report, Pesticide Residues in Tea Samples from India, August 2014. https://www.greenpeace. org/...india/Global/india/image/2014/cocktail/download/Trouble.

Gunnell, D., Eddleston, M., 2003. Suicide by intentional ingestion of pesticides: a continuing tragedy in developing countries. International Journal of Epidemiology 32 (6), 902–909.

Han, Y., et al., 2017. Pesticide residues in nut-planted soils of China and their relationship between nut/soil. Chemosphere 180, 42–47.

Harinathareddy, A., Prasad, N.B.L., Devi, L.K., 2014. Pesticide residues in vegetable and fruit samples from Andhra Pradesh, India. Journal of Biological and Chemical Research 31, 1005–1015.

Harrison, S.A., 1990. The Fate of Pesticides in the Environment. Agrochemical Fact Sheet # 8. Penn, USA.

Hayward, D.G., Wong, J.W., Park, H.Y., 2015. Determinations for pesticides on Black, Green, Oolong, and White Teas by gas chromatography triple-quadrupole mass spectrometry. Journal of Agricultural and Food Chemistry 63 (37), 8116–8124.

Hussain, S., et al., 2009. Impact of pesticides on soil microbial diversity, enzymes, and biochemical reactions. Advances in Agronomy 102, 159–200.

Hussain, S., Masud, T., Ahad, K., 2002. Determination of pesticides residues in selected varieties of mango, Pakistan. Journal of Nutrition 1 (1), 41–42.

Ilyas, M.T., Afzal, S., Hussain, I., 2004. Pesticides in shallow groundwater of Bahawalnagar, Muzafargarh, DG khan and Rajan Pur districts of Punjab, Pakistan. Environment International 30 (4), 471–479.

IRAC, 2010. Resistance Management for Sustainable Agriculture and Improved Public Health. http://www.irac-online.org/.

Jacob, S., Resmi, G., Mathew, P.K., 2014. Environmental pollution due to pesticide application in cardamom hills of Idukki, District, Kerala, India. International Journal of Basic and Applied Science Research (1), 27–34.

Jallow, M.F.A., et al., 2017. Monitoring of pesticide residues in commonly used fruits and vegetables in Kuwait. International Journal of Environmental Research and Public Health 14 (8), 833.

Javaid, M.K., Ashiq, M., Tahir, M., 2016. Potential of biological agents in decontamination of agricultural soil. Scientifica (9).

Jia, K.Z., Cui, Z.L., He, J., Guo, P., Li, S.P., 2006. Isolation and characterization of a denitrifying monocrotophos-degrading *Paracoccus* sp. M-1. FEMS Microbiology Letters 263 (2), 155–162.

Jilani, S., 2013. Comparative assessment of growth and biodegradation potential of soil isolate in the presence of pesticides. Saudi Journal of Biological Sciences 20 (3), 257–264.

Karpouzas, D.G., et al., 2005. Non-specific biodegradation of the organophosphorus pesticides, cadusafos and ethoprophos, by two bacterial isolates. FEMS Microbiology Ecology 53 (3), 369–378.

Karpouzas, D.G., Morgan, J.A.W., Walker, A., 2000. Isolation and characterisation of ethoprophos-degrading bacteria. FEMS Microbiology Ecology 33 (3), 209–218.

Kazemi, M., 2012. Organophosphate pesticides: a general review. Agricultural Science Research Journal 2 (9), 512–522.

Kegley, S.E., et al., 2014. PAN Pesticide Database. Pesticide Action Network. North America. Available online at: http://ww.pesticideinfo.org/.

Kempa, E.S., 1997. Hazardous wastes and economic risk reduction: case study, Poland. International Journal of Environment and Pollution 7 (2), 221–248.

Kim, J.R., Ahn, Y.J., 2009. Identification and characterization of chlorpyrifos-methyl and 3, 5, 6-trichloro-2-pyridinol degrading *Burkholderia* sp. strain KR100. Biodegradation 20 (4), 487–497.

Kim, K., et al., 2009. Genetic and phenotypic diversity of parathion-degrading bacteria isolated from rice paddy soils. Journal of Microbiology and Biotechnology 19 (12), 1679–1687.

Klose, S., Ajwa, H.A., 2004. Enzymes activities in agricultural soils fumigated with methyl bromide alternatives. Soil Biology and Biochemistry 36 (10), 1625–1635.

Kottiappan, M., et al., 2013. Monitoring of pesticide residues in South Indian tea. Environmental Monitoring and Assessment 185 (8), 6413–6417.

Kumari, B., et al., 2003. Magnitude of pesticidal contamination in winter vegetables from Hisar, Haryana. Environmental Monitoring and Assessment 87 (3), 311–318.

Kumari, B., et al., 2005. Monitoring of butter and ghee (clarified butter fat) for pesticidal contamination from cotton belt of Haryana, India. Environmental Monitoring and Assessment 105 (1–3), 111–120.

Kumari, B., Madan, V.K., Kathpal, T.S., 2008. Status of insecticide contamination of soil and water in Haryana, India. Environmental Monitoring and Assessment 136 (1–3), 239–244.

Kumari, B., et al., 2004. Monitoring of pesticidal contamination of farmgate vegetables from Hisar. Environmental Monitoring and Assessment 90 (1–3), 65–71.

Kumar, S., et al., 2018. Microbial degradation of organophosphate pesticides : a Review. Pedosphere 28 (2), 190–208.

Lari, S.Z., et al., 2014. Comparison of pesticide residues in surface water and ground water of agriculture intensive areas. Journal of Environmental Health Science and Engineering 12 (1), 1–7.

Latifi, A.M., et al., 2012. Isolation and characterization of five chlorpyrifos degrading bacteria. African Journal of Biotechnology 11 (13), 3140–3146.

Laura, M., et al., 2013. Pesticide biodegradation: mechanisms, genetics and strategies to enhance the process. Biodegradation – Life of Science. https://doi.org/10.5772/56098.

Lee, S., et al., 2011. Acute pesticide illnesses associated with off-target pesticide drift from agricultural applications: 11 States, 1998–2006. Environmental Health Perspectives 119 (8), 1162.

Liang, Y., et al., 2009. Co-metabolic degradation of dimethoate by *Raoultella* sp . X1. Biodegradation 20 (3), 363−373.

Li, R., et al., 2010. Biochemical degradation pathway of dimethoate by *Paracoccus* sp . Lgjj-3 isolated from treatment wastewater. International Biodeterioration & Biodegradation 64 (1), 51−57.

Liu, Y.H., Chung, Y.C., Xiong, Y., 2001. Purification and characterization of a dimethoate-degrading enzyme of *Aspergillus niger* ZHY256, isolated from sewage. Applied and Environmental Microbiology 67 (8), 3746−3749.

Li, X., et al., 2008. Diversity of chlorpyrifos-degrading bacteria isolated from chlorpyrifos-contaminated samples. International Biodeterioration and Biodegradation 62 (4), 331−335.

Li, X., He, J., Li, S., 2007. Isolation of a chlorpyrifos-degrading bacterium, *Sphingomonas* sp. strain Dsp-2, and cloning of the mpd gene. Research in Microbiology 158 (2), 143−149.

Lopez, L., et al., 2005. Identification of bacteria isolated from an oligotrophic lake with pesticide removal capacities. Ecotoxicology 14 (3), 299−312.

Lorenz, E.S., 2009. Potential health effects of pesticides. Ag Communications and Marketing 1−8.

Lozowicka, B., Abzeitova, E., Sagitov, A., 2015. Studies of pesticide residues in tomatoes and cucumbers from Kazakhstan and the associated health risks. Environmental Monitoring and Assessment 187 (10), 609.

Lu, P., et al., 2013. Biodegradation of chlorpyrifos and 3,5,6-trichloro-2-pyridinol by Cupriavidus sp. DT-1. Bioresource Technology 127, 337−342.

Madhuri, R.J., Rangaswamy, V., 2009. Biodegradation of selected insecticides by *Bacillus* and *Pseudomonas* sp. in ground nut fields. Toxicology International 16 (2), 127−132.

Mandal, K., Singh, B., 2010. Magnitude and frequency of pesticide residues in farmgate samples of cauliflower in Punjab, India. Bulletin of Environmental Contamination and Toxicology 85 (4), 423−426.

Malghani, S., et al., 2009. Isolation and characterization of a profenofos degrading bacterium. Journal of Environmental Sciences 21 (11), 1591−1597.

Ma, J., et al., 2009. Determination of organophosphorus pesticides in underground water by SPE-GC − MS. Journal of Chromatographic Science 47 (2), 110−115.

Marchis, D., et al., 2012. Detection of pesticides in crops: a modified QuEChERS approach. Food Control 25 (1), 270−273.

Masiá, A., et al., 2015. Assessment of two extraction methods to determine pesticides in soils, sediments and sludges. Application to the Túria River Basin. Journal of Chromatography A 1378, 19−31.

Mukherjee, I., 2003. Pesticides residues in vegetables in and around Delhi. Environmental Monitoring and Assessment 86 (3), 265−271.

Maurya, P.K., Malik, D.S., 2016. Bioaccumulation of xenobiotics compound of pesticides in riverine system and its control technique: a critical review. Journal of Industrial Pollution Control 32 (2), 580−594.

Maya, K., et al., 2011. Kinetic analysis reveals bacterial efficacy for biodegradation of chlorpyrifos and its hydrolyzing metabolite TCP. Process Biochemistry 46 (11), 2130−2136.

Mcguinness, M., Dowling, D., 2009. Plant-associated bacterial degradation of toxic organic compounds in soil. International Journal of Environmental Research and Public Health 6 (8), 2226−2247.

Megharaj, M., et al., 2003. Hydrolysis of fenamiphos and its oxidation products by a soil bacterium in pure culture , soil and water. Applied Microbiology and Biotechnology 61 (3), 252−256.

Mohamed, Z.K., et al., 2010. Isolation and molecular characterisation of malathion-degrading bacterial strains from waste water in Egypt. Journal of Advanced Research 1 (2), 145−149.

Mostafalou, S., Abdollahi, M., 2012. Concerns of environmental persistence of pesticides and human chronic diseases. Clinical and Experimental Pharmacology 5, 10−11.

Munoz-Leoz, B., et al., 2011. Tebuconazole application decreases soil microbial biomass and activity. Soil Biology and Biochemistry 43 (10), 2176−2183.

Nguyen, T.D., et al., 2008. A multi-residue method for the determination of 203 pesticides in rice paddies using gas chromatography/mass spectrometry. Analytica Chimica Acta 619 (1), 67−74.

Nguyen, T.D., et al., 2008. A multiresidue method for the determination of 107 pesticides in cabbage and radish using QuEChERS sample preparation method and gas chromatography mass spectrometry. Food Chemistry 110 (1), 207−213.

Ohshiro, K., et al., 1997. Characterization of isofenphos hydrolases from Arthrobacter sp. strain B-5. Journal of Fermentation and Bioengineering 83 (3), 238−245.

Oliveira, B.R., et al., 2015. Biodegradation of pesticides using fungi species found in the aquatic environment. Environmental Science and Pollution Research 22, 11781−11791.

Ortiz-Hernández, M.L., Sánchez-Salinas, E., 2010. Biodegradation of the organophosphate pesticide tetrachlorvinphos by bacteria isolated from agricultural soils in Mexico. Revista Internacional de Contaminación Ambiental 26 (1), 27−38.

Ortiz-Hernández, M.L., et al., 2001. Biodegradation of methyl-parathion by bacteria isolated of agricultural soil. Revista Internacional de Contaminación Ambiental 17 (3), 147–155.

Pampulha, M.E., Oliveira, A., 2006. Impact of an herbicide combination of bromoxynil and prosulfuron on soil microorganisms. Current Microbiology 53 (3), 238–243.

PAN. Germany, 2012. Pesticides and Health Hazards Facts and Figures. Pesticide Action Network, Hamburg Germany.

Pandey, B., Baghel, P.S., Shrivastava, S., 2014. To study the bioremediation of monocrotophos and to analyze the kinetics effect of Tween 80 on fungal growth. Indo American Journal of Pharmaceutical Research 4, 925–930.

Parveen, Z., et al., 2004. Evaluation of multiple pesticide residues in apple and citrus fruits, 1999–2001. Bulletin of Environmental Contamination and Toxicology 73 (2), 312–318.

Parveen, Z., Khuhro, M.I., Rafiq, N., 2005. Monitoring of pesticide residues in vegetables (2000–2003) in Karachi, Pakistan. Bulletin of Environmental Contamination and Toxicology 74 (1), 170–176.

Pawar, K.R., Mali, G.V., 2014. Biodegradation of Quinolphos insecticide by Pseudomonas strain isolated from Grape rhizosphere soils. International Journal of Current Microbiology and Applied Science 3 (1), 606–613.

Paudyal, B.P., 2008. Organophosphorus poisoning. Journal of the Nepal Medical Association 47 (172), 251–258.

Pimentel, D., 1995. Amounts of pesticides reaching target pests: environmental impacts and ethics. Journal of Agricultural and Environmental Ethics 8 (1), 17–29.

Phugare, S.S., Gaikwad, Y.B., Jadhav, J.P., 2012. Biodegradation of acephate using a developed bacterial consortium and toxicological analysis using earthworms (Lumbricus terrestris) as a model animal. International Biodeterioration and Biodegradation 69, 1–9.

Pujeri, U.S., et al., 2011. The status of pesticide pollution in surface water (lakes) of Bijapur. International Journal of Applied Botany and Pharmaceutical Technology 1 (2), 436–441.

Racke, K.D., Coats, J.R., 1988. Comparative degradation of organophosphorus insecticides in soil: specificity of enhanced microbial degradation. Journal of Agricultural and Food Chemistry 36 (1), 193–199.

Ragnarsdottir, K.V., 2000. Environmental fate and toxicology of organophosphate pesticides. Journal of the Geological Society 157 (4), 859–876.

Ramu, S., Seetharaman, B., 2014. Biodegradation of acephate and methamidophos by a soil bacterium Pseudomonas aeruginosa strain Is-6. Journal of Environmental Science and Health 49 (1), 23–34.

Rani, R., Juwarkar, A., 2012. Biodegradation of phorate in soil and rhizosphere of Brassica juncea (L.)(Indian Mustard) by a microbial consortium. International Biodeterioration and Biodegradation 71, 36–42.

Richardson, M., 1998. Pesticides – friend or foe? Water Science and Technology 37 (8), 19–25.

SACONH, Pesticide contamination in select wetlands of Nilgiris district with special reference to sediments and fish, Sálim Ali Centre for Ornithology and Natural History, Coimbatore. Report Submitted to Keystone Foundation, Kotagiri, Nilgiris District. https://indiabiodiversity.org/biodiv/content/projects/project-b03d1ee6-6962.../682.pdf.

Sacramento, C.A., 2008. Department of Pesticide Regulation "What Are the Potential Health Effects of Pesticides?" Community Guide to Recognizing and Reporting Pesticide Problems, pp. 27–29.

Salunkhe, V.P., et al., 2013. Biodegradation of Profenofos by Bacillus subtilis isolated from grapevines (Vitis vinifera). Journal of Agricultural and Food Chemistry 61 (30), 7195–7202.

Sanghi, R., et al., 2003. Organochlorine and organophosphorus pesticide residues in breast milk from Bhopal, Madhya Pradesh, India. Human and Experimental Toxicology 22 (2), 73–76.

Sarnaik, S.S., 2004. Biodegradation of organophosphorus pesticides. Proceeding of the Indian national Science Academy 1, 57–70.

Sasikala, C., et al., 2012. Biodegradation of chlorpyrifos by bacterial consortium isolated from agriculture soil. World Journal of Microbiology and Biotechnology 28 (3), 1301–1308.

Scharf, M.E., et al., 1999. Metabolism of Carbaryl by insecticide-resistant and -susceptible western corn rootworm populations (Coleoptera: Chrysomelidae). Pesticide Biochemistry and Physiology 63 (2), 85–96.

Seenivasan, S., Muraleedharan, N., 2011. Survey on the pesticide residues in tea in south India. Environmental Monitoring and Assessment 176 (1–4), 365–371.

Seo, J.S., et al., 2007. Isolation and characterization of bacteria capable of degrading polycyclic aromatic hydrocarbons (PAHs) and organophosphorus pesticides from PAH-contaminated soil in Hilo, Hawaii. Journal of Agricultural and Food Chemistry 55 (14), 5383–5389.

Shan, X.I.E., et al., 2009. Biodegradation of malathion by Acinetobacter johnsonii MA19 and optimization of cometabolism substrates. Journal of Environmental Sciences 21 (1), 76–82.

Sharma, R.K., Goyal, A.K., 2014. Agro-pesticides and andrology. International Journal of Pharmacy and Pharmaceutical Sciences 6 (10), 12−19.

Sherman, J.D., 1995. Organophosphate pesticides-neurological and respiratory toxicity. Toxicology and Industrial Health 11 (1), 33−39.

Singh, B., Gupta, A., 2002. Monitoring of pesticide residues in farmgate and market samples of vegetables in a semiarid, irrigated area. Bulletin of Environmental Contamination and Toxicology 68 (5), 747−751.

Singh, B., Kaur, J., Singh, K., 2014. Microbial degradation of an organophosphate pesticide, malathion'. Critical Reviews in Microbiology 40 (2), 146−154.

Singh, B., Mandal, K., 2013. Environmental Impact of Pesticides Belonging to Newer Chemistry. Integrated Pest Management, pp. 152−190.

Singh, B.K., Walker, A., 2006. Microbial degradation of organophosphorus compounds. FEMS Microbiology Reviews 30 (3), 428−471.

Singh, P.B., Singh, V., Nayak, P.K., 2008. Pesticide residues and reproductive dysfunction in different vertebrates from north India. Food and Chemical Toxicology 46 (7), 2533−2539.

Sinha, S.N., Rao, M.V.V., Vasudev, K., 2012. Distribution of pesticides in different commonly used vegetables from Hyderabad, India. Food Research International 45 (1), 161−169.

Srivastava, A.K., Trivedi, P., Lohani, M.K.S.M., 2011. Monitoring of pesticide residues in market basket samples of vegetable from Lucknow City, India: QuEChERS method. Environmental Monitoring and Assessment 176 (1−4), 465−472.

Srivastava, S., Narvi, S.S., Prasad, S.C., 2008. Organochlorines and organophosphates in bovine milk samples in Allahabad region. International Journal of Environment Research 2 (2), 165−168.

Tabashnik, B.E., Rensburg, J. B. J. Van, Carrière, Y., 2009. Field-evolved insect resistance to Bt Crops: definition, theory, and data field-evolved insect resistance to Bt Crops: definition, theory. Journal of Insect Physiology 102 (6), 2011−2025.

Tago, K., et al., 2006. Diversity of fenitrothion-degrading bacteria in soils from distant geographical areas. Microbes and Environments 21 (1), 58−64.

Talwar, M.P., Mulla, S.I., Ninnekar, H.Z., 2014. Biodegradation of organophosphate pesticide quinalphos by *Ochrobactrum* sp. strain HZM. Journal of Applied Microbiology 117 (5), 1283−1292.

Tang, J., et al., 2009. Improved degradation of organophosphate dichlorvos by *Trichoderma atroviride* transformants generated by restriction enzyme-mediated integration (REMI). Bioresource Technology 100 (1), 480−483.

Tang, M., You, M., 2012. Isolation, identification and characterization of a novel triazophos-degrading *Bacillus* sp. (TAP-1). Microbiological Research 167 (5), 299−305.

Tariq, M.I., Afzal, S., Hussain, I., 2004. Pesticides in shallow groundwater of bahawalnagar, Muzafargarh, DG Khan and Rajan Pur districts of Punjab, Pakistan. Environment International 30 (4), 471−479.

Tilman, D., et al., 2002. Agricultural sustainability and intensive production practices. Nature 418 (6898), 671.

Vale, A., Lotti, M., 2015. Organophosphorus and carbamate insecticide poisoning. In: Handbook of Clinical Neurology, vol. 131, pp. 149−168.

Van Leeuwen, T., et al., 2010. Acaricide resistance mechanisms in the two-spotted spider mite *Tetranychus urticae* and other important Acari: a review. Insect Biochemistry and Molecular Biology 40 (8), 563−572.

Verma, J.P., Jaiswal, D.K., Sagar, R., 2014. Pesticide relevance and their microbial degradation. Reviews in Environmental Science and Biotechnology 13 (4), 429−466.

Verma, P., Verma, P., Sagar, R., 2013. Forest ecology and management variations in N mineralization and herbaceous species diversity due to sites, seasons, and N treatments in a seasonally dry tropical environment of India. Forest Ecology and Management 297, 15−26.

Vickerman, G.P., 1988. Farm scale evaluation of the long-term effects of different pesticide regimes on the arthropod fauna of winter wheat. In: Fields Methods for the Study of Environmental Effects of Pesticides. Symposium, pp. 127−135.

Vidya Lakshmi, C., Kumar, M., Khanna, S., 2008. Biotransformation of chlorpyrifos and bioremediation of contaminated soil. International Biodeterioration and Biodegradation 62 (2), 204−209.

Wang, L., et al., 2016. Chlorpyrifos exposure in farmers and urban adults: metabolic characteristic, exposure estimation, and potential effect of oxidative damage. Environmental Research 149, 164−170.

Wang, S., Zhang, C., Yan, Y., 2012. Biodegradation of methyl parathion and p-nitrophenol by a newly isolated *Agrobacterium* sp. strain Yw12. Biodegradation 23 (1), 107−116.

Ware, G.W., 1980. Effect of pesticides on non-target organisms. Residue Reviews 76, 173–201.

Wilson, C., Tisdell, C., 2001. Why farmers continue to use pesticides despite environmental, health and sustainability costs. Ecological Economics 39 (3), 449–462.

Worek, F., et al., 2014. Human & Experimental Reappraisal of indications and limitations of oxime therapy in organophosphate poisoning. Human and Experimental Toxicology 16 (8), 466–472.

Worek, F., Diepold, C., 1999. Dimethylphosphoryl-inhibited human cholinesterases: inhibition, reactivation, and aging kinetics. Archives of Toxicology 73 (1), 7–14.

World Health Organisation, 2009. The WHO Recommended Classification of Pesticides by Hazard and Guidelines to Classification. WHO, Geneva. ISSN 1684-1042.

Xu, G., et al., 2007. Mineralization of chlorpyrifos by co-culture of *Serratia* and *Trichosporon* spp. Biotechnology Letters 29 (10), 1469–1473.

Yadav, I.C., Devi, N.L., 2017. Pesticides Classification and its Impact on Human and Environment.

Yadav, M., et al., 2016. Utilization of microbial community potential for removal of chlorpyrifos: a review. Critical Reviews in Biotechnology 36 (4), 727–742.

Yang, C., Li, R., Song, Y., Chen, k., Li, S., Jiang, J., 2011. Identification of the biochemical degradation pathway of triazophos and its intermediate in *Diaphorobacter* sp. TPD-1. Current Microbiology 62 (4), 1294–1301.

Yang, G.L., et al., 2017. Pesticide residues in bayberry (*Myrica rubra*) and probabilistic risk assessment for consumers in Zhejiang, China. Journal of Integrative Agriculture 16 (9), 2101–2109.

Yang, L., et al., 2005. Isolation and characterization of a chlorpyrifos and 3,5,6-trichloro-2- pyridinol degrading bacterium. FEMS Microbiology Letters 251 (1), 67–73.

Zhang, X.H., et al., 2006a. Isolation and characterization of a dichlorvos-degrading strain DDV-1 of *Ochrobactrum* sp. 1. Pedosphere 16 (1), 64–71.

Zhang, Z., et al., 2006b. Isolation of fenitrothion-degrading strain *Burkholderia* sp. FDS-1 and cloning of mpd gene. Biodegradation 17 (3), 275–283.

Zhang, Y.H., et al., 2014. Enhanced degradation of five organophosphorus pesticides in skimmed milk by lactic acid bacteria and its potential relationship with phosphatase production. Food Chemistry 164, 173–178.

Zhongli, C., Shunpeng, L., Guoping, F., 2001. Isolation of methyl parathion-degrading strain M6 and cloning of the methyl parathion hydrolase gene. Applied and Environmental Microbiology 67 (10), 4922–4925.

Zhou, X., et al., 2002. Diagnostic assays based on esterase-mediated resistance mechanisms in western corn rootworms (Coleoptera: Chrysomelidae). Journal of Economic Entomology 95 (6), 1261–1266.

Zhou, Y., Liu, W., Ye, H., 2006. Effects of pesticides metolachlor and S-metolachlor on soil microorganisms in aquisols. II. Soil respiration. Ying Yong Sheng Tai Xue Bao. Journal of Applied Ecology 17 (7), 1305–1309.

Recent trends in the detection and degradation of organic pollutants

Preetismita Borah, Manish Kumar, Pooja Devi

CSIR-Central Scientific Instruments Organisation, Chandigarh, India

1. Introduction

With the increased industrialization, diverse natural activities and adverse uncontrolled human willingness have paved chemical pollutants ways in to soil, water, and atmosphere around the globe (Popek, 2018). Most of these pollutants are classified into two general categories: inorganic and organic. Although metals are natural elements of the earth's crust, the industrial activities have augmented their natural level leading to negative health impacts on humans, flora, and fauna of the earth system. Most of these metals and their associated complexes are enormously environmentally stable owing to their nonbiodegradable nature and these metal complexes can be voyaged at extensive distances by wind and aquatic residue.

On the other hand, organic pollutants such as volatile and semivolatile organic compounds also abbreviated as VOCs and SVOCs are mostly the synthetic chemicals originated from industrial activities and lack of their management. Mostly they are classified into above categories on the basis of their boiling points (WHO, 1989). The classification reflects the differences in the chemical properties that outline their ecological fortune and transport and also the diagnostic methods for their quantitative and qualitative detection. VOCs cover organic compounds (methane, propane, butane, benzene, xylene) having the boiling point in the range of $240°C-260°C$ and hence can be evaporated under normal temperature and pressure condition (Lane et al., 2017). Therefore, they can easily enter into the atmosphere (Sander et al., 2016). Most of the contribution of these VOCs is from agricultural source (mainly methane derivatives), petrochemicals industries, organic solvents, etc., (nonmethane and their derivatives). Because of difference in analysis method, the gases such as methane, ethane, propane, or butane are not categorized as VOCs, while their halogenated derivatives fall under this category. On the other hand, SVOCs, a class of organic compounds having

Abatement of Environmental Pollutants
https://doi.org/10.1016/B978-0-12-818095-2.00003-5

boiling points in the range of 250°C–400°C, generally include a myriad of industrial chemicals such as industrial chemicals and petrochemicals, pesticides, phenols, carbonyl compounds, ethers, aliphatic and aromatic esters, anilines, pyridines, and many others (Diamanti et al., 2009). It is pertinent to note that SVOCs exhibit a definite degree of volatility and hydrophilicity. Being copious and impervious to degradation, SVOCs can be transported with particulate matter. They are found in trace concentration in several matrixes including soil, water, sediments, air, etc. Because of definite persistent stability in environment, bioaccumulation capability, capability of long-range transportability, and environmental poisonousness, they are also internationally determined as persistent organic pollutants (POPs) (Kortenkamp, 2011).

2. Persistent organic pollutants: health effects and environmental chemistry

POPs tend to accumulate in the environment through the food chain because of their high fat solubility. Hence, they are generally found at higher concentration level in fat-containing food such as milks, eggs, fish, and meat, resulting into their subsequent accumulation in human body. POPs can be classified as polycyclic aromatic hydrocarbons (PAHs), perfluoroalkyl and polyfluoroalkyl substances, pesticides (halogenated organic compounds), brominated flame retardants, and so on (Daniel et al., 1998). Fig. 3.1 shows the structure of some POPs and related compounds. The POPs are known to have devastating health effects in terms of abnormal growth, brain malfunction, hormonal imbalance, metabolic disorders, etc. (Noakes et al., 2006; Daniel and Eickhoff, 2005); (EEA) European Environment Agency, 2012 (Karmaus, 2001); WHO (Poon et al., 2003), while the chronic exposure may also lead to immunological effects (Thornton; Exposure and Health Asses, 2000), type 1 immune responses (Santillo et al., 2000), allergy epidemic (Bouwman and Fiedler, 2003), asthma (Wong et al., 2003), cancer, reproductive defects, diabetes, etc. It is, therefore, imperative to have analytical tools for the qualitative and quantitative information in the environment.

On the other hand, atmosphere also acts as a crucial activator in trafficking and insertion of organic pollutants with soil and the aquatic system (Wong et al., 2003). The transportation of POPs to the atmosphere occurs through various physical and chemical processes. It mainly involves phase transformation of compounds from solid to atmospheric region through liquid vapor and then their subsequent dispersion (Toose et al., 2004). For instance, the organic compounds present in glaciers enter in soil and water surface on melting of glaciers, which then make their ways into groundwater. It has been remarked that there is a keen relationship between air, water, animals, and humans in the surroundings. The effects of these identities intentionally or unintentionally disturb the environment. The POPs pollution degrades ecosystem balance; environmental threatening; and health impact on all living organisms also. Many POPs are easily volatile and get accumulated on dust and aerosol, and these readily attack our environment and are easily inhaled to our body (Omar et al., 2018). Walker and his co-workers demonstrated that the inhalation and ingestion of dust and air are the two major sources of POPs (Walker, 2012).

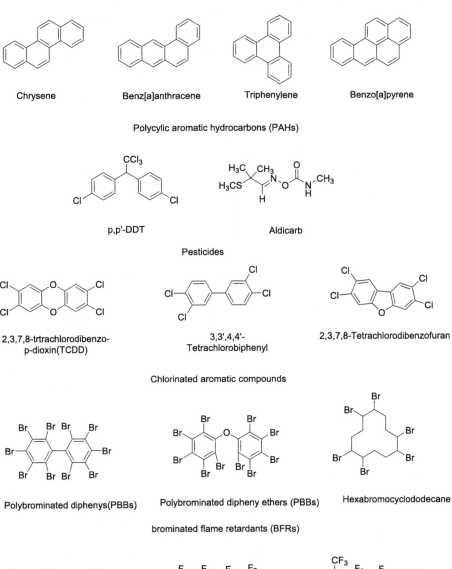

FIGURE 3.1 Structure of selected persistent organic pollutants and related compounds.

3. Method of POPs analysis (soil and water)

The recent development of analytical techniques for POPs degradation and analysis provides a definitive value for their natural or biological existence and plays a vital role in the analysis of their diffusion, environmental fates, and potential sources. This type of quantitative-based analysis technique helps shareholders to share responsibility, as well as provides the imperative information required by regulators.

Different techniques for the analysis of POPs have been identified based on gas chromatography (GC) and liquid chromatography (LC). GC provides an outstanding advantage for compatible analytes because of the high-resolution separations that are intrinsic to the method. For compounds targeted for analysis through GC should have enough volatility and thermal stability. For the analysis of polar or thermally labile compounds, LC is recommended as best technique. Recently, Moriwaki and his co-workers suggested the use of LC for PAHs, organochlorine compounds, and perfluorinated acids in environmental samples. Mass spectrometry (MS) is found to be ubiquitous detection technique for environmental analysis. Several LC-MS techniques have been used for POPs identification, such as quadrupole, ion trap, and time-of-flight mass analyzers, and MS sources such as electrospray ionization, atmospheric pressure photoionization, atmospheric pressure chemical ionization, and direct electron ionization. Besides, several environmental and agricultural additives have been succeeded during the past several decades for the POPs analysis in soil and water. The United Nation Environmental Protection organization has developed complete analysis procedure for POPs. The standard procedure includes either several matrices or individual matrix. Further different novel techniques for specific matrix readily available from the National Environmental Methods Index show that there is no complete or step-by-step analytical procedure suggested for specific POPs. As a result, performance-based methods are used by analytical laboratories. Such workability allows analytical chemists to choose the most suitable technique depending on their applications and drawbacks (Xua et al., 2013).

3.1 Samples collection, extraction, storage, and preparation

The development of soil matrix makes the characterization, identification, and quantification of organic pollutants (OPs) a true scientific dispute. A different mode of analytical techniques, encompassing sample preparation, analytes separation, and detection, is needed to attain the best recovery rates, detection limits, and analytical reliability for OPs under study.

The sample preparation of OPs from soil samples is the most crucial and moderate step of analysis, which may use up to 80% of the total time required for analysis. Generally, in the sample preparation, large volume of organic reagents, hazardous chemicals are used, and low recovery of the analytes with sample contamination is found to be the most relevant problem. Beginning of 21st century, sample preparation method has been done using the classical or traditional techniques, such as Soxhlet extraction, shaking extraction, ultrasonic-assisted extraction (UAE), and liquid–liquid extraction (LLE). In Soxhlet extraction, extract is found by the condensation of the organic solvent under continuous reflux of soil sample in the extractor. The Soxhlet method has high extraction capacity, but it is labor and time-consuming and requires the use of huge amount of organic solvents, which is one of the main disadvantages of this technique. Besides, shaking extraction requires manual or mechanical shaking of soil sample dispersed in a suitable solvent for a definite time period. The UAE requires ultrasound (US) radiation to carry out the extraction of soil sample, which is also dissolved in a definite solvent for a specific time period. The proper function of US is to improve the spiking of the solvent into the soil matrix so as to enhance the extraction time and efficiency. Tadeo et al. (2010) described the use of UAE in the analysis of soil OPs. The use of ultrasonic radiation in a water bath is the cheapest and widely used method. As a result of poor reliability with this technique, the use of more efficient and powerful modules is currently studied. In the above-mentioned techniques, it is necessary to apply a filtration which may tend to lower the contamination level. The LLE technique is also considered as one of the classical extraction techniques used in analysis of soil samples. Nevertheless, in this technology the soil samples require initial extraction of OPs. Therefore, this technique is generally used as a cleaning procedure for purification of soil matrix in combination with other extraction techniques. Among the traditional/classical procedures, shaking and UAE techniques are found to be the most widely used for extraction of OPs from soil samples.

Nowadays, more advanced extraction techniques are extensively used, such as microwave-assisted extraction (MAE), pressurized fluid extraction (PFE), supercritical fluid extraction (SFE), and solid-phase extraction. In the MAE method, the microwave (MW) energy is used to heat the sample added with the solvent. This method takes care of two necessary criteria time required for the extraction and use of less amounts of hazardous chemical and solvents. Despite the advantages of MAE technique, in many applications, it requires step-by-step filtration and cleaning. Wang et al. described an in-depth review of the most important and recent advances of MAE applied to OP analysis in soil and water. On the other hand, in the PFE, which is also known as accelerated solvent extraction (ASE), the use of high pressure allows the use of solvents at temperatures beyond their boiling point. Because of using high temperature and pressure, the solvent can be penetrated deep in to the samples. SFE describes the properties of a supercritical fluid such as carbon dioxide to extract OPs from soil samples. Supercritical fluids have properties intermediate between those of gas and a liquid, i.e., their viscosity is lower than that of liquids, while the diffusion coefficients and permeability are higher, which allow solvating many OPs (Duarte et al., 2018).

Nowadays, because of maximum use of more sophisticated and sensitive-detection techniques in OPs analysis, a tendency is developing to miniaturize sample preparation. Miniaturization provides a green extraction method as very less amounts of organic solvents are used, which generally yield low amounts of unwanted product and are most cost-effective. This is a tremendous advantage over the traditional extraction methods described above, which require large volumes of organic solvents and laborious clean-up and preconcentration steps. For examples, miniaturization of sample preparation is liquid-phase microextraction and solid-phase microextraction. Bruzzoniti et al. (Bruzzoniti et al., 2014) described quick easy cheap effective rugged safe (QuEChERS) method for sample preparation, which is basically used for OPs analysis in soil matrices. According to him, the overall recoveries for different classes of OPs obtained by using QuEChERS are higher than those obtained by other extraction methods such as ASE, UAE, liquid/solid extraction, and Soxhlet extraction.

The collected samples are preserved in a bottle containing 1M sodium thiosulfate solution. The temperature must be maintained at 4°C during the collection and stored at 4°C until the analyst is ready for the extraction process. After collection of the samples, the extraction should be done as soon as possible and can be stored for up to 14 days maintaining these conditions. Once the sample extraction is over, it is necessary to cleanup to remove the unwanted substances (Nagwa et al., 2015).

3.2 Conventional techniques

The conventional methods depend on lab-based reliable instrumentation, which requires experience of expert scientists, numerous chemicals, and simple sample preparation strategies. These are highly important techniques and therefore, they are used extensively for POPs quantifications in various analyses including soil, water, air, and food.

After the completion of sample preparation, the next step is the identification and quantification of OPs that are present in the soil and water samples. The technique used for identification and quantification depends on both the soil extracts and soil and water OPs. Typically, advanced analytical techniques such as GC or LC are used for identification and separation of OPs in complex soil extracts before their both qualitative and quantitative compositions. The GC is one of the major chromatographic methods for both volatile and semi-volatile OPs. Numerous varieties of detectors are available, which are combined to GC separation equipment. Some of these detectors are called electron capture, flame ionization, flame photometric, atomic emission, and nitrogen phosphorus detector, but MS and tandem MS (MS/MS) are the most common detectors performed for this purpose (Kim et al., 2016).

3.3 Analytical techniques for POPs quantification

3.3.1 UV-Vis spectroscopy

Ultraviolet-visible spectroscopy introduces to absorption spectroscopy or reflectance spectroscopy in the ultraviolet-visible region of the spectrum. Generally, this technique uses light in the near-UV and near-infrared ranges. The absorption or reflection directly affects the original color of the chemicals in the visible range. In this region of the electromagnetic spectrum, atoms and molecules undergo electronic transitions. There is utilization of an optical system based on UV spectrophotometry for the detection of organic compounds in water. The system is found to be capable of estimating organic carbon content in water. Furthermore, Kim and his co-workers demonstrated the determination of the organic compound concentration in water by measurement of the UV absorption at a wavelength of 250–300 nm (Dan et al., 2018).

3.3.2 Surface-enhanced Raman scattering

Surface-enhanced Raman scattering (SERS) is a powerful technique for pollutant detection, and magnetic nanoparticles (MNPs) are ubiquitous substrates for SERS detection. The applications of MNPs in SERS are used in the detection of environmental pollutants, such as OPs, heavy metal ions, pathogens, and other pollutants. Nowadays, OPs have been widely disposed to soil and groundwater. Because of its fingerprinting signals and ultrahigh sensitivity, SERS is found to be an ideal technique. Using MNPs as substrates, this technique has been applied in the fruitful detection of different organic pollutants, such as PAHs, explosives, and others. Most of these OPs can be directly detected with characteristic Raman peaks (Perelo, 2010).

4. Methods for POPs degradation

Traditional caring approaches such as incineration, chemical metal reduction, solvent extraction, ground filling, and stabilization or solidification are applied to remove the POPs from soil and sand residues. It is very challenging to manage sophisticated infrastructure and skilled staff with expertise to operate analytic instruments because of its high cost. Furthermore, these approaches are not fully able to eliminate the toxic organic pollutants (Megharaj et al., 2011). One method, i.e., biodegradation, can get rid of the toxic organic compounds more effectively from water and soil through applying microorganisms (Gadd, 2010). The biodegradation system works without affecting the other parameters and does pure the environment. The naturally existing microorganism degrades the typical chemical substances into smaller nonharmful substances through a process called bioconversion (Joutey et al., 2013; Nisha et al., 2018). The technique offers excellent outcomes in terms of eco-friendliness, prominence, effectiveness, and economical acceptability over previous traditional techniques.

4.1 Biological

The degradation of pollutants from organic substances is based on approachability, bioreach, and microorganism's metabolic potential to detoxify. Several chemical activities such as hydrogenation, O_2 double-bond oxidation, dehalogenation, amino groups oxidation, nitro groups reduction, sulfur conversion to O_2, addition of hydroxyl groups to aromatic compounds (benzene), etc., are among the catalytic bioreactants, which are formed throughout the biodegradation process (Singh and Ward, 2004).

4.1.1 Microbial degradation

Biodegradation technique not only possesses the capability to degrade living microorganisms but also immobilizes undesirable substances such as chemical acids, POPs, hydrocarbon, and their derivatives for their subsequent degradation (Haritash and Kaushik, 2009). A number of microorganisms such as bacteria, fungi, algae, etc., are reported for same, wherein Bacillus has found significant role in transformation of remediation of toxic organic pollutants (Hinchee and Leeson, 1996). A set of techniques such as biostimulation, biosparging, bioaugmentation, and bioventing have been used in accordance with organic contaminant to be eliminated. **Bioventing** is an approach of aerating the soil and water for the subsequent biodegradation of organic substances (Hyman and Dupont, 2001; Singh et al., 2011). **Biostimulation** is the conversion of pollutants substances and addition of other nutrients to the medium to maintain pH ration and balance C:N:P ratio to support soil microorganism activities (Stroo et al., 2012). **Bioaugmentation** is a method of insertion of microbial group, i.e., bacteria or fungi and other biocatalyst, i.e., gene or enzyme to degradation of organic/inorganic toxic compounds (Mehboob, 2010). In bioremediation, the aerobic and anaerobic conditions are applied for pollutants degradation. The aerobic conditions need molecular O_2 like mono- and dioxygenase to act as electron acceptor and cosubstrate. The anaerobic approach requires only electron presence for microorganisms. They transform and degrade molecule assembly into CO_2, H_2O, and inorganic compounds (salts). The hard-to-degrade chlorinated compounds are generally subjected to anaerobic condition for their degradation

(Muyzer and Stams, 2008). Metabolic coordination is a method which gets particularly in the coordinated microbial group, and biodegradation includes conversion of the solid surfaces and substances. Sometimes the fungi and bacteria originated enzymes are opted to degrade organic pollutants in place of whole cell microbial species.

4.1.1.1 Bacterial degradation

Microorganism compounds imply for the degrading of organic pollutants and act as an essential sustainable growth of environment. The degradation of contaminants in environment has been supported by hydrocarbon-degrading bacteria. The naturally occurring anaerobic processes by associated facultative microorganism make use of natural environment such as available nitrate, sulfates, iron, carbon dioxide, metal ions, etc., as electron acceptor or donor for the conversion of toxic organic pollutants into nonharmful by-products. For instance, the high level of anaerobic degradation in marine environment is supported by presence of high amount of sulfate in seawater. Fermentative strains have capability of complete reduction when supported with sulfate-reducing bacteria. Carmona et al. demonstrated dechlorination reduction approach for degradation of chlorinated compounds (Carmona et al., 2009). Similarly, consortiums of bacterial strains are reported for complete degradation of POPs including hydrocarbons, pesticides, dioxins, furans, etc. The bacterial consortia isolated from mangrove sediments are reported for several PAH degradation (Wongwongsee et al., 2013; Pino and Peñuela, 2011). Kong and co-workers showed that degradation of chlorinated pesticides with Alcaligenes fecal is JBW4 (Kong et al., 2013; Cuozzo et al., 2012).

4.1.1.2 Fungal degradation

Fungi, an omnipresent microorganism, are eukaryotic bacteria that have the capability to develop on various surfaces and are competent for longer duration and harsh condition. Also, fungi do not have fast growth; therefore, they are generally prefaced onto substrate for their cometabolic system. A diverse community, including molds, yeast, and filamentous fungi, are mainly utilized in POPs degradation. They exhibit strength to remove industrial pollutants (acids, aromatic compounds, etc.) and various kinds of materials such as wood, plastic, paper, textile, and leather. Mycodegradation technology uses the variety of fungi for chemicals or compounds degradation. The fungal cultures (white-rot) are able to degrade and remove PAHs (Bhatt et al., 2002). For example, *Irpex lacteus* and *Pleurotus ostreatus* isolated from contaminated soil are reported to degrade PAHs. Some fungal strains make use of various enzymes for degradation of these POPs such as DDT mineralization with lignin peroxidases (Bhattacharya et al., 2012). The mineralization rate is further dependent on carbon source available. Fungi could also increase the soil permeability, ion exchange strength, and hence can further decontaminate the soil from the pollutants. A number of fungal species, including *Fusarium oxysporum, Trametes versicolor, Phanerochaete chrysosporium*, etc., are reported for degradation of persistent toxic pollutants including DDT. They generally use oxidation/reduction mechanism to achieve this. For example, endosulfan degradation is catalyzed by *Trichoderma harzianum* through oxidation to endosulfan sulfate and then hydrolysis to endosulfan diol. The addition of nicotinamide adenine dinucleotide phosphate (NADPH) increases the metabolism rate of the approach, while the inhibition of endosulfan sulfate by cytochrome P450 increases the hydrolysis process. In other similar study, mycoremediation of PAH of contaminated oil is demonstrated through white-rot fungal.

The observation revealed that the fungi are relevant for biotreating PAH-contaminated oil-based drill cuttings (Okparanma et al., 2013). Ligninolytic enzyme derived from fungi has been shown for benzo-(α) pyrene degradation, supporting the use of specific strain in PAHs degradation and modifications (Cerniglia, 1997).The mechanism of PAHs degradation by lignolytic fungi is proposed as similar to nonphenolic lignin degradation (Peng et al., 2008). The further exploration of various types of mycelia's is shown for oxidation of aromatic compounds and their derivatives including pyrene, halogenated pyrenes, etc.

4.2 Chemical

The traditional applied chemical approaches are the Fenton process, in which a mixture of a soluble iron (II) salt and H_2O_2, known as the Fenton's reagent, is applied to degrade and destroy pollutants. The mechanism of the Fenton's process involves the degradation of the H_2O_2 to produce strongly reactive hydroxyl radicals, which generally degrades the POPs and OPs from water and wastewater sediments (Haag and Yao, 1992). While with advancement in chemical processes, the new technological approaches such as MW and high-intense US are generally utilized for chemical degradation of the POPs. The organic pollutants are degraded or converted into little quantity of inorganic ions, nonharmful chemicals, and mineralized to CO_2, H_2O, etc. Under US or MW heating, fast degradation in pollutant water, having aromatic halides, halogenated phenols, and polychlorinated biphenyls, is performed at neutral pH. In acids condition, acidification with sulfuric acid of pH 1.7–2.0 facilitates full degradation. Cravotto and his staff extracted the dichloromethane (CH_2Cl_2) with hexane. The concentration of emergent organic pollutants in the organic extracts measured using gas chromatography with HP five cross-linked 5% phenyl methyl siloxane column. GC was maintained with injector temperature 240°C; temperature program between 90°C (1 min) and 240°C at 2°C min^{-1} and helium is used as a carrier gas at 1.2 mL min^{-1} (Cravotto et al., 2005). On the other hand, 0.1 or 0.5 N of aqueous alkaline solution of NaOH or KOH was used for the degradation and elimination of POPs and other humic substances from soil matrices (Schnitzer et al., 1978). Furthermore, Hartmann and his co-workers used a hydrolysis method for the extraction of PAHs and OPs from soil. By using the methanolic extraction procedure, the organic contaminates readily degrade from the soil samples. Generally the hydrolysis procedure has been used with the extraction method to identify the organic chemicals in the soil matrices (Hartmann, 1996).

4.3 Advanced oxidation approaches

Advanced oxidation approaches are foremost, favorable, effective, and eco-friendly methods which remove the POPs from each type of water. Normally these approaches are depended on the capable oxidative such as hydroxyl radicals (•OH) (Mehmet and Aaron, 2014). It is suggested that these methods are applied for degradation and conversion of organic pollutants by treatment of water or wastewater. Glaze et al. (Glaze et al., 1987) gave a definition of advanced oxidation approaches; it works as water treatment method at room temperature and normal pressure and produces oxidizing agent such as hydroxyl radicals (•OH). Some significant research works are reported in this direction

(Andreozzi et al., 1999; Herrmann et al., 1999; Tarr, 2003; Gogate and Pandit, 2004; Parsons, 2004; Brillas et al., 2006; Laine and Cheng, 2007; Zaviska et al., 2009). Various advanced oxidation approaches are depended on different chemical, photochemical, or electrochemical reactions. Besides, they provide an optimistic, novel, and environment-friendly technique to separate the POPs from water. Based on the reactions such as chemical, electrochemical, and photolytic via the formation of intermediate, hydroxyl radicals, different types of AOP methods are known. The most widely used AOP is the Fenton method, which is specifically used for degradation and elimination of POPs. However, the efficiency and utilization of this method is significantly upgraded by the use of ultraviolet or by sunlight (Andreozzi et al., 1999). Brillas et al. demonstrated the use of electrochemistry to generate the hydroxyl radical for the elimination of POPs and carcinogenic organic compounds from water (Tarr, 2003).

5. Conclusions

The chapter briefly summarizes the POPs, their classification, sources, and negative health impact, with a focus on need of their analysis and quantifications. A range of conventional and other analytical (optical/electrochemical) techniques are presented and discussed for their deployment in POPs quantification. Besides, a trend toward using biological and chemical processes for POPs degradation is covered and discussed in brief.

Acknowledgments

The sincere support and encouragement by Director, CSIR-Central Scientific Instruments Organisation, Chandigarh, is acknowledged.

References

Andreozzi, R., Caprio, V., Insola, A., Marotta, R., 1999. Advanced oxidation processes (AOP) for water purification and recovery. Catalysis Today 53 (1), 51–59.

Bhatt, M., Cajthaml, T., Šašek, V., 2002. Mycoremediation of PAH-contaminated soil. Folia Microbiologica 47 (3), 255–258.

Bhattacharya, S., Angayarkanni, J., Das, A., Palaniswamy, M., 2012. Mycoremediation of benzo[a]pyrene by pleurotusostreatus isolated from Wayanad district in Kerala, India. International Journal of Pharma and Bio Sciences 2 (2), 84–93.

Bouwman, H., 2003. POPs in Southern Africa. In: Fiedler, H. (Ed.), The Hand Book of Environmental Chemistry, Persistent Organic Pollutants, third ed. Springer-Verlag, Berlin/Heidelberg (Chapter 11).

Brillas, E., Arias, C., Cabot, P.L., Centellas, F., Garrido, J.A., Rodríguez, R.M., 2006. Degradation of organic contaminants by advanced electrochemical oxidation methods. Portugaliae Electrochimica Acta 24 (2), 159–189.

Bruzzoniti, M.C., LC, R.M., Carlo, D., Orlandini, S., Rivoira, L., Bubba, M.D., 2014. QuEChERS sample preparation for the determination of pesticides and other organic residues in environmental matrices: a critical review. Analytical and Bioanalytical Chemistry 406, 4089–4116.

Carmona, M., Zamarro, M.T., Blázquez, B., Durante-Rodríguez, G., Juárez, J.F., Valderrama, J.A., Barragán, M.J., García, J.L., Díaz, E., 2009. Anaerobic catabolism of aromatic compounds: a genetic and genomic view. Microbiology and Molecular Biology Reviews 73 (1), 71–133.

Cerniglia, C.E., 1997. Fungal metabolism of polycyclic aromatic hydrocarbons: past, present and future applications in bioremediation. Journal of Industrial Microbiology and Biotechnology 19 (5), 324–333.

Cravotto, G., Carlo, S. Di, Tumiatti, V., Roggero, C., Bremner, H.D., 2005. Degradation of persistent organic pollutants by Fenton's reagent facilitated by microwave or high-intensity ultrasound. Environmental Tech 26, 721–724.

Cuozzo, S.A., Fuentes, M.S., Bourguignon, N., Benimeli, C.S., Amoroso, M.J., 2012. Chlordane biodegradation under aerobic conditions by indigenous Streptomyces strains, 1274 International Biodeterioration and Biodegradation 66 (1), 19–24.

Dan, S., Rong, Y., Feng, L., Anna, Z., 2018. Applications of magnetic nanoparticles in surface-enhanced Raman scattering (SERS) detection of environmental pollutants. Journal of Environmental Sciences. https://doi.org/10.1016/j.jes.2018.07.004.

Daniel, F.H., Eickhoff, C.S., 2005. Type 1 immunity provides both optimal mucosal and systemic protection against a mucosally invasive, intracellular pathogen. Infection and Immunity 73 (8), 4934–4940.

Daniel, T.C., Sharpley, A.N., Lemunyon, J.L., 1998. Agricultural phosphorus and eutrophication: a symposium overview. Journal of Environmental Quality 27, 251–257.

Diamanti, K.E., Bourguignon, J.P., Giudice, L.C., Hauser, R., Prins, G.S., Soto, A.M., Zoeller, R.T., Gore, A.C., 2009. Endocrine-disrupting chemicals: an endocrine society scientific statement. Endocrine Reviews 30 (4), 293–42.

Duarte, R.M.B.O., Matos João, T.V., Senesi, N., 2018. Organic pollutants in soils. Soil Pollution 103–126.

Exposure and Health Assessment for 2,3,7,8-Tetrachlorobenzo-p-dioxine(TCDD) and Related Compounds, Part 3 Integrated Summary and Risk Characterization for 2,3,7,8-Tetrachlorobenzo-p-dioxin (TCDD) and Related Compounds, 2000. US EPA.

Gadd, G.M., 2010. Metals, minerals and microbes: geomicrobiology and bioremediation. Microbiology 156 (3), 609–643.

Glaze, W.H., Kang, J.W., Chapin, D.H., 1987. The chemistry of water treatment processes involving ozone, hydrogen peroxide and ultraviolet radiation. Ozone Science and Engineering 9 (4), 335–352.

Gogate, P.R., Pandit, A.B., 2004. A review of imperative technologies for wastewater treatment I: oxidation technologies at ambient conditions. Advances in Environmental Research 8 (3–4), 501–551.

Haag, W.R., Yao, C.C.D., 1992. Rate constants for reaction of hydroxyl radicals with several drinking water contaminants. Environmental Science and Technology 26, 1005–1013.

Haritash, A., Kaushik, C., 2009. Biodegradation aspects of polycyclic aromatic hydrocarbons (PAHs): a review. Journal of Hazardous Materials 169 (1), 1–15.

Hartmann, R., 1996. Polycyclic aromatic hydrocarbons (PAHs) in forest soils: critical evaluation of a new analytical procedure. InterNational Journal of Environmental Analytical Chemistry 62, 161–173.

Herrmann, J.M., Guillard, C., Arguello, M., Aguera, A., Tejedor, A., Piedra, L., Fernandez Alba, A., 1999. Photocatalytic degradation of pesticide pirimiphosmethyl. Determination of the reaction pathway and identification of intermediate products by various analytical methods. Catalysis Today 54 (2–3), 353–367.

Hinchee, R.E., Leeson, A., 1996. Soil Bioventing: Principles and Practice. Taylor & Francis.

Hyman, M., Dupont, R.R., 2001. Groundwater and Soil Remediation: Process Design and Cost Estimating of Proven Technologies. ASCE Press.

Joutey, N.T., Bahafid, W., Sayel, H., Ghachtouli, N.E., 2013. Biodegradation: Involved Microorganisms and Genetically Engineered Microorganisms (Chapter).

Karmaus, W., 2001. Article in Archives of Environmental Health and International Journal vol. 56 (6), 485–492.

Kim, C., Joo, B.E., Soyoun, J., Taeksoo, J., 2016. Detection of organic compounds in water by an optical absorbance method. Sensors 16 (1), 61 (1-7).

Kong, L., Zhu, S., Zhu, L., Xie, H., Su, K., Yan, T., Wang, J., Wang, J., Wang, F., Sun, F., 2013. Biodegradation of organochlorine pesticide endosulfan by bacterial strain Alcaligenes faecalis JBW4. Journal of Environmental Sciences 25 (11), 2257–2264.

Kortenkamp, A., 2011. Are cadmium and other heavy metal compounds acting as endocrine disrupters? Metal Ions in Life Sciences 8, 305–317.

Laine, D.F., Cheng, I.F., 2007. The destruction of organic pollutants under mild reaction conditions: a review. Microchemical Journal 85 (2), 183–193.

Lane, C., Sander, M., Schantz, M., Stephen, A., 2017. Environmental Analysis: Persistent Organic Pollutants. Elsevier BV.

Megharaj, M., Ramakrishnan, B., Venkateswarlu, K., Sethunathan, N., Naidu, R., 2011. Bioremediation approaches for organic pollutants: a critical perspective. Environment International 37 (8), 1362–1375.

Mehboob, F., 2010. Anaerobic Microbial Degradation of Organic Pollutants with Chlorate as Electron Acceptor. PhD Thesis. Wageningen University, Wageningen.

Mehmet, A.O., Aaron, J.J., 2014. Advanced oxidation processes in water/wastewater treatment: principles and applications. A Review 44 (23), 2577−2641.

Muyzer, G., Stams, A.J.M., 2008. The ecology and biotechnology of sulphate-reducing bacteria. Nature Reviews Microbiology 6, 441.

Nagwa, A.B.O., ELMaali, A., Yehia, W., 2015. Gas chromatography-mass spectrometric method for simultaneous separation and determination of several POPs with health hazards effects. Modern Chemistry and Applications 3, 4, 1000167.

Nisha, G., Korrapati, N., Pydisetty, Y., 2018. Recent advances in the bio-remediation of persistent organic pollutants and its effect on environment. Journal of Cleaner Production 118, 1602−1631.

Noakes, M., Foster, P.R., Keogh, J.B., James, A.P., Mamo, J.C., Clifton, P.M., 2006. Comparison of isocaloric very low carbohydrate/high saturated fat and high carbohydrate/low saturated fat diets on body composition and cardiovascular risk. Nutrition and Metabolism 3, 7.

Okparanma, R.N., Ayotamuno, J.M., Davis, D.D., Allagoa, M., 2013. Mycoremediation of polycyclicaromatic hydrocarbons (PAH)-contaminated oil-based drill-cuttings. African Journal of Biotechnology 10 (26), 5149−5156.

Omar, M.L., Alharbi, A.A.B., Rafat, A.K., Imran, A., 2018. Health and environmental effects of persistent organic pollutants. Journal of Molecular Liquids 263, 442.

Parsons, S.A. (Ed.), 2004. Advanced Oxidation Processes for Water and Wastewater Treatment. IWA Publishing, Alliance House, London.

Peng, R.H., Xiong, A.S., Xue, Y., Fu, X.Y., Gao, F., Zhao, W., Tian, Y.S., Yao, Q.H., 2008. Microbial biodegradation of polyaromatic hydrocarbons. FEMS Microbiology Reviews 32 (6), 927−955.

Perelo, L.W., 2010. Review: in situ and bioremediation of organic pollutants in aquatic sediments. Journal of Hazardous Materials 177, 81−89.

Pino, N., Peñuela, G., 2011. Simultaneous degradation of the pesticides methyl parathion and chlorpyrifos by an isolated bacterial consortium from a contaminated site. International Biodeterioration and Biodegradation 65 (6), 827−831.

Poon, C.S., Azhar, S., Anson, M., Wong, Y.L., 2003. Performance of metakaol in concrete at elevated temperatures. Cement and Concrete Composites 25, 83.

Popek, E., 2018. Sampling and Analysis of Environmental Chemical Pollutants, Chapter: 2 Environmental Chemical Pollutants, pp. 13−69.

Sander, L.C., Schantz, M.M., Wise, S.A., 2016. Environment Analysis: Persistent Organic Pollutants, pp. 401−449.

Santillo, R., Stringer, L., Johnston, P., 2000. The Global Distribution of PCBs, Organochlorine Pesticides, Polychlorinated Dibenzo-p-dioxins and Polychlorinated Dibenzofurans Using Butter as an Interactive Matrix. Green Peace.

Schnitzer, M., 1978. Humic substances: chemistry and reactions. In: Schnitzer, M., Khan, S.U. (Eds.), Soil Organic Matter. Elsevier Science, New York.

Singh, A., Ward, O.P., 2004. Biodegradation and Bioremediation. Springer.

Singh, A., Parmar, N., Kuhad, R.C., 2011. Bioaugmentation, Biostimulation and Biocontrol. Springer.

Stroo, H.F., Leeson, A., Ward, C.H., 2012. Bioaugmentation for ground water remediation. In: Environmental Remediation Technology. Springer.

Tadeo, J.L., Sánchez-Brunete, C., Albero, B., 2010. Application of ultrasound-assisted extraction to the determination of contaminants in food and soil samples. Journal of Chromatography A 1217, 2415−2440.

Tarr, M.A., 2003. Chemical Degradation Methods for Wastes and Pollutants: Environmental and Industrial Applications. Marcel Dekker, New York, NY.

Thornton, P.E., Biome-Bgc: Modeling Effects of Disturbance and Climate, ORNL DAAC, Oak Ridge, Tennessee, USA.

Toose, L., Woodfine, D.G., MacLeod, M., Mackay, D., Gouin, J., 2004. BETR-World: a geographically explicit model of chemical fate: application to transport of α-HCH to the Arctic. Environmental Pollution 128 (1), 223−240.

Walker, C.H., 2012. Organic Pollutants: An Ecotoxicological Perspective. CRC Press.

WHO, 1989. Indoor Air Quality: Organic Pollutants. EURO Reports and Studies No. 111. WHO Reg. Office for Europe, Copenhagen.

Wong, M., Poon, B., 2003. Sources, fates and effects of persistent organic pollutants in China, with emphasis on the Pearl River Delta. In: Fiedler, H. (Ed.), The Hand Book of Environmental Chemistry, Persistent Organic Pollutants, third ed. Springer-Verlag, Berlin/Heidelberg (Chapter 13).

Wongwongsee, W., Chareanpat, P., Pinyakong, O., 2013. Abilities and genes for PAH biodegradation of bacteria isolated from mangrove sediments from the central of Thailand. Marine Pollution Bulletin 74 (1), 95–104.

Xua, W., Wanga, X., Cai, Z., 2013. Analytical chemistry of the persistent organic pollutants identified in the Stockholm convention. Analytica Chimica Acta 6 (1–13), 790.

Zaviska, F., Drogui, P., Mercier, G., Blais, J.F., 2009. Proc'ed'es d'oxydationavanc'ee dans le traitement des eaux et des effluents industriels: application 'ala d'egradation des polluants r'efractaires. Revue des Sciences de l'Eau 22 (4), 535–564.

Phytoremediation of organic pollutants: current status and future directions

Sachchidanand Tripathi[1], Vipin Kumar Singh[7], Pratap Srivastava[2,3], Rishikesh Singh[4], Rajkumari Sanayaima Devi[1], Arun Kumar[5], Rahul Bhadouria[2,6]

[1]Deen Dayal Upadhyaya College (University of Delhi), New Delhi, India; [2]Department of Botany, Institute of Science, Banaras Hindu University, Varanasi, India; [3]Shyama Prasad Mukherjee Government Degree College, Phaphamau, Prayagraj, India; [4]Institute of Environment & Sustainable Development, Banaras Hindu University, Varanasi, India; [5]Bihar Agricultural University, Sabour, Bhagalpur, India; [6]Department of Botany, University of Delhi, Delhi, India; [7]Center of Advanced Study in Botany, Institute of Science, Banaras Hindu University, Varanasi, India

1. Introduction

Phytoremediation is the technique to remove/remediate contaminated soil or water with the help of plant (directly or indirectly) (Pradhan et al. 1998). Soil, sediments, and water (surface and groundwater) are becoming contaminated because of human activities such as agriculture, industrialization, mining, construction, and other developmental activities. Phytoremediation is a cost-effective, nonintrusive, and effective way of remediating soils, water, and sediments.

Phytoremediation is more cost-effective than mechanical and chemical methods of remediating hazardous compounds from soil, sediments, surface, and groundwater environment (Bollag et al., 1994). Furthermore, it is natural, a novel plant-based, aesthetically pleasing technology (Pradhan et al. 1998).

Abatement of Environmental Pollutants
https://doi.org/10.1016/B978-0-12-818095-2.00004-7

The goal of phytoremediation is to completely mineralize organic pollutants into relatively nontoxic constituent, such as CO_2, nitrate, chlorine, and ammonia (Cunningham et al., 1996). Phytoremediation is defined as the use of green plants to remove pollutants from site of contamination or render them harmless. This technology makes use of naturally occurring processes by which plants and their microbial rhizosphere flora degrade and/or sequester organic and inorganic pollutants (Pradhan et al., 1998). Inorganic pollutants occur as natural elements in the Earth's crust or atmosphere; however, their release into natural ecosystem through human activities such as mining, intensive application of agrochemicals in agriculture, and industrial processes that are causing hazards to human health and environment is well-documented (Nriagu, 1979). Wide diversity of plants can be successfully exploited to decontaminate atmospheric pollutants such as SO_2 and NOx as well as halogenated hydrocarbons of volatile nature (Jeffers and Liddy, 2003). Environmental contaminants can be detoxified through number of physicochemical and biological processes including surface binding, movement inside cellular system, biotransformation, biodegradation, and internal storage. The intracellular sequestration thus helps in mitigating pollutant toxicity to plant system. The process of accumulation may be enhanced up to multiple folds by development of transgenic plants through enhanced expression of a particular gene from a plant or any bacterial system. This chapter is focused to have details on mechanisms of phytoremediation, role of plant-associated microbes in organic contaminant removal, genetically engineered organisms for phytoremediation, and lastly the challenges to currently used phytoremediation techniques. Removal of different organic pollutants through plants is discussed under different sections.

2. The process of phytoremediation

Organic contaminants found in the natural environmental conditions are attacked and mineralized by multiple biological activities. Plants can remediate organic pollutants via immobilization, volatilization, transformation to different extents (even mineralization), or a combination of all, depending on the compound structure, environmental factors, and plant genotypes. In general, various phytoremediation technologies are used to remediate organic pollutants such as the following: (1) phytoextraction (phytoaccumulation, phytoabsorption, phytosequestration) refers to the accumulation of soil contaminants within plant's tissues (rhizofiltration, however, denotes the removal of contaminants from polluted water system) (Ali et al., 2013; Mesjasz- Przyby lowicz et al., 2004); (2) phytodegradation/rhizodegradation deals with the transformation of toxic environmental contaminants into nontoxic forms using plants and associated microorganisms followed by either accumulation or secretion in the vicinity (Wiszniewska et al., 2016); (3) phytovolatilization describes the phenomenon of pollutant uptake from contaminated sites and their release into atmosphere in volatile forms (Wiszniewska et al., 2016); (4) phytostabilization discusses with the process of contaminant entrapment on a suitable matrix through adsorption (Wiszniewska et al., 2016); and (5) phytodesalination refers to the use of halophytic plants for removal of excess salts from saline soils (Ali et al., 2013; Wiszniewska et al., 2016).

The mobilization of a particular organic contaminant is largely determined by physical and chemical characteristics such as volatility, miscibility in water, electric charge, density, size, and interactions with surrounding matrix system. In a soil ecosystem, important parameters such as pH, porosity, bulk density, microbial communities, nutrient content, and organic matter availability affect the transport and immobilization of a contaminant at a greater extent.

A number of issues must be taken into account before application of full-scale phytoremediation including (1) detailed site characterization; (2) selection of specific plant species for phytoremediation in targeted site; (3) cost—benefit analysis including seedling growth, plantation, and management; and (4) the time estimated for remediation, which depends largely on the pollutant's uptake or elimination rate, is also an important parameter.

The above-mentioned issues have been analyzed through pilot experiments for better process optimization (McCutcheon and Schnoor, 2004). Some additional important factors are the compilation of biomass harboring the contaminant and safe disposal after harvesting. Furthermore, the contaminant sequestered into below-ground biomass may pose a significant challenge for their removal and may prove time-consuming and expensive process, restricting their field-scale feasibility.

3. Physiological and biochemical aspects of phytoremediation

To date, lots of researchers have done extensive investigations to unveil the process of pollutant absorption from contaminated sites, their upward movement, and detoxification of accumulated substances. Most of the contaminants present in the plant's surrounding are accumulated by roots followed by leaves (Wang and Liu, 2007). Uptake through leaves mostly happens after intensive spraying of agrochemicals; however, significant content of volatile molecules may also be accumulated (Burken et al., 2005). Generally, contaminants get access to roots by the process of diffusion, finally finding their ways into xylem stream. As plant systems are devoid of specifically designed transporters for the movement of xenobiotics, the rate of transport within plant system is greatly influenced by their physical and chemical properties. As the process of transport is energy independent (passive) and is of physical nature, the transport rate may be evaluated and typical models for movement can be developed (Fujisawa, 2002).

Some of the agricultural activities may also improve the absorption of contaminants by plants. For instance, roots growth may be promoted in the surroundings of contaminant availability. Furthermore, watering at specific locations, addition of fertilizers, and introduction of oxygen into the root zone may facilitate the uptake process. Similarly, the plants having specific root organization and development of plants exhibiting engineered root system may serve the purpose of efficient contaminant uptake. Plants such as *Populus* spp. and Salix with elaborated root architecture and increased rate of water loss through transpiration could be considered for effective contaminant removal (Jansson and Douglas, 2007). After entry inside the xylem stream, contaminants may be transformed through oxidoreduction and hydrolytic reactions and eventually may conjugate with cellular metabolites such as reduced glutathione, sugars, and organic acids. Binding with cellular metabolites improves the contaminant solubility and facilitates further binding with appropriate biological catalysts, transport proteins, or some other relevant molecules of proteinaceous nature (Dietz and Schnoor, 2001; Pilon-Smits, 2005). Thus, bound contaminant may be stored as an ingredient of cell wall matrix or may be accumulated in vacuoles for further metabolic actions for their biological degradation (Pilon-Smits, 2005). Moreover, cellular compartmentalization appears to be a major and vital aspect for the detoxification of hazardous contaminants (Mezzari et al., 2005).

The formation of contaminant-glutathione complex is well recognized in most of the living beings. The process is facilitated by enzyme systems glutathione-S-transferases, possessing the active sites for binding of hydrophobic molecules. The contaminant-enzyme complex then associates with γ-glutamyl transpeptidases (GGTs) and other enzymes of the pathway for subsequent mineralization. Ohkama-Ohtsu et al. (2007) have described the presence of a GGT in *Arabidopsis thaliana* (L.) Heynh. equipped with the ability to direct the primary and inevitable events in the degradation of contaminants complexed with the reduced glutathione. Still very less information with respect to other enzymes such as glucosyltransferases or malonyltransferases participating in degradation of glutathione-contaminant complex is documented (Brazier-Hicks and Edwards, 2005). Noteworthy, ATP-binding cassette transporters play a significant role in the transport of enzyme-contaminant complex from cytoplasmic fluids to specific locations such as vacuole or mobilization into the apoplast (Klein et al., 2006) and thus may be selected as important candidate for their genetic engineering to enhance transport of such complexes.

Multiple enzymes from plants including mono- and dioxygenases, dehydrogenases, hydrolases, peroxidases, nitroreductases, nitrilases, dehalogenases, phosphatases, and carboxylesterases have been suggested to participate in degradation of organic contaminants (Singer et al., 2003; Wolfe and Hoehamer, 2003; Pilon-Smits, 2005). Current world scientists especially biochemists and molecular biotechnologist have recognized the immense potential of these biocatalysts in enhancing the contaminant degradation ability of a plant species. Few of the aforementioned classes of enzymes have been described for their natural secretions into the soil environment and subsequent degradation of myriads of organic contaminants (Singer et al., 2003). The possible application of genetically modified *Arabidopsis* with the capability of enhanced secretion of enzyme laccase into the soil resulting from the gene overexpression may be considered for contaminant degradation (ex planta remediation) and has been evidenced (Wang et al., 2004). The activity of different plant enzymes and various transporters toward a number of man-made compounds (xenobiotics) may be considered essentially as an accidental phenomenon because the inherent functions of these enzymes are well defined in cellular system. In this context, cytochrome P450 monooxygenases could be considered as a suitable enzyme for degradation of environmental contaminants (Morant et al., 2003). The oxidation reactions performed by the enzyme are documented in secondary metabolic activities.

The plant P450, rather than performing only the hydroxylation and epoxidation, is also able to direct others reactions such as phenol integration, ring development, and transformations or reactions leading to evolution of CO_2 from substrate molecule. The possible involvement of P450 in plants survivability at higher concentration of herbicide and its removal from contaminated site is mentioned. *Helianthus tuberosus* CYP76B1 and *Glycine max* (L) Merr. CYP71A10 are reported as the first biological catalysts of plant origin participating in efficient degradation of herbicide (Robineau et al., 1998; Siminszky et al., 1999). Thereafter, numerous P450s of plant origin were demonstrated to be linked with the mineralization of different organic contaminants including persistent organic pollutants (POPs). However, in most of the cases, interestingly, the genetically engineered plants overexpressing P450 genes for the purpose of efficient phytoremediation were of mammalian origin. For instance, Doty et al. (2000) developed genetically engineered tobacco plants having the potential to synthesize mammalian CYP2E1 with significantly enhanced efficiency in absorption and mineralization of trichloroethylene (TCE) and ethylene bromide (EtBr). Similarly, investigations demonstrating considerable increase in tolerance against hazardous herbicides through development of genetically engineered potato (*Solanum tuberosum* L.) (Inui et al., 2001) and rice (*Oryza sativa* L.) (Kawahigashi et al., 2007) have been described.

The extensive research in this direction, however, must be accelerated to make the process of phytoremediation potential of engineered plants efficient and feasible at larger scale through incorporation of P450 genes. Critical investigations have indicated that P450s obtained from mammalian systems may act on larger numbers of xenobiotics as compared with their counterparts of plant origin. The important progress in the development of genetically engineered *Arabidopsis* and tobacco (*Nicotiana tabaccum* L.) harboring the P450 gene of plant origin with improved herbicide tolerance was demonstrated by Didierjean et al. (2002).

4. Strategies of phytoremediation of organic pollutants

Most of the organic pollutants in the environment are man-made and xenobiotic to organisms. Majority of organic pollutants are toxic and some of them are reported as carcinogenic. The main anthropogenic sources of organic pollutants in environment are agricultural activities (pesticides, herbicides, chemical fertilizers), industries (chemical), military activities (explosives, chemical weapons), spills (oil, solvents), urbanization, wood treatment, etc. Most important organic pollutants, which are targeted for phytoremediation, are polychlorinated biphenyls (PCBs) (Harms et al., 2003), polycyclic aromatic hydrocarbons (PAHs) (Hughes et al., 1996), and linear halogenated hydrocarbons (Burken and Schnoor, 1997; Shang et al., 2003).

4.1 Direct uptake (direct phytoremediation)

The direct absorption of contaminants by plant is restricted by the availability and mechanism of transport involved (Salt et al., 1998). In general, transport of organic substances such as herbicides and pesticides happens through soluble portions of soil as largely evidenced for plant systems (Paterson et al., 1994; Topp et al., 1986). The important features deciding the transport of a particular contaminant within plants are physicochemical nature such as concentrations, acidity, octanol:water partition coefficient, and pKa value (Wenzel et al., 1999). Organic substances generally transported into plant systems are known to exhibit mild hydrophobic characteristics with octanol—water partition coefficients within range of 0.5—3 (Ryan et al., 1988; Wenzel et al., 1999). Apart from the parameters governing the availability of organic contaminants, there are significant differences in the absorption and transport of contaminants among different plant species as documented in the case of nitrobenzene

(Farlane et al., 1990) and atrazine (Anderson and Walton, 1995; Burken and Schnoor, 1996). Most importantly, considerable disparity in rate of transpiration could have noticeable impact on differential uptake and transport of organic contaminants transport into plants (Burken and Schnoor, 1996).

Because the availability of contaminant in soil system is a major challenge for efficient removal of organic contaminants (Cunningham et al., 1996; Salt et al., 1998), the introduction of soil amendments could have a major breakthrough in the field of "induced" (in contrast to the "continuous") phytoremediation approaches. The employment of chemically (Triton X-100, SDS) and naturally synthesized (rhamnolipids) surfactants to increase the solubility of target xenobiotics and improvement in bacterially catalyzed degradation is recognized (Van Dyke et al., 1993; Desai and Banat, 1997; Providenti et al., 1995). In addition, the incorporation of cyclodextrins has been evidenced with the enhanced solubility of organic contaminants (Brusseau et al., 1997). Significant improvement in solubility of organic contaminant after introduction of surfactant of either chemical or biological origin and biopolymers such as cyclodextrins (Miller, 1995; Nivas et al., 1996) could have promising future application for the removal of mixed types of pollutants and thereby management of contaminated sites.

While practicing for phytoremediation of organic pollutant contaminated sites, it is necessary to analyze and have thorough knowledge about the fate of parent compound to be removed as well as their metabolic by-products (Wenzel et al., 1999). Furthermore, the accumulation of organic contaminants in root and shoot portion is largely determined by their physical and chemical attributes.

4.2 Phytoremediation explanta

Explanta phytoremediation involves the exploitation of root secreted metabolites to facilitate the multiplication and metabolic processes of large pool of numerous microbial communities inhabiting in the rhizospheric region to remediate the contaminated sites (Anderson et al., 1993; Shimp et al., 1993). Few of the organic substances (i.e., phenolics, organic acids, alcohols, proteins) secreted by roots may be utilized as valuable source of carbon and nitrogen to favor the rapid multiplication and prolonged survivability of diverse microorganisms responsible for catabolism of hazardous organic contaminants. For example, number of nitrogen-fixing microbes may be up to fourfold higher in the vicinity of rhizosphere along with high rate of microbial metabolisms as compared with those occurring in nonrhizospheric region (Anderson et al., 1993; Salt et al., 1998). It is well-documented that constituents of root exudates and their rate of secretion vary significantly in different plant species (Rao, 1990; Salt et al., 1998). Investigations in this direction has fascinated the researchers worldwide to work selectively on the plants possessing the ability to secrete metabolic products in form of root exudates enriched with phenolics to intensify the rapid multiplication of bacteria capable of PCB biodegradation (Fletcher and Hedge, 1995; Salt et al., 1998). As the phenolics present in exudates differ significantly from species to species (Rao, 1990), it may be assumed that exudates of some plants may sufficiently support the fast multiplication of contaminant-degrading bacteria in rhizosphere (Wenzel et al., 1999). Fletcher and Hedge (1995) had worked out to explore the possibility of phenolic components present in root exudates from 17 different plants that could favor the multiplication of PCB-degrading microbes and concluded that phenolics from mulberry (*Morus rubra* L.) possessed multiple promising features for its exploitation in bioremediation. Furthermore, the process of remediation is also aggravated by rhizosphere-inhabiting microbes through volatilization of organic contaminants such as PAHs or through enhanced synthesis of humic compounds from contaminants of organic origin (Cunningham et al., 1996). Similarly, removal of TCE and TNT (trinitrotoluene) through volatilization and increased synthesis of humics has also been documented (Anderson and Walton, 1995; Foth, 1990; Wolfe et al., 1993).

5. Role of enzymes

Enzymes have a great potential of transformation and detoxification of organic pollutants as they have been recognized to be able to transform pollutants and are suitable for restoration of polluted environment (Rao et al., 2010; Schwitzguébel, 2017; Wang et al. 2004). Most of the studies have revealed that enzymes may act as a good alternative in phytoremediation for overcoming most hindrances related to the use of microorganisms (Gianfreda and Bollag, 2002; Gianfreda and Rao, 2004).

Enzymes are able to act on number of available substrate molecules. In general, microbial enzymes favor the mineralization of easily degradable contaminants. The contaminants are considered to be more transportable as compared with the microbial cells because of little size Gianfreda and Bollag (2002). The application of microbial enzymes or any other for contaminant degradation processes relying on enzymes is an environment-friendly approach to clean the hazardous pollutants. The enzymatic actions may occur intracellularly (within producing cells) or extracellularly (outside producing cells). The enzymes acting extracellularly may perform their specified work either in soluble or in immobilized form and the catalytic activity would be homogeneous and heterogeneous, respectively (Gianfreda and Rao, 2004).

Apart from the synthesis and secretion of organic molecules favoring the multiplication and metabolic activities of rhizosphere-inhabiting microorganisms, plant systems are also provided with the ability to secrete myriads of biocatalysts into terrestrial and aqueous environment with the enormous capability to degrade organic pollutants (Salt et al., 1998; Wenzel et al., 1999). The important plant-secreted enzymes present in soil environment are laccases, dehalogenases, nitroreductases, nitrilases, and peroxidases (Carreira and Wolfe, 1996; Schnoor et al., 1995). The application of plant-secreted nitroreductases and laccases under field conditions has shown considerable degradation of different nitrogen containing hazardous chemicals (Wolfe et al., 1993; Wang et al. 2004). Similarly, Boyajian and Carreira (1997) have described the mineralization of different nitroaromatic contaminants through the deployment of enzyme nitroreductase. In another experimental investigation, the involvement of nitrilase for catabolism of 4-chloro-benzonitrile and participation of halogenases in mineralization of hexachloroethane and TCE is well demonstrated (Wenzel et al., 1999).

Biotransformation of organic contaminants through enzymes has been carried out extensively under in vitro conditions. Many of the enzymes performing the biodegradations can be easily identified through different assays (Whiteley and Lee, 2006). However, the real field tests are not gone through much detail. Degradation of ester, amide, and peptide bonds through the enzymatic actions of esterases, amidases, and proteases culminating into the generation of byproducts with minimal or no toxicity may be observed. For example, hydrolases of bacterial species belonging to *Achromobacter, Pseudomonas, Flavobacterium, Nocardia, and Bacillus cereus* have shown promising results and thereby future opportunities for the effective biological conversion of contaminants such as carbofuran and carbaryl or parathion, diazinon, and coumaphos (Coppella et al., 1990; Mulbry and Eaton, 1991; Sutherland et al., 2002).

Although the extent to which the enzymes are secreted into the soil environment is less explored, the determination of half-life of released enzymes has provided the fact that they may actively participate in mineralization of contaminants up to several days once released from plants into the soil system (Schnoor et al., 1995). Therefore, multiple future opportunities exist for the exploitation of plant-secreted enzymes for efficient degradation of hazardous man-made compounds (Salt et al., 1998; Schwitzguébel, 2017).

6. Role of plant-associated microflora

Plants live in association with millions of other microorganisms, such as fungi and bacteria. These associated microorganisms are reported to support plants to cope with abiotic and biotic stresses, to assist their host in nutrient and water uptake, and to produce plant hormones, siderophores, and inhibitory allelochemicals. It is well-documented that plant—microbe interactions play important role in phytoremediation and plant growth (Doty et al. 2017; Deng and Cao, 2017; Feng et al. 2017; Jambon et al. 2018).

Complex interactions of plants and its associated microbes are under investigation (Feng et al. 2017). Endophytic bacteria are well-known for colonizing the internal tissues of plants without infection on their host (Schulz and Boyle, 2007; Schwitzguébel, 2017). Most of the studies confirmed that the main entry point of endophytic bacteria in plants is root (Pan et al. 1997; Germaine et al. 2004). Interaction of endophytic bacteria with their host is closer as compared with rhizosphere and phyllosphere bacteria. Under this close plant—endophyte interactions, host plant and bacteria, both are benefitted as plants provide nutrients for bacteria, and plant also get benefitted directly or indirectly to improve their growth and health (Mastretta et al. 2009; Syranidou et al. 2018).

Apart from their advantageous impacts on plant growth and development, endophytes possess significant biotechnological opportunities for the phytoremediation of contaminated sites (Feng et al. 2017; Thijs et al. 2018). Significantly large numbers of endophytic microbes inhabiting within the plants are reported to date. Following the entry of organic contaminants into plants, endophytic microbes association may be instrumental in biodegradation of targeted compounds. Hence, the intimate interactions of microbes with plants may improve the contaminant mineralization ability along with giving a remarkable opportunity for rapid exchange of different relevant genes.

The usefulness of plant—microbe associations in the removal of environmental pollutants of organic nature has been evidenced for rhizosphere (Ho et al. 2007; Kidd et al. 2008), phyllosphere (Sandhu et al. 2007), and within plants (Barac et al. 2004, 2009; Taghavi et al. 2005).

The combined use of plants and bacteria has a great potential for the remediation of soil and water contaminated with POPs. Plants provide residence and nutrients to their associated rhizosphere and endophytic bacteria, while the bacteria support plant growth by the degradation and detoxification of POPs.

Various phytoremediation strategies for organic contaminants management have been reported by McCutcheon and Schnoor (2003). Porteous Moore et al., 2006 have evaluated the diversity of endophytic bacterial communities inhabiting within the poplar plants surviving at BTEX polluted sites and concluded that many of the endophytic bacterial species were efficient in mineralization of targeted compounds. In this connection, Barac et al. (2009) working on same contaminated site have reported that the disappearance of targeted compound or their presence at nondetectable level because of plant—microbe interaction resulted into the loss of contaminant degradation capability of associated bacterial communities. Analysis of diverse endophytic bacterial communities inhabiting within the English oak and common ash surviving at TCE polluted sites has revealed the tolerance to contaminant in most of the isolates and biodegradation ability in few of the isolates (Weyens et al. 2009).

Because of smaller genome size, the bacterial species may be engineered comparatively with an ease in contrast to plants. Furthermore, there are lots of opportunities of genetic exchange between similar types of endophytic bacterial species. To date, few experimental investigations regarding the exploitation of wild or engineered bacterial species intimately associated with plants roots have been reported. Very often, organic contaminants with high degree of lipophilic nature may enter the xylem stream before it is partially or completely degraded by rhizospheric microorganisms (Trapp et al. 2000). As the organic contaminants may stay in xylem stream up to 2 days (McCrady et al. 1987), the application of genetically engineered endophytic bacteria surviving within xylem cells could have promising potential in degradation of organic contaminants and may alleviate the toxicity of parent compound or degradation by-products as well as mitigate the loss of hazardous organic contaminants into natural environment through evapotranspiration. In case the endophytic bacterial species inefficient in contaminant degradation are expected, the development of engineered endophytic bacterial strains through recombinant DNA technology, possessing the desired metabolic pathway for degradation of a particular pollutant, could be inoculated again with host plant to improve the phytoremediation. The exploitation of genetically engineered microbes (GEMs) for enhancing phytoremediation potential lies in to the improvement in overall metabolic potential of selected plant. Evidences in support were given by transfer of toluene-metabolizing bacterial endophytes to yellow lupine plants (Barac et al. 2004) and poplar (Taghavi et al. 2005), indicating an overall reduction in toluene toxicity to host plants and reduced contaminant loss through evapotranspiration. As soil-inhabiting bacteria are described to possess number of catabolic pathways, generally directed through the genes present on extracellular genetic material (plasmids) or transposons, the engineered bacteria with desired traits could be synthesized in vitro to improve their contaminant degradation potential. The transfer and expression of such catabolic genes within closely related endophytic bacterial strains would not be a major process while considering for the preparation of bacterial strains efficient in contaminant degradation. Interestingly, the heterologous expression of gene of interest, apart from improving the phytoremediation potential, could also be instrumental in minimizing the level of contaminants accumulation enriched in crop plants (Table 4.1).

TABLE 4.1 List of some plants reported for phytoremediation of organic contaminants.

S.N.	Country	Plants used for phytoremediation	Pollutant	Techniques for measurement of contaminant degradation	Remarks	Reference
1	Iran	Festuca arundinacea Schreb. and Festuca pratensis Huds.	Water-soluble phenols, total petroleum hydrocarbon (TPH), and polycyclic aromatic hydrocarbon (PAH) contents	Gas chromatography (GC) equipped with flame ionization detector and GC-MS (mass spectrometer)	Grasses infected with endophytic fungi may be exploited for remediation of TPH-contaminated soils.	Soleimani et al. (2010)
2	France	Alfalfa (Medicago sativa L.)	Petroleum hydrocarbons	Soil acetone-hexane extraction was followed by purification and analysis by GC-FID	Alfalfa-Pseudomonas aeruginosa association improved petroleum hydrocarbon remediation	Agnello et al. (2016)
3	China	Kandelia obovata, Bruguiera gymnorrhiza, and Avicennia marina	PAHs	GC and GC-MS	Among three mangrove plants, Bruguiera gymnorrhiza was the most effective in removing PAHs	Wang et al. (2014)
4	Spain	Maize (Zea mays L.)	Atrazine	HPLC analysis	Enhanced atrazine removal by electric current assisted phytoremediation with maize	Sánchez et al. (2018)
5	Malaysia	Scirpus grossus	Petroleum hydrocarbons	Liquid–liquid extraction with dichloromethane followed by GC	Combined action of plants and bacterial community along with oxygenation improved the remediation of contaminated site	Al-Baldawi et al. (2013)
6	China	Fire phoenix (Festuca arundinacea, Festuca elata, Festuca gigantea)	PAHs	GC-MS	Phoenix promoted the growth of Gordonia sp. as the major bacteria and thus improved PAH degradation	Liu et al. (2014)
7	Romania	Achillea millefolium	Petroleum hydrocarbons	A series of extraction, evaporation, and weight measurement and calculation based on formula	Achillea millefolium could be selected for removal of petro hydrocarbons	Masu et al. (2014)
8	China	Impatiens balsamina L.	TPHs	Column chromatography followed by the gravimetry	Impatiens balsamina L. could be an effective plant for efficient remediation of petroleum hydrocarbon contaminated soils	Cai et al. (2010)

No.	Country	Plant species	Pollutant	Method	Finding	Reference
9	Canada	*Salix miyabeana* SX67 and *S. miyabeana* SX64	Mixture of organics pollutants	MA 400-HAP 1.1 and BPC 1.0 methods	*Salix miyabeana* showed a significantly potential to lower the content of organic contaminants	Guidi et al. (2012)
10	China	*Echinacea purpurea, Festuca arundinacea* Schred, *Medicago sativa* L.	PAHs	GC-MS	Organic contaminants in the soil was closely related to specific characteristics of particular plant species	Liu et al. (2015)
11	Canada	Wild rye (*Elymus angustus*) and alfalfa (*Medicago sativa*)	Hydrocarbon	GC-MS	Plant-specific exudates may considerably affect phytoremediation	Phillips et al. (2012)
12	South Korea	*Pinus densiflora, Populus tomentiglandulosa,* and *Thuja orientalis*	TPHs	Not mentioned	The enzymatic hydrolysis of the woody biomass with the *Armillaria gemina* enzymes enhanced the phytoremediation	Jagtap et al. (2014)
13	Italy	*Paspalum vaginatum, Tamarix gallica, Spartium junceum*	TPHs	Gravimetric analysis	Stimulation of hydrolase enzymes and microbial metabolism activated the phytoremediation of organic contaminants	Masciandaro et al. (2014)
14	Australia	*Poa foliosa* (Hook. f.)	Hydrocarbons	GC-FID	*Poa foliosa* may be exploited for reducing high levels of hydrocarbon	Bramley-Alves et al. (2014)
15	Thailand	*Vetiveria zizanioides* (L.) Nash	Phenol	Colorimetric determination based on 4-amino-antipyrine	Combination of phytochemical and rhizomicrobiological processes eliminated phenol in wastewater in less than 32 days	Phenrat et al. (2017)
16	Mexico	*Melilotus albus*	TPHs	GC-MS	Diesel negatively affected both growth and nutritional status of *Melilotus albus* and AMF combination significantly enhanced the degradation of TPH in the rhizosphere	Hernández-Ortega et al. (2012)
17	Canada	*Lolium perenne, Festuca arundinacea, Hordeum vulgare* treated with PGPR *Hordeum vulgare*	Petroleum hydrocarbons	Gravimetric analysis	Increased plant biomass under the influence of plant growth–promoting bacteria improved the phytoremediation efficiency	Gurska et al. (2009)
18	Brazil	*Helianthus annuus*	Persistent organic pollutants (POPs)	GC-MS	Sunflower-assisted phytoremediation enhanced the remediation of heptachlor, aldrin, heptachlor epoxide, trans-chlordane, chlordane, dieldrin, DDE, DDT, methoxychlor, mirex, and decachlorobiphenyl	de Almeida et al. (2018)

(Continued)

TABLE 4.1 List of some plants reported for phytoremediation of organic contaminants.—cont'd

S.N.	Country	Plants used for phytoremediation	Pollutant	Techniques for measurement of contaminant degradation	Remarks	Reference
19	Italy	*Salix matsudana, Populus deltoides, Populus nigra, Eucalyptus camaldulensis, Helianthus annuus*	n-Alkanes, PAHs, and PCBs	GC-MS	The selected plant species improved phytoremediation of organic contaminants. The highest removal for n-alkanes was noticed for *Salix matsudana* and *Helianthus annuus*	Nissim et al. (2018)
20	Pakistan	*Leptochloa fusca* and *Brachiaria mutica*	Petroleum hydrocarbons	Gravimetric analysis	Use of endophytes with plant could serve as an effective strategy for the cleanup of oil-contaminated soil under field conditions	Fatima et al. (2018)
21	Czech Republic	*Brassica napus, Sorghum bicolor, drummondii*	Petroleum hydrocarbons	GC × GC-TOFMS	*Brassica napus* plants showed high resistance and suitability for PAH accumulation	Petrová et al. (2017)
22	Italy	*Zea mays* L.	Trichloroethylene (TCE)	GC-MS and LC-MS	*Zea mays* may serve as good candidate for the efficient remediation of TCE-contaminated sites	Moccia et al. (2017)
23	China	*Acorus calamus*	PAHs	HPLC	*Acorus calamus* may be exploited for efficient phytoremediation soil contaminated with PAHs	Jeelani et al. (2017)
24	Brazil	*Medicago sativa* cultivar Crioula	Phenanthrene, pyrene, and anthracene	HPLC	The PAH degradation was dependent on initial concentration of selected hydrocarbon; approximately 85% of selected contaminant was degraded in 20 days	Alves et al. (2018)
25	China	Ryegrass (*Lolium perenne*), white clover (*Trifolium repens*), and celery (*Apium graveolens*)	PAH	GC–MS	Multispecies mixtures facilitated the enhanced phytoremediation of PAH contaminated soils	Meng et al. (2011)
26	Iran	*Leucanthemum vulgare*	TPHs	Gravimetric analysis	Significant reduction in level of crude oil contaminants through phytoremediation strategy	Noori et al. (2018)

The utilization of bacterial endophytes to minimize the content of hazardous herbicides in crops is recognized (Germaine et al. 2006). Transfer of endophytes equipped with the property of 2,4-D degradation into *Pisum sativum* has demonstrated the significant removal of chemical in question and most notably the inoculated plant neither stored 2,4-D nor displayed phytotoxicity (Germaine et al. 2006). As there are enormous promising potential of exploitation of genetically engineered bacteria associated with plants to enhance the phytoremediation efficiency, the field-scale possibility could be achieved only after addressing some of the important challenges associated with the process (Newman and Reynolds, 2005). One of the challenging concern is the long-term existence and stability of inoculated GEMs associated with plants, apart from their biodegradation potential under given natural environmental conditions. The presence of bacterial community performing the degradation of a particular organic contaminant (Barac et al. 2009) would be favored only in the presence of targeted compound. Elimination of selection pressure could lead to the survivability of effective degraders in question. Yet, it could not be ascertained that inoculated endophytic strain would be accepted as an integral part of naturally existing bacterial communities in host plant. Furthermore, rather than inoculating the effective endophytic bacterial strains, the naturally surviving microbial species could also be selected for horizontal gene transfer to make the phytoremediation more effective. The employment of horizontal gene transfer process to confer adaptation to microbial communities existing either in rhizosphere (Ronchel et al. 2000; Devers et al. 2005) or endosphere (Taghavi et al. 2005) under different environmental constraints is documented. This process could lead to the significant benefit as there would be no additional requirement for successful integration/association of endophytic inoculants with the host plant.

7. Fate and transport of organic contaminants in phytoremediation

Once entered into plants, contaminants could have several fates. They may be accumulated in different regions of plants (Schroll et al., 1994; Fellows et al., 1996), may undergo volatilization, may be mineralized partially or absolutely (Goel et al., 1997; Newman et al., 1997; Wolfe et al., 1993), may be converted into comparatively lesser toxic forms, or may be stored in plant tissues to unavailable forms (Field and Thurman, 1996). In one of study, authors have indicated the absence of translocation of hexachlorobenzene and octachlorodibenzo-p-dioxin from one location to another after their entry through roots and leaves (Schroll et al., 1994). Contrary to this, the reciprocal translocation of chlorobenzene and trichloroacetic after absorption through root and leaves was described. Some compounds, however, may be preferentially accumulated to different regions of plant body. The significant accumulation of TNT and aniline in roots and much favored storage of hexahydro-1,3,5-trinitro-1,3,5-triazine (RDX), phenol, and quinoline into the shoot after contaminant transport through root is well recognized (Fellows et al., 1996; Wenzel et al., 1999).

Nearly all organic contaminants are modified at some extent before their accumulation inside the vacuole or association with cellular molecules such as lignin (Salt et al., 1998). Nevertheless, some of the chemical entities may be absolutely degraded to water and carbon dioxide, but the complete mineralization at specified locations is represented by very small fraction of original molecule (Newman et al., 1997). In this context, the plant systems are less preferred candidate for organic contaminant remediation in contrast to bacterial cells dealing efficiently with the organic contaminants. Furthermore, the possible generation of comparatively more toxic by-products after contaminant degradation poses regulatory issues to be overcome before large-scale field application could be made feasible.

8. Genetically engineered organisms for phytoremediation

The improved phytoremediation potential could be achieved through genetic modification of host plants (Doty, 2008; Abhilash et al., 2009; Legault et al. 2017; Hussain et al. 2018) and microbial communities (Basu et al. 2018; Kuffner et al. 2008; Hussain et al. 2018) intimately associated with plants. The genetic engineering of microbial communities harbored by host plants to enhance the phytoremediation potential involves (1) isolation of bacteria and incorporation of genes directing the synthesis of particular enzyme responsible for contaminant degradation, followed by transfer of modified organism (Valls and De Lorenz, 2002) and (2) enhancement in number of responsible microbial species within host system (Abou-Shanab et al. 2007; Li et al. 2007).

In spite of several regulatory issues and public concerns regarding the exploitation of GEMs for field-scale phytoremediation, many other important challenges might restrict the full fledged and efficient contaminant degradation ability of GEMs under natural environmental conditions. Generally, the GEMs inoculated under field conditions do not grow as fast as under in vitro conditions, therefore outcompeted by native rhizosphere microbial communities, and eventually their number falls to a level that could not support the process of efficient contaminant degradation (Macek et al. 2000; Gilbertson et al. 2007). In general, the mineralization of organic pollutants is facilitated by a set of metabolic activities governed by different enzymes, and it could not be normally feasible to introduce all the desired genes into a single organism to achieve the absolute degradation of targeted contaminant (Macek et al. 2000). Preserving the stability of transferred gene in host is not an easier task and there are evidences that recombinant organism displaying a particular trait, say for example, the degradation efficiency for a particular contaminant is very often disappeared (Jussila et al., 2007).

9. Research and development in phytoremediation

9.1 Current status

Current developments in the field of biochemistry, plant physiology, and molecular biology together with the technological innovations to understand vital plant metabolic processes, genetic control, and elucidation of protein functioning have provided a new arena for the better management practices to deal with the environmental contaminants through phytoremediation approaches. With the aid of better instrumentation techniques, the more mechanistic details of phytoremediation process could be made possible.

9.2 Biotechnological approaches

The potential application of plant enzymes to enhance the phytoremediation potential is greatly acknowledged by current world biotechnologists. Few of the enzymes participating in degradation of contaminants with diverse chemical nature have been shown to be released into soil ecosystem (Singer et al., 2003). Experimental evidences in support of soil contaminant degradation through the cotton laccase enzyme secreted externally from transgenic *Arabidopsis* (Wang et al., 2004) have been reported. In recent years, more integrated phytoremediation has been tried to be achieved through advances in next-generation sequencing technologies called meta-omics (metagenomics, metatranscriptomics, metaproteomics, metabolomics) (Bell et al. 2014). These technologies have substantially enhanced our understanding of the function of plants, microorganisms, and the interactions between them occurring naturally (Thijs et al., 2017). The lowering costs of high-throughput sequencing technologies have enabled a further extensive integration (Bell et al. 2014), offering a wide range of opportunities for optimization and a better understanding of phytoremediation (Thijs et al. 2017).

9.3 Protein engineering

The desirable enzymes or channel proteins from different sources participating in contaminant degradation or mobilization could be altered through the advanced molecular biology techniques before its enhanced synthesis in genetically modified plants. Some of the important enzymes that could be considered for such modifications include specified oxygenases and reductases. The protein engineering could also be performed for aromatic-ring hydroxylating dioxygenases and ring-fission dioxygenases participating in primary steps of organic contaminants degradation (Furukawa, 2006). Constructing protein molecules with novel features or improved cellular functions needs thorough understanding of structure and activity relations. Keeping this fact in mind, the native 3D confirmations of numerous enzymes have been analyzed, and approaches based on mutations have been performed to mark the appropriate changes in amino acid residues to enhance the enzymatic actions and/or association with substrate molecules (Leungsakul et al., 2006; Zielinski et al., 2006; Camara et al., 2007). Furthermore, with the aid of chemical and physical sciences, studies were conducted to get comprehensive insights into the molecular arrangement of environmental contaminants, determining their reaction and charge characteristics (Pacios and Gómez, 2006). The enzyme biphenyl dioxygenase (*BphA*) governing the initial steps in single pathway reported till date for oxygenic degradation of PCBs could be targeted for engineering (Furusawa et al., 2004). Production of engineered protein performing the oxygen addition to 4-chlorobiphenyl through site-specific mutation in genetically modified tobacco has been demonstrated (Mohammadi et al., 2007). A more detailed account on the relevant genes with efficient contaminant degradation activity and approaches exploited for their engineering has been nicely presented by Whiteley and Lee (2006).

10. Advantages and limitations of phytoremediation

Phytoremediation provides many advantages over other remedial strategies (Susarla et al. 2002; Chaudhry et al. 2005; Pilon-Smits and Freeman, 2006). Conventional *ex situ* methods of phytoremediation includes excavation of soil, off-site storage, soil washing, and *in situ* covering for stabilization, which are more expensive than *in situ* phytoremediation

(Pilon-Smits and Freeman, 2006). Phytoremediation protocols for *in situ* are easy to implement and maintenance costs are low. Phytoremediation has a great potential as a natural, solar energy—driven *in situ* strategy to treat soils and sites moderately polluted over large surfaces, provided the plants have been carefully chosen and the adequate agronomic methods are applied to manage correctly the phytoavailability of organic contaminants (Schwitzguébel, 2017). Furthermore, organic materials, nutrients, and oxygen are added to soil via plant and microbial metabolic processes in *in situ* phytoremediation, by this quality and texture of remediated sites will improve (Schwitzguébel, 2017; Wiszniewska et al., 2016). Wind and water erosion will also be checked by plants used for phytoremediation as they provide groundcover, and their roots help to stabilize soil. Phytoremediation can be applied at any site in any geographical area, where plants can grow. High public acceptance and an attractive option for industry and regulators is an additional advantage of phytoremediation. Furthermore, appropriate plant species, ecotypes, varieties, or cultivars can be selected, tailored, and exploited for the remediation of polluted soils, sites, and brownfields, as well as for the phytotreatment of domestic or industrial wastewater (Schwitzguébel, 2017).

However, there are numerous technical inherent limitations of phytoremediation (Schwitzguébel, 2017) such as (1) pollutants must be within the root zones of plants that are used for phytoremediation. This may create stress conditions to plants for water, depth, nutrient, and other resources, which are essential for plant growth. (2) Adequate size of site to ensure application of farming techniques.

11. Emerging challenges to phytoremediation

The most important restriction of phytoremediation strategies such as excavation and *ex situ* treatment is slower rate of remediation. Plant survival and growth is restricted by contaminated soil, thereby limiting catabolism of contaminants (Harvey et al., 2002; Huang et al., 2004, 2005). Various abiotic (temperature, light, nutrient, precipitation, atmospheric CO_2) and biotic (plant pathogens, competition with grass and herbs, herbivory by insects or animals) factors may also restrict survival and growth of plant (used for phytoremediation) in field (Nedunuri et al. 2000). These biotic or abiotic factors may restrict plant growth and ultimately negatively affect phytoremediation process. For example, attempt of phytoremediation for soil and ground water at a hydrocarbon burn facility, NASA Kennedy Space Center in Florida was failed because of water stress condition and competition with herbs (phytoremediation at the hydrocarbon burn facility at NASA Kennedy Space Center in Florida). However, various studies suggested that weeds/herbs may also remediate organic pollutants, in this project remediation by weeds was not investigated, and intensive weed completion was reported for project failure. Furthermore, physical challenges may also limit the efficiency of phytoremediation. For instance, raising the moisture availability at contaminated sites polluted with organic contaminants may aggravate the problem after being moistureless eventually leading to inhibition of seedling growth and development. Another important constraint of phytoremediation technology is related with the depth to which soil is polluted with a particular contaminant of interest as in most of the cases; the average depth gained by plant roots is usually 50 cm (Pilon-Smits, 2005). This is important to excavate such type of challenge before phytoremediation. Deeper roots of trees facilitate remediation at greater depth (Pilon-Smits, 2005). Dendroremediation of explosive showed great promise for explosives in case of soil and ground remediation (Meagher, 2000; Susarla et al. 2002; Doucette et al. 2003). Furthermore, survival, establishment, and growth are major problems in contaminated soil. Furthermore, trees need several years to attain sufficient biomass for efficient phytoremediation. Furthermore, after completion of remediation, there may be problem of disposal, if roots and woods be contaminated.

In contrast to greenhouse experiments and laboratory conditions, where soils are well mixed to uniformity in their characteristics, there is heterogeneity in distribution of contaminants in fields sites, which is major challenge in phytoremediation (Ferro et al. 1999; Van Dillewijn et al., 2007). In field conditions, it is impossible to achieve uniformity in soil conditions, even if the site is properly tilled before planting. Such type of heterogeneity in soil conditions makes data of phytoremediation scatter. Furthermore, factors such as root architecture, soil morphology, nutrient composition pH, moisture availability, and microbiological actions commonly display considerable spatial and temporal differences (Nedunuri et al. 2000; Corwin et al. 2006).

Application of genetically engineered organisms for contaminant removal has gained remarkable success in vitro and in green house conditions; however, the regulatory issues concerned have restricted their evaluation under natural field environment (Pilon-Smits and Freeman, 2006). In the absence of any suitable containment systems, there is much possibility of transfer of gene of interest from engineered organisms to native organisms. Furthermore, the modified strains harboring the antibiotic genes could not be considered for their release in natural environments (Villacieros et al. 2005). If the problem of gene transfer from engineered to native one could be solved, the low public acceptance may emerge as the major problem of release of engineered organisms as reported for Europe (INRA, 2000). Although the rejection of genetically engineered organisms by a public is relied on irrelevant fear, instead of scientific understanding, regulatory laws in most of such cases during the course of policy development seek for public response.

12. Conclusion

Phytoremediation is an emerging technology to deal with the ever-rising hazardous contaminants in the natural environment. It has been regarded as suitable green technology and good alternative to costly physicochemical approaches being utilized for contaminant degradation. Plants with efficient remediation strategies could be selected for the management of contaminated sites. The plants very often have been reported for having diverse endophytic microbial population that may be very much promising in enhancing the pollutant removal efficiency of plant species. The laboratory studies have been reported to be failed under natural field condition probably because of significant changes in property of soil and microbial community from one place to another place. Some of the challenges of phytoremediation, however, could be overcome through development of genetically engineered plant as well as endophytic microbes equipped with the property of overexpressing the enzymes capable of contaminant degradation.

Acknowledgments

The authors are thankful to UGC and CSIR for providing fellowship. Rahul Bhadouria is thankful to UGC (BSR/BL/ 17–18/0067) for providing Dr. D. S. Kothari fellowship for postdoctoral research.

References

Abhilash, P.C., Jamil, S., Singh, N., 2009. Transgenic plants for enhanced biodegradation and phytoremediation of organic xenobiotics. Biotechnology Advances 27 (4), 474–488.

Abou-Shanab, R.A.I., Angle, J.S., Van Berkum, P., 2007. Chromate-tolerant bacteria for enhanced metal uptake by *Eichhornia crassipes* (Mart.). International Journal of Phytoremediation 9 (2), 91–105.

Agnello, A.C., Huguenot, D., Hullebusch Van, E.D., Esposito, G., 2016. Citric acid-and Tween® 80-assisted phytor-emediation of a co-contaminated soil: alfalfa (Medicago sativa L.) performance and remediation potential. Environmental Science and Pollution Research 23 (9), 9215–9226.

Al-Baldawi, I.A., Abdullah, S.R.S., Suja, F., Anuar, N., Mushrifah, I., 2013. Effect of aeration on hydrocarbon phytore-mediation capability in pilot sub-surface flow constructed wetland operation. Ecological Engineering 61, 496–500.

Ali, H., Khan, E., Sajad, M.A., 2013. Phytoremediation of heavy metals—concepts and applications. Chemosphere 91 (7), 869–881.

Almeida, M.V.D., Rissato, S.R., Galhiane, M.S., Fernandes, J.R., Lodi, P.C., Campos, M.C.D., 2018. In vitro phytore-mediation of persistent organic pollutants by *Helianthus annuus* L. plants. Química Nova 41 (3), 251–257.

Alves, W.S., Manoel, E.A., Santos, N.S., Nunes, R.O., Domiciano, G.C., Soares, M.R., 2018. Phytoremediation of poly-cyclic aromatic hydrocarbons (PAH) by cv. Crioula: a Brazilian alfalfa cultivar. International Journal of Phytore-mediation 20 (8), 747–755.

Anderson, T.A., Walton, B.T., 1995. Comparative fate of [14C] trichloroethylene in the root zone of plants from a former solvent disposal site. Environmental Toxicology and Chemistry: An International Journal 14 (12), 2041–2047.

Anderson, T.A., Guthrie, E.A., Walton, B.T., 1993. Bioremediation in the rhizosphere. Environmental Science and Technology 27 (13), 2630–2636.

Barac, T., Taghavi, S., Borremans, B., Provoost, A., Oeyen, L., Colpaert, J.V., Vangronsveld, J., Van Der Lelie, D., 2004. Engineered endophytic bacteria improve phytoremediation of water-soluble, volatile, organic pollutants. Nature Biotechnology 22 (5), 583.

Barac, T., Weyens, N., Oeyen, L., Taghavi, S., van der Lelie, D., Dubin, D., Split, M., Vangronsveld, J., 2009. Application of poplar and its associated microorganisms for the *in situ* remediation of a BTEX contaminated ground-water plume. International Journal of Phytoremediation 11, 416–424.

Basu, S., Rabara, R.C., Negi, S., Shukla, P., 2018. Engineering PGPMOs through gene editing and systems biology: a solution for phytoremediation? Trends in Biotechnology 36, 499–510.

Bell, T.H., Joly, S., Pitre, F.E., Yergeau, E., 2014. Increasing phytoremediation efficiency and reliability using novel omics approaches. Trends in Biotechnology 32 (5), 271–280.

Bollag, J.M., Mertz, T., Otijen, L., 1994. Role of microorganisms in soil remediation. In: Anderson, T.A., Coats, J.R. (Eds.), Bioremediation through Rhizosphere Technology. ACS. Symp. Ser.563. Am. Chem. Soc. Maple press, York, PA, pp. 2–10.

Boyajian, G.E., Carreira, L.H., 1997. Phytoremediation: a clean transition from laboratory to marketplace? Nature Biotechnology 15 (2), 127.

Bramley-Alves, J., Wasley, J., King, C.K., Powell, S., Robinson, S.A., 2014. Phytoremediation of hydrocarbon contam-inants in subantarctic soils: an effective management option. Journal of Environmental Management 142, 60–69.

Brazier-Hicks, M., Edwards, R., 2005. Functional importance of the family 1 glucosyltransferase UGT72B1 in the metabolism of xenobiotics in *Arabidopsis thaliana*. The Plant Journal 42 (4), 556–566.

Brusseau, M.L., Wang, X., Wang, W.Z., 1997. Simultaneous elution of heavy metals and organic compounds from soil by cyclodextrin. Environmental Science and Technology 31 (4), 1087–1092.

Burken, J.G., Schnoor, J.L., 1996. Phytoremediation: plant uptake of atrazine and role of root exudates. Journal of Environmental Engineering 122 (11), 958–963.

Burken, J.G., Schnoor, J.L., 1997. Uptake and metabolism of atrazine by poplar trees. Environmental Science and Technology 31 (5), 1399–1406.

Burken, J.G., Ma, X., Struckhoff, G.C., Gilbertson, A., 2005. Volatile organic compound fate in phytoremediation applications: natural and engineered systems. Zeitschrift fur Naturforschung. C, Journal of Biosciences 60 (3–4), 208–215.

Cai, Z., Zhou, Q., Peng, S., Li, K., 2010. Promoted biodegradation and microbiological effects of petroleum hydrocarbons by Impatiens *balsamina* L. with strong endurance. Journal of Hazardous Materials 183 (1–3), 731–737.

Cámara, B., Seeger, M., González, M., Standfuß-Gabisch, C., Kahl, S., Hofer, B., 2007. Generation by a widely applicable approach of a hybrid dioxygenase showing improved oxidation of polychlorobiphenyls. Applied and Environmental Microbiology 73 (8), 2682–2689.

Carreira, L.H., Wolfe, N.L., 1996, May. Isolation of a sediment nitroreductase, antibody production, and identification of possible plant sources. In: IBC Symp. On Phytoremediation, Arlington, VA.

Chaudhry, M.S., Batool, Z., Khan, A.G., 2005. Preliminary assessment of plant community structure and arbuscular mycorrhizas in rangeland habitats of Cholistan desert, Pakistan. Mycorrhiza 15 (8), 606–611.

Coppella, S.J., DelaCruz, N., Payne, G.F., Pogell, B.M., Speedie, M.K., Karns, J.S., Sybert, E.M., Connor, M.A., 1990. Genetic engineering approach to toxic waste management: case study for organophosphate waste treatment. Biotechnology Progress 6 (1), 76–81.

Corwin, D.L., Hopmans, J., De Rooij, G.H., 2006. From field-to landscape-scale vadose zone processes: scale issues, modeling, and monitoring. Vadose Zone Journal 5 (1), 129–139.

Cunningham, S.D., Anderson, T.A., Schwab, A.P., Hsu, F.C., 1996. Phytoremediation of soils contaminated with organic pollutants. Advances in Agronomy 56 (1), 55–114.

Deng, Z., Cao, L., 2017. Fungal endophytes and their interactions with plants in phytoremediation: a review. Chemosphere 168, 1100–1106.

Desai, J.D., Banat, I.M., 1997. Microbial production of surfactants and their commercial potential. Microbiology and Molecular Biology Reviews 61 (1), 47–64.

Devers, M., Henry, S., Hartmann, A., Martin-Laurent, F., 2005. Horizontal gene transfer of atrazine-degrading genes (atz) from *Agrobacterium tumefaciens* St96-4 pADP1:: Tn5 to bacteria of maize-cultivated soil. Pest Management Science: Formerly Pesticide Science 61 (9), 870–880.

Didierjean, L., Gondet, L., Perkins, R., Lau, S.M.C., Schaller, H., O'Keefe, D.P., Werck-Reichhart, D., 2002. Engineering herbicide metabolism in tobacco and *Arabidopsis* with CYP76B1, a cytochrome P450 enzyme from *Jerusalem artichoke*. Plant Physiology 130 (1), 179–189.

Dietz, A.C., Schnoor, J.L., 2001. Advances in phytoremediation. Environmental Health Perspectives 109, 163–168.

Doty, S.L., 2008. Enhancing phytoremediation through the use of transgenics and endophytes. New Phytologist 179 (2), 318–333.

Doty, S.L., Freeman, J.L., Cohu, C.M., Burken, J.G., Firrincieli, A., Simon, A., Khan, Z., Isebrands, J.G., Lukas, J., Blaylock, M.J., 2017. Enhanced degradation of TCE on a superfund site using endophyte-assisted poplar tree phytoremediation. Environmental Science and Technology 51 (17), 10050–10058.

Doty, S.L., Shang, T.Q., Wilson, A.M., Tangen, J., Westergreen, A.D., Newman, L.A., Strand, S.E., Gordon, M.P., 2000. Enhanced metabolism of halogenated hydrocarbons in transgenic plants containing mammalian cytochrome P450 2E1. Proceedings of the National Academy of Sciences 97 (12), 6287–6291.

Doucette, W.J., Bugbee, B.G., Smith, S.C., Pajak, C.J., Ginn, J.S., 2003. Uptake, metabolism, and phytovolatilization of trichloroethylene by indigenous vegetation: impact of precipitation. Phytoremediation: Transformation and Control of Contaminants 561–588.

Farlane, C.M., Pfleeger, T., Fletcher, J., 1990. Effect, uptake and disposition of nitrobenzene in several terrestrial plants. Environmental Toxicology and Chemistry: An International Journal 9 (4), 513–520.

Fatima, K., Imran, A., Amin, I., Khan, Q.M., Afzal, M., 2018. Successful phytoremediation of crude-oil contaminated soil at an oil exploration and production company by plants-bacterial synergism. International Journal of Phytoremediation 20 (7), 675–681.

Fellows, R.J., Harvey, S.D., Ainsworth, C.C., Cataldo, D.A., 1996, May. Biotic and abiotic transformation of munitions materials (TNT, RDX) by plants and soils. Potentials for attenuation and remediation of contaminants. In: IBC Int. Conf. Phytoremed., Arlington, VA.

Feng, N.X., Yu, J., Zhao, H.M., Cheng, Y.T., Mo, C.H., Cai, Q.Y., Li, Y.W., Li, H., Wong, M.H., 2017. Efficient phytoremediation of organic contaminants in soils using plant–endophyte partnerships. The Science of the Total Environment 583, 352–368.

Ferro, A.M., Rock, S.A., Kennedy, J., Herrick, J.J., Turner, D.L., 1999. Phytoremediation of soils contaminated with wood preservatives: greenhouse and field evaluations. International Journal of Phytoremediation 1 (3), 289–306.

Field, J.A., Thurman, E.M., 1996. Glutathione conjugation and contaminant transformation. Environmental Science and Technology 30 (5), 1413–1418.

Fletcher, J.S., Hegde, R.S., 1995. Release of phenols by perennial plant roots and their potential importance in bioremediation. Chemosphere 31 (4), 3009–3016.

Foth, H.D., 1990. Fundamentals of Soil Science, eighth ed. Wiley, New York.

Fujisawa, T., 2002. Model of the uptake of pesticides by plant. Journal of Pesticide Science-Pesticide Science Society of Japan-Japanese Edition- 27 (3), 279–286.

Furukawa, K., 2006. Oxygenases and dehalogenases: molecular approaches to efficient degradation of chlorinated environmental pollutants. Bioscience Biotechnology and Biochemistry 70 (10), 2335–2348.

Furusawa, Y., Nagarajan, V., Tanokura, M., Masai, E., Fukuda, M., Senda, T., 2004. Crystal structure of the terminal oxygenase component of biphenyl dioxygenase derived from *Rhodococcus* sp. strain RHA1. Journal of Molecular Biology 342 (3), 1041–1052.

Germaine, K., Keogh, E., Garcia-Cabellos, G., Borremans, B., Van Der Lelie, D., Barac, T., Oeyen, L., Vangronsveld, J., Moore, F.P., Moore, E.R., Campbell, C.D., 2004. Colonisation of poplar trees by gfp expressing bacterial endophytes. FEMS Microbiology Ecology 48 (1), 109–118.

Germaine, K.J., Liu, X., Cabellos, G.G., Hogan, J.P., Ryan, D., Dowling, D.N., 2006. Bacterial endophyte-enhanced phytoremediation of the organochlorine herbicide 2, 4-dichlorophenoxyacetic acid. FEMS Microbiology Ecology 57 (2), 302–310.

Gianfreda, L., Rao, M.A., 2004. Potential of extra cellular enzymes in remediation of polluted soils: a review. Enzyme and Microbial Technology 35 (4), 339–354.

Gianfreda, L., Bollag, J.-M., 2002. Isolated enzymes for the transformation and detoxification of organic pollutants. In: Burns, R.G., Dick, R. (Eds.), Enzymes in the Environment: Activity, Ecology and Applications. Marcel Dekker, New York, pp. 491–538.

Gilbertson, A.W., Fitch, M.W., Burken, J.G., Wood, T.K., 2007. Transport and survival of GFP-tagged root-colonizing microbes: implications for rhizodegradation. European Journal of Soil Biology 43 (4), 224–232.

Goel, A., Kumar, G., Payne, G.F., Dube, S.K., 1997. Plant cell biodegradation of a xenobiotic nitrate ester, nitroglycerin. Nature Biotechnology 15 (2), 174.

Guidi, W., Kadri, H., Labrecque, M., 2012. Establishment techniques to using willow for phytoremediation on a former oil refinery in southern Quebec: achievements and constraints. Chemistry and Ecology 28 (1), 49–64.

Gurska, J., Wang, W., Gerhardt, K.E., Khalid, A.M., Isherwood, D.M., Huang, X.D., Glick, B.R., Greenberg, B.M., 2009. Three year field test of a plant growth promoting rhizobacteria enhanced phytoremediation system at a land farm for treatment of hydrocarbon waste. Environmental Science and Technology 43 (12), 4472–4479.

Harms, H., Bokern, M., Kolb, M., Bock, C., 2003. Transformation of organic contaminants by different plant systems. Phytoremediation. Transformation and Control of Contaminants 285–316.

Harvey, P.J., Campanella, B.F., Castro, P.M., Harms, H., Lichtfouse, E., Schäffner, A.R., Smrcek, S., Werck-Reichhart, D., 2002. Phytoremediation of polyaromatic hydrocarbons, anilines and phenols. Environmental Science and Pollution Research 9 (1), 29–47.

Hernández-Ortega, H.A., Alarcón, A., Ferrera-Cerrato, R., Zavaleta-Mancera, H.A., López-Delgado, H.A., Mendoza-López, M.R., 2012. Arbuscular mycorrhizal fungi on growth, nutrient status, and total antioxidant activity of *Melilotus albus* during phytoremediation of a diesel-contaminated substrate. Journal of Environmental Management 95, S319–S324.

Ho, C.H., Applegate, B., Banks, M.K., 2007. Impact of microbial/plant interactions on the transformation of polycyclic aromatic hydrocarbons in rhizosphere of *Festuca arundinacea*. International Journal of Phytoremediation 9 (2), 107–114.

Huang, X.D., El-Alawi, Y., Gurska, J., Glick, B.R., Greenberg, B.M., 2005. A multi-process phytoremediation system for decontamination of persistent total petroleum hydrocarbons (TPHs) from soils. Microchemical Journal 81 (1), 139–147.

Huang, X.D., El-Alawi, Y., Penrose, D.M., Glick, B.R., Greenberg, B.M., 2004. A multi-process phytoremediation system for removal of polycyclic aromatic hydrocarbons from contaminated soils. Environmental Pollution 130 (3), 465–476.

Hughes, J.B., Shanks, J., Vanderford, M., Lauritzen, J., Bhadra, R., 1996. Transformation of TNT by aquatic plants and plant tissue cultures. Environmental Science and Technology 31 (1), 266–271.

Hussain, I., Aleti, G., Naidu, R., Puschenreiter, M., Mahmood, Q., Rahman, M.M., Wang, F., Shaheen, S., Syed, J.H., Reichenauer, T.G., 2018. Microbe and plant assisted-remediation of organic xenobiotics and its enhancement by genetically modified organisms and recombinant technology: a review. The Science of the Total Environment 628, 1582–1599.

Inra (Europe)-ECOSA, March 15, 2000. Eurobarometer 52.1, the Europeans and Biotechnology Report. http://ec.europa.eu/public_opinion/archives/ebs/ebs_134_en.pdf.

Inui, H., Shiota, N., Motoi, Y., Ido, Y., Inoue, T., Kodama, T., Ohkawa, Y., Ohkawa, H., 2001. Metabolism of herbicides and other chemicals in human cytochrome P450 species and in transgenic potato [*Solanum tuberosum*] plants co-expressing human CYP1A1, CYP2B6 and CYP2C19. Journal of Pesticide Science 26, 28–40.

Jagtap, S.S., Woo, S.M., Kim, T.S., Dhiman, S.S., Kim, D., Lee, J.K., 2014. Phytoremediation of diesel-contaminated soil and saccharification of the resulting biomass. Fuel 116, 292–298.

Jambon, I., Thijs, S., Weyens, N., Vangronsveld, J., 2018. Harnessing plant-bacteria-fungi interactions to improve plant growth and degradation of organic pollutants. Journal of Plant Interactions 13 (1), 119–130.

Jansson, S., Douglas, C.J., 2007. Populus: a model system for plant biology. Annual Review of Plant Biology 58, 435–458.

Jeelani, N., Yang, W., Xu, L., Qiao, Y., An, S., Leng, X., 2017. Phytoremediation potential of *Acorus calamus* in soils co-contaminated with cadmium and polycyclic aromatic hydrocarbons. Scientific Reports 7 (1), 8028.

Jeffers, P.M., Liddy, C.D., 2003. Treatment of atmospheric halogenated hydrocarbons by plants and fungi. Phytoremediation: Transformation and Control of Contaminants 787–804.

Jussila, M.M., Zhao, J., Suominen, L., Lindström, K., 2007. TOL plasmid transfer during bacterial conjugation in vitro and rhizoremediation of oil compounds in vivo. Environmental Pollution 146 (2), 510–524.

Kawahigashi, H., Hirose, S., Ohkawa, H., Ohkawa, Y., 2007. Herbicide resistance of transgenic rice plants expressing human CYP1A1. Biotechnology Advances 25 (1), 75–84.

Kidd, P.S., Prieto-Fernández, A., Monterroso, C., Acea, M.J., 2008. Rhizosphere microbial community and hexachlorocyclohexane degradative potential in contrasting plant species. Plant and Soil 302 (1–2), 233–247.

Klein, M., Burla, B., Martinoia, E., 2006. The multidrug resistance-associated protein (MRP/ABCC) subfamily of ATP-binding cassette transporters in plants. FEBS Letters 580 (4), 1112–1122.

Kuffner, M., Puschenreiter, M., Wieshammer, G., Gorfer, M., Sessitsch, A., 2008. Rhizosphere bacteria affect growth and metal uptake of heavy metal accumulating willows. Plant and Soil 304 (1–2), 35–44.

Legault, E.K., James, C.A., Stewart, K., Muiznieks, I., Doty, S.L., Strand, S.E., 2017. A field trial of TCE phytoremediation by genetically modified poplars expressing cytochrome P450 2E1. Environmental Science and Technology 51 (11), 6090–6099.

Leungsakul, T., Johnson, G.R., Wood, T.K., 2006. Protein engineering of the 4-methyl-5-nitrocatechol monooxygenase from Burkholderia sp. strain DNT for enhanced degradation of nitroaromatics. Applied and Environmental Microbiology 72 (6), 3933–3939.

Li, W.C., Ye, Z.H., Wong, M.H., 2007. Effects of bacteria on enhanced metal uptake of the Cd/Zn-hyperaccumulating plant, *Sedum alfredii*. Journal of Experimental Botany 58 (15–16), 4173–4182.

Liu, R., Dai, Y., Sun, L., 2015. Effect of rhizosphere enzymes on phytoremediation in PAH-contaminated soil using five plant species. PLoS One 10 (3), e0120369.

Liu, R., Xiao, N., Wei, S., Zhao, L., An, J., 2014. Rhizosphere effects of PAH-contaminated soil phytoremediation using a special plant named Fire Phoenix. The Science of the Total Environment 473, 350–358.

Macek, T., Mackova, M., Káš, J., 2000. Exploitation of plants for the removal of organics in environmental remediation. Biotechnology Advances 18 (1), 23–34.

Masciandaro, G., Di Biase, A., Macci, C., Peruzzi, E., Iannelli, R., Doni, S., 2014. Phytoremediation of dredged marine sediment: monitoring of chemical and biochemical processes contributing to sediment reclamation. Journal of Environmental Management 134, 166–174.

Mastretta, C., Taghavi, S., Van Der Lelie, D., Mengoni, A., Galardi, F., Gonnelli, C., Barac, T., Boulet, J., Weyens, N., Vangronsveld, J., 2009. Endophytic bacteria from seeds of *Nicotiana tabacum* can reduce cadmium phytotoxicity. International Journal of Phytoremediation 11 (3), 251–267.

Masu, S., Albulescu, M., Balasescu, L.C., 2014. Assessment on Phytoremediation of Crude Oil Polluted Soils with *Achillea millefolium* and Total Petroleum Hydrocarbons Removal Efficiency. http://hdl.handle.net/123456789/585.

McCrady, J.K., McFarlane, C.R.A.I.G., Lindstrom, F.T., 1987. The transport and affinity of substituted benzenes in soybean stems. Journal of Experimental Botany 38 (11), 1875–1890.

McCutcheon, S.C., Schnoor, J.L., 2003. Overview of phytotransformation and control of wastes. In: McCutcheon, S.C., Schnoor, J.L. (Eds.), Phytoremediation: Transformation and Control of Contaminants. Willey, p. 358.

McCutcheon, S.C., Schnoor, J.L., 2004. Phytoremediation: Transformation and Control of Contaminants, vol. 121. John Wiley and Sons.

Meagher, R.B., 2000. Phytoremediation of toxic elemental and organic pollutants. Current Opinion in Plant Biology 3 (2), 153–162.

Meng, L., Qiao, M., Arp, H.P.H., 2011. Phytoremediation efficiency of a PAH-contaminated industrial soil using ryegrass, white clover, and celery as mono-and mixed cultures. Journal of Soils and Sediments 11 (3), 482–490.

Mesjasz-Przybyłowicz, J., Nakonieczny, M., Migula, P., Augustyniak, M., Tarnawska, M., Reimold, W.U., Koeberl, C., Przybyłowicz, W., Głowacka, E., 2004. Uptake of cadmium, lead nickel and zinc from soil and water solutions by the nickel hyperaccumulator *Berkheya coddii*. Acta Biologica Cracoviensia Series Botanica 46, 75–85.

Mezzari, M.P., Walters, K., Jelínkova, M., Shih, M.C., Just, C.L., Schnoor, J.L., 2005. Gene expression and microscopic analysis of *Arabidopsis* exposed to chloroacetanilide herbicides and explosive compounds. A phytoremediation approach. Plant Physiology 138 (2), 858–869.

Miller, R.M., 1995. Biosurfactant-facilitated remediation of metal-contaminated soils. Environmental Health Perspectives 103 (Suppl. 1), 59–62.

Moccia, E., Intiso, A., Cicatelli, A., Proto, A., Guarino, F., Iannece, P., Castiglione, S., Rossi, F., 2017. Use of *Zea mays* L. in phytoremediation of trichloroethylene. Environmental Science and Pollution Research 24 (12), 11053–11060.

Mohammadi, M., Chalavi, V., Novakova-Sura, M., Laliberté, J.F., Sylvestre, M., 2007. Expression of bacterial biphenyl-chlorobiphenyl dioxygenase genes in tobacco plants. Biotechnology and Bioengineering 97 (3), 496–505.

Morant, M., Bak, S., Møller, B.L., Werck-Reichhart, D., 2003. Plant cytochromes P450: tools for pharmacology, plant protection and phytoremediation. Current Opinion in Biotechnology 14 (2), 151–162.

Mulbry, W.W., Eaton, R.W., 1991. Purification and characterization of the N-methylcarbamate hydrolase from *Pseudomonas* strain CRL-OK. Applied and Environmental Microbiology 57 (12), 3679–3682.

Nedunuri, K.V., Govindaraju, R.S., Banks, M.K., Schwab, A.P., Chen, Z., 2000. Evaluation of phytoremediation for field-scale degradation of total petroleum hydrocarbons. Journal of Environmental Engineering 126 (6), 483–490.

Newman, L.A., Reynolds, C.M., 2005. Bacteria and phytoremediation: new uses for endophytic bacteria in plants. Trends in Biotechnology 23 (1), 6–8.

Newman, L.A., Strand, S.E., Choe, N., Duffy, J., Ekuan, G., Ruszaj, M., Shurtleff, B.B., Wilmoth, J., Heilman, P., Gordon, M.P., 1997. Uptake and biotransformation of trichloroethylene by hybrid poplars. Environmental Science and Technology 31 (4), 1062–1067.

Nissim, W.G., Cincinelli, A., Martellini, T., Alvisi, L., Palm, E., Mancuso, S., Azzarello, E., 2018. Phytoremediation of sewage sludge contaminated by trace elements and organic compounds. Environmental Research 164, 356–366.

Nivas, B.T., Sabatini, D.A., Shiau, B.J., Harwell, J.H., 1996. Surfactant enhanced remediation of subsurface chromium contamination. Water Research 30 (3), 511–520.

Nriagu, J.O., 1979. Global inventory of natural and anthropogenic emissions of trace metals to the atmosphere. Nature 279 (5712), 409.

Noori, A., Zare Maivan, H., Alaie, E., Newman, L.A., 2018. Leucanthemum vulgare Lam. crude oil phytoremediation. International Journal of Phytoremediation 20 (13), 1292–1299.

Ohkama-Ohtsu, N., Zhao, P., Xiang, C., Oliver, D.J., 2007. Glutathione conjugates in the vacuole are degraded by γ-glutamyl transpeptidase GGT3 in *Arabidopsis*. The Plant Journal 49 (5), 878–888.

Pacios, L.F., Gómez, L., 2006. Conformational changes of the electrostatic potential of biphenyl: a theoretical study. Chemical Physics Letters 432 (4–6), 414–420.

Pan, M.J., Rademan, S., Kunert, K., Hastings, J.W., 1997. Ultrastructural studies on the colonization of banana tissue and *Fusarium oxysporum* f. sp. cubense race 4 by the endophytic bacterium *Burkholderia cepacia*. Journal of Phytopathology 145 (11-12), 479–486.

Paterson, S., Mackay, D., McFarlane, C., 1994. A model of organic chemical uptake by plants from soil and the atmosphere. Environmental Science and Technology 28 (13), 2259–2266.

Petrová, Š., Rezek, J., Soudek, P., Vaněk, T., 2017. Preliminary study of phytoremediation of brownfield soil contaminated by PAHs. The Science of the Total Environment 599, 572–580.

Phenrat, T., Teeratitayangkul, P., Prasertsung, I., Parichatprecha, R., Jitsangiam, P., Chomchalow, N., Wichai, S., 2017. Vetiver plantlets in aerated system degrade phenol in illegally dumped industrial wastewater by phytochemical and rhizomicrobial degradation. Environmental Science and Pollution Research 24 (15), 13235–13246.

Phillips, L.A., Greer, C.W., Farrell, R.E., Germida, J.J., 2012. Plant root exudates impact the hydrocarbon degradation potential of a weathered-hydrocarbon contaminated soil. Applied Soil Ecology 52, 56–64.

Pilon-Smits, E., 2005. Phytoremediation. Annual Review of Plant Biology 56, 15–39.

Pilon-Smits, E.A., Freeman, J.L., 2006. Environmental cleanup using plants: biotechnological advances and ecological considerations. Frontiers in Ecology and the Environment 4 (4), 203–210.

Porteous Moore, F., Barac, T., Borremans, B., Oeyen, L., Vangronsveld, J., Van Der Lelie, D., Campbell, C.D., Moore, E.R., 2006. Endophytic bacterial diversity in poplar trees growing on a BTEX-contaminated site: the characterisation of isolates with potential to enhance phytoremediation. Systematic and Applied Microbiology 29 (7), 539–556.

Pradhan, S.P., Conrad, J.R., Paterek, J.R., Srivastava, V.J., 1998. Potential of phytoremediation for treatment of PAHs in soil at MGP sites. Journal of Soil Contamination 7 (4), 467–480.

Providenti, M.A., Flemming, C.A., Lee, H., Trevors, J.T., 1995. Effect of addition of rhamnolipid biosurfactants or rhamnolipid-producing Pseudomonas aeruginosa on phenanthrene mineralization in soil slurries. FEMS Microbiology Ecology 17 (1), 15–26.

Rao, A.S., 1990. Root flavonoids. The Botanical Review 56 (1), 1–84.

Rao, M.A., Scelza, R., Scotti, R., Gianfreda, L., 2010. Role of enzymes in the remediation of polluted environments. Journal of Soil Science and Plant Nutrition 10 (3), 333–353.

Robineau, T., Batard, Y., Nedelkina, S., Cabello-Hurtado, F., LeRet, M., Sorokine, O., Didierjean, L., Werck-Reichhart, D., 1998. The chemically inducible plant cytochrome P450 CYP76B1 actively metabolizes phenylureas and other xenobiotics. Plant Physiology 118 (3), 1049–1056.

Ronchel, M.C., Ramos-Díaz, M.A., Ramos, J.L., 2000. Retrotransfer of DNA in the rhizosphere. Environmental Microbiology 2 (3), 319–323.

Ryan, J.A., Bell, R.M., Davidson, J.M., O'connor, G.A., 1988. Plant uptake of non-ionic organic chemicals from soils. Chemosphere 17 (12), 2299–2323.

Salt, D.E., Smith, R.D., Raskin, I., 1998. Phytoremediation. Annual Review of Plant Biology 49 (1), 643–668.

Sánchez, V., López-Bellido, F.J., Cañizares, P., Rodríguez, L., 2018. Can electrochemistry enhance the removal of organic pollutants by phytoremediation? Journal of Environmental Management 225, 280–287.

Sandhu, A., Halverson, L.J., Beattie, G.A., 2007. Bacterial degradation of airborne phenol in the phyllosphere. Environmental Microbiology 9 (2), 383–392.

Schnoor, J.L., Light, L.A., McCutcheon, S.C., Wolfe, N.L., Carreia, L.H., 1995. Phytoremediation of organic and nutrient contaminants. Environmental Science and Technology 29 (7), 318A–323A.

Schroll, R., Bierling, B., Cao, G., Dörfler, U., Lahaniati, M., Langenbach, T., Scheunert, I., Winkler, R., 1994. Uptake pathways of organic chemicals from soil by agricultural plants. Chemosphere 28 (2), 297–303.

Schulz, B., Boyle, C., 2007. What are endophytes?. In: Schulz, B.J., Boyle, C.J., Sieber, T.N. (Eds.), Microbial Root Endophytes, vol. 9. Springer Science and Business Media, pp. 1–13.

Schwitzguébel, J.P., 2017. Phytoremediation of soils contaminated by organic compounds: hype, hope and facts. Journal of Soils and Sediments 17 (5), 1492–1502.

Shang, T.Q., Newman, L.A., Gordon, M.P., 2003. Fate of trichloroethylene in terrestrial plants. Phytoremediation: Transformation and Control of Contaminants 529–560.

Shimp, J.F., Tracy, J.C., Davis, L.C., Lee, E., Huang, W., Erickson, L.E., Schnoor, J.L., 1993. Beneficial effects of plants in the remediation of soil and groundwater contaminated with organic materials. Critical Reviews in Environmental Science and Technology 23 (1), 41–77.

Siminszky, B., Corbin, F.T., Ward, E.R., Fleischmann, T.J., Dewey, R.E., 1999. Expression of a soybean cytochrome P450 monooxygenase cDNA in yeast and tobacco enhances the metabolism of phenylurea herbicides. Proceedings of the National Academy of Sciences 96 (4), 1750–1755.

Singer, A.C., Crowley, D.E., Thompson, I.P., 2003. Secondary plant metabolites in phytoremediation and biotransformation. Trends in Biotechnology 21 (3), 123–130.

Soleimani, M., Hajabbasi, M.A., Afyuni, M., Mirlohi, A., Borggaard, O.K., Holm, P.E., 2010. Effect of endophytic fungi on cadmium tolerance and bioaccumulation by Festuca arundinacea and Festuca pratensis. International Journal of Phytoremediation 12 (6), 535–549.

Susarla, S., Medina, V.F., McCutcheon, S.C., 2002. Phytoremediation: an ecological solution to organic chemical contamination. Ecological Engineering 18 (5), 647–658.

Sutherland, T., Russell, R., Selleck, M., 2002. Using enzymes to clean up pesticide residues. Pesticide Outlook 13 (4), 149–151.

Syranidou, E., Thijs, S., Avramidou, M., Weyens, N., Venieri, D., Pintelon, I., Vangronsveld, J., Kalogerakis, N., 2018. Responses of the endophytic bacterial communities of Juncus acutus to pollution with metals, emerging organic pollutants and to bioaugmentation with indigenous strains. Frontiers of Plant Science 9.

Taghavi, S., Barac, T., Greenberg, B., Borremans, B., Vangronsveld, J., van der Lelie, D., 2005. Horizontal gene transfer to endogenous endophytic bacteria from poplar improves phytoremediation of toluene. Applied and Environmental Microbiology 71 (12), 8500–8505.

Thijs, S., Sillen, W., Weyens, N., Vangronsveld, J., 2017. Phytoremediation: state-of-the-art and a key role for the plant microbiome in future trends and research prospects. International Journal of Phytoremediation 19 (1), 23–38.

Thijs, S., Weyens, N., Gkorezis, P., Vangronsveld, J., 2018. Plant-endophyte partnerships to assist petroleum hydrocarbon remediation. In: Consequences of Microbial Interactions with Hydrocarbons, Oils, and Lipids: Biodegradation and Bioremediation, pp. 1–34.

Topp, E., Scheunert, I., Attar, A., Korte, F., 1986. Factors affecting the uptake of 14C-labeled organic chemicals by plants from soil. Ecotoxicology and Environmental Safety 11 (2), 219–228.

Trapp, S., Zambrano, K.C., Kusk, K.O., Karlson, U., 2000. A phytotoxicity test using transpiration of willows. Archives of Environmental Contamination and Toxicology 39 (2), 154–160.

Valls, M., De Lorenzo, V., 2002. Exploiting the genetic and biochemical capacities of bacteria for the remediation of heavy metal pollution. FEMS Microbiology Reviews 26 (4), 327–338.

Van Dillewijn, P., Caballero, A., Paz, J.A., González-Pérez, M.M., Oliva, J.M., Ramos, J.L., 2007. Bioremediation of 2, 4, 6-trinitrotoluene under field conditions. Environmental Science and Technology 41 (4), 1378–1383.

Van Dyke, M.I., Gulley, S.L., Lee, H., Trevors, J.T., 1993. Evaluation of microbial surfactants for recovery of hydrophobic pollutants from soil. Journal of Industrial Microbiology 11 (3), 163–170.

Villacieros, M., Whelan, C., Mackova, M., Molgaard, J., Sánchez-Contreras, M., Lloret, J., de Cárcer, D.A., Oruezábal, R.I., Bolanos, L., Macek, T., Karlson, U., 2005. Polychlorinated biphenyl rhizoremediation by Pseudomonas fluorescens F113 derivatives, using a Sinorhizobium meliloti nod system to drive bph gene expression. Applied and Environmental Microbiology 71 (5), 2687–2694.

Wang, C.J., Liu, Z.Q., 2007. Foliar uptake of pesticides—present status and future challenge. Pesticide Biochemistry and Physiology 87 (1), 1–8.

Wang, G.D., Li, Q.J., Luo, B., Chen, X.Y., 2004. Ex planta phytoremediation of trichlorophenol and phenolic allelochemicals via an engineered secretory laccase. Nature Biotechnology 22 (7), 893.

Wang, Y., Fang, L., Lin, L., Luan, T., Tam, N.F., 2014. Effects of low molecular-weight organic acids and dehydrogenase activity in rhizosphere sediments of mangrove plants on phytoremediation of polycyclic aromatic hydrocarbons. Chemosphere 99, 152–159.

Wenzel, W.W., Adriano, D.C., Salt, D., Smith, R., 1999. Phytoremediation: a plant-microbe-based remediation system. Bioremediation of Contaminated Soils, Agronomy Monograph 37, 457–508.

Weyens, N., Taghavi, S., Barac, T., Van Der Lelie, D., Boulet, J., Artois, T., Carleer, R., Vangronsveld, J., 2009. Bacteria associated with oak and ash on a TCE-contaminated site: characterization of isolates with potential to avoid evapotranspiration of TCE. Environmental Science and Pollution Research 16 (7), 830–843.

Whiteley, C.G., Lee, D.J., 2006. Enzyme technology and biological remediation. Enzyme and Microbial Technology 38 (3–4), 291–316.

Wiszniewska, A., Hanus-Fajerska, E., Muszyńska, E., Ciarkowska, K., 2016. Natural organic amendments for improved phytoremediation of polluted soils: a review of recent progress. Pedosphere 26 (1), 1–12.

Wolfe, N.L., Hoehamer, C.F., 2003. Enzymes used by plants and microorganisms to detoxify organic compounds. In: Phytoremediation: Transformation and Control of Contaminants. Wiley, NY, USA, pp. 159–187.

Wolfe, N.L., Ou, T.Y., Carreira, L., 1993. Biochemical Remediation of TNT Contaminated Soils. US Army Corps Eng., Washington, DC.

Zielinski, M., Kahl, S., Standfuß-Gabisch, C., Cámara, B., Seeger, M., Hofer, B., 2006. Generation of novel-substrate-accepting biphenyl dioxygenases through segmental random mutagenesis and identification of residues involved in enzyme specificity. Applied and Environmental Microbiology 72 (3), 2191—2199.

Further reading

Gerhardt, K.E., Huang, X.D., Glick, B.R., Greenberg, B.M., 2009. Phytoremediation and rhizoremediation of organic soil contaminants: potential and challenges. Plant Science 176 (1), 20—30. Why silence is not an option, Nature Biotechnology 24 (2006) 1177.

Kärenlampi, S., Schat, H., Vangronsveld, J., Verkleij, J.A.C., van der Lelie, D., Mergeay, M., Tervahauta, A.I., 2000. Genetic engineering in the improvement of plants for phytoremediation of metal polluted soils. Environmental Pollution 107 (2), 225—231.

Lebeau, T., Braud, A., Jézéquel, K., 2008. Performance of bioaugmentation-assisted phytoextraction applied to metal contaminated soils: a review. Environmental Pollution 153 (3), 497—522.

Matzke, A.J., Matzke, M.A., 1998. Position effects and epigenetic silencing of plant transgenes. Current Opinion in Plant Biology 1 (2), 142—148.

Mrozik, A., Miga, S., Piotrowska-Seget, Z., 2011. Enhancement of phenol degradation by soil bioaugmentation with Pseudomonas sp. JS150. Journal of Applied Microbiology 111 (6), 1357—1370.

Newman, L.A., Doty, S.L., Gery, K.L., Heilman, P.E., Muiznieks, I., Shang, T.Q., Siemieniec, S.T., Strand, S.E., Wang, X., Wilson, A.M., Gordon, M.P., 1998. Phytoremediation of organic contaminants: a review of phytoremediation research at the university of Washington. Journal of Soil Contamination 7 (4), 531—542.

Bioremediation of dyes from textile and dye manufacturing industry effluent

Razia Khan[1], Vipul Patel[2], Zeenat Khan[3]

[1]Department of Microbiology, Girish Raval College of Science, Gujarat University, Gandhinagar, India; [2]Environment Management Group, Center for Environment Education, Ahmedabad, India; [3]Environmental Genomics and Proteomics Lab, BRD School of Biosciences, Satellite Campus, Sardar Patel University, Vallabh Vidyanagar, India

1. Introduction

Highly colored wastewater is a major environmental barrier for the growth of the textile and dye manufacturing industry, besides the other minor issues such as solid waste management. These industries collectively consume several hundred thousand gallons of water each day and proportionally produce a huge volume of wastewater. Wastewaters generating from these industries contain high concentrations of inorganic and organic pollutants. Textile and dye manufacturing industry wastewaters contain dyes, salts, surfactants, acids, binders, reducing agents, thickeners, etc. Diverse kind of synthetic dyes is present in wastewaters of these industries, as uptake of these dyes by the fabric is very poor. Such wastewater is characterized in terms of strong color, salinity, temperature, pH, biological oxygen demand (BOD), chemical oxygen demand (COD), total dissolved solids, and nonbiodegradable organic compounds. It also contains trace heavy metals such as chromium, arsenic, copper, and zinc. The concentrations of nutrients varying in textile wastewaters are source-dependent. In addition, COD and BOD also vary because of the dyes used and their metabolites produced in wastewater because different dyes inherit different structure. This highly colored textile wastewater severely affects the photosynthetic function of receiving waterbodies. It also has an influence on aquatic life because of low light penetration and oxygen depletion. It may also be lethal to certain forms of marine life because of the occurrence of component metals and chlorine present in the synthetic dyes (Sarayu and Sandhya, 2012). Therefore, this kind of colored wastewater must be treated before their discharge.

Abatement of Environmental Pollutants
https://doi.org/10.1016/B978-0-12-818095-2.00005-9

Many combined physicochemical and biochemical procedures and other cutting-edge innovations have been created to treat the effluents from textile and dye manufacturing industries. Physical (adsorption and filtration) and chemical treatment (coagulation, photocatalytic oxidation, Fenton process, ozonation) techniques are viable for decolorization yet use more chemical and energy. Likewise, these techniques gather the contaminants into solid or liquid side streams and require extratreatment strategies for disposal. Biological treatment strategies including bacteria, fungi, and algae can totally mineralize pollutants and are typically less expensive. Biological treatment techniques alone may not be adequate to deal with the nonbiodegradable substrate present in the textile and dye manufacturing industry effluent.

Enzyme-based remediation technique is currently generally utilized for the decolorization of dye-containing effluent. However, this strategy is additionally confronting a few issues, for example, the expense of enzymes, stability of enzyme, and product inhibition. Then again, biological procedures give an ease and proficient option for simultaneous color and organic matter removal. However, it is challenging to attain complete mineralization only using aerobic biological processes (Méndez-Paz et al., 2005). Anaerobic treatment might be a possible option for the treatment of wastewater containing dye. The fundamental disadvantage of textile dye reduction under anaerobic conditions is the generation of aromatic amines, which cannot be degraded under anaerobic conditions (Van der Zee et al., 2001). Likewise, biological decolorization through anaerobic reductive procedures is typically moderate and requires an electron donor or co-substrate to create necessary reductive conditions (Dos Santos et al., 2005).

Among biological remediation system, constructed wetland has evolved into a reliable technology and has been successfully applied for the treatment of various types of industrial effluents including dye-containing wastewater. The thought of transforming waste to energy has attained great interest worldwide as it can directly solve global environmental issues with a more sustainable approach. Recently, the exploration of sustainable industrial effluent treatment technology has driven the development of microbial fuel cell (MFC) integrated with constructed wetland. Both the systems consist of an anaerobic and aerobic regions or chambers for oxidation and reduction processes, respectively. Such kind of systems are still in their developmental stage; more research works need to be conducted to overcome or solve the limitation and challenges that exist for the treatment of industrial effluent containing textile dyes. Hence, it is equally important to determine the mechanistic information of biological remediation in constructed wetland system and electricity generation in MFC simultaneously to overcome the limitation. In this chapter, we have focused on different factors influencing biological dye removal, microorganisms involved, and mechanism behind dye bioremediation process. Application of enzymes, MFC, constructed wetlands, and genetically engineered cultures in biological dye removal have also been discussed.

2. Importance of characterization of dye-containing wastewater

So as to design a powerful and effective treatment process, characterization of industrial effluent is maybe the most basic and critical step. The industrial effluent generated at different industrial units differs regarding their quality and quantity. The characteristics of this effluent are critical to comprehend as the ensuing treatment alternatives are directly related to these characteristics. The effluent characterization studies constitute an important prerequisite for the success of any treatment facility, such as common effluent treatment plants (CETPs). Subsequently, this investigation ought to be finished by a certified proficient. Furthermore, it should utilize certified analytical research facility. The certified analytical research centers are required to meet least execution measures and must pass periodic proficiency tests. The scientific strategies utilized are likewise vital as the decision of technique influences the outcomes. The variation acquired because of the technique pursued may interfere with the reliability and in the term it's summed up capacity.

3. Factors affecting biological removal of textile dyes

The factors affecting biological dye removal from industrial effluents include molecular structure of dyes, pH, temperature, dye concentration, agitation, etc. Biological degradation of textile dyes using microbial cultures has been explained in Fig. 5.1. The molecular configuration of dyes is found to have an influence on the extent of dye bioremoval. Wong and Yu (1999) reported that dye biodecolorization by *Trametes versicolor* was dependent on dye structures. Spadaro et al. (1992) reported that aromatic rings with hydroxyl, amino, or nitro functional groups were mineralized to a greater extent than unsubstituted rings in dye removal by *Phanerochaete chrysosporium*. Dye concentration additionally influences the color removal efficiency. A decline in decolorization efficiency (from 100% to 80%) by *Coriolus versicolor* was seen when the dye concentration was increased from 100–500 mg/L to 700–1200 mg/L (Kapdan et al., 2000). The ionic forms of the dye in solution and the surface electrical charge of the biomass depend on pH of the solution. Consequently, pH impacts both the fungal biomass surface dye binding sites and the dye chemistry in the medium. Fu and Viraraghavan (2000) mentioned that the effective initial pH of dye solution was 6 and 4, respectively, for Basic Blue 9 and Acid Blue 29. At pH of 2.0, no biosorption happened for Basic Blue 9 because of the high concentration of protons, while at pH of 12, no biosorption happened for Acid Blue 29. Arica and Bayramoglu (2007) revealed that as the pH was decreased, the biosorption of Reactive Red 120 dye on the fungal biomass *Lentinus sajor-caju* enhanced. Thus, O'Mahony et al. (2002) reported the maximum removal of reactive

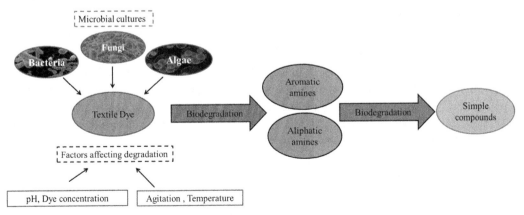

FIGURE 5.1 Mechanism for biological degradation of textile dyes.

dye Remazol Black B in the range of 1—2 with a sharp drop off at higher values. At low pH, the fungal biomass will have a net positive charge. These sites become available for binding anionic groups of reactive dyes. Most textile and other colored effluents are generated at generally high temperatures and henceforth temperature will be a critical factor in the real application of biosorption in future. Arica and Bayramoglu (2007) reported biological sorption of dye by *L. sajor-caju* increased with an increase in temperature from 5 to 35 °C. Aksu and Cagatay (2006) found an increase in biosorption with increase in temperature up to 45 °C for *Rhizopus arrhizus* culture. Dyeing processes consume a lot of salt and consequently the dyeing wastewater contains high salt concentration. Consequently, the ionic strength is an imperative factor. Zhou and Banks (1991, 1993) revealed that high ionic strength prompted high biosorption of humic acid by *R. arrhizus*. However, adsorption capacities of *L. sajor-caju* biomass demonstrated no noteworthy impact with increasing NaCl from 0 to 0.5 M concentration (Arica and Bayramoglu, 2007).

4. Microorganisms and mechanism involved in dye bioremediation process

Microbial remediation of textile dyes can be accomplished using microbial cultures, each of which has their own advantages and disadvantages in terms of dye removal efficiency and system suitability. Wide varieties of microbial strains possess the ability of extracellular enzyme synthesis that can remediate toxic contaminants from polluted sites (Table 5.1).

4.1 Bacteria

Bacteria have been a most studied organism for the remediation of textile dyes from industrial effluents. Most of the research involving bacterial strains has focused on studying the biodecolorization and degradation as compared with biosorption (application of dead bacterial biomass) of dyes (Forgacs et al., 2004). Biodegradation processes involving bacterial cells may be anaerobic, aerobic, or involve a combination of the two. The mode of dye uptake by dead/inactive cells in the biosorption process is extracellular; the chemically functional groups of the bacterial cell wall play vital roles in the process. Won et al. (2005) reported *C. glutamicum* as a potential biosorbent of Reactive Red 4 dye.

Generally, decontamination of textile dyes in industrial effluent by bacterial culture involves breaking of azo bond with the help of different enzymes and electron donating systems (Brüschweiler and Merlot, 2017). Hence, application of a bacterial culture or bacterial consortium is very common. Recently, 93% of Reactive black 5 dye with 500 mg/L of dye concentration has been remediated using a *Pseudomonas entomophila* culture after 120 h of incubation time (Khan and Malik, 2015). A bacterial consortium of *Providencia rettgeri* and *Pseudomonas* sp. culture was reported to mineralize four textile azo dyes (Reactive Black 5, Direct Red 81, Reactive Orange 16, and Disperse Red 78) (Lade et al., 2015). Complete decolorization of these azo dyes was accomplished in microaerophilic, sequential microaerophilic/aerobic, and aerobic/microaerophilic environments with slight difference in incubation time, perhaps because of their difference in chemical structure. It is well-known that the subsequent breakdown products like aromatic amines from the degradation of dyes experience

TABLE 5.1 Reports on microbial cultures for dye removal.

Sr. No.	Microbial culture	Dye	Mechanism	References
Bacteria				
1	*Lysinibacillus* sp.	Remazol Red	Biodegradation	Saratale et al. (2013)
2	*Pseudomonas desmolyticum*	Direct Blue 6	Biodegradation	Kalme et al. (2007)
3	*Micrococcus glutamicus*	Reactive Green 19A	Biodegradation	Saratale et al. (2009)
4	*Pseudomonas putida*	Crystal violet	Biodegradation	Chen et al. (2007)
5	*Pseudomonas pulmonicola*	Malachite green	Biodegradation	Chen et al. (2009)
6	*Brevibacillus parabrevis*	Congo red	Biodegradation	Talha et al. (2018)
Fungi				
1	*Aspergillus bombycis*	Reactive Red 31	Biodegradation	Khan and Fulekar (2017)
2	*Penicillium ochrochloron*	Cotton blue	Biodegradation	Shedbalkar et al. (2008)
3	*Aspergillus* sp.	Brilliant green	Biodegradation	Kumar et al. (2012)
4	*Coriolopsis* sp.	Crystal Violet Cotton Blue	Biodegradation	Munck et al. (2018)
5	*Peyronellaea prosopidis*	Scarlet RR	Biodegradation	Bankole et al. (2018)
6	*Myceliophthora vellerea*	Reactive Blue 220	Biodegradation	Patel et al. (2013)
Algae				
1	*Cyanobacterium Phormidium*	Indigo blue	Biodegradation and biosorption	Dellamatrice et al. (2017)
2	*Chlorella vulgaris*	Remazol Black B	Biosorption	Aksu and Tezer (2005)
3	*Nostoc linckia*	Methyl red	Biodegradation	El-Sheekh et al. (2009)
4	*Oscillatoria rubescens*	Basic Fuchsin	Biodegradation	El-Sheekh et al. (2009)

further mineralization by oxygenase and hydroxylase enzymes secreted by bacterial culture (Xiang et al., 2016). However, structurally complex textile dyes and their degraded products minimize the application of bacterial cultures for bioremediation use. For instance, naphthylamine sulfonic acids are refractory as they are difficult to degrade using bacterial cultures because of the nonpermeability of a strongly charged anionic species, viz. sulfonyl group, through the bacterial membranes (Gao et al., 2014).

4.2 Fungi

Fungi have emerged as effective biological tools for the degradation and mineralization of structurally rigid textile dyes owing to their strong enzymatic system and diverse metabolism (Ahmad et al., 2015; Rahimnejad et al., 2015). They possess the strong ability to degrade complex organic pollutants by producing extracellular ligninolytic enzymes including laccase, lignin peroxidase, and manganese peroxidase. A wide range of fungal strains is capable of removal of a wide range of textile azo dyes (Fu and Viraraghavan, 2001). Many fungi strains have been applied in either living or inactivated form. The use of white-rot fungi such as *P. chrysosporium* in textile wastewater remediation has been widely reported in the literature (Gomaa et al., 2008; Sharma et al., 2009; Faraco et al., 2009). Apart from white-rot fungi, other fungi such as *Aspergillus niger* (Fu and Viraraghavan, 2002), *R. arrhizus* (Zhou and Banks, 1991), and *Rhizopus oryzae* (Gallagher et al., 1997) can also decolorize and biosorb diverse range of dyes. *Penicillium oxalicum* strain was able to biodegrade Acid Red 183, Direct Red 75, and Direct Blue 15 dyes (Saroj et al., 2014). Other fungal cultures, such as *Magnusiomyces* and *Candida*, were also recognized for their capability to completely degrade textile dyes (Brüschweiler and Merlot, 2017). Complete decolorization of a red azo dye (1000 mg/L dye concentration) was carried out using *A. niger* culture at 9 pH (Mahmoud et al., 2017). The mechanism of fungal remediation involves adsorption, enzymatic degradation, or a combination of both. The main aim of fungal-based remediation is to decolorize and detoxify the dye-contaminated industrial effluents.

4.3 Algae

Algal cultures have been proved to be potential candidates for the removal of dyestuff compounds from colored effluents. They are effective biosorbents and bio-coagulants because of their availability in both fresh and marine water. The biosorption capability of algal culture relies on their relatively high surface area and high binding affinity (Donmez and Aksu, 2002). Algal cell wall characteristics possess an important part in biosorption. Electrostatic attraction and complexation are known to take place during algal biosorption (Satiroglu et al., 2002). Functional groups such as carboxylate, hydroxyl, and amino found on the surface of algal cells are considered to be responsible for removal of pollutants from wastewater. The dye remediation, focusing on algal applications, may be due to the accumulation of dye ions on the surface of algal biopolymers and further to the diffusion of the dye particles from aqueous system onto the solid surface of the biopolymer (Ozer and Benet-Martinez, 2006). Before using microalgal cells as the biosorbent or bio-coagulator, their growth tolerance in presence of various dyes should be taken into consideration.

5. Application of enzymes as biocatalyst in dye bioremediation

Microorganisms are living factories that own the ability to utilize or degrade organic pollutants or contaminants in different forms. Microbial cultures produce complex chemicals such as enzymes. Complex organic pollutants are converted into simpler ones by these enzymes. These enzymes show high specificity and are capable to differentiate between substrate molecules with minute difference. Therefore, directly applying these enzymes may result in decline in the microbial production. Since last 10 years, these biocatalysts have been exposed to wide areas of research which includes biofuel cells (Zhang et al., 2016), biodiesel generation (Taher and Al-Zuhair, 2017), biosensor applications (Sharma and Leblanc, 2017), and biodecolorization of synthetic dyes (Bilal et al., 2017). Most of the enzymes are well-known as biological catalysts (Bilal et al., 2017). They possess a very high ability for the biodegradation of wide variety of complex contaminants including synthetic textile dyes. The manufacturing rate for microbial enzymes in bulk can be decreased using advanced methods such as molecular biology techniques, and the application of immobilized enzyme can enhance their effectiveness and lifetime (Chatha et al., 2017). Biological dye removal can be carried out using various approaches as mentioned earlier. However, the azo dye reduction using bacterial cultures (single, mixture, and consortium) can give rise to colorless and highly toxic and recalcitrant aromatic amine generation (Brüschweiler and Merlot, 2017; Xiang et al., 2016). Also instead of that, direct application of enzymes such as laccase enzyme was observed to give good results for effective decolorization of industrial wastewater via enzymatic degradation of textile dyes (Chatha et al., 2017). Main step involved in the application of biocatalysts for effective remediation of textile dye in a biological reactor can be mentioned as follows: (1) immobilization of the enzyme and (2) determination of biocatalyst potential for its repeated applications.

5.1 Immobilization of biological catalysts

Highly colored or contaminated industrial effluent can subsequently reduce the catalytic activity and stability of enzymes in free state. Immobilized biocatalysts are appropriate to solve these issues which rely on their durability and stability (Jiang et al., 2014). Various kind of support materials comprising beads and gels are reported for the application in immobilization of enzyme. It was reported that such immobilization process also facilitated to increase and improve the repeated application of enzymes. Research on support-free immobilization methods, on the other hand, has attracted the attention of investigators around the world, as enzymes bound to carrier surface are liable for the less stability and leaching in an aqueous system (Bilal et al., 2017). Three kinds of techniques have principally been applied for enzyme encapsulation/immobilization: (1) cross-linking of the enzyme, (2) enzyme molecules encapsulation in a polymer, and (3) attachment with an inorganic/organic polymer (Jiang et al., 2011). During the procedure of cross-linking enzyme aggregate formation, in the first stage, the enzyme is precipitated using precipitants, with the aim of the resulting aggregates cross-linked with less effective cross-linkers, e.g., glutaraldehyde (Bilal et al., 2017). Such technique is equally cost-effective and ecologically sustainable. In case of covalent immobilization, the supporting material needs selective activation and coupling procedures (Blamey et al., 2017). Jiang et al. (2011) immobilized an enzyme, horseradish peroxidase in phospholipid-templated TiO_2 particles.

5.2 Potential of biocatalysts for reusability

Major benefits of enzymatic system application for dye bioremediation are that they can be recycled. Considering large-scale applications, reusability of enzymes has great benefit in decreasing handling expenses (Taher and Al-Zuhair, 2017; Sharma and Leblanc, 2017). Immobilized microbial cells can easily be collected by shifting them into fresh medium, whereas free microbial cells can be collected using centrifugation (Chen and Lin, 2007). Recycling of enzymes has attracted the attention of various scientists all over the globe. In a study carried out by Shaheen et al. (2017), calcium alginate encapsulated LiP enzyme was reused for seven consecutive cycles for the reactive dye decolorization and 41% of color removal efficiency was maintained. In the situation of free microbial cells, enzyme showed absence of any activity after seven repetitive cycles of application. Therefore, the color removal efficiency of encapsulated system was found to be comparatively more stable as compared with free cultures after repetitive applications. Kunamneni et al. (2008) reported effective maintenance of 41% of the preliminary enzyme activity of immobilized laccase enzyme after five repetitive cycles for color removal of methyl green dye. Nevertheless, the motive behind the decline in enzyme activity could be that the biocatalysts are secreted from the support system as the immobilization beads are washed after each cycle (Daâssi et al., 2014).

6. Advancements in bioreactor systems for dye remediation

A biological reactor is considered as the basic part of the biological processes engaged for the removal of diverse contaminants. Design and development of an appropriate bioreactor for a specific treatment process requires broad study on the biosystem, including cell growth, genetic manipulation, and metabolism, and establishing proper experimental factors such as operational stability, O_2 transfer, scale-up, establishment cost, and so on (Zhong, 2010). Biological remediation of synthetic dyes can be accomplished by means of anaerobic and aerobic methods, as well as a blend of both systems. Nonetheless, whichever technique is applied, biological degradation or sorption ought to be done in a specially designed biological reactor to improve the overall efficiency of whole process. These kinds of biological reactors are of wide variety and incorporate membrane-based reactors, air-pulsed bioreactors, aerobic—anaerobic sequential reactors, MFCs, hybrid bioreactors, and semicontinuous bioreactors. Various investigations presently exist on biodegradation methods for textile dyes in effluents. These include activated sludge, trickling filters, aerobic granular sludge, lagoons, oxidizing beds, constructed wetlands, aerobic filters, membrane bioreactors, and so on. These strategies include the utilization of redox mediators, anaerobic, aerobic, and enzymatic treatments.

A 5 L airlift biological reactor was utilized for the decolorization of indigo carmine dye in the optimized environment with dye concentration of 25 mg/L of and an airflow rate of 4 L/min (Teerapatsakul et al., 2017). In that report, 100% biological dye removal was attained with just a single enzymatic supplementation, i.e., laccase. An anaerobic biofilter combined with anoxic—aerobic membrane-based reactor was applied simultaneously for the biological remediation of an orange dye of Reactive Orange 16 (Spagni et al., 2010). The researchers revealed the total decolorization and degradation of Reactive Orange 16 dye because of the breakdown of azo linkages, aromatic amines, and sulfonated aromatic amines through the collective procedure. An additional advantage reported in that study was the production of 60—70% of methane. Similarly, an aerobic system combined with fixed-film biological reactor using a continuous stirred-tank bioreactor and fixed-film biological reactor was likewise used to remediate indigo carmine dye (Khelifi et al., 2008). Such arrangement was likewise exceedingly effective at COD decrease (97.5%) and dye decolorization (97.3%) under ideal environment of 7.5 pH and an early COD value of 1185 mg/L.

A membrane bioreactor utilizing *T. versicolor* combined with reverse osmosis was effective for decolorization of colored effluent (Kim et al., 2004). A wood-decaying fungal strain F29 decolorized 95%—99% Orange II in a continuous packed bed and fluidized bed bioreactor frameworks (Zhang et al., 1999). *Irpex lacteus* immobilized in a pine wood reactor removed Remazol Brilliant Blue R more quickly than a polyurethane foam-based bioreactor

(Kasinath et al., 2003). Borras et al. (2008) achieved a high degree of dye decolorization utilizing *T. versicolor* culture as pellet; notwithstanding, the dye removal efficiency was not compared with other morphological forms. Prominently, compared with the morphologies experienced in suspended cultures, immobilized cultures have an inclination to accomplish a more elevated amount of enzymatic action and greater adaptability to ecological perturbations, for example, shear damage and pH/toxic shock (Rodríguez-Couto et al., 2009). For instance, Zheng et al. (1999) reported better decolorization of an azo dye by alginate-immobilized fungus than dispersed filaments in various reactor designs; nonetheless, coordinate correlation with pellet morphology was not performed in their examination.

7. Treatment of dye-containing industrial effluents using genetically modified microorganisms or enzymes

Biological remediation is an eco-friendly way for the treatment of dye manufacturing and textile industry effluent, but the physicochemical nature of the effluents, including pH, high NaCl concentration, temperature, and the presence of toxicants, can result in the deactivation of enzymes and microbial cells. Therefore, it is essential to have more active and versatile enzymes and microbial cultures with high stability and low cost that is suitable to fulfill the criteria for the treatment of colored effluents.

There are molecular biology approaches, such as cloning, heterologous expression, random mutagenesis, site-directed mutagenesis, gene recombination techniques, directed evolution, rational design, and metagenomics, to accelerate the evolution processes in such a way that the bioremediation process is enhanced. Additionally, advances in molecular genetics and genetic engineering have made it possible to clone and express virtually any gene in a targeted microbial host. The Lac gene from white-rot fungi *Trametes* sp. was cloned in *Pichia pastoris*, and the purified recombinant Lac possesses a stronger capacity for decolorizing different dyes, such as Methyl Orange and Bromophenol Blue, compared with some other known laccases (Fan et al., 2011). A Lac gene from *Ganoderma lucidum* was synthesized using optimized codons and a PCR-based two-step DNA synthesis method; the resulting recombinant Lac was overexpressed in *P. pastoris* and showed high levels of Methyl Orange biodegradation (Sun et al., 2012). The purified recombinant Lac obtained from the heterologous synthesis of *Pycnoporus sanguineus* Lac in *P. pastoris* could effectively decolorize textile dyes in the absence of mediators (Lu et al., 2009). Among 2300 randomly mutated variants of *Pleurotus ostreatus*, two mutants showed higher stability in a variety of environmental conditions and higher ability to decolorize azo dyes than the wild-type Lac; the mutant also proved to be highly stable at both acidic and alkaline pH. In contrast, fungal Lac showed optimal dye decolorization at acidic pH and in the presence of redox mediators (Pereira et al., 2009).

8. Current status of bioreactor application in CETPs of industrial areas for dye removal

The chemical industry has been one of the driving forces for industrial growth in India and has contributed enormously to the economic and social development of the country.

However, it has also been responsible for damage to the environment and natural ecosystems of the country. The small-scale dyes chemical manufacturing clusters have increased significantly in the last couple of decades and currently, manufacture a wide range of textile dyes and its intermediates, which find application in many sectors of industry such as dyes and pigments, textile, printing ink, plastic, surface coating, etc.

The manufacturing of synthetic textile dyes and intermediate involves a number of stages and production is based on various unit processes and unit operations and generates colored effluent of diverse composition mostly at all stages. These results in the generation of process waste of varying quantities and characteristics. The production capacity in individual units is being small, and it is difficult and uneconomical to treat the effluents before final disposal in the environment. Hence, CETP concepts have come in practice at the cluster level. In current scenario, the discharge of treated effluent from varied industrial clusters to the natural ecosystem from CETPs is receiving increasing attention as a reliable method for pollution control and water resource. The idea of CETP was originally encouraged by the Ministry of Environment and Forests (MOEF) in 1984 for the treatment of effluent from a large number of small- and medium-scale industries.

A significant interest in applying various treatment strategies to effluent from the CETP has been reported. Raj (2007) reported the treatment of effluent from the CETP located in Hyderabad city using an extended aeration system. Combined chemical and biooxidation processes were applied by Manekar et al. (2014) to remediate a high strength textile process effluent at CETP located in western India. However, Moosvi and Madamwar (2007) have reported an integrated process for the treatment of CETP effluent, using chemical coagulation, anaerobic, and aerobic process studies on a CETP, which receives effluent from small-scale industries. Pophali et al. (2003) have reported that CETP is effectively applied to treat combined effluent from cotton and synthetic textile industries in Rajasthan by mixed activated sludge process. A study was conducted by Rathi (2013) on the management of eight CETP located in different industrial estates in Gujarat, India, manufacturing a wide range of chemicals.

9. Microbial fuel cell: a novel system for the remediation of colored wastewater

MFC is a biological system that converts chemical energy present in the organic pollutants into electrical energy by using microorganisms as a biocatalyst. They possess great potential for the remediation of industrial effluents containing dyes and simultaneous electricity generation with the help of microorganisms as biocatalysts. Recently, they have been well explored for their pioneering features and environmental benefits. The concept of MFC has already been well-established for electricity generation; however, not much work has been reported regarding dye removal with simultaneous electricity generation using MFC.

In an MFC system, the electrochemically active microorganisms oxidize various organic components of colored effluents in the anode compartment and generate protons and electrons that transport to the cathode compartment to reduce O_2 to H_2O. Most of the systems have a membrane to separate the compartments of the anode and the cathode (Li et al., 2014). The electricity generated can be easily collected by an external resistor placed between the anode and the cathode. The main disadvantage of this system is its application on large scale because of the lower production of power and higher material cost. Tremendous work has been made over the past decade to improve the electricity generation in MFC.

9.1 Microorganisms used in microbial fuel cells

Mostly, microbial cultures are electrochemically inactive. Electrochemically active bacterial species are applied to transfer electrons to the electrode in MFC. A list of microbes is shown in Table 5.2 along with their substrates. Different bacterial cultures of genus *Geobacter, Enterobacter, Shewanella,* and *Bacillus* (Richter et al., 2008; Rezaei et al., 2009; Nimje et al., 2009; Watson and Logan, 2010) have been reported with respect to their generation or optimization of the power output in the system. Recently, MFC populated by mixed microbial communities has garnered much attention owing to their stability, robustness because of nutrient adaptability, stress resistance, and general tendency to produce higher current densities than those obtained using pure cultures. But the study performed by Kiely et al. (2010) showed that a *Shewanella* strain produced greater electricity generation as compared with mixed culture. This may be because of different electron transfer mechanisms used by *Shewanella* strain in comparison with the mechanisms of mixed cultures.

9.2 Microbial fuel cell configuration and operation

Several MFC designs for the remediation of industrial effluents have been studied. Broadly they can be partitioned into two classes, for example, double chamber and single chamber. Double-chambered MFC includes an anaerobic anode chamber and an aerobic cathode chamber, which are normally separated by a proton exchange membrane. The substrate is oxidized by microorganisms producing electrons and protons at the anode chamber. The protons going through the layer of the membrane and the electrons going through the outer circuit are joined with electron acceptors at the cathode chamber. The anode is inoculated with a blended solution of anaerobic sludge and substrate like glucose. Then again, the cathode is inoculated with aerobic sludge. The final pH of the medium in both the chambers is adjusted to 7 (Li et al., 2010). Microbial cultures applied in various kinds of MFC for dye removal are mentioned in Table 5.3.

TABLE 5.2 Microbial cultures and their substrates for microbial fuel cell applications.

Sr. No.	Microbial cultures	Substrates	References
1	*Clostridium butyricum*	Starch, glucose, lactose, molasses	Niessen et al. (2004)
2	*Aeromonas hydrophila*	Acetate	Pham et al. (2003)
3	*Escherichia coli*	Glucose and sucrose	Ieropoulos et al. (2005)
4	*Pseudomonas aeruginosa*	Glucose	Narayanasamy and Jayaprakash (2018)
5	*Actinobacillus succinogenes*	Glucose	Park et al. (1999)
6	*Desulfovibrio desulfuricans*	Sucrose	Ieropoulos et al. (2005)

TABLE 5.3 Microbial cultures applied in different types of microbial fuel cell (MFC) for dye removal.

Sr. no.	Microbial culture	Dye and dye concentration	Type of MFC	References
1	*Klebsiella pneumoniae*	Methyl Orange 0.016 mg/L	Single-chambered system	Liu et al. (2009)
2	Sulfate-reducing mixed communities	Acid Red 114 (100–1000 mg/L)	Dual-compartment system	Miran et al. (2018)
3	Enterobacter cancerogenus	Reactive Green 19	Single-chambered membrane-less system	Hsueh et al. (2014)
4	Microbial consortium	Acid orange 7 (0.06–0.24 mg/L)	Two equal rectangular Perspex frames	Mu et al., 2009
5	Microbial consortium	Amaranth (75 mg/L)	Dual-chambered system	Fu et al., 2010
6	Enterobacter cancerogenus	Reactive green 19, Reactive blue160	Single-chambered system	Chen et al., 2016

Single-chambered MFC compromises of simpler design and cost savings. It commonly comprises of an anode chamber with a microfiltration layer air-cathode. The cathode was presented to air on one side and water on the opposite side (inside). There is no proton exchange membrane. The whole cathode is secured with a thick plexiglass cover with gaps to permit O_2 to reach the cathode. They are inoculated with a blend of aerobic and anaerobic sludge. Amid electricity generation, azo bonds of dye are cleaved utilizing protons and electrons from substrate oxidized by anodic microbes, resulting in the formation of colorless intermediates (Hou et al., 2011; Sun et al., 2009).

10. Potential of constructed wetlands for the treatment of dye-contaminated effluents

Constructed wetlands are designed frameworks that have been planned and developed to use the characteristic procedures including wetland vegetation, soils, and their related microbial communities for wastewater treatment. It is a potential natural treatment technology for industrial wastewater treatment because of low operational and maintenance costs (Wang, 2017). They are intended to impersonate similar procedures that happen in natural wetlands yet in an increasingly controlled condition. A portion of these frameworks have been structured and worked with the sole motivation behind treating wastewater, while others have been actualized in light of various use destinations, for example, utilizing treated wastewater effluent as a water source for the creation and rebuilding of wetland living space for wildlife use, for reuse in farming or ecological improvement. Synonymous terms to "constructed" incorporate man-made, designed, and fake wetlands.

Vertical stream pilot-scale constructed wetlands were utilized to expel two textile azo dyes—Acid Blue 113 and Reactive Blue 171—from a synthetic wastewater simulating textile effluent. The wetlands were loaded up with layers of gravel of different sizes and planted with *Phragmites australis*. For both dyes, 98% removal was accomplished. It was accounted for that most of dye (57%) was removed in the initial 12 h by adsorption onto charged surfaces in the substratum. The performance was significantly improved by the expansion of peat or another appropriate natural substratum. Davies et al. (2005) utilized a vertical stream pilot-scale constructed wetland planted with *P. australis* to expel an azo dye Acid Orange 7. At the organic loading rates somewhere in the range of 21 and 105 g COD m^2/d, the system could be able to remove 11−67 g COD m^2/d, and the removal of COD, TOC, and color added up to 64%, 71%, and 74%, individually. It was seen that *Phragmites* played an active role in Acid Orange 7 degradation. In addition to Acid Orange 7 degradation, *Phragmites* also degraded aromatic amines discharged amid Acid Orange 7 breakdown.

Bulc (2006) and Bulc and Ojstršek (2008) reported the utilization of pilot vertical flow− horizontal flow constructed wetland for the treatment of real textile effluent. The framework comprised of two parallel vertical stream beds (20 m^2 each) trailed by a single horizontal stream bed (40 m^2). All beds were planted with *P. australis*. At the flow rate of 1 m^3/d, 84% of COD, 66% of BOD, 93% of TSS, 88% of sulfate, and 90% of color were removed in first vertical stream bed. Mbuligwe (2005) designed and developed horizontal stream constructed wetland framework loaded up with sand, planted with *Typha* sp. and *Colocasia esculenta* for the treatment of highly colored effluent from "tie-and-die" unit. The initial pH (10.7) of effluent was reduced to 7.8 and 7.9 in *Typha* and *Colocasia* beds separately after treatment. At an HLR of 45 cm/d, 72%, 77%, and 15% of color, 53%, 59%, and 25% of sulfate, and 68%, 72%, and 51% of COD were removed in *Typha*, *Colocasia*, and unplanted units. The outcomes uncovered the marginally better performance of *Colocasia* unit over the *Typha* unit and much better removal in vegetated units when compared with the unvegetated unit. Bulc and Ojstršek (2008) reported that low BOD$_5$/COD proportion of textile effluent shows the nonbiodegradable nature of textile effluent and hence high BOD$_5$ removal cannot be expected. Ong et al. (2009) reported upstream vertical constructed wetlands for the treatment of azo dye (Acid Orange 7) containing industrial effluent. The developed system was loaded up with gravel and planted with *P. australis* and *Zizania aquatica*. Color removal was observed to be 96% in both planted and unplanted units. This demonstrates the vast majority of the color was removed in the anaerobic lower portion of the developed constructed wetland.

Biodegradability of industrial effluent and input loadings are basic components for accomplishing higher removal efficiency utilizing wetland frameworks. Media are a vital wetland segment, which gives attachment sites and fixings to biofilm development; likewise, media additionally support the development and growth of wetland plants. Because of novel capacities for supporting physicochemical and biological removals, diverse kinds of waste materials, for instance, gravels, sand, plant biomass, alum sludge, blast furnace slag, coal ash, and so forth, have been utilized as wetland media to date (Bai et al., 2014; Bavandpour et al., 2015; Saeed and Sun, 2012). These investigations revealed satisfactory removal performances in constructed wetlands, demonstrating potential change of waste materials into assets. Effluent distribution and air circulation had been accounted for enhancing remediation proficiency in constructed wetlands (Torrijos et al., 2016).

11. Conclusion and suggestions

The aim of the effluent treatment plants in the textile and dye manufacturing industrial clusters is to actualize innovations giving minimum or negligible water contamination. These effluent treatment units in the textile and dye manufacturing industrial clusters are the most acknowledged methodologies toward achieving environmental safety. In any case, sadly, no particular treatment approach/technique is suitable or all around appropriate for a wide range of textile and dye manufacturing industry effluents. Along these lines, the treatment/remediation of textile and dye manufacturing industry effluent is performed by combining several methods or techniques, which contain physical, chemical, and biological frameworks relying on the sort and quantum of contamination load and contaminants included. In this section we have examined a few methodologies that can be adjusted to treat the dye in textile and dye manufacturing industry effluents and to decrease the pollution load. Physical and oxidation strategies are compelling for the removal of dye compounds in dye-containing effluents only if the colored effluent volume is small. This restricts the use of physical and chemical techniques. Cost of utilization of membrane filtration confines its utilization in large-scale treatment. These are genuine even in lab-scale studies. Hence, they are not used in large-scale studies. Effluent treatment plants using biological treatment approaches, instead of chemical methods, guarantee that their inclination is because of low generation of inorganic sludge, low expenses, and complete mineralization of dye compounds in the biological method. By and large, textile and dye manufacturing industry effluents after biological treatment are not in consistence with the wastewater release standards. So as to meet effluent release standards and to lessen the impact of harmful or inhibitory substances on microorganisms, initially, recalcitrant compounds and dye compounds ought to be oxidized by chemical oxidation or advanced oxidation technique to change over it to biodegradable constituents previously exposing the wastewater to bacterial treatment is favored. Cavitation can be utilized to obliterate microbial life in water assuming any. The treated water after removal of microorganisms can be reused for cleaning purpose. Uses of eco-friendly and cost-effective frameworks, for example, constructed wetlands for the treatment of dye-containing industrial effluents, have been all around examined. These systems have been turned out to be effective in treating expansive volume of textile industry effluents. Treatment approaches including MFC for the treatment of industrial effluents served double advantage of remediation and energy generation in the form of bioelectricity. Presently onward, more scientists should concentrate on the kinetic study of decolorization/degradation and modeling of a biological reactor for the combined biological and chemical system as a pre- or posttreatment of textile and dye manufacturing industry effluents. The research studies on pollution control for the textile and dye manufacturing industry ought to likewise concentrate on the quantitative depiction of combination processes in spite of only qualitative discussion. Rigorous research has been done to determine the involvement of the different microorganisms in the remediation of azo-based water-soluble dyes. Barely any investigations have been accounted for on real colored industrial effluents. Along these lines, in future, it is essential to conduct more research that will concentrate on or include the remediation of real textile and dye manufacturing industry effluents with the help of integrated approaches. Such methodology may upgrade the biodegradability of textile and dye manufacturing industry effluent containing different groups of textile dyes.

References

Ahmad, A., Mohd-Setapar, S.H., Chuong, C.S., Khatoon, A., Wani, W.A., Kumar, R., Rafatullah, M., 2015. Recent advances in new generation dye removal technologies: novel search for approaches to reprocess wastewater. RSC Advances 5 (39), 30801–30818.

Aksu, Z., Çağatay, S.S., 2006. Investigation of biosorption of Gemazol Turquise Blue-G reactive dye by dried *Rhizopus arrhizus* in batch and continuous systems. Separation and Purification Technology 48 (1), 24–35.

Aksu, Z., Tezer, S., 2005. Biosorption of reactive dyes on the green alga *Chlorella vulgaris*. Process Biochemistry 40 (3–4), 1347–1361.

Arica, M.Y., Bayramoğlu, G., 2007. Biosorption of Reactive Red-120 dye from aqueous solution by native and modified fungus biomass preparations of *Lentinus sajor-caju*. Journal of Hazardous Materials 149 (2), 499–507.

Bai, Y., Mora-Sero, I., De Angelis, F., Bisquert, J., Wang, P., 2014. Titanium dioxide nanomaterials for photovoltaic applications. Chemical Reviews 114 (19), 10095–10130.

Bankole, P.O., Adekunle, A.A., Obidi, O.F., Chandanshive, V.V., Govindwar, S.P., 2018. Biodegradation and detoxification of Scarlet RR dye by a newly isolated filamentous fungus, *Peyronellaea prosopidis*. Sustainable Environment Research 28 (5), 214–222.

Bavandpour, R., Karimi, M.H., Beitollahi, H., Taghavi, M., Ghaemy, M., 2015. A nanostructure based carbon paste electrode as a voltammetric sensor for determination of chlorpromazine.

Bilal, M., Asgher, M., Parra-Saldivar, R., Hu, H., Wang, W., Zhang, X., Iqbal, H.M., 2017. Immobilized ligninolytic enzymes: an innovative and environmental responsive technology to tackle dye-based industrial pollutants—a review. The Science of the Total Environment 576, 646–659.

Blamey, J.M., Fischer, F., Meyer, H.P., Sarmiento, F., Zinn, M., 2017. Enzymatic biocatalysis in chemical transformations: a promising and emerging field in green chemistry practice. Biotechnology of Microbial Enzymes 347–403.

Borràs, E., Blánquez, P., Sarrà, M., Caminal, G., Vicent, T., 2008. *Trametes versicolor* pellets production: low-cost medium and scale-up. Biochemical Engineering Journal 42 (1), 61–66.

Brüschweiler, B.J., Merlot, C., 2017. Azo dyes in clothing textiles can be cleaved into a series of mutagenic aromatic amines which are not regulated yet. Regulatory Toxicology and Pharmacology 88, 214–226.

Bulc, T.G., Ojstršek, A., 2008. The use of constructed wetland for dye-rich textile wastewater treatment. Journal of Hazardous Materials 155 (1–2), 76–82.

Bulc, T.G., 2006. Long term performance of a constructed wetland for landfill leachate treatment. Ecological Engineering 26 (4), 365–374.

Chatha, S.A.S., Asgher, M., Iqbal, H.M., 2017. Enzyme-based solutions for textile processing and dye contaminant biodegradation—a review. Environmental Science and Pollution Research 24 (16), 14005–14018.

Chen, C.C., Liao, H.J., Cheng, C.Y., Yen, C.Y., Chung, Y.C., 2007. Biodegradation of crystal violet by *Pseudomonas putida*. Biotechnology Letters 29 (3), 391–396.

Chen, C.Y., Kuo, J.T., Cheng, C.Y., Huang, Y.T., Ho, I.H., Chung, Y.C., 2009. Biological decolorization of dye solution containing malachite green by *Pandoraea pulmonicola* YC32 using a batch and continuous system. Journal of Hazardous Materials 172 (2–3), 1439–1445.

Chen, J.P., Lin, Y.S., 2007. Decolorization of azo dye by immobilized *Pseudomonas luteola* entrapped in alginate–silicate sol–gel beads. Process Biochemistry 42 (6), 934–942.

Chen, B.Y., Ma, C.M., Han, K., Yueh, P.L., Qin, L.J., Hsueh, C.C., 2016. Influence of textile dye and decolorized metabolites on microbial fuel cell-assisted bioremediation. Bioresource Technology 200, 1033–1038.

Daâssi, D., Rodríguez-Couto, S., Nasri, M., Mechichi, T., 2014. Biodegradation of textile dyes by immobilized laccase from *Coriolopsis gallica* into Ca-alginate beads. International Biodeterioration and Biodegradation 90, 71–78.

Davies, L.C., Carias, C.C., Novais, J.M., Martins-Dias, S., 2005. Phytoremediation of textile effluents containing azo dye by using *Phragmites australis* in a vertical flow intermittent feeding constructed wetland. Ecological Engineering 25 (5), 594–605.

Dellamatrice, P.M., Silva-Stenico, M.E., de Moraes, L.A.B., Fiore, M.F., Monteiro, R.T.R., 2017. Degradation of textile dyes by cyanobacteria. Brazilian Journal of Microbiology 48 (1), 25–31.

Dönmez, G., Aksu, Z., 2002. Removal of chromium (VI) from saline wastewaters by *Dunaliella* species. Process Biochemistry 38 (5), 751–762.

Dos Santos, A.B., Traverse, J., Cervantes, F.J., Van Lier, J.B., 2005. Enhancing the electron transfer capacity and subsequent color removal in bioreactors by applying thermophilic anaerobic treatment and redox mediators. Biotechnology and Bioengineering 89 (1), 42–52.

El-Sheekh, M.M., Gharieb, M.M., Abou-El-Souod, G.W., 2009. Biodegradation of dyes by some green algae and cyanobacteria. International Biodeterioration and Biodegradation 63 (6), 699–704.

Fan, F., Zhuo, R., Sun, S., Wan, X., Jiang, M., Zhang, X., Yang, Y., 2011. Cloning and functional analysis of a new laccase gene from *Trametes* sp. 48424 which had the high yield of laccase and strong ability for decolorizing different dyes. Bioresource Technology 102 (3), 3126–3137.

Faraco, V., Pezzella, C., Miele, A., Giardina, P., Sannia, G., 2009. Bio-remediation of colored industrial wastewaters by the white-rot fungi *Phanerochaete chrysosporium* and *Pleurotus ostreatus* and their enzymes. Biodegradation 20 (2), 209–220.

Forgacs, E., Cserhati, T., Oros, G., 2004. Removal of synthetic dyes from wastewaters: a review. Environment International 30 (7), 953–971.

Fu, Y., Viraraghavan, T., 2000. Removal of a dye from an aqueous solution by the fungus *Aspergillus niger*. Water Quality Research Journal 35 (1), 95–112.

Fu, Y., Viraraghavan, T., 2001. Fungal decolorization of dye wastewaters: a review. Bioresource Technology 79 (3), 251–262.

Fu, Y., Viraraghavan, T., 2002. Dye biosorption sites in *Aspergillus niger*. Bioresource Technology 82 (2), 139–145.

Fu, L., You, S.J., Zhang, G.Q., Yang, F.L., Fang, X.H., 2010. Degradation of azo dyes using in-situ Fenton reaction incorporated into H_2O_2-producing microbial fuel cell. Chemical Engineering Journal 160 (1), 164–169.

Gallagher, K.A., Healy, M.G., Allen, S.J., 1997. Biosorption of synthetic dye and metal ions from aqueous effluents using fungal biomass. Studies in Environmental Science 66, 27–50.

Gao, D.W., Hu, Q., Yao, C., Ren, N.Q., 2014. Treatment of domestic wastewater by an integrated anaerobic fluidized-bed membrane bioreactor under moderate to low temperature conditions. Bioresource Technology 159, 193–198.

Gomaa, O.M., Linz, J.E., Reddy, C.A., 2008. Decolorization of Victoria blue by the white rot fungus, *Phanerochaete chrysosporium*. World Journal of Microbiology and Biotechnology 24 (10), 2349–2356.

Hou, B., Sun, J., Hu, Y., 2011. Effect of enrichment procedures on performance and microbial diversity of microbial fuel cell for Congo red decolorization and electricity generation. Applied Microbiology and Biotechnology 90 (4), 1563–1572.

Hsueh, C.C., Wang, Y.M., Chen, B.Y., 2014. Metabolite analysis on reductive biodegradation of reactive green 19 in *Enterobacter cancerogenus* bearing microbial fuel cell (MFC) and non-MFC cultures. Journal of the Taiwan Institute of Chemical Engineers 45 (2), 436–443.

Ieropoulos, I.A., Greenman, J., Melhuish, C., Hart, J., 2005. Comparative study of three types of microbial fuel cell. Enzyme and Microbial Technology 37 (2), 238–245.

Jiang, L., Wang, J., Liang, S., Cai, J., Xu, Z., Cen, P., Yang, S., Li, S., 2011. Enhanced butyric acid tolerance and bioproduction by *Clostridium tyrobutyricum* immobilized in a fibrous bed bioreactor. Biotechnology and Bioengineering 108 (1), 31–40.

Jiang, Y., Liu, X., Chen, Y., Zhou, L., He, Y., Ma, L., Gao, J., 2014. Pickering emulsion stabilized by lipase-containing periodic mesoporous organosilica particles: A robust biocatalyst system for biodiesel production. Bioresource technology 153, 278–283.

Kalme, S.D., Parshetti, G.K., Jadhav, S.U., Govindwar, S.P., 2007. Biodegradation of benzidine based dye Direct Blue-6 by *Pseudomonas desmolyticum* NCIM 2112. Bioresource Technology 98 (7), 1405–1410.

Kapdan, I.K., Kargia, F., McMullan, G., Marchant, R., 2000. Effect of environmental conditions on biological decolorization of textile dyestuff by *C. versicolor*. Enzyme and Microbial Technology 26 (5–6), 381–387.

Kasinath, A., Novotný, Č., Svobodová, K., Patel, K.C., Šašek, V., 2003. Decolorization of synthetic dyes by *Irpex lacteus* in liquid cultures and packed-bed bioreactor. Enzyme and Microbial Technology 32 (1), 167–173.

Khan, R., Fulekar, M.H., 2017. Mineralization of a sulfonated textile dye Reactive Red 31 from simulated wastewater using pellets of *Aspergillus bombycis*. Bioresources and Bioprocessing 4 (1), 23.

Khan, S., Malik, A., 2015. Degradation of Reactive Black 5 dye by a newly isolated bacterium *Pseudomonas entomophila* BS1. Canadian Journal of Microbiology 62 (3), 220–232.

Khelifi, E., Gannoun, H., Touhami, Y., Bouallagui, H., Hamdi, M., 2008. Aerobic decolourization of the indigo dye-containing textile wastewater using continuous combined bioreactors. Journal of Hazardous Materials 152 (2), 683–689.

Kiely, P.D., Call, D.F., Yates, M.D., Regan, J.M., Logan, B.E., 2010. Anodic biofilms in microbial fuel cells harbor low numbers of higher-power-producing bacteria than abundant genera. Applied Microbiology and Biotechnology 88 (1), 371–380.

Kim, T.H., Lee, Y., Yang, J., Lee, B., Park, C., Kim, S., 2004. Decolorization of dye solutions by a membrane bioreactor (MBR) using white-rot fungi. Desalination 168, 287−293.

Kumar, C.G., Mongolla, P., Joseph, J., Sarma, V.U.M., 2012. Decolorization and biodegradation of triphenylmethane dye, brilliant green, by *Aspergillus* sp. isolated from Ladakh, India. Process Biochemistry 47 (9), 1388−1394.

Kunamneni, A., Camarero, S., García-Burgos, C., Plou, F.J., Ballesteros, A., Alcalde, M., 2008. Engineering and applications of fungal laccases for organic synthesis. Microbial Cell Factories 7 (1), 32.

Lade, H., Kadam, A., Paul, D., Govindwar, S., 2015. Biodegradation and detoxification of textile azo dyes by bacterial consortium under sequential microaerophilic/aerobic processes. EXCLI Journal 14, 158.

Li, W.W., Yu, H.Q., He, Z., 2014. Towards sustainable wastewater treatment by using microbial fuel cells-centered technologies. Energy and Environmental Science 7 (3), 911−924.

Li, Z., Zhang, X., Lin, J., Han, S., Lei, L., 2010. Azo dye treatment with simultaneous electricity production in an anaerobic−aerobic sequential reactor and microbial fuel cell coupled system. Bioresource Technology 101 (12), 4440−4445.

Liu, L., Li, F.B., Feng, C.H., Li, X.Z., 2009. Microbial fuel cell with an azo-dye-feeding cathode. Applied Microbiology and Biotechnology 85 (1), 175.

Lu, L., Zhao, M., Liang, S.C., Zhao, L.Y., Li, D.B., Zhang, B.B., 2009. Production and synthetic dyes decolourization capacity of a recombinant laccase from Pichia pastoris. Journal of Applied Microbiology 107 (4), 1149−1156.

Mahmoud, M.S., Mostafa, M.K., Mohamed, S.A., Sobhy, N.A., Nasr, M., 2017. Bioremediation of red azo dye from aqueous solutions by *Aspergillus niger* strain isolated from textile wastewater. Journal of Environmental Chemical Engineering 5 (1), 547−554.

Manekar, P., Patkar, G., Aswale, P., Mahure, M., Nandy, T., 2014. Detoxifying of high strength textile effluent through chemical and bio-oxidation processes. Bioresource Technology 157, 44−51.

Mbuligwe, S.E., 2005. Comparative treatment of dye-rich wastewater in engineered wetland systems (EWSs) vegetated with different plants. Water Research 39 (2−3), 271−280.

Méndez-Paz, D., Omil, F., Lema, J.M., 2005. Anaerobic treatment of azo dye Acid Orange 7 under fed-batch and continuous conditions. Water Research 39 (5), 771−778.

Miran, W., Jang, J., Nawaz, M., Shahzad, A., Lee, D.S., 2018. Sulfate-reducing mixed communities with the ability to generate bioelectricity and degrade textile diazo dye in microbial fuel cells. Journal of Hazardous Materials 352, 70−79.

Moosvi, S., Madamwar, D., 2007. An integrated process for the treatment of CETP wastewater using coagulation, anaerobic and aerobic process. Bioresource Technology 98 (17), 3384−3392.

Mu, W., Liu, F., Jia, J., Chen, C., Zhang, T., Jiang, B., 2009. 3-Phenyllactic acid production by substrate feeding and pH-control in fed-batch fermentation of Lactobacillus sp. SK007. Bioresource technology 100 (21), 5226−5229.

Munck, C., Thierry, E., Gräßle, S., Chen, S.H., Ting, A.S.Y., 2018. Biofilm formation of filamentous fungi *Coriolopsis* sp. on simple muslin cloth to enhance removal of triphenylmethane dyes. Journal of Environmental Management 214, 261−266.

Narayanasamy, S., Jayaprakash, J., 2018. Improved performance of *Pseudomonas aeruginosa* catalyzed MFCs with graphite/polyester composite electrodes doped with metal ions for azo dye degradation. Chemical Engineering Journal 343, 258−269.

Niessen, J., Schröder, U., Scholz, F., 2004. Exploiting complex carbohydrates for microbial electricity generation−a bacterial fuel cell operating on starch. Electrochemistry Communications 6 (9), 955−958.

Nimje, V.R., Chen, C.Y., Chen, C.C., Jean, J.S., Reddy, A.S., Fan, C.W., Pan, K.Y., Liu, H.T., Chen, J.L., 2009. Stable and high energy generation by a strain of *Bacillus subtilis* in a microbial fuel cell. Journal of Power Sources 190 (2), 258−263.

O'mahony, T., Guibal, E., Tobin, J.M., 2002. Reactive dye biosorption by *Rhizopus arrhizus* biomass. Enzyme and Microbial Technology 31 (4), 456−463.

Ong, S.A., Uchiyama, K., Inadama, D., Yamagiwa, K., 2009. Simultaneous removal of color, organic compounds and nutrients in azo dye-containing wastewater using up-flow constructed wetland. Journal of Hazardous Materials 165 (1−3), 696−703.

Ozer, D.J., Benet-Martinez, V., 2006. Personality and the prediction of consequential outcomes. Annual Review of Psychology 57, 401−421.

Park, D.H., Laivenieks, M., Guettler, M.V., Jain, M.K., Zeikus, J.G., 1999. Microbial utilization of electrically reduced neutral red as the sole electron donor for growth and metabolite production. Applied and Environmental Microbiology 65 (7), 2912—2917.

Patel, V.R., Bhatt, N.S., Bbhatt, H., 2013. Involvement of ligninolytic enzymes of *Myceliophthora vellerea* HQ871747 in decolorization and complete mineralization of Reactive Blue 220. Chemical Engineering Journal 233, 98—108.

Pereira, L., Coelho, A.V., Viegas, C.A., Ganachaud, C., Iacazio, G., Tron, T., Robalo, M.P., Martins, L.O., 2009. On the mechanism of biotransformation of the anthraquinonic dye acid blue 62 by laccases. Advanced Synthesis and Catalysis 351 (11-12), 1857—1865.

Pham, C.A., Jung, S.J., Phung, N.T., Lee, J., Chang, I.S., Kim, B.H., Yi, H., Chun, J., 2003. A novel electrochemically active and Fe (III)-reducing bacterium phylogenetically related to *Aeromonas hydrophila*, isolated from a microbial fuel cell. FEMS Microbiology Letters 223 (1), 129—134.

Pophali, G.R., Kaul, S.N., Mathur, S., 2003. Influence of hydraulic shock loads and TDS on the performance of large-scale CETPs treating textile effluents in India. Water Research 37 (2), 353—361.

Rahimnejad, M., Adhami, A., Darvari, S., Zirepour, A., Oh, S.E., 2015. Microbial fuel cell as new technology for bioelectricity generation: a review. Alexandria Engineering Journal 54 (3), 745—756.

Raj, K., 2007. Relocating Modern Science: Circulation and the Construction of Knowledge in South Asia and Europe. Springer, pp. 1650—1900.

Rathi, A.K.A., 2013. Common environmental infrastructure: case study on the management of common effluent treatment plants. International Journal of Environmental Engineering 5 (1), 93—110.

Rezaei, F., Xing, D., Wagner, R., Regan, J.M., Richard, T.L., Logan, B.E., 2009. Simultaneous cellulose degradation and electricity production by *Enterobacter cloacae* in a microbial fuel cell. Applied and Environmental Microbiology 75 (11), 3673—3678.

Richter, H., McCarthy, K., Nevin, K.P., Johnson, J.P., Rotello, V.M., Lovley, D.R., 2008. Electricity generation by *Geobacter sulfurreducens* attached to gold electrodes. Langmuir 24 (8), 4376—4379.

Rodríguez-Couto, S., Osma, J.F., Toca-Herrera, J.L., 2009. Removal of synthetic dyes by an eco-friendly strategy. Engineering in Life Sciences 9 (2), 116—123.

Saeed, T., Sun, G., 2012. A review on nitrogen and organics removal mechanisms in subsurface flow constructed wetlands: dependency on environmental parameters, operating conditions and supporting media. Journal of Environmental Management 112, 429—448.

Saratale, R.G., Gandhi, S.S., Purankar, M.V., Kurade, M.B., Govindwar, S.P., Oh, S.E., Saratale, G.D., 2013. Decolorization and detoxification of sulfonated azo dye CI Remazol Red and textile effluent by isolated *Lysinibacillus* sp. RGS. Journal of Bioscience and Bioengineering 115 (6), 658—667.

Saratale, R.G., Saratale, G.D., Chang, J.S., Govindwar, S.P., 2009. Ecofriendly degradation of sulfonated diazo dye CI Reactive Green 19A using *Micrococcus glutamicus* NCIM-2168. Bioresource Technology 100 (17), 3897—3905.

Sarayu, K., Sandhya, S., 2012. Current technologies for biological treatment of textile wastewater—a review. Applied Biochemistry and Biotechnology 167 (3), 645—661.

Saroj, S., Kumar, K., Pareek, N., Prasad, R., Singh, R.P., 2014. Biodegradation of azo dyes acid red 183, direct blue 15 and direct red 75 by the isolate Penicillium oxalicum SAR-3. Chemosphere 107, 240—248.

Satiroğlu, N., Yalçınkaya, Y., Denizli, A., Arıca, M.Y., Bektas, S., Genç, Ö., 2002. Application of NaOH treated *Polyporus versicolor* for removal of divalent ions of Group IIB elements from synthetic wastewater. Process Biochemistry 38 (1), 65—72.

Shaheen, R., Asgher, M., Hussain, F., Bhatti, H.N., 2017. Immobilized lignin peroxidase from Ganoderma lucidum IBL-05 with improved dye decolorization and cytotoxicity reduction properties. International Journal of Biological Macromolecules 103, 57—64.

Sharma, P., Singh, L., Dilbaghi, N., 2009. Biodegradation of Orange II Dye by *Phanerochaete chrysosporium* in Simulated Wastewater.

Sharma, S.K., Leblanc, R.M., 2017. Biosensors based on β-galactosidase enzyme: recent advances and perspectives. Analytical Biochemistry 535, 1—11.

Shedbalkar, U., Dhanve, R., Jadhav, J., 2008. Biodegradation of triphenylmethane dye cotton blue by Penicillium ochrochloron MTCC 517. Journal of Hazardous Materials 157 (2—3), 472—479.

Spadaro, J.T., Gold, M.H., Renganathan, V., 1992. Degradation of azo dyes by the lignin-degrading fungus *Phanerochaete chrysosporium*. Applied and Environmental Microbiology 58 (8), 2397—2401.

Spagni, A., Grilli, S., Casu, S., Mattioli, D., 2010. Treatment of a simulated textile wastewater containing the azo-dye reactive orange 16 in an anaerobic-biofilm anoxic–aerobic membrane bioreactor. International Biodeterioration and Biodegradation 64 (7), 676–681.

Sun, J., Hu, Y.Y., Bi, Z., Cao, Y.Q., 2009. Simultaneous decolorization of azo dye and bioelectricity generation using a microfiltration membrane air-cathode single-chamber microbial fuel cell. Bioresource Technology 100 (13), 3185–3192.

Sun, J., Peng, R.H., Xiong, A.S., Tian, Y., Zhao, W., Xu, H., Liu, D.T., Chen, J.M., Yao, Q.H., 2012. Secretory expression and characterization of a soluble laccase from the *Ganoderma lucidum* strain 7071-9 in *Pichia pastoris*. Molecular Biology Reports 39 (4), 3807–3814.

Taher, H., Al-Zuhair, S., 2017. The use of alternative solvents in enzymatic biodiesel production: a review. Biofuels, Bioproducts and Biorefining 11 (1), 168–194.

Talha, M.A., Goswami, M., Giri, B.S., Sharma, A., Rai, B.N., Singh, R.S., 2018. Bioremediation of Congo red dye in immobilized batch and continuous packed bed bioreactor by *Brevibacillus parabrevis* using coconut shell bio-char. Bioresource Technology 252, 37–43.

Teerapatsakul, C., Parra, R., Keshavarz, T., Chitradon, L., 2017. Repeated batch for dye degradation in an airlift bioreactor by laccase entrapped in copper alginate. International Biodeterioration and Biodegradation 120, 52–57.

Torrijos, V., Gonzalo, O.G., Trueba-Santiso, A., Ruiz, I., Soto, M., 2016. Effect of by-pass and effluent recirculation on nitrogen removal in hybrid constructed wetlands for domestic and industrial wastewater treatment. Water Research 103, 92–100.

Van der Zee, F.P., Lettinga, G., Field, J.A., 2001. Azo dye decolourisation by anaerobic granular sludge. Chemosphere 44 (5), 1169–1176.

Wang, H.F., 2017. Theory of Linear Poroelasticity with Applications to Geomechanics and Hydrogeology. Princeton University Press.

Watson, V.J., Logan, B.E., 2010. Power production in MFCs inoculated with *Shewanella oneidensis* MR-1 or mixed cultures. Biotechnology and Bioengineering 105 (3), 489–498.

Won, S.W., Choi, S.B., Yun, Y.S., 2005. Interaction between protonated waste biomass of *Corynebacterium glutamicum* and anionic dye Reactive Red 4. Colloids and Surfaces A: Physicochemical and Engineering Aspects 262 (1–3), 175–180.

Wong, Y., Yu, J., 1999. Laccase-catalyzed decolorization of synthetic dyes. Water Research 33 (16), 3512–3520.

Xiang, D., Wang, X., Jia, C., Lee, T., Guo, X., 2016. Molecular-scale electronics: from concept to function. Chemical Reviews 116 (7), 4318–4440.

Zhang, E., Xie, Y., Ci, S., Jia, J., Wen, Z., 2016. Porous Co_3O_4 hollow nanododecahedra for nonenzymatic glucose biosensor and biofuel cell. Biosensors and Bioelectronics 81, 46–53.

Zhang, G., Campbell, E.A., Minakhin, L., Richter, C., Severinov, K., Darst, S.A., 1999. Crystal structure of *Thermus aquaticus* core RNA polymerase at 3.3 Å resolution. Cell 98 (6), 811–824.

Zheng, Z., Levin, R.E., Pinkham, J.L., Shetty, K., 1999. Decolorization of polymeric dyes by a novel *Penicillium* isolate. Process Biochemistry 34 (1), 31–37.

Zhong, J.J., 2010. Recent advances in bioreactor engineering. Korean Journal of Chemical Engineering 27 (4), 1035–1041.

Zhou, J.L., Banks, C.J., 1991. Removal of humic acid fractions by *Rhizopus arrhizus*: uptake and kinetic studies. Environmental Technology 12 (10), 859–869.

Zhou, J.L., Banks, C.J., 1993. Mechanism of humic acid colour removal from natural waters by fungal biomass biosorption. Chemosphere 27 (4), 607–620.

6

Mycoremediation of polycyclic aromatic hydrocarbons

Shalini Gupta, Bhawana Pathak

School of Environment and Sustainable Development, Central University of Gujarat, Gandhinagar, India

1. Introduction

The pervasive exploitation, transportation, and consumption of crude oil and petroleum products have enticed concern over adverse effect of hazardous compounds. Crude oil contains a complex mixture of aliphatic and aromatic compounds such as polycyclic aromatic hydrocarbons (PAHs). PAHs are hazardous organic compounds and major concern to human health and environment. PAHs possess mutagenic, carcinogenic threat to microorganisms, plants, and animals because of persistence and low aqueous solubility (Williams et al., 2013; Cao et al., 2009; Marchand et al., 2017). In spite of several advance available treatment technologies, there are many drawbacks related to generation of secondary toxic pollutants from breakdown of these complex hydrocarbon compounds. Thus, moving toward environmentally sound treatment technology, bioremediation studies have been focused. Bioremediation is a biological approach where living microbes are used for removal of hazardous compounds. Research studies on the bioremediation competences of bacteria to transform PAHs into nontoxic compounds are well-documented. Bacteria transform PAHs using intracellular enzymatic system which helps in mineralization of complex hydrocarbon. Though there are few studies that deal with indigenous fungi in PAH-contaminated sites, the mechanisms and pathways convoluted in the breakdown of PAHs (Ritz and Young, 2004; Godoy et al., 2016). Fungal bioremediation may also be called as mycoremediation because it is a promising approach for the efficient breakdown of PAHs because of its extracellular and intracellular enzymatic system. PAH-contaminated sites are commonly characterized by restricted microbial diversity. Fungi are extremely tolerant under antagonistic environmental conditions, possibly because of high selective pressure, and constitute an influential device for pollutant bioconversion (Margesin and Schinner, 2001).

Abatement of Environmental Pollutants
https://doi.org/10.1016/B978-0-12-818095-2.00006-0

Fungi degrading PAHs generally belong to the division Ascomycota and Basidiomycota (Hibbet et al., 2007; Harms et al., 2011; Godoy et al., 2016).

Extensively studied lignin degrading white-rot fungi are well recognized to be capable to mineralize PAHs through ligninolytic enzymatic activity (Cerniglia, 1993). However, nonligninolytic fungi, such as *Cunninghamella*, *Aspergillus,* and *Penicillium*, have been found to degrade aromatic compounds, aliphatic hydrocarbons, and PAHs through cytochrome P450 monooxygenases enzymes (Pinedo- Rivilla et al., 2009; Marco-Urrea et al., 2015; Godoy et al., 2016). Hence this chapter includes fundamental of mycoremediation, mechanism of fungal degradation, ligninolytic and nonligninolytic fungal enzymatic system, biosurfactant production capacity, and factors affecting fungal growth.

1.1 PAHs: environmental concern

PAHs are worldwide environmental concern because of their recalcitrance and adverse effect on environment and human health. PAHs are produced by natural emissions such as wood fires or volcano eruptions. Anthropogenic emissions mainly derive from combustion processes: industrial processes, refining processes such as coking for coal and cracking for petroleum products such as tar, waxes, oils, and indoor sources such as fireplaces, tobacco smoke. PAHs are natural constituents of fossil fuels, coal and petroleum comprising 0.2% and 7% PAHs (National Research Council, 2003). PAHs adsorb to dust or soot particles and enter into the atmosphere and transported over long ranges because of their persistence. In a cyclic process, PAHs return to the surface of the earth via rain or fog, deposit on soil and plants, and percolate in surface waters (Quiroz et al., 2010). PAHs mainly enter into the environment through dusts produced by coal mining, vehicular exhaust, transportation, and drilling of oil. Stock piles and tailings also consist PAHs and contaminate soil, water, and groundwater (Nikitha et al., 2017). There are list of 16 PAH compounds classified as mutagenic and carcinogenic by the USEPA (Table 6.1).

1.2 Effect of PAHs exposure on environment and human health

PAHs are ubiquitously present in the environment mainly evaporated into the atmosphere. PAHs undergo photolysis in presence of sunlight, mainly when adsorbed to dust particles. Oxidation of PAHs can break down the complex compound structure in days or week (Santodonato, 1981). PAH compounds are hydrophobic, immiscible in water and adsorbed on dust, and precipitate in sediments of aquatic water bodies, else they are miscible in any hydrophobic matter which may contaminate aquatic body. Microbes inhabited in terrestrial and water system possess adaptability to degrade and mineralize PAHs over longer or shorter time duration (ATSDR, 2010).

The PAH metabolites are usually more toxic in the existence of UV light. PAHs in soil are improbable to employ toxicity influence on terrestrial invertebrates (Peter, 2003). PAHs get absorbed to plants from roots from soils and translocate contaminant to rest of the plant parts. Mobility of these contaminants is commonly ruled by dose, solubility, and other physicochemical properties such as nature of soil. Some plant species contains constituents

TABLE 6.1 Properties of polycyclic aromatic hydrocarbons (PAHs).

S.no	PAH compound	Density	Melting point	Boiling point
1	Acenaphthylene	0.8987 g cm−3	91.8 °C (197.2°F; 364.9 K)	280 °C (536°F; 553 K)
2	Fluoranthene	1.252 g/cm3 (0 °C), solid	110.8 °C (231.4°F; 383.9 K)	375 °C (707°F; 648 K)
3	Indeno[1,2,3-cd]pyrene	NA	320−325°F/164°C	997 degrees F at 760 mm Hg/536°C
4	Pyrene	1.271 g/mL	145−148 °C (293−298°F; 418−421 K)	404 °C (759°F; 677 K)
5	Naphthalene	1.145 g/cm3 (15.5 °C) 1.0253 g/cm3 (20 °C) 0.9625 g/cm3 (100 °C)	78.2 °C (172.8°F; 351.3 K) 80.26 °C (176.47°F; 353.41 K) at 760 mmHg	217.97 °C (424.35°F; 491.12 K) at 760 mmHg
6	Benz[a]anthracene	1.19 g/cm3	158 °C (316°F; 431 K)	438 °C (820°F; 711 K)
7	Acenaphthylene, 1,2-dihydro-	NA	93-95°C	NA
8	Dibenz[a,h]anthracene	1.232 g/cm3	262 °C (504°F; 535 K	NA
9	Phenanthrene	1.18 g/cm3	101 °C (214°F; 374 K)	332°C (630°F; 605 K)
10	Anthracene	1.28 g/cm3 (25 °C) 0.969 g/cm3 (220 °C)	215.76 °C (420.37°F; 488.91 K) at 760 mmHg	339.9 °C (643.8°F; 613.0 K) at 760 mmHg
11	Benz[e] acephenanthrylene	1.286 g/cm3	166 °C (331°F; 439 K)	481 °C (898°F; 754 K)
12	9H-Fluorene	1.202 g/ML	116−117 °C (241−243°F; 389−390 K)	295 °C (563°F; 568 K)
13	Benzo[k]fluoranthene	1.286 g/cm3	217 °C (423°F; 490 K)	NA
14	Chrysene	1.274 g/cm3	254 °C (489°F; 527 K)	448 °C (838°F; 721 K)
15	Benzo[ghi]perylene	1.378 g/cm3	278 °C (532°F; 551 K)	500 °C (932°F; 773 K)
16	Benzo[a]pyrene	1.24 g/cm3 (25 °C)	179 °C (354°F; 452 K)	495 °C (923°F; 768 K)

NA, not available.

which may guard against toxic consequence of PAHs; however, some plants could synthesize PAHs and perform as growth hormones (ATSDR, 2010; Beyer et al., 2010). PAH bioaccumulation has been observed in terrestrial invertebrates because of persistence and longer half-life shellfish expected to consist much higher concentration of PAH than in the environment. Nevertheless, metabolism of PAHs is sufficient to prevent biomagnifications (Tudoran and Putz, 2012; Inomata et al., 2012; Abdel-Shafy and Mansour, 2016; Borosky, 1999). Organisms are adversely affected because of tumors, reproduction, growth development, and immunity. PAH absorption occurs in mammals by inhalation, dermal contact, and ingestion (Dong et al., 2012; Veltman et al., 2012; Beyer et al., 2010).

1.3 Bioremediation approach

Bioremediation is an ecologically viable technique that aids in cleanup of contaminated site by using natural biological (microbes) activity. It is a relatively low-cost, low-technology technique and can often be carried out on site. For the successful implementation of bioremediation process, a brief assessment of a contaminated site is prerequisite to augment environmental settings to accomplish a desirable effect. Bioremediation technique could be upgraded through better knowledge and skill, and no suspicion on bioremediation process having immense potential for allocating with different contaminated sites. But the principles, advantages, and disadvantages of this technology are not extensively known (King et al., 1997; NRC, 1993; Norris et al., 1993; Hinchee et al., 1995), and it is required to look at relevant field application case histories of bioremediation (Flathman et al., 1993; Vidali, 2001). Bacteria-mediated PAH degradation is mainly owing to intracellular dioxygenase enzymatic system (Johnsen et al., 2005). Bacteria utilizes PAH compounds as source for the proliferation and results into breakdown of PAH compound (Kästner et al., 1994; Haderlein et al., 2006). Bacteria-mediated PAH degradation is begun through oxygenation of complex aromatic compound and formation of cis-dihydrodiol followed by dehydrogenation reaction which results into formation of dihydroxylated metabolites. Bacterial enzymatic route for PAH degradation and genes encoding for respective enzymes are identified. Varied groups of Gram-positive and Gram-negative bacteria are identified for PAHs degradation, and specific genes encoding specific enzymes are fairly different (Cebron et al., 2008; Erika et al., 2014); *Burkholderia* (Gram-negative bacteria) can degrade 2- or 3-ring PAH such as naphthalene and anthracene, whereas *Mycobacterium* (Gram-positive bacteria) are more capable in degrading complex PAHs, for example, flouroanthene, benzo-alpha-pyrene (Johnsen et al., 2005; Erika et al., 2014).

2. Mycoremediation: intact potential

Mycoremediation PAH breakdown using fungi is still less explored than bacteria. Fungi are naturally exiled microorganisms in competition with bacteria and are more apt to the polluted site. Fungi share 75% of the total soil microbial biomass; bacteria and fungi are immobilized at <-50 kPa matric potential, and continuous water phases are not required by fungi for active dispersal. Fungal hyphae develop through air—water interfaces, connecting soil pores with a diameter ranges from 2 to 7 µm (Bornyasz et al., 2005). Mycelia morphology reveals an operative rummaging approach; fungi represent massive growth and linking scattered extension in poor nutrient conditions also (Ritz and Young, 2004). Mycelia affect soil structure by electrostatic, adhesive, entrapment mechanisms, and organic matter decomposition (Rillig and Mummey, 2006; Read and Perez-Moreno, 2003). Fungi are exudation of oxidative exoenzymes at their continuously expansion of hyphal tips. Saprobic fungi using cellulose and lignin play significant roles in biogeochemical cycling (Bebber et al., 2007; Tero et al., 2010). Both diffusive and active translocation of compounds occurs in fungal hyphae (Govindarajulu et al., 2005; Darrah et al., 2006). The aquatic and terrestrial inhabitant fungi are heavily exposed to hazardous chemicals, resemble as heterotrophic chemical substrates, and metals may be utilized as micronutrients for fungal growth in natural conditions.

Fungal-mediated PAH metabolism is classified into two major classes: nonligninolytic and ligninolytic fungi. Nonligninolytic eukaryotes metabolize PAH compounds, using P_{450} monooxygenase, to hydroxylization of metabolites can be ejected directly as conjugates and convert into more polar water soluble molecules (Gibson and Subramanian, 1984). For example, *Cunninghamella* sp. utilize cytochrome P450 systems to oxidize naphthalene, phenanthrene, anthracene, benz[a]anthracene, 3-methylcholanthrene, and benzo[a]pyrene (Cerniglia and Gibson, 1979; Cerniglia and Yang, 1984; Gibson and Subramanian, 1984). Aliphatic hydrocarbons, chlorophenols, PAHs, and 2,4,6-trinitrotoluene (TnT) are degraded by *Aspergillus* sp. and *Penicillium* sp (Cerniglia and Sutherland, 2010; Pinedo-Rilla et al., 2009; Hofrichter et al., 1994; Scheibner et al., 1997; Prince, 2010). Ligninolytic fungi are active to wide array of organic compounds because of release of extracellular lignin transforming enzymes and act on various organic compounds which broadly show resemblance to lignin (Adenipekun and Lawal, 2012; Rhodes, 2014). Enzymes degrading lignin comprise mainly lignin peroxidase (LiP), manganese peroxidase (MnP), versatile peroxidases, and laccase (Kirk and Farrell, 1987; Rhodes, 2014). *Phanerochaete chrysosporium* (white-rot basidiomycete) transform phenanthrene to trans-dihydrodiols and phenanthrol conjugates cultivated under nonligninolytic conditions in nutrient-enriched medium (Sutherland et al., 1991; Kenneth and Hammel, 1995). Partial knowledge of the techniques and environmental approaches vital for adequate fungal biomass and enzymatic activity in PAH contaminated sites is still a pronounced intrusion to mycoremediation process (Harms et al., 2011).

2.1 Ligninolytic fungi

Ligninolytic fungi are found to be more efficient in the degradation of PAHs as compared with bacteria (Davis et al., 1993). Ligninolytic fungi are taxonomically heterogeneous higher fungi characterized by exceptional ability to depolymerize lignin. They comprise wood and soil inhabiting basidiomycetes and ascomycetes (Natalia, 2017). In ligninolytic conditions, white-rot fungi can oxidize PAHs via generating hydroxyl free radicals and donation of one electron results into PAH quinones and acids (Sutherland et al., 1995). Degradation of PAHs by ligninolytic fungi (Table 6.2, Figs 6.1 and 6.2) has reported that PAHs probably degraded by an incorporation of epoxide hydrolases, cytochrome P450 monooxygenases, and ligninolytic enzymes, which led to complete mineralization of the hazardous compound (Bezalel et al., 1997). Ability of white-rot fungi to degrade and mineralize diversified recalcitrant compounds such as organochlorines, polychlorinated biphenyls, PAHs, synthetic dyes, wood preservatives, and synthetic polymers because of the nonspecificity of their enzyme machinery (Pointing, 2001). White-rot fungi degrade lignin by the action of ligninolytic enzymes that are nonspecific and activate through radical reactions (Hatakka, 2001; Tuomela and Hatakka, 2011). The ligninolytic enzymes are extracellular in nature and they play vital role in the breakdown of complex PAHs in soil, resulting into more water-soluble intermediates, end products, and its bioavailability (Sack et al., 1997a; Harms et al., 2011). The metabolites formed by the activity of extracellular enzymes could either be used as substrate by many bacteria or be passed to intracellular enzymes CP-450 monooxygenase for further degradation (Sack et al., 1997a; Pozdnyakova, 2012). *Irpex lacteus* showed the degradation of several PAH compounds and resultant metabolites by the action of extracellular lignin-degrading enzymes and intracellular monooxygenase (Cajthaml et al., 2002, 2006).

TABLE 6.2 Polycyclic aromatic hydrocarbon (PAH) compounds degradation by ligninolytic fungi.

Ligninolytic fungi	PAH Compound	Metabolite	Reference
Pleurotus ostreatus	Phenanthrene	3% CO_2, 52% trans-9,10-dihydroxy-9, 10-dihydrophenanthrene (phenanthrene trans-9,10-dihydrodiol) (28%), 2,2*-diphenic acid (17%), and unidentified metabolites (7%). 35% Nonextractable metabolites	Bezalel et al. (1996)
Phanerochaete chrysosporium	[14C]-phenanthrene	14CO$_2$.	John and Bumpus (1989)
Trametes versicolor	Fluoranthene, anthracene, and phenanthrene	Quinone	Pozdnyakova et al. (2018)
Ganoderma lucidum	Anthracene, benzo[a]pyrene, fluorine, acenaphthene, and benzo[a]anthracene	NA	Hunsa et al. (2009)
Nematoloma frowardii	Phenanthrene, anthracene, pyrene, fluoranthene, benzo[a]pyrene	NA	Sack et al. (1997a), b

NA, not available.

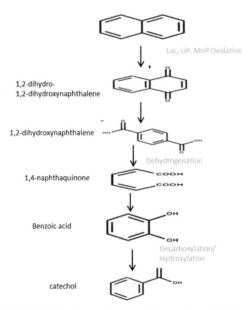

FIGURE 6.1   Degradation pathway of naphthalene.

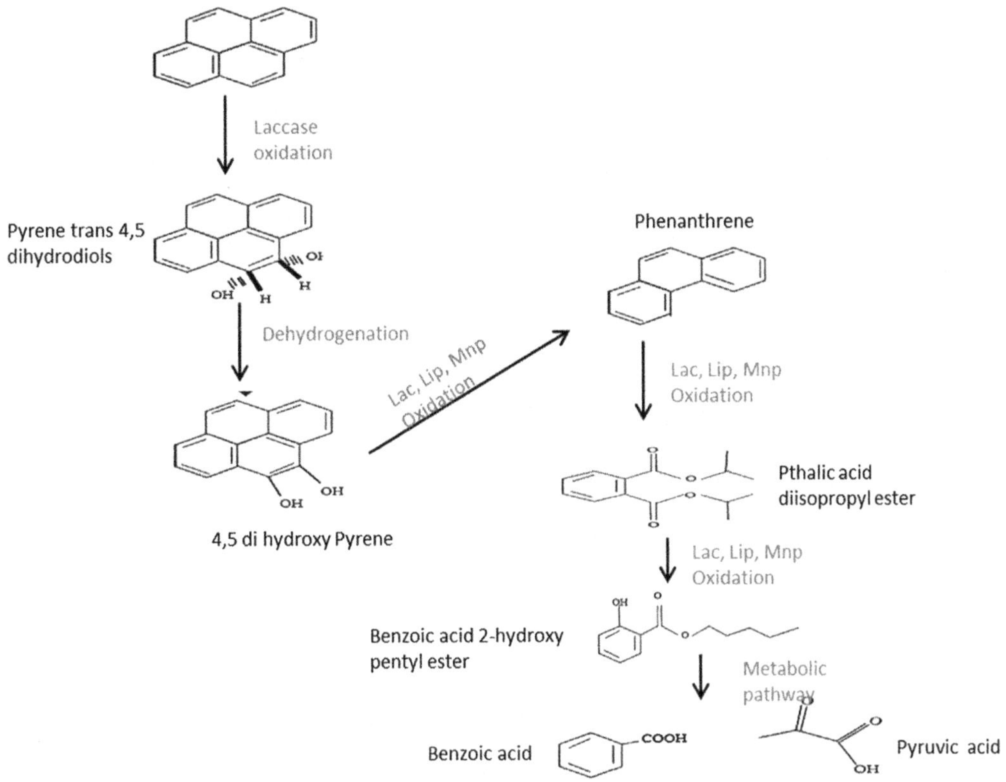

FIGURE 6.2   Degradation pathway of pyrene.

Benzo[a]pyrene and pyrene degraded by *Pleurotus ostreatus* and *P. chrysosporium* with combination of nonligninolytic and ligninolytic enzymes probably are the keys to the complete mineralization intractable compounds (Bezalel et al., 1997). Combination of white-rot fungi and brown-rot fungi, *Antrodia vaillantii* and *P. ostreatus,* could be used to degrade larger ring structured PAH compounds (Andersson et al., 2003; Al-Hawash et al., 2018).

2.2 Nonligninolytic fungi

Nonligninolytic fungi are also involved in PAH metabolism by oxidation of PAHs via intracellular P450 monooxygenase enzyme—stimulated response for configuration of arene oxide similar to mammalian PAH metabolism (Table 6.3; Fig 6.3, Sutherland et al., 1995). The monooxygenase enzyme involves incorporation of single oxygen molecule in compound for configuration of epoxide hydrolase and transform to trans-dihydrodiols (Jerina and Brodie, 1983). Nonenzymatic reorganization of parent compound as substrates leads to formation of phenol derivatives and further methylation, sulfation, conjugation with glucuronic acid forms xylose, or glucose (Cerniglia and Sutherland, 2009). Nonligninolytic fungi

TABLE 6.3   Polycyclic aromatic hydrocarbon (PAH) compounds degradation by nonligninolytic fungi.

Nonligninolytic	PAH compound	Metabolites	References
Trichoderma harzianum	Pyrene	NA	Saraswathy and Rolf (2002)
Aspergillus fumigatus	Anthracene	Phthalic anhydride, anthrone, and anthraquinone	Ye et al. (2010)
Penicillium sp.	Naphthalene (15.0%), acenaphthene	NA	Govarthanana et al. (2017)
Penicillium oxalicum	Naphthalene	NA	Kannangara et al. (2016)
Trichoderma asperellum H15	Mixture of phenanthrene, pyrene, and benzo[a]pyrene	NA	Zafra et al. (2015)
Mucor sp	benzo[a]pyrene	NA	Su et al. (2006)
Fusarium solani	benzo[a]pyrene	NA	Rafin et al. (2006)

NA, not available.

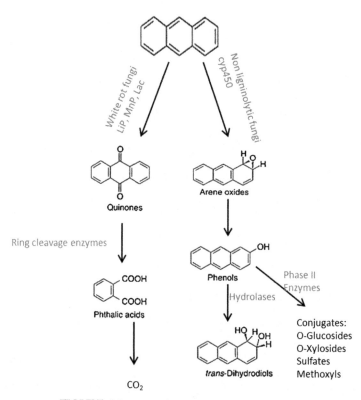

FIGURE 6.3   Degradation pathway of anthracene.

such as *Aspergillus niger, Chrysosporium pannorum*, and *Cunninghamella elegans* involve cytochrome P450 monooxygenase enzyme—mediated oxidative route for phenanthrene degradation (Al-Hawash et al., 2018). Fluorene and anthracene metabolized by *C. elegans* transformed to 9-fluorenol, 9-fluorenone, 2-hydroxy-9-fluorenone, and anthracene *trans*-1,2-dihydrodiol enantiomer and 1-anthryl sulfate, respectively (Pothuluri et al., 1993; Cerniglia et al., 1982; Cerniglia and Yang, 1984; Cerniglia and Sutherland, 2009). Phenanthrene transformation by *C. elegans* and *Syncephalastrum racemosum* result into phenanthrene *trans*-1,2-dihydrodiol, *trans*-9,10-dihydrodiol, and phenanthrene *trans*-3,4-dihydrodiol, respectively (Cerniglia and Yang, 1984; Casillas et al., 1996; Cernigilia and Sutherland, 2009). Fluoranthene degradation by *C. elegans* results into 3-fluoranthene-β-glucopyranoside, 3(8-hydroxy-fluoranthene), β-glucopyranoside, and fluoranthene *trans*-2,3-dihydrodiol metabolites considerably lower toxic than fluoranthene (Pothuluri et al., 1996).

3. Major enzymes

All fungi are heterotrophic and depend on carbon compounds synthesized by other living organisms. Fungi possess intracellular and extracellular enzymes. For breaking down of larger complex compounds such as cellulose, hemicellulose, lignin, starch, and pectin, fungi secrete extracellular enzymes. The main extracellular enzymes participating in lignin degradation are heme-containing lignin peroxidase, manganese peroxidase, and Cu-containing laccase. The extracellular enzymatic system includes nonspecific and oxidative enzymes of fungal assistance in catalysis and degradation of lignin (Dashtban et al., 2010). Lignin peroxidase, manganese peroxidase, versatile peroxidase, and laccase are extracellular enzymes produce by fungi. These nonspecific oxidative enzymes act on many persistent aromatic compounds and led to complete removal of toxic compounds (Tables 6.4 and 6.5). Enzyme synthesis is not suppressed by the concentrations of compounds. The enzymes can degrade even low concentrations of pollutants. Ligninolytic enzymes tend to produce polar and water-soluble products by catalytic action, which are more accessible for both fungal metabolism and further degradation by the natural soil microflora (Natalia, 2017). Fungal strains also possessed hydrolytic enzymes, peroxidases, and proteases (Alves et al., 2002) as follows:

3.1 Hydrolases

Hydrolytic enzymes consist of important roles in reducing the toxicity of complex organic compounds. The mechanism of hydrolytic enzymes involves disruption of major chemical bond of toxic compounds such as hydrocarbons, organophosphate, carbamate, and insecticides. Certain hazardous organic compounds like such as DDT, heptachlor are stable in aerobic conditions though get readily degraded in anaerobic condition where in hydrolases catalyzes these compounds via condensations and alcoholysis (Williams, 1977; Vasileva-Tonkova and Galabova, 2003; Lal and Saxena, 1982). Hydrolases are readily available and are tolerant to hydrophilic solvents. Hydrolases classified as proteases, cellulases, and lipases are discussed below.

TABLE 6.4 Role of major enzymes during polycyclic aromatic hydrocarbon (PAH) degradation by ligninolytic fungi.

PAH compound	Ligninolytic fungi	Enzymes	Metabolites	References
Phenanthrene	*Phanerochaete chrysosporium*	Monooxygenase; epoxide hydrolase	PHE-trans-9,10-dihydrodiol; PHE-trans-3,4-dihydrodiol; 9-phenantrol, 3-phenanthrol; 4-phenanthrol; 9-phenanthryl-D-glucopyranoside, CO_2	Sutherland et al. (1991); Pozdnyakova (2012)
Phenanthrene	*Phanerochaete sordida*	Manganese peroxidase (MnP)	NA	Lee et al. (2010); Pozdnyakova (2012)
Phenanthrene	*Ganoderma lucidum*	LAC	NA	Ting et al. (2011)
Anthracene	*Pleurotus ostreatus*	MnP; LAC	9,10-Anthraquinone	(Johannes et al., 1996; Collins et al., 1996; Schützendübel et al., 1999; Vyas et al., 1994; Pozdnyakova, 2012)
Benzo[a]pyrene	*P. chrysosporium*	Lignin peroxidase; MnP	Quinone metabolite; CO_2	Steffen et al. (2003)
Fluoranthene	*P. sordida*	MnP	NA	Lee et al. (2010); Pozdnyakova (2012)
Anthracene	*Stropharia coronilla*	MnP	NA	Steffen et al. (2003)

NA, not available.

TABLE 6.5 Role of major enzymes during polycyclic aromatic hydrocarbon (PAH) degradation by nonligninolytic fungi.

PAH Compound	Nonligninolytic Fungi	Enzymes	Metabolites	References
Pyrene	*Penicillium janthinellum*	Cytochrome P450 monooxygenase	1-Pyrenol, followed by 1,6- and 1,8-pyrene quinones	Launen et al. (1995)
Naphthalene, acenaphthene, and Benzo[a]pyrene	*Penicillium* sp. CHY-2	Manganese peroxidase	NA	Govarthanan et al. (2017)
Naphthalene	*Penicillium fastigiata* and *Penicillium digitatum*	Laccase	NA	Simanjorang and Subowo (2018)
Phenanthrene, anthracene, pyrene, fluorene, and fluoranthene	*Penicillium* sp. M 1	Extracellular peroxidases	NA	Sack and Günther (1993)

NA, not available.

3.1.1 Proteases

Proteases belong to group of hydrolases in which it aids in breakdown of peptide bonds in aqueous phase and synthesis in nonaqueous phase. Proteases cover wide group of industrial applications such as in food, leather, detergent, and pharmaceutical fields (Singh, 2003; Beena and Geevarghese, 2010). A great number of fungal strains have been used to produce proteases belonging to the genera *Aspergillus, Penicillium, Rhizopus, Mucor, Trichoderma reesei* QM9414, among others (Rajmalwar and Dabholkar, 2009; Germano et al., 2003; Andrade et al., 2002; Dienes et al., 2007; de Souza et al., 2015).

3.1.2 Cellulases

Some organisms produce membrane-bound extracellular cellulases. Extracellular cellulases are constitutively present at very low concentrations by some microorganisms (Rixon et al., 1992; Bennet et al., 2002; Adriano-Anaya et al., 2005). Cellulases are involved in the hydrolysis process, mainly (1) endoglucanase acts on the regions of low crystallinity in the cellulose fiber and produces free chain ends; (2) exoglucanase degrades the cellulose molecule and removes cellobiose units from the free chain ends; (3) β-glucosidase hydrolyzes cellobiose to glucose (Adriano-Anaya et al., 2005). Some studied cellulolytic fungi are *Aspergillus, Penicillium, Chaetomium, Trichoderma, Fusarium, Stachybotrys, Cladosporium* (Wood, 1985; Mehrotra and Aneja, 1990; Sajith et al., 2016).

3.1.3 Lipases

From the recent research works, lipase showed its association with organic contaminants present in the soil system. Lipase activity responses for the reduction of total hydrocarbon compounds (Margesin et al., 1999; Riffaldi et al., 2006). Lipase catalyzes various reactions such as hydrolysis, esterification, alcoholysis, and aminolysis (Prasad and Manjunath, 2011). Lipase activity is the most useful indicator for estimation of hydrocarbon degradation in soil (Margesin et al., 1999; Riffaldi et al., 2006). Lipases have analytical usage in bioremediation and industrial applications such as food, chemical, detergent manufacturing, and cosmetic. Lipase production by microbes is more versatile because of potential industrial application (Sharma et al., 2011; Joseph et al., 2006). Fungal species also known to produce lipases are *Candida rugosa, Candida Antarctica, Aspergillus, A. niger, Penicillium* sp. (Liu et al., 2015; Amoah et al., 2016; Wolski et al., 2008; Mehta et al., 2017).

3.2 Versatile peroxidases

Versatile peroxidase (VP) is a heme-containing ligninolytic peroxidase with mixed molecular structure associated with different oxidation-active sites (Perez-Boada et al., 2005; Ruiz-Dueñas Francisco et al., 2001; Abdel-Hamid et al., 2013). VPs are able to directly oxidize Mn2+, methoxybenzenes, and phenolic aromatic substrates and have extraordinary broad substrate specificity (Ruiz-Dueñas et al., 2007; Karigar; Rao, 2011).

3.3 Ligninolytic enzymes

3.3.1 *Laccase*

Laccase are multicopper oxidases and blue multicopper oxidases (Couto and Toca-Herrera, 2006). Four copper (Cu) molecules present in the active sites of laccases participate in oxygen declination and water production (Dias et al., 2007). Laccase is considered as eco-friendly enzyme and also called as blue enzyme (Riva, 2006). Laccase could aid in degradation of hazardous organic compounds, for instance, phenols, chlorophenols, and PAHs (Dec and Bollag, 1995; Riva, 2006; Wu et al., 2008; Li et al., 2010). Laccase is widely distributed in plants and fungi effective in degrading phenol and lignin. The activity and stability of laccase produced by white-rot fungi, *P. ostreatus*, enhanced by addition of copper, while cadmium, silver, mercury, and lead prevented the enzyme activity (Bhattacharya, 2014; Ali et al., 2017).

3.3.2 *Heme peroxidases*

Heme peroxidases are categorized as LiP and MnP under ligninolytic enzymes. These enzymes are heme proteins consisting of protoporphyrin IX as a prosthetic group, principally reported in the batch cultures of *P. chrysosporium* (Dias et al., 2007; Piontek et al., 1993; Sundaramoorthy et al., 1994; Angel, 2002; Choinowski et al., 1999). LiP activates through oxidation of aromatic rings by providing electron substitutes and common peroxidases involved in the catalysis of aromatic substrates (amine and hydroxyl) (Khindaria et al., 1996).

MnP are glycoproteins, and its structure has two domains with the heme ionic group in the middle, 10 major helixes, a minor helix, and 5 disulfide linkages (Martin, 2002). MnP possess specific linkage stakes in the manganese-bonding site. This is a distinctive feature of MnP from other peroxidases (Plácido et al., 2015).

4. Biosurfactant production by fungi and its application in bioremediation

Biosurfactants are surface active compounds consisting of both hydrophobic and hydrophilic fragments which lower the surface tension between individual molecules (Lang, 2002; Satpute et al., 2010). Microbes producing biosurfactant can be used in the bioremediation technologies such as solubilization, removal of oil from contaminated site, and sludge in oil storage tank. Biosurfactants are ecologically safe and can be applied in bioremediation processes (Banat, 1995). Microbial-enhanced oil recovery is a worthy example of utilization of biosurfactant in petroleum recovery process (Sen, 2008). Numerous bacteria and limited fungi are reported to produce biosurfactant (Satpute et al., 2010b; Patil and Chopade, 2001). Fungi have been reported for higher yields of 120 and 40 g/L of surfactants using carbon sources such as tallow fatty acid residues, glycerol, and oleic acid (Felse et al., 2007; Deshpande and Daniels, 1995; Kim et al., 2002). Fungi yield higher amount of biosurfactant compared with bacteria because of rigid cell wall in fungi (Kim et al., 1999; Bhardwaj et al., 2013) (Table 6.6). Hydrocarbon utilization aided by developed cell contact and capability to adhere

TABLE 6.6 Type of biosurfactant produced by different fungi.

Fungi	Biosurfactant type	References
Candida tropicalis and *Candida albicans*	NA	Padmapriya et al. (2013)
Candida bombicola Atcc 22214	Sophorolipids	Minucelli et al. (2016)
Aspergillus ustus	Glycolipoprotein	Kiran et al. (2009)
Trichosporon ashii	Sophorolipid	Chandran and Das (2010); Da Silva et al. (2017)
Torulopsis petrophilum	Glycolipid	Folch et al. (1957); Da Silva et al. (2017)
Fusarium fujikuroi Ufsm-Bas-01	Trehalolipids	Reis et al. (2018)
Fusarium proliferatum	Enamide	Bhardwaj et al. (2015)
Penicillium chrysogenum Snp5	Lipopeptide	Gautam et al. (2014)
Aspergillus niger	Glycolipid	Kannahi and Sherley (2012)

NA, not available.

on hydrocarbon is associated with cell surface hydrophobicity. Surfactants can boost the attraction between a microbial cell and hydrocarbon (Franzeeti et al., 2008). There are several methods for screening of biosurfactant produced by microbes such as drop collapse test and oil spreading technique. The drop collapse method depends on immiscible liquids interface and principally drop of a liquid containing a bioemulsifier. Biosurfactant usually exhibits emulsifying biosurfactant, which collapse completely over oil surface (Jain et al., 1991; Ewa et al., 2011). The oil spreading method measures the diameter of clear zones by group of surface active agent caused when a drop of a biosurfactant synthesized by a variety of bacteria, yeasts, and fungi and placed on oil—water surface (Morikawa et al., 2000; Padmapriya et al., 2013). These methods can be used for rapid screening of large number of microbial samples (Bodour et al., 2003).

Biosurfactants comprise of potential application in bioremediation because of inherent degradability. Certainly, microbial populations degrading petroleum hydrocarbons produce biosurfactants to escalate substrate bioavailability; therefore, increased rate of biodegradation could be accomplished (Muller-Hurting et al., 1993; Padmapriya et al., 2013).

5. Factors affecting growth of fungi

Growth of fungi and its ability to perpetuate under given environmental condition affects the mineralization of toxic organic compound. Various abiotic factors may influence the fungal growth and biodegradation process as well, such as temperature, pH, carbon and nitrogen sources, aeration, humidity, light intensity, and trace elements, which are discussed below. Along with these environmental factors, genomic characteristics and different categories of enzymes and proteins produced by the fungi are also contributing factors.

5.1 Temperature

Temperature is a key factor for stimulating and hindering the fungal growth. *Entomophthora* sp. showed most dynamic growth in the optimum temperature ranges of 24–27°C (Hall and Bell, 1961). Optimum temperature range showed stimulating effect on the growth of mycelium of *Beauveria, Metarhizium*, and *Paecilomyces* sp. (Halsworth and Magan, 1999; Piatkowski and Krzyzewska, 2007). The effect of temperature and pH on the radial growth rate and biomass yield of two strain of *Trichoderma* (Td85 and Td50) with high biocontrol potential was studied. Four incubation temperature (15°C, 25°C, 30°C, 35°C) and both strains were grown better at 25–30°C and slow growth was observed at 15 °C (Cristina et al., 2016).

5.2 Humidity

Humidity or moisture content employs a direct impact on fungi. Both vegetative cells and spores of fungi require optimum humidity and oscillate within a wide range depending on species involved. *Entomophthora aphidis* and *Entomophthora thaxteriana* discharge spores only when humidity ranges between 70% and 90% (Wilding, 1973). Various fungal genera germination is suppressed even at 90% or lower humidity range (Uziel and Kennet, 1991; Piatkowski and Krzyzewska, 2007).

5.3 pH

pH is an important factor which may accelerate or prevent the fungal growth and biodegradation process. Two strains of *Trichoderma* (Td85 and Td50) tested with different pH ranges 4.5; 5.5; 7.5; 8.5 and selected strains were able to tolerate a different range of pH levels, but growth was reduced on alkaline media between 7.5 and 8.5 (Cristina et al., 2016). Likewise, *Fusarium aqueducturn* and *Trichosporon cutaneurn* were capable of growing over the wide pH range 4–9. *Geotrichum* sp. were also able to tolerate a wide range (pH 3–10) and grew well over the range pH 3–9; best growth observed at pH 3, though no growth occurs at pH 2. *Sepedonium* sp. tolerated the pH range 4–10. The pH 5.5 was best for the growth of *Rhizopus stolonifer*, *A. niger* and best growth of *Alternaria alternata* and *Phytophthora nicotianae* found at 6.5 while pH 7.0 was optimum for the growth of *Fusarium oxysporum and Fusarium lycopersici* (Hassan et al., 2017). Growth rate of fungal species against different pH ranges primarily depends on the habitat and isolation site.

5.4 Light

Light also effects the growth and sporulation of fungi. Shehu and Bello (2011) reported highest mycelium growth diameter of 84.80 mm found in *A. alternata* under exposure to alternating light and dark, whereas lowest growth (14.40 mm) of *A. alternate* occurred at 8 h dark condition. Similarly Hubballi et al. (2010) reported that alternate exposure cycles of light and dark resulted in maximum fungal growth than continuous light and dark exposure, while *F. oxysporum* and *F. lycopersici* showed maximum growth under dark and hindered growth at alternate cycles of light and dark (Hassan et al., 2017).

5.5 Trace elements

Trace elements or metals are required in a very low amount but are essential for growth and proliferation of microorganisms. Some trace metals such as zinc, manganese, iron, and copper required by *Sepedonizcm* sp. have frequently been reported to be necessary for the growth of other fungi also. Calcium has rarely been reported to be essential. However, growth of various fungi was inhibited to varying degrees in Ca-deficient media reported by Steinberg (1948); Painter (1954).

5.6 Aeration

Metabolism of lignin in manure fiber by *P. chrysosporium* was influenced by aeration (Rosenberg and Wilke, 1979; Yang et al., 1980). Similarly *Ganoderma lucidum* showed maximum production of biomass and polysaccharides under high agitation and aeration (Agudelo-Escobar et al., 2017).

6. Conclusion and future perspective

Mycoremediation using fungi could prove as potential biomass degraders for complex organic compounds because of their ability to survive in harsh environmental conditions, resulting into production of versatile extracellular ligninolytic enzymes and biosurfactant production. Mycoremediation for any contaminated site could be implemented by using several strategies depending on site conditions and physicochemical properties of chemical pollutants, tolerance ability of fungal biomass for the degradation, and complete removal of targeted pollutants. For successful bioremediation, greater understanding of the capabilities of the selected fungi is required. During application of fungi for the bioremediation of PAH-contaminated sites, the environmental and nutritional factors must be considered as important factors that may influence the rate of biodegradation. More research is required for exploration and identification of PAH intermediates and enzymatic pathways responsible for biodegradation process. Genes encrypting specific enzymes involved in degradation of PAHs must be emulated, sequenced, and described in a direction to expose a novel epoch in mycoremediation for the detoxification of wide range of persistent organic compounds.

References

Abdel-Shafy Hussein, I., Mansour Mona, S.M., 2016. A review on polycyclic aromatic hydrocarbons: source, environmental impact, effect on human health and remediation. Egyptian Journal of Petroleum 25, 107–123. https://doi.org/10.1016/j.ejpe.2015.03.011.

Andersson, B.E., Lundstedt, S., Tornberg, K., Schnürer, Y., Öberg, L.G., Mattiasson, B., 2003. Incomplete degradation of polycyclic aromatic hydrocarbons in soil inoculated with wood-rotting fungi and their effect on the indigenous soil bacteria. Environmental Toxicology and Chemistry 22, 1238–1243.

Al-Hawash, A.B., Alkooranee, J.T., Zhang, X., Ma, F., 2018. Fungal degradation of polycyclic aromatic hydrocarbons. International Journal of Pure and Applied Bioscience 6 (2), 8–24. https://doi.org/10.18782/2320-7051.6302.

Adriano-Anaya, M., Salvador-Figueroa, V., Ocampo, J.A., García-Romera, I., 2005. Plant cell-wall degrading hydrolytic enzymes of *Gluconacetobacter diazotrophicus*. Symbiosis 40 (3), 151–156.

Agudelo-Escobar, L.M., Gutiérrez-López, Y., Urrego-Restrepo, S., 2017. Effects of aeration, agitation and pH on the production of mycelial biomass and exopolysaccharide from the filamentous fungus *Ganoderma lucidum*. DYNA 84 (200), 72–79.

Agency for Toxic Substance and Disease Registry (ATSDR), 2010. Public Health Statement. August 1995.

Angel, T.M., 2002. Molecular biology and structure-function of lignindegrading heme peroxidases. Enzyme and Microbial Technology 30, 425–444. https://doi.org/10.1016/S0141-0229(01)00521-X.

Amoah, J., Ho, S.H., Hama, S., Yoshida, A., Nakanishi, A., Hasunuma, T., Ogino, C., Kondo, A., 2016. Lipase cocktail for efficient conversion of oils containing phospholipids to biodiesel. Bioresource Technology 211, 224–230.

Andrade, V.S., Sarubbo, L.A., Fukushima, K., 2002. Production of extracellular proteases by *Mucor circinelloides* using D-glucose as carbon source/substrate. Brazilian Journal of Microbiology 33, 106–110.

Alves, M.H., Campos-Takaki, G.M., Porto, A.L.F., Milanez, A.I., 2002. Screening of *Mucor* spp. for the production of amylase, lipase, polygalacturonase and protease. Brazilian Journal of Microbiology 33, 325–330.

Ali, A., Guo, D., Mahar, A., Ping, W., Feng, S., Ronghua, L., Zengqiang, Z., 2017. Mycoremediation of potentially toxic trace elements—a biological tool for soil cleanup: a review. Pedosphere 27 (2), 205–222. https://doi.org/10.1016/S1002-0160(17)60311-4. ISSN 1002-0160/CN 32-1315/P, (Soil Science Society of China, Published by Elsevier B.V. and Science Press).

Abdel-Hamid, A.M., Solbiati, J.O., Cann, I.K., 2013. Insights into lignin degradation and its potential industrial applications. Advances in Applied Microbiology 82, 1–28. https://doi.org/10.1016/B978-0-12-407679-2.00001-6.

Adenipekun, C.O., Lawal, R., 2012. Uses of mushrooms in bioremediation: a review. Biotechnology and Molecular Biology Reviews 7 (3), 62–68.

Banat, I.M., 1995. Biosurfactants production and possible uses in microbial enhanced oil recovery and oil pollution remediation: a review. Bioresource Technology 51, 1–12.

Bebber, D.P., Hynes, J., Darrah, P.R., Boddy, L., Fricker, M.D., 2007. Biological solutions to transport network design. Proceedings of the Royal Society 274, 2307–2315.

Beena, A.K., Geevarghese, P.I., 2010. A solvent tolerant thermostable protease from a psychrotrophic isolate obtained from pasteurized milk. Developmental Microbiology and Molecular Biology 1, 113–119.

Bennet, J.W., Wunch, K.G., Faison, B.D., 2002. Use of Fungi Biodegradation. ASM Press, Washington, DC, USA.

Beyer, J., Jonsson, G., Porte, C., Krahn, M.M., Ariese, F., 2010. Environmental Toxicology and Pharmacology 30, 224–244.

Bezalel, L., Hadar, Y., Cerniglia, C.E., 1997. Enzymatic mechanisms involved in phenanthrene degradation by the white rot fungus *Pleurotus ostreatus*. Applied and Environmental Microbiology 63, 2495–2501.

Bezalel, L., Hadar, Y., Fu Peter, P., Freeman James, P., Cerniglia Carl, E., 1996. Metabolism of phenanthrene by the white rot fungus *Pleurotus ostreatus*. Applied and Environmental Microbiology 62 (7), 2547–2553.

Bhardwaj, H., Gupta, R., Tiwari, A., 2013. Communities of microbial enzymes associated with biodegradation of plastics. Journal of Polymers and the Environment 21, 575–579.

Bhardwaj, G., Cameotra, S.S., Chopra, H.K., 2015. Isolation and purification of a new enamide biosurfactant 1 from *Fusarium proliferatum* using rice-bran. RSC Advances (67).

Bhattacharya, S., Das, A., Prashanthi, K., Palaniswamy, M., Angayarkanni, J., 2014. Mycoremediation of Benzo[a]pyrene by Pleurotus ostreatus in the presence of heavy metals and mediators. 3 Biotechnology 4 (2), 205–211. https://doi.org/10.1007/s13205-013-0148-y.

Bodour, A.A., Drees, K.P., Maier, R.M., 2003. Distribution of biosurfactant-producing bacteria in undisturbed and contaminated arid southwestern soils. Applied and Environmental Microbiology 69, 3280–3287.

Bornyasz, M.A., Graham, R.C., Allen, M.F., 2005. Ectomycorrhizae in a soil-weathered granitic bedrock regolith: linking matrix resources to plants. Geoderma 126, 141–160.

Borosky, G.L., 1999. Theoretical study related to the carcinogenic activity of polycyclic aromatic. hydrocarbon derivatives. Journal of Organic Chemistry 64, 7738–7744.

Cerniglia, C.E., Freeman, J.P., Mitchum, R.K., 1982. Glucuronide and sulfate conjugation in the fungal metabolism of aromatic hydrocarbons. Applied and Environmental Microbiology 43, 1070–1075.

Cerniglia, C.E., Yang, S.K., 1984. Stereoselective metabolism of anthracene and phenanthrene by the fungus *Cunninghamella elegans*. Applied and Environmental Microbiology 47, 119–124.

Cebron, A., Norini, M.P., Beguiristain, T., Leyval, C., 2008. Real-time PCR quantification of PAH-ring hydroxylating dioxygenase (PAH-RHDα) genes from Gram positive and Gram negative bacteria in soil and sediment samples. Journal of Microbiological Methods 73, 148–159.

Cristina, P., Alexandru, P., Florica, C., 2016. Influence of abiotic factors on *in vitro* growth of *Trichoderma* strains. Proceedings of the Romanian Academy. Series B 18 (1), 11−14.

Choinowski, T., Blodig, W., Winterhalter, K.H., Piontek, K., 1999. The crystal structure of lignin peroxidase at 1.70 Å resolution reveals a hydroxy group on the Cβ of tryptophan 171: a novel radical site formed during the redox cycle. Journal of Molecular Biology 286, 809−827. https://doi.org/10.1006/jmbi.1998.2507.

Chandran, P., Das, N., 2010. Biosurfactant production and diesel oil degradation by yeast species *Trichosporon asahii* isolated from petroleum hydrocarbon contaminated soil. International Journal of Engineering Science and Technology 2, 6942−6953.

Cernigilia, C.E., Gibson, D.T., 1979. Oxidation of benzo[a]pyrene by the filamentous fungus cunninghamella elegans. Journal of Biological Chemistry 254, 12174−12180.

Cerniglia Carl, E., Sutherland John, B., 2009. Bioremediation of polycyclic aromatic hydrocarbons by ligninolytic and non-ligninolytic fungi. In: GADD, G.M. (Ed.), Fungi in Bioremediation. Cambridge University Press, pp. 136−186. ISBN-13 978-0-511-40917-2.

Casillas, R.P., Crow Jr., S.A., Heinze, T.M., Deck, J., Cerniglia, C.E., 1996. Initial oxidative and subsequent conjugative metabolites produced during the metabolism of phenanthrene by fungi. Journal of Industrial Microbiology 16, 205−215.

Cao, B., Nagarajan, K., Loh, K.C., 2009. Biodegradation of aromatic compounds: current status and opportunities for biomolecular approaches. Applied Microbiology and Biotechnology 85, 207−228.

Cerniglia, C.E., 1993. Biodegradation of polycyclic aromatic hydrocarbons. Current Opinion in Biotechnology 4, 331−338.

Cerniglia, C.E., Sutherland, J.B., 2010. In: Timmis, K.N., McGenity, T., van der Meer, J.R., de Lorenzo, V. (Eds.), Handbook of Hydrocarbon and Lipid Microbiology, pp. 2079−2110.

Cajthaml, T., Erbanová, P., Sasek, V., Möder, M., 2006. Breakdown products on metabolic pathway of degradation of benz[a]anthracene by a ligninolytic fungus. Chemosphere 64, 560−564.

Cajthaml, T., Möder, M., Kacer, P., Sasek, V., Popp, P., 2002. Study of fungal degradation products of polycyclic aromatic hydrocarbons using gas chromatography with ion trap mass spectrometry detection. Journal of Chromatography A 974, 213−222.

Couto, S., Toca-Herrera, J., 2006. Lacases in the textile industry. Biotechnology and Molecular Biology Reviews 1, 115−120.

Collins, P.J., Kotterman, M.J.J., Field, J.A., Dobson, A.D.W., 1996. Oxidation of anthracene and benzo[a]pyrene by laccases from *Trametes versicolor*. Applied and Environmental Microbiology 62 (12), 4563−4567.

de Souza, P.M., Bittencourt, M.L., Caprara, C.C., de Freitas, M., de Almeida, R.P., Silveira, D., Fonseca, Y.M., Ferreira Filho, E.X., Pessoa Junior, A., et al., 2015. A biotechnology perspective of fungal proteases. Publication of the Brazilian Society for Microbiology Brazilian Journal of Microbiology 46 (2), 337−346. https://doi.org/10.1590/S1517-838246220140359.

Dienes, D., Börjesson, J., Hägglund, P., 2007. Identification of a trypsin-like serine protease from *Trichoderma reesei* QM9414. Enzyme and Microbial Technology 40, 1087−1094.

Dashtban, M., Schraft, H., Syed, T.A., Qin, W., 2010. Fungal biodegradation and enzymatic modification of lignin. International Journal of Biochemistry and Molecular Biology 1 (1), 36−50.

Davis, M.W., Glaser, J.A., Evans, J.W., Lamar, R.T., 1993. Field evaluation of the lignindegrading fungus *Phanerochaete sordida* to treat creosote-contaminated soil. Environmental Science and Technology 27, 2572−2576.

Dec, J., Bollag, J.M., 1995. Effect of various factors on dehalogenation of chlorinated phenols and anilines during oxidative coupling. Environmental Science and Technology 29, 657−663.

Dong, C., Chen, C., Chen, C., 2012. International Journal of Environmental Research and Public Health 9, 2175−2188.

Da Silva, A.C.S., Dos Santos, P.N., Lima e Silva, T.A., Silva Andrade, R.F., Takaki Galba, M.C., 2017. Biosurfactant production by fungi as a sustainable alternative. Agricultural Microbiology/Review Article. https://doi.org/10.1590/1808-165700050.

Deshpande, M., Daniels, L., 1995. Evaluation of sophorolipid biosurfactant production by *Candida bombicola* using animal fat. Bioresource Technology 54, 143−150.

Darrah, P.R., Tlalka, M., Ashford, A., Watkinson, S.C., Fricker, M.D., 2006. The vacuole system is a significant intracellular pathway for longitudinal solute transport in basidiomycete fungi. Eukaryotic Cell 5, 1111−1125.

Dias, A., Sampaio, A., Bezerra, R., 2007. Environmental applications of fungal and plant systems: decolourisation of textile wastewater and related dyestuffs. In: Singh, S., Tripathi, R. (Eds.), Environmental Bioremediation Technologies. Springer Berlin Heidelberg, pp. 445−463.

Ewa, K., Moszynska, S., Olszanowski, A., 2011. Modification of cell surface properties of *Pseudommonas alcaligenes* S22 during hydrocarbon biodegradation. Biodegradation 22, 359–366.

Erika, W., Katarina, B., Schultz, E., Markus, R., Kalle, S., Festus, A., Tomás, C., Steffen Kari, T., Jørgensen Kirsten, S., Marja, T., 2014. Bioremediation of PAH-contaminated soil with fungi from laboratory to field scale. International Biodeterioration & Biodegradation 86, 238–247.

Franzeeti, A., Di Gennaro, P., Bestetti, G., Lasagni, M., Pitea, D., Collina, E., 2008. Selection of surfactants for enhancing diesel hydrocarbons – contaminated media bioremediation. Journal of Hazardous Materials 152, 1309–1316.

Flathman, P.E., Jerger, D., Exner, J.E., 1993. Bioremediation: Field Experience. Lewis, Boca Raton, FL.

Folch, J., Lees, M., Stanly, S.G.H., 1957. A simple method for the isolation and quantification of total lipids from animal tissues. Journal of Biological Chemistry 226, 497–509.

Felse, P.A., Shah, V., Chan, J., Rao, K.J., Gross, R.A., 2007. Sophorolipid biosynthesis by *Candida bombicola* from industrial fatty acid residues. Enzyme and Microbial Technology 40, 316–323.

Govindarajulu, M., et al., 2005. Nitrogen transfer in the arbuscular mycorrhizal symbiosis. Nature 435, 819–823.

Godoy, P., Reina, R., Calderón, A., Wittich, R.-M., García-Romera, I., Aranda, E., 2016. Exploring the potential of fungi isolated from PAH-polluted soil as a source of xenobiotics-degrading fungi. Environmental Science & Pollution Research. https://doi.org/10.1007/s11356-016-7257-1. ISSN 0944-1344.

Gautam, G., Mishra, V., Verma, P., Pandey, A.K., Negi, S., 2014. A cost effective strategy for production of bio-surfactant from locally isolated *Penicillium chrysogenum* SNP5 and its applications. Journal of Bioprocessing and Biotechniques 4 (6), 1–7. https://doi.org/10.4172/2155-9821.1000177.

Gibson, D.T., Subramanian, V., 1984. Microbial degradation of aromatic hydrocarbons. In: Gibson, D.T. (Ed.), Microbial Degradation of Organic Compounds. Marcell Dekker. Inc., NewYork, pp. 181–252.

Govarthanan, M., a Fuzisawa, S., b Hosogaia, T., Chang, Y.-C., 2017. Biodegradation of aliphatic and aromatic hydrocarbons using the filamentous fungus *Penicillium* sp. CHY-2 and characterization of its manganese peroxidase activity. RSC Advances 7, 20716–20723.

Germano, S., Pandey, A., Osaku, C.A., 2003. Characterization and stability of proteases from *Penicillium* sp. produced by solid-state fermentation. Enzyme and Microbial Technology 32, 246–251.

Hall, M., Bell, J.V., 1961. Further studies on the effect of temperature on the growth of entomophthorous fungi. Journal of Insect Pathology 3, 289–296.

Hallsworth, J.E., Magan, N., 1999. Water and temperature relations of growth of the entomogenous fungi beauveria bassiana, metarhizium anisopliae, and paecilomyces farinosus. Journal of Invertebrate Pathology 74 (3), 261–266.

Harms, H., Dietmar, S., Wick Lukas, Y., 2011. Untapped potential: exploiting fungi in bioremediation of hazardous chemicals, applied and industrial microbiology. Nature Reviews Microbiology 177–192.

Hatakka, A., 2001. Biodegradation of Lignin. In: Hofrichter, M., Steinbuchel, A. (Eds.), Biopolymers, Vol. 1, Lignin, Humic Substances and Coal. Wiley-VCH, Weinheim, Germany, 129-1.

Hubballi, M., Nakkeeran, S., Raguchander, T., Anand, T., Samiyappan, R., 2010. Effect of environmental conditions on growth of *Alternaria alternata* causing leaf blight of Noni. World Journal of Agricultural Sciences 6 (2), 171–177.

Hibbett, D.S., Binder, M., Bischoff, J.F., Blackwell, M., Cannon, P.F., Eriksson, O.E., et al., 2007. A higher-level phylogenetic classification of the fungi. Mycological Research 111, 509–547.

Hassan, N., Shafiq, R., Haleema, B., 2017. Environmental factors affecting growth of pathogenic fungi causing fruit rot in tomato (Lycopersicon Esculentum). International Journal of Engineering Research and Technology 6 (02). ISSN: 2278-0181, IJERTV6IS020357.

Hinchee, R.E., Means, J.L., Burrisl, D.R., 1995. Bioremediation of Inorganics. Battelle Press, Columbus, OH.

Haderlein, A., Legros, R., Ramsay, B.A., 2006. Pyrene mineralization capacity increases with compost maturity. Biodegradation 17, 293–302.

Hunsa, P., Prasongsuk, S., Messner, K., Danmek, K., Lotrakul, P., 2009. Polycyclic aromatic hydrocarbons (PAHs) degradation by laccase from a tropical white rot fungus *Ganoderma lucidum*. African Journal of Biotechnology 8 (21), 5897–5900, 2.

Hofrichter, M., Bublitz, F., Fritsche, W., 1994. Unspecific degradation of halogenated phenols by the soil fungus *Penicillium frequentans* Bi 7/2. Journal of Basic Microbiology 34, 163–172.

Inomata, Y., Kajino, M., Sato, K., Ohara, T., Kurokawa, J.I., 2012. Environmental Science and Technology 46, 4941–4949.

Johannes, C., Majcherczyk, A., Hüttermann, A., 1996. Degradation of anthracene by laccase of *Trametes versicolor* in the presence of different mediator compounds. Applied Microbiology and Biotechnology 46 (3), 313—317.

Jain, D.K., Collins-Thompson, D., Lee, H., 1991. A drop-collapsing test for screening surfactant. Journal of Microbiological Methods 13, 271—279.

Johnsen, A.R., Wick, L.Y., Harms, H., 2005. Principles of microbial PAH-degradation in soil. Environmental Pollution 133, 71e84.

Jerina, D.M., Brodie Bernard, B., 1983. Metabolism of Aromatic Hydrocarbons by the Cytochrome P-450 System and Epoxide Hydrolase, Drug Metabolism and Disposition: The Biological Fate of Chemicals, vol. 11, pp. 1—4.

JOHN, A., BUMPUS, Jan. 1989. Biodegradation of polycyclic aromatic hydrocarbons by phanerochaete chrysosporium. Applied And Environmental Microbiology 55 (1), 154—158 (American Society for Microbiology).

Joseph, B., Ramteke, P.W., Kumar, P.A., 2006. Studies on the enhanced production of extracellular lipase by *Staphylococcus epidermidis*. Journal of General and Applied Microbiology 52 (6), 315—320.

Kenneth, E., Hammel, 1995. Microbial Transformation and Degradation of Toxic Organic Chemicals. Wiley-Liss. Inc, pp. 331—346.

Kim, H.S., Jeon, J.W., Lee, H.W., Park, Y.I., Seo, W.T., et al., 2002. Extracellular production of a glycolipid biosurfactant, mannosylerythritol lipid, from *Candida antarctica*. Biotechnology Letters 24, 225—229.

Kim, H.S., Yoon, B.D., Choung, D.H., Oh, H.M., Katsuragi, T., et al., 1999. Characterization of a biosurfactant mannosylerythritol lipid produced from *Candida* sp. SY 16. Applied Microbiology and Biotechnology 52, 713—721.

King, R.B., Long, G.M., Sheldon, J.K., 1997. Practical Environmental Bioremediation: The Field Guide, second ed. Lewis, Boca Raton, FL.

Kannahi, M., Sherley, M., 2012. Biosurfactant production by *Pseudomonas putida* and *Aspergillus niger* from oil contaminated site. International Journal of Chemical and Pharmaceutical Sciences 3 (4), 37—43.

Kiran, G.S., Hema, T.A., Gandhimathi, R., Selvin, J., Thomas, T.A., et al., 2009. Optimization and production of a biosurfactant from the sponge-associated marine fungus *Aspergillus ustus* MSF3. Colloids and Surfaces B 73, 250—256.

Kästner, M., Breuer-Jammali, M., Mahro, B., 1994. Enumeration and characterization of soil microflora from hydrocarbon contaminated soil sites able to mineralize polycyclic aromatic hydrocarbons (PAH). Applied Microbiology and Biotechnology 41, 267—273.

Karigar Chandrakant, S., Rao Shwetha, S., 2011. Role of microbial enzymes in the bioremediation of pollutants: a review. Enzyme Research Volume 11. https://doi.org/10.4061/2011/805187. Article ID 805187.

Khindaria, A., Yamazaki, I., Aust, S.D., 1996. Stabilization of the veratryl alcohol cation radical by lignin peroxidase. Biochemistry 35, 6418—6424. https://doi.org/10.1021/bi9601666.

Kannangara, S., Undugoda, L., Rajapaksha, N., Abeywickrama, K., 2016. Depolymerizing activities of aromatic hydrocarbon degrading Phyllosphere fungi in Sri Lanka. Journal of Bioremediation and Biodegradation 7, 6. https://doi.org/10.4172/2155-6199.1000372.

Kirk, T.K., Farrell, R.L., 1987. Enzymatic combination— microbial degradation of lignin. Annual Review of Microbiology 41, 465—505.

Lal, R., Saxena, D.M., 1982. Accumulation, metabolism and effects of organochlorine insecticides on microorganisms. Microbiological Reviews 46 (1), 95—127.

Li, X., Lin, X., Zhang, J., Wu, Y., Yin, R., Feng, Y., Wang, Y., 2010. Degradation of polycyclic aromatic hydrocarbons by crude extracts from spent mushroom substrate and its possible mechanisms. Current Microbiology 60, 336—342.

Lee, H., Choi, Y.S., Kim, M.J., 2010. Degradation ability of oligocyclic aromates by *Phanerochaete sordida* selected via screening of white-rot fungi. Folia Microbiologica 55 (5), 447—453.

Launen, L., Pinto, L., Wiebe, C., Kiehlmann, E., Moore, M., 1995. The oxidation of pyrene and benzo[a]pyrene by nonbasidiomycete soil fungi. Canadian Journal of Microbiology 41, 477—488.

Lang, S., 2002. Biological amphiphiles (microbial biosurfactants). Current Opinion in Colloid and Interface Science 7, 12—20.

Liu, G., Hu, S., Li, L., Hou, Y., 2015. Purification and characterization of a lipase with high thermostability and polar organic solvent tolerance from *Aspergillus niger* AN0512. Lipids 50, 1155—1163.

Mehta, A., Bodh, U., Gupta, R., 2017. Fungal lipases: a review. ISSN: 1944-3285 Journal of Biotech Research 8, 58—77, 58.

Martin, H., 2002. Review: lignin conversion by manganese peroxidase (MnP). Enzyme and Microbial Technology 30, 454—466. https://doi.org/10.1016/S0141-0229(01)00528-2.

Minucelli, T., Riberio-Viana, R.M., Borsato, D., Andrade, G., Cely, M.V.T., De Oliveira, M.R., Baldo, C., Celligoi, M.A.P.C., 2016. Sophorolipid production by candida bombicola ATCC 22214 and its potential application in soil bioremediation. Waste Biomass Valor. https://doi.org/10.1007/s12649-016-9592-3.

Muller-Hurting, R., Wanger, F., Blaszeyk, R., Kosaric, N., 1993. Biosurfactants for environmental control. In: Kosaric, N. (Ed.), Biosurfactants. Production, Properties, Applications. Marcel Dekker, New York, pp. 447–469.

Marchand, C., St-Arnaud, M., Hogland, W., Bell Terrence, H., Hijri, M., 2017. Petroleum biodegradation capacity of bacteria and fungi isolated from petroleum-contaminated soil. International Biodeterioration and Biodegradation 116, 48–57.

Margesin, R., Zimmerbauer, A., Schinner, F., 1999. Soil lipase activity—a useful indicator of oil biodegradation. Biotechnology Techniques 13 (12), 859–863.

Marco-Urrea, E., García-Romera, I., Aranda, E., 2015. Potential of nonligninolytic fungi in bioremediation of chlorinated and polycyclic aromatic hydrocarbons. New Biotechnology 32, 620–628.

Margesin, R., Schinner, F., 2001. Biodegradation and bioremediation of hydrocarbons in extreme environments. Applied Microbiology and Biotechnology 56, 650–663.

Morikawa, M., Hirata, Y., Imanaka, T., 2000. A study on the structure function relationship of the lipopeptide biosurfactants. Biochimica et Biophysica Acta 1488, 211–218.

Mehrotra, R., Aneja, K., 1990. An Introduction to Mycology: New Age International.

National Research Council (NRC), 1993. In Situ Bioremediation: When Does It Work? National Academy Press, Washington, DC.

Norris, R.D., Hinchee, R.E., Brown, R., McCarty, P.L., Semprini, L., Wilson, J.T., Kampbell, D.H., Reinhard, M., Bouwer, E.J., Borden, P.C., Vogel, T.M., Thomas, J.M., Ward, C.H., 1993. Handbook of Bioremediation. Lewis, Boca Raton, FL.

National Research Council, 2003. Oil in the Sea III: Inputs, Fates, and Effects.

Natalia, P., 2017. Ligninolytic fungi: their degradative potential and the prospect for the development of environmentally significant biotechnologies. In: 2nd International Conference on, Environmental Health & Global Climate Change, Occup Med Health Aff 2017, 5:2 (Suppl). https://doi.org/10.4172/2329-6879-C1-030.

Nikitha, T., Satyaprakash, M., Vani, S.S., Sadhana, B., Padal, S.B., 2017. A review on polycyclic aromatic hydrocarbons: their transport, fate and biodegradation in the environment. International Journal of Current Microbiology and Applied Sciences 6 (4), 1627–1639.

Plácido, J., Capareda, S., Plácido, Capareda, 2015. Ligninolytic enzymes: a biotechnological alternative for bioethanol production. Bioresources and Bioprocessing 2, 23. https://doi.org/10.1186/s40643-015-0049-5.

Pérez-Boada, M., Ruiz-Dueñas, F.J., Pogni, R., Basosi, R., Choinowski, T., Martínez, M.J., Piontek, K., Martínez, A.T., 2005. Versatile peroxidase oxidation of high redox potential aromatic compounds: site-directed mutagenesis, spectroscopic and crystallographic investigation of three long-range electron transfer pathways, 25 Journal of Molecular Biology 354 (2), 385–402 (Epub).

Pozdnyakova, N.N., Balandina, S.A., Dubrovskaya, E.V., Golubev, C.N., Turkovskaya, O.V., 2018. Ligninolytic basidiomycetes as promising organisms for the mycoremediation of PAH-contaminated environments. IOP Conference Series: Earth and Environmental Science 107, 012071. https://doi.org/10.1088/1755-1315/107/1/012071.

Piatkowski, J., Krzyzewska, A., 2007. Influence of some physical factors on the growth and sporulation of entomopathogenic fungi. Acta Mycological 42 (2), 255–265.

Padmapriya, B., Suganthi, S., Anishya, R.S., 2013. Screening, optimization and production of biosurfactants by Candida species isolated from oil polluted soils. American-Eurasian Journal of Agricultural and Environmental Sciences 13 (2), 227–233. https://doi.org/10.5829/idosi.aejaes.2013.13.02.2744. ISSN 1818-6769.

Piontek, K., Glumoff, T., Winterhalter, K., 1993. Low pH crystal structure of glycosylated lignin peroxidase from Phanerochaete chrysosporium at 2.5 Å resolution. FEBS Letters 315, 119–124. https://doi.org/10.1016/0014-5793(93) 81146-Q.

Pinedo-Rivilla, C., Aleu, J., Collado, I.G., 2009. Pollutants biodegradation by fungi. Current Organic Chemistry 13, 1194–1214.

Prince, R.C., 2010. In: Timmis, K.N., McGenity, T., van der Meer, J.R., de Lorenzo, V. (Eds.), Handbook of Hydrocarbon and Lipid Microbiology, pp. 2065–2078.

Pointing, S.B., 2001. Feasibility of bioremediation by white-rot fungi. Applied Microbiology and Biotechnology 57, 20–33.

Pothuluri, J.V., Evans, F.E., Heinze, T.M., Cerniglia, C.E., 1996. Formation of sulfate and glucoside conjugates of benzo[e]pyrene by Cunninghamella elegans. Applied Microbiology and Biotechnology 45, 677–683. https://doi.org/10.1007/s002530050747.

Prasad, M.P., Manjunath, K., 2011. Comparative study on biodegradation of lipid-rich wastewater using lipase producing bacterial species. Indian Journal of Biotechnology 10 (1), 121–124.

Peter, H.A., 2003. Petroleum and individual polycyclic aromatic hydrocarbon. In: Hoffman, D.J., Rattner, B.A., Buston, G.A., Cairns, J. (Eds.), Handbook of Ecotoxicology. Lewis Publisher, pp. 342–359.

Pozdnyakova, N.N., 2012. Involvement of the ligninolytic system of white-rot and litter-decomposing fungi in the degradation of polycyclic aromatic hydrocarbons. Biotechnology Research International 243217.

Pothuluri, J.V., Freeman, J.P., Evans, F.E., Cerniglia, C.E., 1993. Biotransformation of fluorene by the fungus Cunninghamella elegans. Applied and Environmental Microbiology 59, 1977–1980.

Patil, J.R., Chopade, B.A., 2001. Studies on bioemulsifier production by Acinetobacter strains isolated from healthy human skin. Journal of Applied Microbiology 91, 290–298.

Painter, H.A., 1954. Factors affecting the growth of some fungi associated with sewage purification. Journal of General Microbiology 10, 177–190.

Quiroz, R., Grimalt, J.O., Fernandez, P., 2010. Toxicity assessment of polycyclic aromatic hydrocarbons in sediments from European high mountain lakes. Ecotoxicology and Environmental Safety 73, 559–564.

Rhodes Christopher, J., 2014. Mycoremediation (bioremediation with fungi) – growing mushrooms to clean the earth. Chemical Speciation and Bioavailability 26 (3), 196–198. https://doi.org/10.3184/09542291 4X14047407349335.

Riva, S., 2006. Laccases: blue enzymes for green chemistry. Trends Biotechnology 24, 219–226.

Ruiz-Dueñas Francisco, J., Susana, C., Pérez-Boada, M., Angel, T.M., Maria, J.M., 2001. A new versatile peroxidase from Pleurotus. Biochemical Society Transactions 29 (Pt 2), 116–122. https://doi.org/10.1042/BST0290116.

Reis, D.L.B.C., Morandini, B.M.L., Bevilacqua, C.B., Bublitz, F., Ugalde, G., Mazutti, A.M., Jacques, S.J.R., 2018. First report of the production of a potent biosurfactant with α, β-trehalose by Fusarium fujikuroi under optimized conditions of submerged fermentation, biotechnology and industrial microbiology. Brazilian Journal of Microbiology 49S, 185–192.

Ritz, K., Young, I.M., 2004. Interactions between soil structure and fungi. Mycologist 18, 52–59.

Rajmalwar, S., Dabholkar, P.S., 2009. Production of protease by Aspergillus sp. using solidstate fermentation. African Journal of Biotechnology 8, 4197–4198.

Rixon, J.E., Ferreira, L.M.A., Durrant, A.J., Laurie, J.I., Hazlewood, G.P., Gilbert, H.J., 1992. Characterization of the gene celD and its encoded product 1,4-β-D-glucan glucohydrolase D from Pseudomonas fluorescens subsp. cellulosa. Biochemical Journal 285 (3), 947–955.

Ruiz-Dueñas, F.J., Morales, M., Pérez-Boada, M., 2007. Manganese oxidation site in Pleurotus eryngiiversatile peroxidase: a site-directed mutagenesis, kinetic, and crystallographic study. Biochemistry 46 (1), 66–77.

Riffaldi, R., Levi-Minzi, R., Cardelli, R., Palumbo, S., Saviozzi, A., 2006. Soil biological activities in monitoring the bioremediation of diesel oil-contaminated soil. Water, Air, and Soil Pollution 170 (1–4), 3–15.

Rillig, M.C., Mummey, D.L., 2006. Mycorrhizas and soil structure. New Phytologist 171, 41–53.

Read, D.J., Perez-Moreno, J., 2003. Mycorrhizas and nutrient cycling in ecosystems – a journey towards relevance? New Phytologist 157, 475–492.

Rosenberg, S.L., Wilke, C.R., 1979. In: Kirk, T.K., Chang, H.-rn, Higuchi, T. (Eds.), Lignin Biodegradation: Microbiology. Chemistry, and Potential Applications. CRC Press, West Palm Beach, FL.

Rafin, C., Potin, O., Veignie, E., Sahraoui, L.-H.A., Sancholle, M., 2006. Degradation of benzo[a]pyrene as sole carbon source by a non white rot fungus, Fusarium solani. Journal Polycyclic Aromatic Compounds 21 (1–4), 311–329.

Sundaramoorthy, M., Kishi, K., Gold, M.H., Poulos, T.L., 1994. The crystal structure of manganese peroxidase from Phanerochaete chrysosporium at 2.06-A resolution. Journal of Biological Chemistry 269, 32759–32767.

Sack, U., Hofrichter, M., Fritsche, W., 1997a. Degradation of polycyclic aromatic hydrocarbons by manganese peroxidase of Nematoloma frowardii. FEMS Microbiology Letters 152, 227–234.

Sack, U., Heinze, T.M., Deck, J., Cerniglia, C.E., Cazau, M.C., Fritsche, W., 1997b. Novel metabolites in phenanthrene and pyrene transformation by Aspergillus niger. Applied and Environmental Microbiology 63, 2906–2909.

Su, D.1, Li, P.J., Frank, S., Xiong, X.Z., 2006. Biodegradation of benzo[a]pyrene in soil by Mucor sp. SF06 and Bacillus sp. SB02 co-immobilized on vermiculite. Journal of Environmental Sciences 18 (6), 1204–1209.

Sen, R., 2008. Biotechnology in petroleum recovery: the microbial EOR. Progress in Energy and Combustion Science 34, 714–724.

Satpute, S.K., Banat, I.M., Dhakephalkar, P.K., Banpurkar, A.G., Chopade, B.A., 2010b. Biosurfactants, bioemulsifiers and exopolysaccharides from marine microorganisms. Biotechnology Advances 28, 436–450.

Sharma, D., Sharma, B., Shukla, A.K., 2011. Biotechnological approach of microbial lipase: a review. Biotechnology 10 (1), 23–40.

Steinberg, R.A., 1948. Essentiality of calcium in the nutrition of fungi. Science 107, 423.

Shehu, K., Bello, M.T., 2011. Effect of environmental factors on the growth of *Aspergillus* species associated with stored millet grains in Sokoto. Nigerian Journal of Basic and Applied Science 19 (2), 218–223.

Sajith, S., Priji, P., Sreedevi, S., Benjamin, S., 2016. An overview on fungal cellulases with an industrial perspective. Journal of Nutrition and Food Sciences 6, 461. https://doi.org/10.4172/2155-9600.1000461.

Sutherland, J.B., Selby, A.L., Freeman, J.P., Evans, F.E., Cerniglia, C.E., 1991. Metabolism of phenanthrene by *Phanerochaete chrysosporium*. Applied and Environmental Microbiology 57 (11), 3310–3316 ([PMC free article] [PubMed]).

Schützendübel, A., Majcherczyk, A., Johannes, C., Hüttermann, A., 1999. Degradation of fluorene, anthracene, phenanthrene, fluoranthene, and pyrene lacks connection to the production of extracellular enzymes by *Pleurotus ostreatus* and *Bjerkandera adusta*. International Biodeterioration and Biodegradation 43 (3), 93–100.

Scheibner, K., Hofrichter, M., Herre, A., Michels, J., Fritsche, W., 1997. Screening for fungi intensively mineralizing 2,4,6-trinitrotoluene. Applied Microbiology and Biotechnology 47, 452–457.

Sutherland, J.B., Rafii, F., Khan, A.A., Cerniglia, C.E., 1995. Mechanisms of Polycyclic Aromatic Hydrocarbon Degradation, Microbial Transformation and Degradation of Toxic Organic Chemicals, 15, p. 269.

Singh, C.J., 2003. Optimization of an extracellular protease of *Chrysosporium keratinophilum* and its potential in bioremediation of keratinic wastes. Mycopathologia 156 (3), 151–156.

Satpute, S.K., Banpurkar, A.G., Dhakephalkar, P.K., Banat, I.M., Chopade, B.A., 2010. Methods for investigating biosurfactants and bioemulsifiers: a review. Critical Reviews in Biotechnology 30, 127–144.

Sack, U., Günther, T., 1993. Metabolism of PAH by fungi and correlation with extracellular enzymatic activities. Journal of Basic Microbiology 33 (4), 269–277.

Simanjorang, S.W., Subowo, Y.B., 2018. The ability of soil-Borne fungi to degrade polycyclic aromatic hydrocarbon (PAH). AIP Conference Proceedings 030019. https://doi.org/10.1063/1.5062743.

Saraswathy, A., Hallberg, R., 2002. Degradation of pyrene by indigenous fungi from a former gasworks site. FEMS Microbiology Letters 210, 227–232.

Santodonato, J., 1981. Polycyclic organic matter. Journal of Environmental Pathology & Toxicology 5, 1.

Steffen, K., Hatakka, A., Hofrichter, M., 2003. Removal and mineralization of polycyclic aromatic hydrocarbons by litter-decomposing basidiomycetous fungi. Applied Microbiology and Biotechnology 60 (1–2), 212–217.

Ting, W.T.E., Yuan, S.Y., Wu, S.D., Chang, B.V., 2011. Biodegradation of phenanthrene and pyrene by *Ganoderma lucidum*. International Biodeterioration & Biodegradation 65 (1), 238–242.

Tudoran, M.A., Putz, M.V., 2012. Chemical Bulletin of Politehnica University of Timisoara 57 (71), 1.

Tuomela, M., Hatakka, A., 2011. Oxidative fungal enzymes for bioremediation. In: Moo-Young, M., Agathos, S. (Eds.), Comprehensive Biotechnology, second ed. Elsevier, Spain, p. 183e196.

Tero, A., et al., 2010. Rules for biologically inspired adaptive network design. Science 327, 439–442.

Uziel, A., Kenneth, R.G., 1991. Survival of primary conidia and capilliconidia at different humidities in Erynia (subgen. *Zoophthora*) spp. and in *Neozygites freseni* (Zygomycotina, Entomophthorales) with special emphasis on *Erynia radicans*. Journal of Invertebrate Pathology 58, 118–126.

Veltman, K., Huijbregts, M.A.J., Rye, H., Hertwich, E.G., 2012. Integrated Environmental Assessment and Management 7, 678–686.

Vidali, M., 2001. Bioremediation. An overview. Pure and Applied Chemistry 73 (7), 1163–1172, 2001. © 2001 IUPAC 1163.

Vasileva-Tonkova, E., Galabova, D., 2003. Hydrolytic enzymes and surfactants of bacterial isolates from lubricant-contaminated wastewater. Zeitschrift fur Naturforschung 58 (1–2), 87–92.

Vyas, B.R.M., Bakowski, S., Sasek, V., Matucha, M., 1994. Degradation of anthracene by selected white rot fungi. FEMS Microbiology Ecology 14 (1), 65–70.

Williams, E.S., Mahler, B.J., Van Metre, P.C., 2013. Cancer risk from incidental ingestion exposures to PAHs associated with coal-tar-sealed pavement. Environmental Science and Technology 47, 1101–1109.

Williams, P.P., 1977. Metabolism of synthetic organic pesticides by anaerobic microorganisms. Residue Reviews 66, 63–135.

Wilding, N., 1973. The survival of *Entomophthora* spp. in mummified aphids at different temperatures and humidities. Journal of Invertebrate Pathology 21, 309–311.

Wolski, E., Menusi, E., Mazutti, M., Toniazzo, G., Rigo, E., Cansian, R.L., Mossi, A., Oliveira, J.V., Luccio, M.D., Oliveira, J.V., Luccio, M.D., Oliveira, D., Treichel, H., 2008. Response surface methodology for optimization of lipase production by an immobilized newly isolated *Penicillium* sp. Industrial and Engineering Chemistry Research 47, 9651–9657.

Wu, Y., Teng, Y., Li, Z., Liao, X., Luo, Y., 2008. Potential role of polycyclic aromatic hydrocarbons (PAHs) oxidation by fungal laccase in the remediation of an aged contaminated soil. Soil Biology and Biochemistry 40, 789–796.

Wood, T.M., 1985. Properties of cellulolytic enzyme systems. Biochemical Society Transactions 13, 407–410.

Yang, H.H., Effland, M.J., Kirk, T.K., 1980. Factors influencing fungal degradation of lignin in a representative ligno-cellulosic, thermomechanical pulp. Biotechnology and Bioengineering 22 (1), 65–77. https://doi.org/10.1002/bit.260220106.

Ye, J.-S., Yin, H., Jing, Q., Peng, H., Qin, H.-M., Zhang, N., He, B.-Y., 2010. Biodegradation of anthracene by *Aspergillus fumigatus*. Journal of Hazardous Materials 185 (1), 174–181. https://doi.org/10.1016/j.jhazmat.2010.09.015.

Zafra, G., Montaño, A., Angel, A., Diana, C.-E., 2015. Degradation of polycyclic aromatic hydrocarbons in soil by a tolerant strain of *Trichoderma asperellum*. Environmental Science and Pollution Research 22 (2), 1034–1042. https://doi.org/10.1007/s11356-014-3357-y.

Further reading

Baldrian, P., Gabriel, J., 2002. Copper and cadmium increase laccase activity in *Pleurotus ostreatus*. FEMS Microbiology Letters 206, 69–74.

Bhattacharya, S., Das, A., G M, K V, J S., 2011. Mycoremediation of congo red dye by filamentous fungi. Brazilian Journal of Microbiology 42 (4), 1526–1536. https://doi.org/10.1590/S1517-838220110004000040.

Cernigilia, C.E., Yang, S.K., 1979. Stereoselective metabolism of anthracene and phenanthrene by the fungus cunninghamella elegans. Applied and Environmental Microbiology 47, 119–124.

Plant growth—promoting rhizobacteria and their functional role in salinity stress management

Akanksha Gupta[1], Sandeep Kumar Singh[2], Manoj Kumar Singh[3], Vipin Kumar Singh[2], Arpan Modi[4], Prashant Kumar Singh[4], Ajay Kumar[1,5]

[1]Institute of Environment & Sustainable Development, Banaras Hindu University, Varanasi, India; [2]Center of Advanced Study in Botany, Institute of Science, Banaras Hindu University, Varanasi, India; [3]Department of Chemistry, Indian Institute of Technology Delhi, Hauzkhas, India; [4]Agriculture Research Organization, Ministry of Agriculture and Rural Development Volcani Centre, Rishon LeZion, Israel; [5]Agriculture Research Organization (ARO), Volcani Center, Rishon LeZion, Israel

1. Introduction

Currently, the whole world is facing high risk of a series of biotic and abiotic stress factors. Biotic factors include the effect of living entities such as phytopathogens to the plants, whereas abiotic stress factors include draught, salinity, high pH, xenobiotic compounds, and accumulation of heavy metals. Currently, some of the abiotic factors such as salinity and draught appear as most severe problem for the growing population of world because of adverse effect on the growth yields, productivity of plant, soil, and environment (Wani et al., 2016). Besides growing populations of the world, rapid industrialization and use of huge amounts of pesticides in the crop fields generate a large amount of toxic, heavy metals, xenobiotics in the environment or agriculture field that severally affect the health, growth, productivity of plant, soil, or humans as well.

In all the abiotic factors, salinity, drought, and heat have been appeared as the most influencing or destructive factor that limits the growth, yield, and productivity of crops at each stages of life cycle (Mittler, 2006; Prasad et al., 2011). These stress factors influences normal physiology of the plants and affect their optimum productivity, yields, etc. Besides physiology, stress factors also influence the normal interactions between plant and pests, plant and beneficial microbes, etc. (Scherm and Coakley, 2003). The growth and productivity of plant depends on many factors in which soil characteristics are one of the important. The disturbances in soil properties by abiotic and biotic factors lead to affect the yield and productivity of the plants. Currently, 6% or 800 million hectares of total available land are used for the agricultural purposes only, and large part of the area have been affected by salinity, which affect the growth, yields, and productivity of the crops (Munns and Tester, 2008). Recently, FAO (Food and Agricultural Organization) has mentioned about availability of agricultural land and stated half of the total available agricultural land had been lost because of various stress factors in which salinity is one of the leading factors (Ilangumaran and Smith, 2017).

Generally, saline soil is defined as the soil having electrical conductivity (EC) of the saturation extract (ECe) more than $4\,dS\,mL^{-1}$ (approximately 40 mM NaCl) in the root zone at 25°C and exchangeable Na 15% (Shrivastava and Kumar, 2015). At this ECe and in certain crops below this ECe, hamper the growth and productivity of many crops, which ultimately affect the yields of crops (Munns, 2005; Jamil et al., 2012). The thresholds of salt stress vary with crop species, and most of the crops having low salinity thresholds such as maize have salinity tolerance up to $2\,dSm^{-1}$ and for wheat $6\,dSm^{-1}$ of tolerance (Saeed et al., 2001). In the arid and semiarid region of the world, it covers around 15% of the area and in the irrigated land covers up to 40% (Orhan, 2016). On the whole, 20% of total cultivated and 33% of irrigated agricultural lands are affected by high salinity worldwide. Furthermore, the level of salinity increasing at a rate of 10% annually and estimated that by the year of 2050, it covers more than 50% of the arable land (Jamil et al., 2012). The primary causing factor of the salinity in the soil and water surface is the accumulation of natural salt in the long-term process. Natural process includes breakdown of natural salts from the rocks, containing Na^+, Ca^{2+}, and Mg^{2+}, deposition of sea salt through the wind, surface runoff water, etc., whereas the human-induced anthropogenic activities are the main secondary process involved in causing salinity in the soil (Shrivastava and Kumar, 2015).

Under the saline conditions, plant faces irregularities in the metabolic pathways because of ion imbalance and their toxic effect. The high contents of ion normal osmosis give rise to water stress; all these phenomena lead to cellular toxicity in plants. In this stress condition, antioxidant system is like a boon to plants. In the salinity stress conditions, the root systems of the plant are largely influenced by the stress factors because of absorption of water and minerals absorbed from the surrounding (Jung and McCouch, 2013). In legumes, they not only limit the plant growth but also the nodulation and biomass of the plants (Han and Lee, 2005).

Recently, from last few decades, many researchers reported various studies about how to cope up with the salinity stress on plants. Genetic engineering emerges as a solution, but the difficulties in their implementation and hardness of the techniques make them unpopular (Xu et al., 2015; Liang et al., 2018). In this aspect, from last few decades, various beneficial microbes are used as plant or soil inoculants to withstand the salinity stress. The use of plant growth–promoting bacteria (PGPB) in the salinity stress management is not only sustainable approach to the plant soil or environment but also a cost-effective method in respect to other conventional management practices.

2. Plant growth—promoting rhizobacteria

Rhizosphere of the plant is the hot spot of microbial communities such as fungi, bacteria, cyanobacteria, and nematode microorganism. In this zone, large number of microbial communities interacts with the root exudates of the plant, some of them effectively colonized with the plant and help in growth promotion via direct or indirect ways (Kumar et al., 2014, 2015a,b,c). The microorganism which directly or indirectly associated with the plants for their growth promotion is called plant growth—promoting microorganism and the bacterial species, which resides around the rhizosphere of the plant and helps in growth promotion, is called plant growth—promoting rhizobacteria (PGPR). In the recent past, PGPR have been broadly used as microbial inoculants for the growth enhancement as a bio fertilizer, as biocontrol agent to inhibit the growth of various phytopathogens, or in the management of various biotic or abiotic stress (Kumar et al., 2016a,b, 2017, 2018; Singh et al., 2017, 2019). Many researchers have reported the ability of PGPR strains in reducing the effects of salinity and helping plants in tolerating the stress conditions. Some of the PGPR commonly used in mitigating salinity conditions are *Azospirillum, Azotobacter, Arthrobacter, Enterobacter, Burkholderia, Bacillus*, and *Pseudomonas* (Kumar et al., 2016a,b; Singh et al., 2018). Studies have shown that these bacterial genera have potential to increase the tolerance of salinity stress in many agricultural lands and enhance the productivity of crops. Some of them have been listed in Table 7.1.

3. Plant growth—promoting rhizobacteria in salinity stress

The generation of reactive oxygen species (ROS) in plant cell is the result of various metabolic stresses. According to Choudhary et al. (2007), when the plant is facing any kind of stress such as water, temperature, exposure to metals, the level of ROS increases with the level of severity of the stress; increase in ROS produces another oxidative stress to the plant cell. Oxidation of membrane proteins, lipids, or DNA is prevented by scavenging enzymes including catalase, superoxide dismutase, and ascorbate peroxidase. The ionic imbalance generated various kinds of free radicals during the ETS cycle. In another report by Noctor et al. (2002a,b), 70% of the H_2O_2 production is due to photorespiration, which is caused by the high salt concentration. Recently, many reports are published regarding production of ROS in plants by various plant growth—promoting microorganism endophytes (Hamilton et al., 2012).

3.1 Functional aspects of PGPR under salt stress

The aftermath of high salt condition on plants physiology and their development can be reduced by the use of PGPR. The approach of PGPR in salinity tolerance is proven as a green method to increase growth and productivity (Venkateswarlu et al., 2008; Grover et al., 2011; Coleman-Derr and Tringe, 2014). Many pathways are known in relation to understand how PGPR works in abiotic stress condition; these pathways are interlinked to each other but

TABLE 7.1   An overview of plant growth—promoting bacteria mediated salt tolerance.

PGPB	Experiments	Results	References
Pseudomonas fluorescens, Pseudomonas migulae	Two ACCD + bacterial endophytes were tested on tomato plants under salt stress	Plants pretreated with wild-type ACC deaminase containing endophytic strains were healthier and grew to a much larger size	Ali et al. (2014)
Azospirillum	Pot experiments; nitrogenase activity (in situ) was assayed, acetylene reduction assay was used	Restricted Na + uptake and increased K^+ and Ca^{2+} uptake	Hamdia et al. (2004)
Pseudomonas syringae, Pseudomonas fluorescens, Enterobacter aerogenes	Pot and jar experiments, Pot experiments, characterization of strains	Improving the growth and yield under salinity stress condition, it also increases total chlorophyll content (a, b, and carotenoids)	Nadeem et al. (2007))
Pseudomonas fluorescens	Serial dilution technique used to isolate pseudomonad strains, DNA extraction and amplification of *Pseudomonas* sp., amplification of ACC deaminase, isolation, cloning, and sequencing of 16S rDNA of *Pseudomonas fluorescens* strain TDK1	Enhanced ACC deaminase activity	Saravanakumar and Samiyappan (2007)
Pseudomonas mendocina	Seeds were grown for 15 days in peat substrate under nursery conditions, without any fertilization treatment, CAT activity was determined	ACC deaminase activity and enhanced uptake of essential nutrients	Kohler et al. (2009)
Rhizobium, Pseudomonas	Pot experiments; determination of phosphate-solubilizing bacteria was done	Decreased electrolyte leakage and increase in proline production, maintenance of relative water content of leaves, and selective uptake of K^+ ion	Bano and Fatima (2009)
Pseudomonas putida	Bacterial characterization, 16S rDNA identification, extraction and estimation of the IAA, plant growth promotion by Rs-198 in the fields	Increase the absorption of the Mg^{2+}, K^+, and Ca^{2+} and decrease the uptake of the Na_2	Yao et al. (2010)
Streptomyces sp.	Screening of actinobacterial isolates for NaCl tolerance, assessment of tomato growth promotion under NaCl stress	ACC deaminase activity and IAA production and phosphate, solubilization	Palaniyandi et al. (2014)

slight complicated in understanding. Removal of free radicals generated during high salt conditions in plant cell needs to be removed and performed by various enzymatic mechanisms of the plant cell. The power to withstand hardship of the microbes in the soil is very much dependent on their beneficial role in plant for growth and promotion during stress conditions (Dimkpa et al., 2009). According to Radhakrishnan and Baek (2017), PGPR triggers the ROS removal mechanisms by increasing the level of antioxidants and other compounds which act as ROS scavengers; this elevation in ROS reduction automatically increases the skilfulness of photosynthesis. Neutralization of free radicals also performed by the production of catalase enzyme as it is secreted by PGPR in significant amount. The search for the superoxide radicals is carried out by superoxide dismutase (SOD) whose efficiency can be improved by the application of PGPR.

Tripathi et al. (2002) reported under saline condition the activity and synthesis of nitrogenase enzyme is affected in the case of *Azospirillum* sp. *Azospirillum* sp. regulate the osmotic adaptation by depositing some compounds which regulate the process, e.g., proline, glutamate, trehalose, and glycinebetaine in the stress condition. In a study, Tripathi et al. (1998) inoculated *Azospirillum* strain in sorghum plants and found drastic change in the amount of proline, as the glutamate was also converted to proline due to the inoculation; the amount of water with its characteristics such as water potential is elevated and the canopy temperature has been declined compared to standard plants growing in the salinity stress condition. In another study, Gururani et al. (2013) inoculated two strains of *Bacillus* in to the potato plants and found that the gene related to free radical scavenging enzymes is expressed significantly that leads to the enhancement in photosynthetic ability of the treated plants.

Liu et al. (2017) used *Bacillus amyloliquefaciens* strain to identify the genes involved in tolerance of salinity stress. They have prepared the transcriptome profiles of *Arabidopsis* upper part (above ground), which was growing in saline condition and studied Illumina sequencing. After RNA sequencing, positive feedback to some mechanisms had been found because of more tolerance to salinity stress through the removal of free radicals, increase in the efficiency of photosynthesis, and accumulation of osmoregulation-related compounds. When the *Dietzia natronolimnaea* STR1 was inoculated in wheat plants growing in salt stress, it increases the expression level of TaWRKY10 in plants. According to Bharti et al. (2016), WRKY TFs play a role in regulation of abiotic stress by regulating removal of free radicals, some other stress-related genes and also by osmoregulation in cells. Jha et al. (2011) inoculated rice plant with *Pseudomonas pseudoalcaligenes* and *Bacillus pumilus* and found that the strains lowered the toxic effects of free radicals by lowering the apoptosis rate, plant cell membrane index, and the activity of cell caspase such as protease; these activities ultimately increase the viability of plant cells. These microbe-inoculated rice plants have shown lowered activities of APX and SOD, which can be interpreted as both the enzymes are involved in the stress tolerance of high salt content in GJ17 variety.

Plants inoculated with PGPB under high salt conditions produces antioxidant genes that involved in maintaining levels of ROS and confirm the role of PGPB in free radicals scavenging under salinity stress condition (Upadhyay et al., 2012). Gururani et al. (2013) reported some plants growing under the high salt conditions, after inoculation with *Pseudomonas frederiksbergensis* strain, showed decline in the levels of activities of APX and SOD enzymes this drop down can be interpreted as the enzymes are involved in the neutralization of the free radicals so the amount of free enzyme, which leads to the lowering in enzyme production, and also have plant growth promotion activity. The molecular approach of PGPR to the plants is a multiway pathway and mutualistic in nature (Singh et al., 2018).

4. PGPR and ACC deaminase activity

Ethylene hormone negatively affects the growth of plants, so the reduction in the level of ethylene is necessary function in saline stress condition to improvising the growth of plants (Nadeem et al., 2010). ACC deaminase producing PGPR promote the growth of plants through breakdown of ethylene under abiotic and biotic stresses (Glick et al., 2007). PGPB accumulate the ACC, a precursor of ethylene, and decrease its level by simply breaking down its precursor. In a study, Barnawal et al. (2017) monitored the level of ethylene between PGPB-inoculated plants having ACC deaminase activity and uninoculated plants growing under saline condition and observed that PGPB inoculation reduced the production level of ethylene compared with uninoculated plants. The production of ethylene is a two-step process, firstly conversion of S-adenosyl-L-methionine to ACC by the activity of ACC synthase and secondly conversion of ACC to ethylene by enzyme ACC oxidase. When plants are facing salinity stress, the level of ACC synthase rises in the absence of microbes showing ACC deaminase activity.

Singh et al. (2015) reported high salinity tolerance in wheat plants after inoculation with *Klebsiella* sp., which also showed ACC deaminase activity. Similarly, Li and Glick (2001) reported an increased growth in canola plants under salt stress when inoculated with *Pseudomonas putida* that have also ACC deaminase activity. The plants facing salinity stress, inoculated with *Enterobacter cloacae* HSNJ4, showed prominent tolerance level to the stress and lower the level of ethylene because of the bacteria-mediated digestion of ACC into ammonia and α-ketobutyrate, and ultimately lower the ethylene production in plant roots (Ali et al., 2014; Saleem et al., 2015). In a study, *Pseudomonas stutzeri*, *Pseudomonas fluorescens*, and *Pseudomonas aeruginosa* were isolated from the rhizospheric zone of tomato growing under salinity stress caused by NaCl; these isolates showed plant growth–promoting activities. These strains help plants to get over from the stress condition by producing various plants hormones including ACC deaminase activity (Bal et al., 2013; Tank and Saraf, 2010). When plants of tomato and peppers were inoculated with ACC deaminase producing strain *Achromobacter piechaudii* ARV8, plants have shown an enhanced level of stress tolerance (drought and salt) (Mayak et al., 2004). When tomato plants grown up to 90 mM concentration of NaCl were inoculated with *Pseudomonas putida* UW4, after 6 weeks of inoculation plants have shown good growth in shoots. Yan et al. (2014) reported enhancement in the activity of a gene from the chloroplast *Toc GTPase*, which import proteins in chloroplast.

5. Conclusion

In the recent scenario of increasing global population and changing climatic conditions, sustainable agricultural methods are of immediate need. From last few decades, various abiotic stresses adversely affect the yield and productivity of crops. Salinity is one of the severe stresses that affect the health, yield, and productivity of plant, soil, and environment. Mitigation of salinity stress through the use of beneficial microbes, including bacteria, fungi, appears as a promising approach in sustainable agriculture because of eco-friendly nature and less side effects. These beneficial microbes are currently used as plant and soil inoculants to withstand the various stresses through various direct or indirect mechanisms.

PGPR helps plants in salt stress via breakdown of enhanced ethylene hormone level with the ACC deaminase, as ethylene negatively affects the plant growth in stress condition. The use of PGPR in agriculture field is a green option to increase productivity and crop yield; but how they work and affect the plant system still needs more research.

The knowledge concerning the molecular signals and pathways driving the beneficial plant—microbe interactions is still limited; and lesser is known regarding the relationship between phytohormones in PGPR inoculated plants and their cumulative response to salinity stress at the whole plant scale. The systems biology approach for unraveling and understanding the intricacies of plant— microbe interactions under salt stressed conditions offers novel possibilities for employing the rhizosphere bacteria as sustainable agents of crop improvement. Research focused toward the utilization of PGPR in salt-affected agricultural fields promotes development of bacterial inoculants as commercial biofertilizers for improved salinity tolerance.

References

Ali, S., Charles, T.C., Glick, B.R., 2014. Amelioration of high salinity stress damage by plant growth-promoting bacterial endophytes that contain ACC deaminase. Plant Physiology and Biochemistry 80, 160—167.

Bal, H.B., Nayak, L., Das, S., Adhya, T.K., 2013. Isolation of ACC deaminase producing PGPR from rice rhizosphere and evaluating their plant growth promoting activity under salt stress. Plant and Soil 366 (1—2), 93—105.

Bano, A., Fatima, M., 2009. Salt tolerance in *Zea mays* (L.) following inoculation with *Rhizobium* and *Pseudomonas*. Biology and Fertility of Soils 45, 405—413.

Barnawal, D., Bharti, N., Maji, D., Chanotiya, C.S., Kalra, A., 2014. ACC deaminase-containing *Arthrobacter protophormiae* induces NaCl stress tolerance through reduced ACC oxidase activity and ethylene production resulting in improved nodulation and mycorrhization in *Pisum sativum*. Journal of Plant Physiology 171 (11), 884—894.

Bharti, N., Pandey, S.S., Barnawal, D., Patel, V.K., Kalra, A., 2016. Plant growth promoting rhizobacteria *Dietzia natronolimnaea* modulates the expression of stress responsive genes providing protection of wheat from salinity stress. Scientific Reports 6.

Choudhary, M., Jetley, U.K., Khan, M.A., Zutshi, S., Fatma, T., 2007. Effect of heavy metal stress on proline, malondialdehyde, and superoxide dismutase activity in the cyanobacterium *Spirulina platensis*-S5. Ecotoxicology and Environmental Safety 66 (2), 204—209.

Coleman-Derr, D., Tringe, S.G., 2014. Building the crops of tomorrow: advantages of symbiont-based approaches to improving abiotic stress tolerance. Frontiers in Microbiology 5.

Dimkpa, C., Weinand, T., Asch, F., 2009. Plant—rhizobacteria interactions alleviate abiotic stress conditions. Plant, Cell and Environment 32 (12), 1682—1694.

Glick, B.R., Cheng, Z., Czarny, J., Duan, J., 2007. Promotion of plant growth by ACC deaminase-producing soil bacteria. In: New Perspectives and Approaches in Plant Growth-Promoting Rhizobacteria Research. Springer, Dordrecht, pp. 329—339.

Grover, M., Ali, S.Z., Sandhya, V., Rasul, A., Venkateswarlu, B., 2011. Role of microorganisms in adaptation of agriculture crops to abiotic stresses. World Journal of Microbiology and Biotechnology 27 (5), 1231—1240.

Gururani, M.A., Upadhyaya, C.P., Baskar, V., Venkatesh, J., Nookaraju, A., Park, S.W., 2013. Plant growth-promoting rhizobacteria enhance abiotic stress tolerance in *Solanum tuberosum* through inducing changes in the expression of ROS-scavenging enzymes and improved photosynthetic performance. Journal of Plant Growth Regulation 32 (2), 245—258.

Hamdia, M.A.E.S., Shaddad, M.A.K., Doaa, M.M., 2004. Mechanisms of salt tolerance and interactive effects of *Azospirillum brasilense* inoculation on maize cultivars grown under salt stress conditions. Plant Growth Regulation 44 (2), 165—174.

Hamilton, C.E., Gundel, P.E., Helander, M., Saikkonen, K., 2012. Endophytic mediation of reactive oxygen species and antioxidant activity in plants: a review. Fungal Diversity 54 (1), 1—10.

Han, S.H., Lee, J.H., 2005. An overview of peak-to-average power ratio reduction techniques for multicarrier transmission. IEEE wireless communications 12 (2), 56—65.

Ilangumaran, G., Smith, D.L., 2017. Plant growth promoting rhizobacteria in amelioration of salinity stress: a systems biology perspective. Frontiers of Plant Science 8.

Jamil, M., Bashir, S., Anwar, S., Bibi, S., Bangash, A., Ullah, F., Rha, E.S., 2012. Effect of salinity on physiological and biochemical characteristics of different varieties of rice. Pakistan Journal of Botany 44, 7–13.

Jha, Y., Subramanian, R.B., Patel, S., 2011. Combination of endophytic and rhizospheric plant growth promoting rhizobacteria in *Oryza sativa* shows higher accumulation of osmoprotectant against saline stress. Acta Physiologiae Plantarum 33 (3), 797–802.

Jung, J.K., McCouch, S., 2013. Getting to the roots of it: genetic and hormonal control of root architecture. Frontiers of Plant Science 4.

Kohler, J., Hernandez, J.A., Caravaca, F., Roldan, A., 2009. Induction of antioxidant enzymes is involved in the greater effectiveness of a PGPR versus AM fungi with respect to increasing the tolerance of lettuce to severe salt stress. Environmental and Experimental Botany 65, 245–252.

Kumar, A., Singh, V.K., Tripathi, V., Singh, P.P., Singh, A.K., 2018. Plant growth-promoting rhizobacteria (PGPR): perspective in agriculture under biotic and abiotic stress. In: Crop Improvement Through Microbial Biotechnology, pp. 333–342. https://doi.org/10.1016/B978-0-444-63987-5.00016-5.

Kumar, A., Vandana, Singh, R., Singh, M., Pandey, K.D., 2015c. Plant growth promoting rhizobacteria (PGPR): a promising approach for disease management. In: Singh, J.S., Singh, D.P. (Eds.), Microbes and Environmental Management. Studium Press, New Delhi, pp. Pp195–209.

Kumar, A., Vandana, Yadav, A., Giri, D.D., Singh, P.K., Pandey, K.D., 2015b. Rhizosphere and their role in plant – microbe interaction. In: Chaudhary, K.K., Dhar, D.W. (Eds.), Microbes in Soil and Their Agricultural Prospects. Nova Science Publisher, Inc., pp. 83–97

Kumar, A., Verma, H., Singh, V.K., Singh, P.P., Singh, S.K., Ansari, W.A., Yadav, A., Singh, P.K., Pandey, K.D., 2017. Role of *Pseudomonas* sp. in sustainable agriculture and disease management. In: Agriculturally Important Microbes for Sustainable Agriculture. Springer, Singapore, pp. 195–215.

Kumar, V., Kumar, A., Pandey, K.D., Roy, B.K., 2015a. Isolation and characterization of bacterial endophytes from the roots of *Cassia tora* L. Annals of Microbiology 65, 1391, 139.

Kumar, A., Singh, R., Giri, D.D., Singh, P.K., Pandey, K.D., 2014. Effect of *Azotobacter chroococcum* CL13 inoculation on growth and curcumin content of turmeric (*Curcuma longa* L.). International Journal of Current Microbiology and Applied Sciences 3 (9), 275–283.

Kumar, A., Singh, R., Yadav, A., Giri, D.D., Singh, P.K., Pandey, K.D., 2016b. Isolation and characterization of bacterial endophytes of Curcuma longa L. 3 Biotechnology 6, 60.

Kumar, A., Vandana, Singh, M., Singh, P.P., Singh, S.K., Singh, P.K., Pandey, K.D., 2016a. Isolation of plant growth promoting rhizobacteria and their impact on growth and curcumin content in Curcuma longa L. Biocatalysis and Agricultural Biotechnology 8, 1–7.

Li, J., Glick, B.R., 2001. Transcriptional regulation of the *Enterobacter cloacae* UW4 1-aminocyclopropane-1-carboxylate (ACC) deaminase gene (acdS). Canadian Journal of Microbiology 47 (4), 359–367.

Liang, W., Ma, X., Wan, P., Liu, L., 2018. Plant salt-tolerance mechanism: a review. Biochemical and Biophysical Research Communications 495 (1), 286–291.

Liu, S., Hao, H., Lu, X., Zhao, X., Wang, Y., Zhang, Y., Wang, R., 2017. Transcriptome profiling of genes involved in induced systemic salt tolerance conferred by *Bacillus amyloliquefaciens* FZB42 in *Arabidopsis thaliana*. Scientific Reports 7 (1), 10795.

Mayak, S., Tirosh, T., Glick, B.R., 2004. Plant growth-promoting bacteria confer resistance in tomato plants to salt stress. Plant Physiology and Biochemistry 42 (6), 565–572.

Mittler, R., 2006. Abiotic stress, the field environment and stress combination. Trends in Plant Science 11 (1), 15–19.

Munns, R., Tester, M., 2008. Mechanisms of salinity tolerance. Annual Review of Plant Biology 59, 651–681.

Munns, R., 2005. Genes and salt tolerance: bringing them together. New Phytologist 167 (3), 645–663.

Nadeem, S.M., Zahir, Z.A., Naveed, M., Arshad, M., 2007. Preliminary investigation on inducing salt tolerance in maize through inoculation with rhizobacteria containing ACC-deaminase activity. Canadian Journal of Microbiology 53, 1141–1149.

Nadeem, S.M., Zahir, Z.A., Naveed, M., Asghar, H.N., Arshad, M., 2010. Rhizobacteria capable of producing ACC-deaminase may mitigate salt stress in wheat. Soil Science Society of America Journal 74 (2), 533–542.

Noctor, G., Gomez, L., Vanacker, H., Foyer, C.H., 2002a. Interactions between biosynthesis, compartmentation and transport in the control of glutathione homeostasis and signalling. Journal of Experimental Botany 53 (372), 1283−1304.

Noctor, G., Veljovic−Jovanovic, S.O.N.J.A., Driscoll, S., Novitskaya, L., Foyer, C.H., 2002b. Drought and oxidative load in the leaves of C3 plants: a predominant role for photorespiration? Annals of Botany 89 (7), 841−850.

Orhan, F., 2016. Alleviation of salt stress by halotolerant and halophilic plant growth-promoting bacteria in wheat (*Triticum aestivum*). Brazilian Journal of Microbiology 47 (3), 621−627.

Palaniyandi, S.A., Damodharan, K., Yang, S.H., Suh, J.W., 2014. Streptomyces sp. strain PGPA39 alleviates salt stress and promotes growth of 'Micro Tom' tomato plants. Journal of Applied Microbiology 117, 766−773.

Prasad, P.V.V., Pisipati, S.R., Momcilovic, I., Ristic, Z., 2011. Independent and combined effects of high temperature and drought stress during grain filling on plant yield and chloroplast EF-Tu expression in spring wheat. Journal of Agronomy and Crop Science 197 (6), 430−441.

Radhakrishnan, R., Baek, K.H., 2017. Physiological and biochemical perspectives of non-salt tolerant plants during bacterial interaction against soil salinity. Plant Physiology and Biochemistry 116, 116−126.

Saeed, M.M., Ashraf, M., Asghar, M.N., Bruen, M., Shafique, M.S., 2001. Root Zone Salinity Management Using Fractional Skimming Wells With Pressurized Irrigation. International Water Management Institute (IWMI), Regional Office for Pakistan, Central Asia and Middle East, Lahore, p. 46.

Saleem, A.R., Bangash, N., Mahmood, T., Khalid, A., Centritto, M., Siddique, M.T., 2015. Rhizobacteria capable of producing ACC deaminase promote growth of velvet bean (*Mucuna pruriens*) under water stress condition. International Journal of Agriculture and Biology 17 (3).

Saravanakumar, D., Samiyappan, R., 2007. ACC deaminase from Pseudomonas fluorescens mediated saline resistance in groundnut (*Arachis hypogea*) plants. Journal of Applied Microbiology 102, 1283−1292.

Seherm, H., Coakley, S.M., 2003. Plant pathogens in a changing world. Australasian Plant Pathology 32 (2), 157−165.

Shrivastava, P., Kumar, R., 2015. Soil salinity: a serious environmental issue and plant growth promoting bacteria as one of the tools for its alleviation. Saudi Journal of Biological Sciences 22 (2), 123−131.

Singh, R.P., Jha, P., Jha, P.N., 2015. The plant-growth-promoting bacterium *Klebsiella* sp. SBP-8 confers induced systemic tolerance in wheat (*Triticum aestivum*) under salt stress. Journal of Plant Physiology 184, 57−67.

Singh, V.K., Singh, A.K., Kumar, A., 2017. Disease management of tomato through PGPB: current trends and future perspective. 3 Biotechnology 7 (4), 255.

Singh, V.K., Singh, A.K., Singh, P.P., Kumar, A., 2018. Interaction of plant growth promoting bacteria with tomato under abiotic stress: a review. Agriculture, Ecosystems and Environment 267, 129−140.

Singh, M., Singh, D., Gupta, A., Pandey, K.D., Singh, P.K., Kumar, A., 2019. Plant Growth Promoting Rhizobacteria: Application in Biofertilizers and Biocontrol of Phytopathogens,. In: PGPR Amelioration in Sustainable Agriculture,. Woodhead Publishing, pp. 41−66.

Tank, N., Saraf, M., 2010. Salinity-resistant plant growth promoting rhizobacteria ameliorates sodium chloride stress on tomato plants. Journal of Plant Interactions 5 (1), 51−58.

Tripathi, A.K., Mishra, B.M., Tripathi, P., 1998. Salinity stress responses in the plant growth promoting rhizobacteria, *Azospirillum* spp. Journal of Biosciences 23 (4), 463−471.

Tripathi, A.K., Nagarajan, T., Verma, S.C., Le Rudulier, D., 2002. Inhibition of biosynthesis and activity of nitrogenase in *Azospirillum* brasilense Sp7 under salinity stress. Current Microbiology 44 (5), 363−367.

Upadhyay, S.K., Singh, J.S., Saxena, A.K., Singh, D.P., 2012. Impact of PGPR inoculation on growth and antioxidant status of wheat under saline conditions. Plant Biology 14 (4), 605−611.

Venkateswarlu, B., Desai, S., Prasad, Y.G., 2008. Agriculturally important microorganisms for stressed ecosystems: challenges in technology development and application. In: Khachatourians, G.G., Arora, D.K., Rajendran, T.P., Srivastava, A.K. (Eds.), Agriculturally Important Microorganisms, vol. 1. Academic World, Bhopal, pp. 225−246.

Wani, S.H., Kumar, V., Shriram, V., Sah, S.K., 2016. Phytohormones and their metabolic engineering for abiotic stress tolerance in crop plants. The Crop Journal 4 (3), 162−176.

Xu, J., Xing, X.J., Tian, Y.S., Peng, R.H., Xue, Y., Zhao, W., Yao, Q.H., 2015. Transgenic *Arabidopsis* plants expressing tomato glutathione S-transferase showed enhanced resistance to salt and drought stress. PLoS One 10 (9), e0136960.

Yan, J., Wang, Q., Wei, T., Fan, Z., 2014. Recent advances in design and fabrication of electrochemical supercapacitors with high energy densities. Advanced Energy Materials 4 (4).

Yao, L., Wu, Z., Zheng, Y., Kaleem, I., Li, C., 2010. Growth promotion and protection against salt stress by *Pseudomonas putida* Rs-198 on cotton. European Journal of Soil Biology 46, 49—54.

Further reading

Farooq, M., Wahid, A., Kobayashi, N., Fujita, D., Basra, S.M.A., 2009. Plant drought stress: effects, mechanisms and management. In: Sustainable Agriculture. Springer, Dordrecht, pp. 153—188.

Han, Y., Wang, R., Yang, Z., Zhan, Y., Ma, Y., Ping, S., Zhang, L., Lin, M., Yan, Y., 2015. 1-Aminocyclopropane-1-carboxylate deaminase from *Pseudomonas stutzeri* A1501 facilitates the growth of rice in the presence of salt or heavy metals. Journal of Microbiology and Biotechnology 25 (7), 1119—1128.

Plant growth—promoting bacteria and their role in environmental management

Divya Singh[1], Sandeep Kumar Singh[1], Vipin Kumar Singh[1], Akanksha Gupta[2], Mohd Aamir[1], Ajay Kumar[3]

[1]Center of Advanced Study in Botany, Institute of Science, Banaras Hindu University, Varanasi, India; [2]Institute of Environment & Sustainable Development, Banaras Hindu University, Varanasi, India; [3]Agriculture Research Organization (ARO), Volcani Centre, Rishon LeZion, Israel

1. Introduction

Modern agricultural and industrialization practices lead to nonjudicious usage of natural resources and excessive production of xenobiotic compounds that give rise to several environmental problems resembling contamination of soil, water, and air, harmful effects on various organism, and interruption of biogeochemical cycles. Xenobiotic compounds (such as pesticides, azo dyes, explosives, industrial solvents, polycyclic aromatic hydrocarbons, alkanes, furans, dioxins, nitro-aromatic compounds, chlorinated aromatic compounds, brominated retardants, and petroleum products, etc.) are the substances synthesized for several agricultural and industrial applications. From last few decades, rapid industrialization and indiscriminate use of chemical fertilizers results in deposition of contaminants or pollutants in the environment. Accumulation of xenobiotics is the severe environmental concern that adversely influences the health and productivity of plant and soil (Gerhardt et al., 2009). Remediation of these pollutants is the prime concern for the researchers, farmers, and policy-makers.

To remediate the problem of pollutants, various physical and chemical technologies are present, but this leads to various complications in the environment because of their degradation products or side effects. In this context use of biological remediation techniques in the mitigation of various pollutants, they are less expensive, suitable, sustainable, eco-friendly

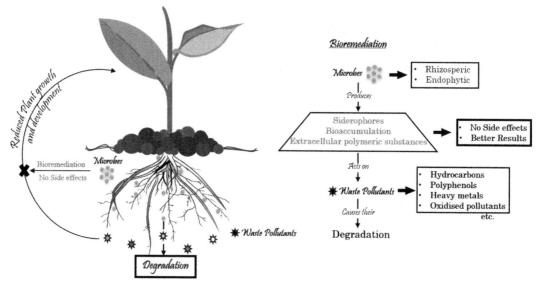

FIGURE 8.1   Overview of plant growth—promoting bacteria (PGPB) in bioremediation.

in nature, and the much attractive choice of remediation. Bioremediation is the remediation techniques used to reclaimate or mitigate the environment contaminants or pollutants by using the beneficial microbes and plants through the specific biological mechanisms (Fig. 8.1).

Currently numerous reports are documented regarding the use of beneficial microbes in the bioremediation of organic pollutants, xenobiotics, etc. (Zhuang et al., 2007; Myresiotis et al., 2012; Roy et al., 2013). The use of rhizosphere-associated microbes in the remediation of various contaminants is termed as rhizoremediation (Schwab and Banks, 1994). It is observed that bioremediation is an eco-friendly and cost-effective method as compared to the conventional expensive and ineffective, physical, or chemical methods (Ayangbenro and Babalola, 2017). The cost-effectiveness of bioremediation is very much low as compared to the conventional physical or chemical methods. Blaylock et al. (1997) reported 50%—65% per Acer low cost in the treatment of Pb-polluted soil through the bioremediation techniques.

2. Plant growth—promoting bacteria

Plant growth—promoting bacteria (PGPB) are a heterogeneous group of bacteria comprising of both rhizospheric and endophytic regions that are directly or indirectly involved in growth promotion. Currently from last few decades various plant growth—promoting microbe used as plant or soil inoculants not only for plant growth promotion but also for rthe management of various biotic and abiotic stresses (Kumar et al., 2017a, 2018). The rhizosphere, a thin layer of soil adhering to the root surface of the plant, is the hotspot of various group of microbes, including bacteria, actinomycetes, fungi, etc. (Kumar et al., 2015a,b; 2017a, 2018). The plant secretes an array of root exudates that contains diverse amount of nutrient such as amino acids, lipids, carbohydrates. These exudates provide nutrient source for the growing microbes at the rhizospheric regions and help in effective colonization. The microbes present at the rhizospheric region are called rhizospheric microorganisms. Some of the rhizospheric microbes enter inside the plants and act as endophytes without causing any external sign of infection. Both the rhizospheric and endophytic bacteria effectively colonized the plants and directly or indirectly involved in growth promotion, disease, and various biotic and abiotic stresses including salinity, heavy metals, pesticides, and pH management (Kumar et al., 2017a,b; Singh, 2017a,b,c; Hayat et al., 2010; Rajkumar et al., 2010).

3. Xenobiotic compounds and their classification

Most of the xenobiotic compounds are characterized by their chemical and thermodynamic stability that makes them ideal for industrial utilization and enhances their commercial importance. These compounds can enter to the environment via different ways: (1) pharmaceutical and chemical industry liberate synthetic polymers, (2) bleaching of paper and pulp releases chlorinated compounds into the environment, (3) mining is the source of entry of heavy metals into biogeochemical cycles, (4) oil spills cause accidental release of fossil fuels into the ecosystem, (5) rigorous agriculture discharges considerable quantity of pesticides and fertilizers into the soil and water recourses.

Many xenobiotic compounds such as polycyclic aromatic hydrocarbons (PAHs) consequent of fossil fuel consumption and exploration, pentachlorophenol (PCP), hexachlorocyclohexane (HCH), and polychlorinated biphenyls (PCBs) are highly resistant to biodegradation. HCH, PCB, and PCP exhibit strong bioaccumulation tendency. Thus, organisms present at higher trophic level have a tendency to accumulate high amount of these toxic compounds. During embryonic development and breastfeeding, these xenobiotic compounds may reach to infant from the mother. Besides bioaccumulation, many xenobiotic compounds may impart toxic effects on humans, animals, and plants (Hassen et al., 2018; Zhuang et al., 2007).

Pesticides: Pesticides are widely used in agriculture and household purpose throughout the world. Pesticides are chemical substances synthesized to control pests, diseases, and weeds. It has been estimated that a very small part (less than 0.1%) of total proportion of pesticides applied in the field reaches to the action sites, a huge amount being wasted as runoff, spray drift, photodegradation, etc.

Heavy metals: Normal concentration of heavy metals is an essential requirement for the growth and functioning of all the living organisms. Their presence plays a significant role in numerous biological activities such as act as cofactor in enzymatic reactions, proteins, stimulator in biological pathway, etc. (da Silva et al., 1991; Appanna et al., 1995). The metals are classified on the basis of their specific weight, metals having specific weight more than $5.0 \, g/cm^3$, classified into heavy metals, which are subcategorized into three classes: toxic metals (e.g., Cr, Pb, Zn, Cu, Ni, Cd, As, Co, Sn, etc.), precious metals (e.g., Ag, Au, Pd, Pt, etc.) and radionuclides (e.g., Ra, Am U, Th, etc.) (Nies, 1999; Bishop, 2002; Ahemad, 2014b).

It is found that accumulation of excess amount of metal ions over the period of time results in the increased toxicity through the biomagnifications process and adversely affects the normal functioning and causes carcinogenic and various types of chronic syndromes in human beings by affecting their normal food chain (Ahemad and Malik, 2011). The excess concentration of heavy metal is the great concern for the researchers and environmentalists because of high persistence, toxicity, and low degradability (Ahmad and Ashraf, 2012). The toxicity of the metal varies with the nature and the concentration of the elements. Elements even change their nature by changing environmental conditions.

Persistent organic pollutants (POPs): They are anthropogenic substances synthesized by humans. United Nations Environment Program (UNEP) has listed 12 chemicals under POPs category including furans, dioxins, and nine organochlorine pesticides. POPs are resistant to natural degradation processes and persists in the environment. They induce toxic effects on living organism by accumulating in the soil and water resources as well as in food chains.

4. Effect of xenobiotics on the health of human beings

With the development of society, xenobiotic compounds have brought high potential risk to human and animals. Earlier studies have shown the adverse effect of these compounds on health (Dobrovolskaia, 2015; Hessel et al., 2016; Moustafa et al., 2016). The immune system is highly susceptible to these compounds that results in the reduced immunity and generation of tumors (Zhu et al., 2017). Immunotoxic xenobiotics can suppress the defense system of body against various pathogens. These compounds also disrupt the mitochondrial functions, resulting in generation of reactive oxygen and nitrogen species and may cause several age-related disorders such as Alzheimer's disease, acute chronic syndrome, cancer, coronary artery disease, diabetes, metabolic syndrome, and Parkinson's disease.

Human has high risk of pesticide exposure during manufacturing, formulation, and application processes. Pesticides induce various toxic effects (such as cytotoxic, genotoxic, hepatotoxic, immunotoxic, neurotoxic, and nephrotoxic) on humans and animals. Long-term exposure of some pesticides has carcinogenic, hormonal, and mutagenic effects. However, exposure of pesticide before pregnancy increases risk of infertility, whereas exposure during pregnancy causes spontaneous abortion, prenatal death, retardation of fetal growth, congenital deformity, and infancy cancer. Pesticides also affect humans and animals indirectly. They can accumulate in vegetables, fruits, and grains to a toxic level and enter in the organism after consumption of those food materials and influence health-related problems (Nicolopoulou-Stamati et al., 2016; Zhu et al., 2017).

Ionic forms of heavy metals are extremely toxic to humans and animals. They induce oxidative stress, impair cell structure, and damage tissues and organs. Cd interferes with calcium metabolism and higher doses of Cd cause itai-itai disease. Arsenate agitates metabolism of phosphate. However, renal damage is prominent in As toxicity, whereas Cd toxicity leads to osteoporosis. Similarly, low exposure of Pd induces intellectual disabilities in children as well as neurotoxicity. Methyl mercury also shows neurotoxic effects (Jaishankar et al., 2014; Clemens and Ma, 2016).

POPs are fat-soluble chemical substances that are highly toxic to living beings. Long-term exposure of POPs causes abnormal development and growth, increased pathogen susceptibility, and impaired neurobehavioral development and functions. Dioxins are responsible for birth defects, tumor production, immunotoxicity, and death. Along with dioxins, many organochlorinated chemicals such as DDT, PCBs, etc., are regarded as endocrine disrupter. POPs affect DNA directly or indirectly. PCBs have negative effects on behavior, immune, hormonal, and reproductive system of animals and human. Furthermore, PCBs are also considered as neurotoxic substances that adversely affect central nervous system and sensory organs (Qing et al., 2006; Sharma et al., 2014).

5. Effects of xenobiotics on the plant growth

Besides various health-related risks to human, contamination of the environment with xenobiotic substances poses high threat to the diversity of ecosystems. Random and injudicious application of pesticides in agriculture has emanated in numerous undesirable consequences such as ecological imbalances, presence of pesticide residues in vegetables, fruits, food, soil, and water, pest rejuvenation, etc. Application of pesticides can inhibit seed germination and growth of plants. Although at low concentrations many pesticides do not show pernicious effects, they adversely affect plant growth and productivity at higher concentrations. Rajashekar and Murthy (2012) have reported that pendamethane drastically reduced seed germination in *Zea mays* L. cv NAAC-6002 at higher concentration (10 ppm). However, reduced shoot/root growth has been reported in different plants after application of imidacloprid, dimethoate, and monocrotophos (Stevens et al., 2008; Mishra et al., 2008; Saraf and Sood, 2002). Pesticide accumulation also results in several metabolic disorders and oxidative damage in plants. They induce plant death by several mechanisms such as inhibition of cell division, photosynthesis, enzyme function, pigment synthesis, interference in protein and DNA synthesis, and promotion of unrestrained growth (Parween et al., 2016).

Heavy metals including Cd, Hg, As, and Pb are extremely toxic for plants when present in ionic forms in the soil. At higher concentration, they impede seed germination, shoot/root growth, production, and alter physiological and metabolic processes in plants. Nickel (Ni) alters protease, amylase and ribonuclease activity in seeds resulting in retardation of germination. In addition, Cu toxicity generates oxidative stress by altering antioxidant defense system of plants. On the other hand, Cd toxicity delays seed germination, inducing membrane damage and nutrient loss and inhibiting invertase and α-amylase enzyme activity (Sethy and Ghose, 2013).

Numerous POP pesticides have been widely used in agricultural field and most of them are absorbed by roots and leaves from soil and air causing severe damage to plants. Polychlorinated dibenzofurans (PCDD) and PCBs can be taken up by the roots and leaves from soil and air, respectively. PCDD influences ultrastructure of cell such as surface of cell wall and chloroplasts, inducing oxidative stress in tobacco (Zhang et al., 2012). PCBs can inhibit several physiological processes in plants including cell division, DNA synthesis, photosynthesis, etc. (Cheng et al., 1984). PCP adversely affects plant survival, growth, and reproduction (https://nature.nps.gov/hazardssafety/toxic/pcp.pdf). It has been reported that PAHs can stimulate plant growth at early developmental stage when applied in low concentration (100 mg/kg); at higher concentration it inhibits the growth in bean, maize, oat, sunflower, and wheat (Maliszewska-Kordybach and Smreczak, 2000).

5.1 Plant growth—promoting bacteria in bioremediation

PGPB are generally used as plant and soil inoculants from last few decades to enhance the growth promotion and disease management in plants. PGPB involve various direct or indirect mechanisms for the growth promotion or to control the growth of phytopathogens (Kumar et al., 2014,2016a,b). Currently plant growth–promoting microbes are well-documented for their ability of biodegradation against various environmental pollutants (Shaheen and Sundari, 2013; Pratibha and Krishna, 2015) as briefly described in Table 8.1. Recent developments have suggested that plant growth–promoting rhizobacteria (PGPR) degrade toxic xenobiotic compounds more rapidly than the endophytic microbes. In this way, PGPR support plant growth in contaminated site with xenobiotic compounds that would alternatively obstruct normal plant growth. Microbes metabolize toxic xenobiotic compounds using as a source of carbon, energy, and micro/macronutrients. Díaz et al. (2010) have reported that some microbes can use xenobiotic compounds as final electron receptor in respiratory system. Complete degradation/mineralization of xenobiotic compounds is essentially required for energy generation and nutrients utilization. In some cases, microbes can transform toxic xenobiotic compound into less toxic organic substrate. Such processes would require the presence of a metabolizable substrate and known as cometabolic biotransformation. The cometabolic biotransformation alters molecular structure of xenobiotic compound, which result in modification or complete loss of characteristic features of the original compound.

Many researchers well-documented the use of PGPB strains in bioremediation of various xenobiotics compounds. Myresiotis et al. (2012) have reported that *Bacillus subtilis* GB03, *Bacillus subtilis* FZB24, *Bacillus amyloliquefaciens* IN937a, and *Bacillus pumilus* SE34 are able to degrade acibenzolar-S-methyl around 5 times higher than the uninoculated samples in the medium. Myresiotis et al. (2012) have conducted an experiment in which they have inoculated four PGPR strains in different medium containing napropamide, thiamethoxam, propamocarb hydrochloride and metribuzin; after 72 h they have recorded that all the medium were devoid of pesticides from 9% to 11%, 11% to 22%, 15% to 36%, and 8% to 18%, respectively, and concluded that the presence of PGPR strains lowered the half-life of acibenzolar-S-methyl. In another study, Yee et al. (1998) have reported that *Pseudomonas fluorescens* degraded trichloroethylene into toluene *o*-monooxygenase, whereas *Pseudomonas fluorescens* F113 can be metabolized PCBs as reported by Villacieros et al. (2005). *Pseudomonas putida* PML2 is capable in direct degradation of PCBs (Narasimhan et al., 2003). Bisht et al. (2015) reviewed remediation ability of PAHs by the use of microbes from the contaminated sites. In another study Tesar et al. (2002) utilized rhizospheric bacterial strains to clean the diesel-contaminated sites. Grass inoculations with *Pseudomonas aeruginosa* strain stimulate the germination of seed by 80% and degrade 2-chlorobenzoic acid up to 20% (Siciliano and Germida, 1997).

5.2 Plant growth–promoting bacteria mechanism of xenobiotics degradation

PGPR play important role in bioremediation of pollutants. It has been reported that PGPR can solubilize and degrade various kind of pollutants including herbicides, hydrocarbons, and explosives, etc. Several PGPR have been isolated from different contaminated locations to study the degradation pathways and enzyme-coding genes (Burd et al., 2000; Germaine et al., 2006; Singh and Cameotra, 2013). Microbes apply two methods for biodegradation of xenobiotic compounds—anaerobic biodegradation and aerobic biodegradation. Mobilization, immobilization, and transformation mechanisms of PGPR resistance reduce toxicity to

TABLE 8.1 Plant growth–promoting bacteria used in bioremediation studies.

Bacteria	Plant	Heavy metal	Conditions	Role of PGPR	References
Ochrobactrum sp., *Bacillus* sp.	*Oryza sativa*	Cd, Pb	Pot experiment	Germination percentage, relative root elongation (RRE), amylase, and protease activities were increased	Pandey et al. (2013)
Pseudomonas putida KNP9	*Vigna radiate*	Cd, Pb	Greenhouse experiment	Reduction of Cd and Pb uptake and enhanced plant growth	Tripathi et al. (2005)
Achromobacter xylosoxidans Ax10	*Brassica juncea*	Cu	Pot experiment	Enhance Cu uptake; increase root, shoot length and fresh, dry weights	Ma et al. (2009)
Pseudomonas sp.; *Pseudomonas fluorescence*; *Bacillus cereus*	*Zea mays*	Pb, Cd, Ni	Pot experiment	Ag-nano particles augmented the PGPR-induced increase in root area and root length	Khan and Bano (2015)
Pseudomonas fluorescens G10; *Microbacterium* sp. G16 (EN)	*Brassica napus*	Pb	Pot experiment	Increase Pb uptake in shoots, increase root length, shoot and dry weight	Sheng et al. (2008)
Methylobacterium oryzae; *Berknolderia* sp.	*Lycopersicon esculentom*	Ni, Cd	Pot culture experiment	Promotes growth of plants	Madhaiyan et al. (2007)
Pseudomonas sp.	Soybean, mung bean; *Triticum aestivum*	Ni, Cd, Cr	Pot experiment	Promotes growth of plants	Gupta et al. (2002)
Mesorhizobium huakuii subsp. rengei B3	*Astragalus sinicus*	Cd	Hydroponic experiment	Expression of PCS_{At} gene increased ability of cells to bind Cd^{2+} approximately 9- to 19-fold	Sriprang et al. (2003)
Xanthomonas sp., RJ3, *Azomonas* sp. RJ4, *Pseudomonas* sp. RJ10, *Bacillus* sp. RJ31	*Brassica napus*	Cd	Pot experiment	Cd accretion elevated and stimulation of growth of plant	Sheng and Xia (2006)
Azotobacter chroococcum HKN-5	*Brassica juncea*	Pb, Zn	Greenhouse experiment	Stimulated plant growth	Wu et al. (2006)
Pseudomonas aeruginosa MKRh3	*Vigna mungo*	Cd	Pot experiment	Plants elaborated lessened accretion, increase root, and stimulation of growth of plant	Ganesan (2008)
Brevibacillus sp.	*Trifolium repens*	Zn	Pot experiment	Improve plant growth, nutrition, and reduced Zn content in plant tissues	Vivas et al. (2006)
Kluyvera ascorbata SUD165	*Brassica juncea*	Nickel, lead, zinc	Pot experiment	Both strains decreased some plant growth inhibition by heavy metals	Burd et al. (2000)

(Continued)

TABLE 8.1   Plant growth—promoting bacteria used in bioremediation studies.—cont'd

Bacteria	Plant	Heavy metal	Conditions	Role of PGPR	References
Rhizobium sp. RP5	Pisum sativum	Ni	Pot experiment	Improved the dry biomass, nodule numbers, seed yield, and grain protein	Wani et al. (2007)
Ochrobactrum; Bacillus Cereus	Vigna radiata	Cr	Pot experiment	Stimulated plant growth and decreased Cr content	Rajkumar et al. (2006)
Bacillus subtilis SJ-101	Brassica juncea	Ni	Pot experiment	Facilitated Ni accumulation	Zaidi et al. (2006)
Bacillus megaterium HKP-1	Brassica juncea	Pb, Zn	Greenhouse experiment	Protected plant from metal toxicity	Wu et al. (2006)
Ochrobactrum intermedium	Helianthus annuus	Cr	Pot experiment	Increased plant growth and decreased Cr uptake	Faisal and Hasnain. (2005)
Microbacterium oxydans AY509223 (RS)	Alyssum murale	Ni	Pot experiment	Aided in phytoextraction of Ni	Abou- Shanab et al. (2006)
Bacillus licheniformis NCCP-59	Oryza sativa	Ni	Pot experiment	Improved seed germination under Ni stress and safeguarded against toxicity	Jamil et al. (2014)
Bacillus weihenstephanensis SM3	Helianthus annuus	Ni, Cu, Zn	Pot experiment	Improved plant biomass and the accretion of Zn and Cu in the root and shoot systems; enhanced the concentrations of soluble Ni, Zn, and Cu in soil with their metal mobilizing potential	Rajkumar et al. (2008)

tolerate heavy metal uptake (Ahemad, 2014b). The major mechanisms applied by these microorganisms are exclusion, complexation, sequestration, and detoxification, etc. Immobilization of heavy metals by extracellular materials can prevent their uptake into bacterial cells. Many heavy metals bind with the anionic functional groups (such as amide, amine, carboxyl, hydroxyl, sulfhydryl, and sulfonate groups) of cell surface. Similarly, extracellular polymers (e.g., humic substances, polysaccharides, and proteins) of the bacterial cell also proficiently bind heavy metals (Ahemad and Kibret, 2013; Ayangbenro and Babalola, 2017). Thus, these substances can detoxify heavy metals by forming complex or effective barrier encircling cell. Furthermore, siderophore production also diminishes bioavailability of metal especially iron (Rajkumar et al., 2010). In *Pseudomonas*, siderophore secretes pyoverdine- and pseudobactin-carrying hydroxamate and catecholate groups that help in chelating process (Pattus and Abdallah, 2000; Boukhalfa and Crumbliss, 2002). Sometimes, production of specific metabolites or bacteria-mediated reactions induces precipitation or crystallization of heavy metals (Diels et al., 2003; Rajkumar et al., 2010). Moreover, rhizobacteria can eliminate heavy metals from the contaminated site by changing their redox state. For instance, bacterial strains detoxify Cr through aerobic and anaerobic reduction (Jing et al., 2007). Although PGPR are used in agriculture to facilitate plant growth, this plant microbe system is proved to be more effective in removal of heavy metals from contaminated sites (Zhuang et al., 2007). Abou-Shanab et al. (2003) observed that inoculation of *Microbacterium arabinogalactanolyticum*, *Microbacterium liquefaciens*, and *Sphingomonas macrogoltabidus* to *Alyssum murale* efficiently increased uptake of Ni in plants in comparison to uninoculated plants. Likewise, Carrillo-Castaneda et al. (2003) demonstrated that seeds of alfalfa *Medicago sativa* could be protected from Cu toxicity by application of *Azospirillum lipoferum* UAP40, UAP154, *Rhizobium leguminosarum* CPMex46, and *Pseudomonas fluorescens* Avm. Inoculation with *Bacillus megaterium* Bm4C and *Pseudomonas* sp. Ps29C significantly reduced Ni toxicity in *Brassica juncea* (Rajkumar and Freitas, 2008). PGPR inoculation with plants may stabilize, revegetate, and remediate the highly contaminated soils, where heavy metal concentration exceeds the tolerance limit of plant.

However, the complete biodegradation of pesticides involves oxidation of pesticide resulting into CO_2 and water that provide energy to microorganisms. Microbial-induced biodegradation process depends on enzyme system, temperature, pH, and availability of nutrients. It has been reported that *Pseudomonas rhizophila* S211 exhibited remediation capability by producing rhamnolipids and daioxygenases. The dioxygenases play important role in modification of numerous recalcitrant compounds including chemical pesticides (Shen et al., 2013). In the rhizosphere region, rhamnolipids can enhance the degradation of pesticides either by increasing substrate availability or enhancing hydrophobicity of bacterial cell surface to facilitate the association of hydrophobic substrates (Pacwa-Płociniczak et al., 2014).

Moreover, plant growth—promoting microbes can also mineralize organic pollutants. Aromatic compounds such as coumarins and flavonoids synthesized by plants are essential for root colonization. These complex aromatic compounds show structural similarity with many organic pollutants such as PAHs, PCBs, and petroleum hydrocarbons providing a way for bioremediation of these contaminants in rhizosphere. Flavonoids and other compounds secreted by roots can induce growth and activity of PAH- and PCB-degrading microbes. Root growth and death elevate aeration in soil that enhances oxidative degradation of rebellious organic contaminants (Gerhardt et al., 2009).

5.3 Microbial degradation of xenobiotic compounds

The degradation or break down of xenobiotic compounds is very typical because of their rebellious character. Decontamination of sites contaminated by these compounds with recent technology is expensive and relatively slow process. As a drawback, these methods can produce more toxic and rebellious intermediates. However, biological approaches have been emerged as an appropriate alternative because of their high efficiency, selectivity, cost-effective, and eco-friendly characteristics (Wu et al., 2016). On the basis of biocatalyst applied in the decontamination process, biological approaches are classified into two categories—microbial bioremediation (using microbes) and phytoremediation (using plants). Moreover, microbial bioremediation method is generally categorized into two classes—bioaugmentation and biostimulation.

Bioaugmentation involves the addition of microbial culture or microbial consortium into the contamination site to speed up the biodegradation process of specific contaminants. However, biostimulation is defined as a process of environmental modification that facilitates the stimulation of indigenous microbes for decontamination of contaminants by adding different forms of rate-limiting nutrients and electron acceptors such as carbon, oxygen, nitrogen, or phosphorus (e.g., in the form of molasses). Although earlier studies have reported the successful trails of microbial bioremediation program for decontamination of xenobiotic compounds at diverse scales, further advancement in these technologies is required for application at commercial scales.

Comparative to microbial bioremediation methods, phytoremediation methods are moderately diverse, including phytoextraction, phytostabilization, phytotransformation, phytovolatilization, and rhizofiltration. In phytoextraction process, plants extract contaminants from soil or water and convert it into harvestable biomass. Phytostabilization involves the immobilization of pollutants on the contamination site by plants and reduces bioavailability of these pollutants to the food chain. Phytotransformation is referred as the enzymatic breakdown or immobilization of specific contaminants sequestered by plants. In phytovolatilization process, plants remove pollutants from the soil and liberate them in the form of volatile products by evapotranspiration mechanism. Rhizofiltration refers to the technique of using plant roots for filtering contaminated water to remove contaminants by absorption and precipitation process.

6. Future prospective

Presently, changing climatic condition is a severe concern at the global level. The changes in climatic conditions abruptly influence the temperature and rainfall to the particular or global level. Researchers and policy-makers are continuously working on the management techniques to control or check the global climatic condition. Rapid industrialization, utilization of huge amount of chemical fertilizers and pesticides, new mining areas, and natural calamities such as volcanic eruptions are some prime factors responsible for the changing natural ecosystem or environmental condition. The accumulation of xenobiotics, industrial effluents, chemical pesticides, and heavy metals in the environment leads to severe chronic diseases in human beings through entering the normal food chains. These environmental pollutants also adversely affect the texture and productivity of plant and soil. The conventional physical and chemical methods applied for the management of xenobiotics or contaminants are the expensive and less effective methods. In this context, use of PGPB strains appears as a best suited choice in the management of environmental contaminants. In sustainable agriculture, many PGPB strains such as *Pseudomonas, Bacillus, Azotobacter*, etc., are broadly used as biofertilizer or biocontrol agents for the growth enhancement and control of phytopathogens to increases the agricultural productivity. Many PGPB strains are also utilized in the management of various abiotic stresses such as salinity, draught, pH, etc., but still there is need to focus on the management strategies of environment contaminants because currently limited number of PGPR or endophytic strains utilized in the environmental management because of less known mechanism of action. Furthermore, there is also need to study about the genetically modified microbes and nanoparticles synthesized by the PGPB strains and their possible uses in the environmental management.

Acknowledgments

The authors thank the University Grants Commission and CSIR, New Delhi, for fellowship in the form of JRF and SRF.

References

Abou-Shanab, R.A., Angle, J.S., Delorme, T.A., Chaney, R.L., van Berkum, P., Moawad, H., Ghanem, K., Ghozlan, H.A., 2003. Rhizobacterial effects on nickel extraction from soil and uptake by *Alyssum murale*. New Phytologist 158, 219–224.

Abou-Shanab, R.A.I., Angle, J.S., Chaney, R.L., 2006. Bacterial inoculants affecting nickel uptake by *Alyssum murale* from low, moderate and high Ni soils. Soil Biology and Biochemistry 38 (9), 2882–2889.

Ahemad, M., 2014b. Remediation of metalliferous soils through the heavy metal resistant plant growth promoting bacteria: paradigms and prospects. Arabian Journal of Chemistry. https://doi.org/10.1016/j.arabjc.2014.11.020.

Ahemad, M., Kibret, M., 2013. Recent trends in microbial biosorption of heavy metals: a review. Biochemistry and Molecular Biology Journal 1, 19–26.

Ahemad, M., Malik, A., 2011. Bioaccumulation of heavy metals by zinc resistant bacteria isolated from agricultural soils irrigated with wastewater. Bacteriology Journal 2, 12–21.

Ahmad, M.S.A., Ashraf, M., 2012. Essential roles and hazardous effects of nickel in plants. In: Reviews of Environmental Contamination and Toxicology. Springer, New York, NY, pp. 125–167.

Appanna, V.D., Finn, H., Pierre, M.S., 1995. Exocellular phosphatidylethanolamine production and multiple-metal tolerance in *Pseudomonas fluorescens*. FEMS Microbiology Letters 131 (1), 53–56.

Ayangbenro, A.S., Babalola, O.O., 2017. A new strategy for heavy metal polluted environments: a review of microbial biosorbents. International Journal of Environmental Research and Public Health 14 (1), 94.

Bishop, P.L., 2002. Pollution Prevention: Fundamentals and Practice. Tsinghua University Press, Beijing, p. 768.

Bisht, S., Pandey, P., Bhargava, B., Sharma, S., Kumar, V., Sharma, K.D., 2015. Bioremediation of polyaromatic hydrocarbons (PAHs) using rhizosphere technology. Brazilian Journal of Microbiology 46 (1), 7–21.

Blaylock, M.J., Salt, D.E., Dushenkov, S., Zakharova, O., Gussman, C., Kapulnik, Y., Ensley, B.D., Raskin, I., 1997. Enhanced accumulation of Pb in Indian mustard by soil-applied chelating agents. Environmental Science and Technology 31 (3), 860–865.

Boukhalfa, H., Crumbliss, A.L., 2002. Chemical aspects of siderophore mediated iron transport. Biometals 15 (4), 325–339.

Burd, G.I., Dixon, D.G., Glick, B.R., 2000. Plant growth promoting bacteria that decrease heavy metal toxicity in plants. Canadian Journal of Microbiology 46, 237–245.

Carrillo-Castaneda, G., Munoz, J.J., Peralta-Videa, J.R., Gomez, E., Gardea-Torresdey, J.L., 2003. Plant growth-promoting bacteria promote copper and iron translocation from root to shoot in alfalfa seedlings. Journal of Plant Nutrition 26, 1801–1814.

Cheng, K.J., Stewart, C.S., Dinsdale, D., Costerton, J.W., 1984. Electron microscopy of bacteria involved in the digestion of plant cell walls. Animal Feed Science and Technology 10 (2–3), 93–120.

Clemens, S., Ma, J.F., 2016. Toxic heavy metal and metalloid accumulation in crop plants and foods. Annual Review of Plant Biology 67, 489–512.

da Silva, Frausto, J.J.R., Williams, R.J.P., 1991. The Biological Chemistry of the Elements. Clarendon, pp. 3–22.

Díaz, F.J., O'Geen, A.T., Dahlgren, R.A., 2010. Efficacy of constructed wetlands for removal of bacterial contamination from agricultural return flows. Agricultural Water Management 97 (11), 1813–1821.

Diels, L., Spaans, P.H., Van Roy, S., Hooyberghs, L., Ryngaert, A., Wouters, H., Walter, E., Winters, J., Macaskie, L., Finlay, J., Pernfuss, B., Woebking, H., Pumpel, T., Tsezos, M., 2003. Heavy metals removal by sand filters inoculated with metal sorbing and precipitating bacteria. Hydrometallurgy 71, 235–241.

Dobrovolskaia, M.A., 2015. Pre-clinical immunotoxicity studies of nanotechnology-formulated drugs: challenges, considerations and strategy. Journal of Controlled Release 220, 571–583.

Faisal, M., Hasnain, S., 2005. Bacterial Cr (VI) reduction concurrently improves sunflower (*Helianthus annuus* L.) growth. Biotechnology Letters 27 (13), 943–947.

Ganesan, V., 2008. Rhizoremediation of cadmium soil using a cadmium-resistant plant growth-promoting rhizopseu-domonas. Current Microbiology 56, 403—407.

Gerhardt, K.E., Huang, X.D., Glick, B.R., Greenberg, B.M., 2009. Phytoremediation and rhizoremediation of organic soil contaminants: potential and challenges. Plant Science 176 (1), 20—30.

Germaine, K.J., Liu, X., Cabellos, G.G., Hogan, J.P., Ryan, D., 2006. Bacterial endophyte-enhanced phytoremediation of the organochlorine herbicide 2,4-dichlorophenoxyacetic acid. FEMS Microbiology Ecology 57, 302—310.

Gupta, A., Meyer, J.M., Goel, R., 2002. Development of heavy metal resistant mutants of phosphate solubilizing *Pseudomonas* sp. NBRI4014 and their characterization. Current Microbiology 45, 323—332.

Hassen, W., Neifar, M., Cherif, H., Najjari, A., Chouchane, H., Driouich, R.C., Salah, A., Naili, F., Mosbah, A., Souissi, Y., Raddadi, N., 2018. *Pseudomonas rhizophila* S211, a new plant growth-promoting rhizobacterium with potential in pesticide-bioremediation. Frontiers in Microbiology 9, 34.

Hayat, R., Ali, S., Amara, U., Khalid, R., Ahmed, I., 2010. Soil beneficial bacteria and their role in plant growth promotion: a review. Annals of Microbiology 60 (4), 579—598.

Hessel, E.V., Ezendam, J., Van Broekhuizen, F.A., Hakkert, B., DeWitt, J., Granum, B., Guzylack, L., Lawrence, B.P., Penninks, A., Rooney, A.A., Piersma, A.H., 2016. Assessment of recent developmental immunotoxicity studies with bisphenol A in the context of the 2015 EFSA t-TDI. Reproductive Toxicology 65, 448—456. https://nature.nps.gov/hazardssafety/toxic/pcp.pdf.

Jaishankar, M., Tseten, T., Anbalagan, N., Mathew, B.B., Beeregowda, K.N., 2014. Toxicity, mechanism and health effects of some heavy metals. Interdisciplinary Toxicology 7 (2), 60—72.

Jamil, M., Zeb, S., Anees, M., Roohi, A., Ahmed, I., Rehman, S., Rha, E.S., 2014. Role of *Bacillus licheniformis* in phy-toremediationof nickel contaminated soil cultivated with rice. International Journal of Phytoremediation 16 (6), 554—571.

Jing, Y., He, Z., Yang, X., 2007. Role of soil rhizobacteria in phytoremediation of heavy metal contaminated soils. Jour-nal of Zhejiang University - Science B 8, 192—207.

Khan, N., Bano, A., 2015. Role of plant growth promoting rhizobacteria and Ag-nano particle in the bioremediation of heavy metals and maize growth under municipal wastewater irrigation. International Journal of Phytoremedia-tion 18 (3), 211—221.

Kumar, A., Singh, A.K., Kaushik, M.S., Mishra, S.K., Raj, P., Singh, P.K., Pandey, K.D., 2017a. Interaction of turmeric (*Curcuma longa* L.) with beneficial microbes: a review. 3 Biotech 7 (6), 357.

Kumar, A., Singh, R., Giri, D.D., Singh, P.K., Pandey, K.D., 2014. Effect of *Azotobacter chroococcum* CL13 inoculation on growth and curcumin content of turmeric (*Curcuma longa* L.). International Journal of Current Microbiology and Applied Sciences 3 (9), 275—283.

Kumar, A., Singh, R., Yadav, A., Giri, D.D., Singh, P.K., Pandey, K.D., 2016a. Isolation and characterization of bacterial endophytes of *Curcuma longa* L. 3 Biotech 6, 60.

Kumar, A., Vandana, Singh, M., Singh, P.P., Singh, S.K., Singh, P.K., Pandey, K.D., 2016b. Isolation of plant growth promoting rhizobacteria and their impact on growth and curcumin content in *Curcuma longa* L. Biocatalysis and Agricultural Biotechnology 8, 1—7.

Kumar, A., Singh, V.K., Tripathi, V., Singh, P.P., Singh, A.K., 2018. Plant growth-promoting rhizobacteria (PGPR): perspective in agriculture under biotic and abiotic stress. In: Crop Improvement through Microbial Biotechnology, pp. 333—342.

Kumar, A., Vandana, R.S., Yadav, A., Giri, D.D., Singh, P.K., Pandey, K.D., 2015b. Rhizosphere and their role in plant—microbe interaction. In: Microbes in Soil and Their Agricultural Prospects. Nova Science Publisher, Inc, New York, pp. 83—97.

Kumar, A., Verma, H., Singh, V.K., Singh, P.P., Singh, S.K., Ansari, W.A., Yadav, A., Singh, P.K., Pandey, K.D., 2017b. Role of *Pseudomonas* sp. in sustainable agriculture and disease management. In: Agriculturally Important Microbes for Sustainable Agriculture. Springer, Singapore, pp. 195—215.

Kumar, V., Kumar, A., Pandey, K.D., Roy, B.K., 2015a. Isolation and characterization of bacterial endophytes from the roots of *Cassia tora* L. Annals of Microbiology 65 (3), 1391—1399.

Ma, Y., Rajkumar, M., Freitas, H., 2009. Improvement of plant growth and nickel uptake by nickel resistant-plant-growth promoting bacteria. Journal of Hazardous Material 166, 1154—1161.

Madhaiyan, M., Poonguzhali, S., Sa, T., 2007. Metal tolerating methylotrophic bacteria reduces nickel and cadmium toxicity and promotes plant growth of tomato (*Lycopersicon esculentum* L.). Chemosphere 69, 220—228.

Maliszewska-Kordybach, B., Smreczak, B., 2000. Ecotoxicological activity of soils polluted with polycyclic aromatic hydrocarbons (PAHs)-effect on plants. Environmental Technology 21 (10), 1099–1110.

Mishra, V., Srivastava, G., Prasad, S.M., Abraham, G., 2008. Growth, photosynthetic pigments and photosynthetic activity during seedling stage of cowpea (*Vigna unguiculata*) in response to UV-B and dimethoate. Pesticide Biochemistry and Physiology 92 (1), 30–37.

Moustafa, G.G., Shaaban, F.E., Hadeed, A.A., Elhady, W.M., 2016. Immunotoxicological, biochemical, and histopathological studies on Roundup and Stomp herbicides in Nile catfish (*Clarias gariepinus*). Veterinary World 9 (6), 638.

Myresiotis, C.K., Vryzas, Z., Papadopoulou-Mourkidou, E., 2012. Biodegradation of soil-applied pesticides by selected strains of plant growth-promoting rhizobacteria (PGPR) and their effects on bacterial growth. Biodegradation 23 (2), 297–310.

Narasimhan, K., Basheer, C., Bajic, V.B., Swarup, S., 2003. Enhancement of plant-microbe interactions using a rhizosphere metabolomics-driven approach and its application in the removal of polychlorinated biphenyls. Plant Physiology 132 (1), 146–153.

Nicolopoulou-Stamati, P., Maipas, S., Kotampasi, C., Stamatis, P., Hens, L., 2016. Chemical pesticides and human health: the urgent need for a new concept in agriculture. Frontiers in Public Health 4, 148.

Nies, D.H., 1999. Microbial heavy-metal resistance. Applied Microbiology and Biotechnology 51 (6), 730–750.

Pacwa-Płociniczak, M., Płaza, G.A., Poliwoda, A., Piotrowska-Seget, Z., 2014. Characterization of hydrocarbon-degrading and biosurfactant producing *Pseudomonas* sp. P-1 strain as a potential tool for bioremediation of petroleum-contaminated soil. Environmental Science and Pollution Research International 21, 9385–9395.

Pandey, S., Ghosh, P.K., Ghosh, S., De, T.K., Maiti, T.K., 2013. Role of heavy metal resistant *Ochrobactrum* sp. and *Bacillus* spp. strains in bioremediation of a rice cultivar and their PGPR like activities. Journal of Microbiology 51 (1), 11–17.

Parween, T., Jan, S., Mahmooduzzafar, S., Fatma, T., Siddiqui, Z.H., 2016. Selective effect of pesticides on plant—a review. Critical Reviews in Food Science and Nutrition 56 (1), 160–179.

Pattus, F., Abdallah, M.A., 2000. Siderophores and iron-transport in microorganisms. Journal of the Chinese Chemical Society 47 (1), 1–20.

Pratibha, Y., Krishna, S.S., 2015. Plant growth promoting Rhizobacteria: an effective tool to remediate residual organophosphate pesticide methyl parathion, widely used in Indian agriculture. Journal of Environmental Research and Development 9 (4), 1138.

Qing Li, Q., Loganath, A., Seng Chong, Y., Tan, J., Philip Obbard, J., 2006. Persistent organic pollutants and adverse health effects in humans. Journal of Toxicology and Environmental Health, Part A. 69 (21), 1987–2005.

Rajashekar, N., Murthy, T.S., 2012. Seed germination and physiological behavior of maize (cv. Nac-6002) seedlings under abiotic stress (Pendimethalin) condition. Asian Journal of Crop Science 4 (2), 80–85.

Rajkumar, M., Ae, N., Prasad, M.N.V., Freitas, H., 2010. Potential of siderophore-producing bacteria for improving heavy metal phytoextraction. Trends in Biotechnology 28 (3), 142–149.

Rajkumar, M., Freitas, H., 2008. Effects of inoculation of plant growth promoting bacteria on Ni uptake by Indian mustard. Bioresource Technology 99, 3491–3498.

Rajkumar, M., Ma, Y., Freitas, H., 2008. Characterization of metal resistant plant-growth promoting *Bacillus weihenstephanensis* isolated from serpentine soil in Portugal. Journal of Basic Microbiology 48, 500–508.

Rajkumar, M., Nagendran, R., Lee, K.J., Lee, W.H., Kim, S.Z., 2006. Influence of plant growth promoting bacteria and Cr6+ on the growth of Indian mustard. Chemosphere 62 (5), 741–748.

Roy, A.S., Yenn, R., Singh, A.K., Boruah, H.P.D., Saikia, N., Deka, M., 2013. Bioremediation of crude oil contaminated tea plantation soil using two *Pseudomonas aeruginosa* strains AS 03 and NA 108. African Journal of Biotechnology 12 (19).

Saraf, M., Sood, N., 2002. Influence of monocrotophos on growth, oxygen uptake and exopolysaccharide production of rhizobium NCIM 2271 on chickpea. Journal of the Indian Botanical Society 82, 157–164.

Schwab, A.P., Banks, M.K., 1994. Biologically Mediated Dissipation of Polyaromatic Hydrocarbons in the Root Zone.

Sethy, S.K., Ghosh, S., 2013. Effect of heavy metals on germination of seeds. Journal of Natural Science, Biology, and Medicine 4 (2), 272–275.

Shaheen, S., Sundari, K.S., 2013. Exploring the applicability of PGPR to remediate residual organophosphate and carbamate pesticides used in agriculture fields. International Journal of Agriculture and Food Science Technology 4 (10), 947–954.

Sharma, B.M., Bharat, G.K., Tayal, S., Nizzetto, L., Čupr, P., Larssen, T., 2014. Environment and human exposure to persistent organic pollutants (POPs) in India: a systematic review of recent and historical data. Environment International 66, 48—64.

Shen, X., Hu, H., Peng, H., Wang, W., Zhang, X., 2013. Comparative genomic analysis of four representative plant growth-promoting rhizobacteria in *Pseudomonas*. BMC Genomics 14, 271.

Sheng, X.F., Xia, J.J., 2006. Improvement of rape (Brassica napus) plant growth and cadmium uptake by cadmium-resistant bacteria. Chemosphere 64, 1036—1042.

Sheng, X.F., Xia, J.J., Jiang, C.Y., He, L.Y., Qian, M., 2008. Characterization of heavy metal-resistant endophytic bacteria from rape (*Brassica napus*) roots and their potential in promoting the growth and lead accumulation of rape. Environmental Pollution 156, 1164—1170.

Siciliano, S.D., Germida, J.J., 1997. Bacterial inoculants of forage grasses that enhance degradation of 2-chlorobenzoic acid in soil. Environmental Toxicology and Chemistry: An International Journal 16 (6), 1098—1104.

Singh, A.K., Cameotra, S.S., 2013. Rhamnolipids production by multi-metalresistant and plant-growth-promoting rhizobacteria. Applied Biochemistry and Biotechnology 170, 1038—1056.

Singh, R., Pandey, K.D., Kumar, A., Singh, M., 2017a. PGPR isolates from the rhizosphere of vegetable crop *Momordica charantia*: characterization and application as biofertilizer. International Journal of Current Microbiology and Applied Sciences 6 (3), 1789—1802.

Singh, V.K., Singh, A.K., Kumar, A., 2017b. Disease management of tomato through PGPB: current trends and future perspective. 3 Biotech 7 (4), 255.

Singh, M., Kumar, A., Singh, R., Pandey, K.D., 2017c. Endophytic bacteria: a new source of bioactive compounds. 3 Biotech 7 (5), 315.

Sriprang, R., Hayashi, M., Ono, H., Takagi, M., Hirata, K., Murooka, Y., 2003. Enhanced accumulation of Cd2+ by a *Mesorhizobium* sp. transformed with a gene from *Arabidopsis thaliana* coding for phytochelatin synthase. Applied and Environmental Microbiology 69, 1791—1796.

Stevens, M.M., Reinke, R.F., Coombes, N.E., Helliwell, S., Mo, J., 2008. Influence of imidacloprid seed treatments on rice germination and early seedling growth. Pest Management Science: Formerly Pesticide Science 64 (3), 215—222.

Tesar, M., Reichenauer, T.G., Sessitsch, A., 2002. Bacterial rhizosphere populations of black poplar and herbal plants to be used for phytoremediation of diesel fuel. Soil Biology and Biochemistry 34 (12), 1883—1892.

Tripathi, M., Munot, H.P., Shouch, Y.J., Meyer, M., Goel, R., 2005. Isolation and functional characterization of siderophore-producing lead and cadmium resistant *Pseudomonas putida* KNP9. Current Microbiology 5, 233—237.

Villacieros, M., Whelan, C., Mackova, M., Molgaard, J., Sánchez-Contreras, M., Lloret, J., de Cárcer, D.A., Oruezábal, R.I., Bolanos, L., Macek, T., Karlson, U., 2005. Polychlorinated biphenyl rhizoremediation by *Pseudomonas fluorescens* F113 derivatives, using a *Sinorhizobium meliloti* nod system to drive bph gene expression. Applied and Environmental Microbiology 71 (5), 2687—2694.

Vivas, A., Biro, B., Ruiz-Lozano, J.M., Barea, J.M., Azcon, J.M., 2006. Two bacterial strains isolated from a Zn-polluted soil enhance plant growth and mycorrhizal efficiency under Zn toxicity. Chemosphere 52, 1523—1533.

Wani, P.A., Khan, M.S., Zaidi, A., 2007. Co-inoculation of nitrogen fixing and phosphate solubilizing bacteria to promote growth, yield and nutrient uptake in chickpea. Acta Agronomica Hungarica 55, 315—323.

Wu, C.H., Wood, T.K., Mulchandani, A., Chen, W., 2006. Engineering plant-microbe symbiosis for rhizoremediation of heavy metals. Applied and Environmental Microbiology 72, 1129—1134.

Wu, H., Shen, J., Wu, R., Sun, X., Li, J., Han, W., Wang, L., 2016. Biodegradation mechanism of 1H-1, 2, 4-triazole by a newly isolated strain *Shinella* sp. NJUST26. Scientific Reports 6, 29675.

Yee, D.C., Maynard, J.A., Wood, T.K., 1998. Rhizoremediation of trichloroethylene by a recombinant, root-colonizing *Pseudomonas fluorescens* strain expressing toluene ortho-monooxygenase constitutively. Applied and Environmental Microbiology 64 (1), 112—118.

Zaidi, S., Usmani, S., Singh, B.R., Musarrat, J., 2006. Significance of *Bacillus subtilis* strain SJ-101 as a bioinoculant for concurrent plant growth promotion and nickel accumulation in *Brassica juncea*. Chemosphere 64 (6), 991—997.

Zhang, B., Zhang, H., Jin, J., Ni, Y., Chen, J., 2012. PCDD/Fs-induced oxidative damage and antioxidant system responses in tobacco cell suspension cultures. Chemosphere 88 (7), 798—805.

Zhu, Y., Boye, A., Body-Malapel, M., Herkovits, J., 2017. The toxic effects of xenobiotics on the health of humans and animals. BioMed Research International. https://doi.org/10.1155/2017/4627872.

Zhuang, X., Chen, J., Shim, H., Bai, Z., 2007. New advances in plant growth-promoting rhizobacteria for bioremediation. Environment International 33 (3), 406—413.

Further reading

Ahemad, M., 2014a. Bacterial mechanisms for Cr(VI) resistance and reduction: an overview and recent advances. Folia Microbiologica. https://doi.org/10.1007/s12223-014-0304-8.

Fungi as potential candidates for bioremediation

Rajesh Kumar Singh[1], Ruchita Tripathi[1], Amit Ranjan[2], Akhileshwar Kumar Srivastava[3]

[1]Department of Dravyaguna, Institute of Medical Sciences, Banaras Hindu University, Varanasi, India; [2]Department of Kayachikitsa Institute of Medical Sciences, Banaras Hindu University, Varanasi, India; [3]The National Institute for Biotechnology in the Negev, Ben-Gurion University of the Negev, Beer-Sheva, Israel

1. Introduction

The condition of environment directly influences the quality of life in the ecosystem on earth. Human beings are most dominant animal on the earth, have developed the science and technology for their comforts, and industrialized at large scale for the fulfillment of their requirements, but unfortunately these produce adverse affect on environment, often resulting to the extinction of several species of the earth. The civilization and industrialization produce large amount of organic and inorganic pollutants, as well as nuclear wastes, which are generally flushed into the river or dumped into the soil. The flushing and dumping of such materials pollute to system, creating a serious problem for the survival of organisms. Such pollutants are disposed by making a hole into land and filling it by waste material or just filling the low-land areas just away from the cities, but continuous increases of the pollutants need new place in near future, which develops other issues such as lack of water-bodies, floods, etc. These methods of disposal are difficult to sustain because of limited space and are also not economic and healthy (Karigar and Rao, 2011). Hence, bioremediation is alternative, economical, and acceptable method to dispose these polluting materials.

Bioremediation is a process in which the waste and hazardous materials are transformed into nonhazardous or less-hazardous substances by microbes such as bacteria, fungi, algae, etc. The applications of fungal microbes in bioremediation have been well-known as fungal bioremediation or mycoremediation. The fungi decompose the biomass residues and chemical pollutants by producing several enzymes such as amylase, protease, lipase, nuclease, etc., which degrade the cell walls, proteins, lipids, DNA, and other organic materials. These microbes are used for biotransformation of organic waste materials and removal of pollutants from environment (Balabanova et al., 2018). Fungi are capable to degrade the organic chemicals, metals, metalloids, and radionuclides, by secreting enzymes and other chemicals through chemical modification and/or influencing chemical bioavailability. It has also adopted to survive in diverse condition continuously. Fungi are competent for metabolize various environmental chemicals and utilize its product for survival without any additional need of nutrition. It degrades pollutants through several extracellular oxidoreductases for decomposing the lignocelluloses along with several pollutants. Fungi metabolize and immobilize the metals, metalloids, and radionuclides in the mycosphere, store in various parts of the cell, or translocate through fungal hyphae. It also degrades the compounds that are not intoxicated by bacteria efficiently such as dioxins and 2,4,6-trinitrotoluene, drugs, etc. These features enable fungi as potential candidate for removing organic or metal contaminants in the soils, water, and air and can be used for bioremediation with cost-effective and nature friendly ways (Harms et al., 2011). This chapter explains the diverse role of fungi group in overcoming the pollutant toxicity through a robust technique bioremediation as summarized in Fig. 9.1.

1.1 Fungal enzymes for bioremediation

The enzymes of many fungi have low specificity, which enables the fungal strain to metabolize several compounds of different pollutants even dissimilar in structure simultaneously, like *Phanerochaete chrysosporium* is a crust fungi, which degrades many harmful chemicals including benzene, ethylbenzene, xylene, toluene, organochlorines, *N*-heterocyclic explosives,

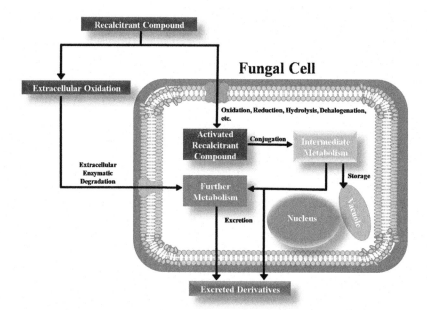

FIGURE 9.1 Describing role of fungi in bioremediation and involvement of fungal enzymes in removal/transformation of materials.

polycyclic aromatic hydrocarbons (PAHs), nitroaromatic compounds, pesticides, synthetic dyes, polychlorinated dibenzo-p-dioxins, and synthetic polymers simultaneously even in mixture of all (Kues, 2015). The major groups of enzymes involved in the degradation of wastes materials and pollutants include extracellular oxidoreductases, cell-bound enzymes, and other transferase enzymes.

1.1.1 Extracellular oxidoreductases

The several extracellular oxidoreductase enzymes are secreted by fungi to degrade the organic compounds such as laccases, tyrosinases, lignin peroxidases, manganese peroxidases, etc. (Harms et al., 2011). These enzymes enabled the fungal strains to grow on recalcitrant substrates such as lignocelluloses that are not efficiently degraded by most bacteria, and fungi could survive on the substrates where bacteria cannot grow. These properties of fungi are advantageous over bacteria for application in bioremediation (Verma et al., 2017).

The prominent copper-containing laccases are mainly produced in *Basidiomycetes* and *Ascomycetes* and used for decolorization and detoxification in textile industries, bleaching of ink in paper industries, degradation and detoxification of recalcitrant compounds in wastewater, hazardous compounds arising from coal processing, and degradation of soil pollutants (Alneyadi et al., 2018; Spina et al., 2015).

Another copper-containing enzyme tyrosinase also oxidizes the phenols and chlorinated phenolic compounds. It has potential for the removal of such chlorinated compounds from the environment. Presently, many peroxidases are characterized in microbes especially from fungi, oxidize pollutants, and convert them into nonhazardous products with high redox potentials (Al-Maqdi et al., 2017).

1.2 Cell-bound enzymes

Many compounds of waste materials and pollutants are permeable to the fungal cells membranes, which are chemically catabolized through several intracellular fungal enzymes such as polyaromatic hydrocarbons and dioxins. Cytochrome P450s also have a vital role in intracellular degradation of the pollutant and involve in fungal metabolism of many drugs such as antiinflammatory drugs, lipid regulator drugs, antiepileptic and antianalgesic drugs, and diphenyl ether herbicides. The pollutants under fungal bioremediation undergo degradation by fungal metabolism till they are mineralized (Harms et al., 2011). The metabolic activities may further also involve in formation of metabolites at different stages of oxidation in the degradation process.

1.3 Transferases

Transferases include mainly aromatic nitroreductases, quinone reductases, etc., which transfer the functional groups and convert the hazardous pollutants into nonhazardous products. The transferase enzymes degrade the pollutants containing hydroxyl groups by converting it into the conjugates. The conjugates are stored, fixed, or secreted into environment in inactive conjugated forms (Morel et al., 2013). The different enzymes and components are involved in remediation of contaminated materials as mentioned in Table 9.1.

TABLE 9.1 The details of enzymes producing fungal taxa and mechanism of action.

S.No.	Enzymes	Fungal taxa	Enzyme occurrence	Mechanism of action	References
1	Laccases	Ascomycota and Basidiomycota	Extracellular	O_2-dependent one-electron oxidation of organic compounds.	(Baldrian, n.d.; Majeau et al., 2010)
2	Tyrosinases	Ascomycota, Basidiomycota, and Mucoromycotina	Intracellular (but sometimes extracellular)	O_2-dependent hydroxylation of monophenols to o-diphenols (cresolase activity) Oxidation of o-diphenols to catechols (catecholase activity).	(Halaouli et al., n.d.; Ullrich and Hofrichter, 2007)
3	Lignin peroxidases	Basidiomycota	Extracellular	H_2O_2-dependent one-electron oxidation of aromatic compounds.	(Hofrichter et al., 2010; Karich et al., 2017; Ruiz-Dueñas et al., 2009)
4	Manganese peroxidases	Basidiomycota	Extracellular	H_2O_2-dependent one-electron oxidation of Mn^{2+} to Mn^{3+}, which subsequently oxidizes organic compounds.	(Hofrichter et al., 2010, n.d.; Ruiz-Dueñas et al., 2009)
5	Versatile peroxidases	Basidiomycota	Extracellular	H_2O_2-dependent direct one-electron oxidation of aromatic compounds H_2O_2-dependent one-electron oxidation of Mn^{2+} to Mn^{3+}, which subsequently oxidizes organic compounds.	(Hofrichter et al., 2010; Martínez, 2002; Ruiz-Dueñas et al., 2009)
6	Coprinopsis cinerea peroxidase	Basidiomycota	Extracellular	H_2O_2-dependent one-electron oxidation of aromatic compounds.	(Hofrichter et al., 2010; Ikehata et al., 2005)
7	Dye-decolorizing peroxidases	Basidiomycota	Extracellular	H_2O_2-dependent one-electron oxidation of organic compounds Additional hydrolyzing activity.	Hofrichter et al. (2010)
8	Caldariomyces fumago haem—thiolate chloroperoxidase	Ascomycota	Extracellular	H_2O_2-dependent halogenation of organic compounds in the presence of halides (one-electron transfer), H_2O_2-dependent one-electron oxidations of phenols and anilines in the absence of halides, and H_2O_2-dependent peroxygenation (two-electron oxidation), leading to epoxidation of (cyclo)alkenes, hydroxylation of benzylic carbon and sulphoxidation of S-containing organic compounds.	Hofrichter et al. (2010)

#	Enzyme	Taxa	Localization	Description	References
9	Haem–thiolate peroxygenases	Basidiomycota	Extracellular	H_2O_2-dependent peroxygenation of aromatic, aliphatic, and heterocyclic compounds, leading to aromatic and alkylic carbon hydroxylation, double-bond epoxidation, ether cleavage, sulphoxidation or N-oxidation reactions (depending on the substrate). H_2O_2-dependent one-electron abstractions from phenols, and H_2O_2-dependent bromination of organic substrates.	Hofrichter et al. (2010)
10	Cytochrome P450 monooxygenases	Ascomycota, Basidiomycota, Mucoromycotina, and Chytridiomycota	Cell bound	Incorporation of a single atom from O_2 into a substrate molecule, with concomitant reduction of the other atom to H_2O.	(Kasai et al., 2010; Subramanian and Yadav, 2009; Ullrich and Hofrichter, 2007; Yadav et al., 2006)
11	Phenol 2-monooxygenases	Ascomycota and Basidiomycota	Cell bound	Incorporation of a single atom from O_2 into a substrate molecule, with concomitant reduction of the other atom to H_2O.	(Hofrichter et al., n.d.; Ullrich and Hofrichter, 2007)
12	Nitroreductases	Ascomycota, Basidiomycota, and Mucoromycotina	Cell bound	NADPH-dependent reduction of nitroaromatics to hydroxylamino and amino(nitro) compounds and of nitro functional groups of N-containing heterocycles.	(Bhushan et al., 2002; Crocker et al., 2006; Esteve-Núñez et al., 2001; Fournier et al., 2004; Scheibner et al., 1997)
13	Reductive dehalogenases	Basidiomycota and perhaps Ascomycota	Cell bound	Two-component system comprising a membrane-bound glutathione S-transferase that produces glutathionyl conjugates with concomitant chlorine removal and a soluble glutathione conjugate reductase that releases reductively dechlorinated compounds.	(Jensen et al., 2001; Nakamiya et al., 2005)
14	Miscellaneous transferases	Ascomycota, Basidiomycota, and Mucoromycotina	Cell bound	Formation of glucoside, glucuronide, xyloside, sulfate, or methyl conjugates from hydroxylated compounds.	Kasai et al. (2010)

2. Fungal bioremediation

2.1 Toxic recalcitrant compound

The poor electron donor compounds such as polychlorinated dibenzo-p-dioxins (PCDDs) and polychlorinated dibenzofurans (PCDFs) occurring at very low concentration are oxidized to the electronegativity of the chlorine atoms by accepting the electrons from the aromatic rings (Hammel, 1995). The compounds with more than one chlorine atom are not employed as a source of carbon and energy by bacteria. Therefore, the biotransformation of the toxic compounds, PCDD 2,3,7,8-tetrachlorodibenzo-*p*-dioxin (2,3,7,8-TCDD), through aerobic and anaerobic bacteria is slow process ranging to months; however, it is not completely biotransformed and finally halted at the stage of less chlorinated aromatic products. Comparatively, the white-rot fungus *Phanerochaete sordid* converts the compound 2,3,7,8-TCDD into chlorocatechols within 10 days and other fungus of same species *P. chrysosporium* mineralizes the intermediate products. The fungus *P. sordida* has no specific to particular compound and it transforms compounds containing different chlorine atoms, namely PCDDs (6—8 chlorines) and PCDFs (4—8 chlorines) (Lipnick et al., 2000). The ascomycete *Cordyceps sinensis* has great potential to remove the highly chlorinated dioxins at fast rate similar to white-rot fungi. The 2,4,6-trinitrotoluene (TNT) is also highly oxidized to the electronegativity of its nitro groups. The transformation of TNT is accomplished mostly through aerobic bacteria into monoamino-dinitrotoluenes, diamino-nitrotoluenes, and hydroxylamino-dinitrotoluenes that acquire to produce mutagenic azoxy-tetranitrotoluenes, although the transformation of TNT through anaerobic bacteria mostly halts at the stage of triaminotoluene, whereas white-rot and litter-decaying basidiomycetes mineralize TNT rapidly (Esteve-Núñez et al., 2001). Researches into use of nonspecific enzymes through biotechnological approaches must elucidate the problems generated during the reactions. The specific fungal transformation products of PAHs have less concentration of mutagenic and carcinogenic metabolites in which most transformed products are lower toxic than its parent compounds. The formations of PCDDs and PCDFs on enzymatic oxidative coupling of chlorophenols are required to assess the structure, biological activity, stability, and environmental behavior (Lamar et al., 2003). The laccase-catalyzed transformation of 1-ethyl-3-(3-dimethylaminopropyl) carbodiimide hydrochloride (EDCs) to oligomeric coupling products reduces the endocrine-disrupting activity of such compounds. Similarly, the toxicity is reduced by laccases and other phenoloxidases for covalently linked chlorophenols, other phenols, and TNT metabolites to components of soil humic substances.

PAHs as good substrates in respect to energetic perspective and its low molecular mass PAHs are commonly used by bacteria. The bioavailability decreases inverse to molecular mass, only a few bacteria grow at high molecular mass PAHs with five or more aromatic rings. The fungal species of Ascomycota, Basidiomycota, and Mucoromycotina having a nonspecific mechanism to detoxify the compounds hydroxylate to PAHs intracellularly and then transfer to water-soluble compounds that could be excreted out. In addition, different fungi utilize the extracellular oxidoreductases in PAH degradation and mineralize to high molecular mass PAHs, such as highly carcinogenic benzo[*a*]pyrene and others (Mao and Guan, 2016).

2.2 Heavy metal

Fungi for remediation of materials (metals, metalloids, and radionuclides) are identified on the basis of morphological, ecological, and biochemical properties. The accumulations of toxic metals in environment are accomplished through different stages of the life cycle of metallic substances, from the mining of metal ores to the final disposal of metal-containing hazardous substances. The mobilization of metals from geogenic sources, agriculture field, and incident like fires are attributed a constant input into waters and the atmosphere and from it further is translocated into soils and aquatic sediments (Sanyal et al., 2005). The contaminated environment with metals could be hazardous for mainly to near of ore mines and smelters or at disposal sites where proper remediation is not adequate. The main risk of metallic products is entirely associated with the nondegradable element. So, biological methods are not capable to reduce metal toxicity in an irreversible way. Such rising risk from such elements could be overcome through biological conversion into less toxic species or by separation from the affected biota. With advancement of technology, it has also drawn much interest toward elimination of metals from wastewater streams for restoration of valuable metals. It accomplishes through such methods such as mobilization and immobilization in the mycosphere, sorption to cell walls and uptake into fungal cells (Ayangbenro and Babalola, 2017). The metals could be chemically converted after incorporation into fungus and accumulated in different parts of the cell or translocated from fungal hyphae into symbiotic plant.

2.3 Municipal solid waste

Nowadays, the municipal solid waste (MSW) in developing countries has been one of the major problems for environment along with public health. For removal of MSW, the process incineration and landfilling are mainly employed, although incineration is a very costly process, hence the landfilling sites produce the secondary environmental pollution such as fouling of air, bad odor, and increased pathogenic activity in soil. In another way, the countries with limited land areas are not fit for composting and landfilling; hence it is essential to expand the land areas. Deshmukh et al. have explained about several modern technologies for resolving of such misplaced resources. It has also explored the composting and biomethanation method through anaerobic process to manage the MSW with production of valuable substances such as volatile fatty acids (VFAs), biogas, and organic residue/compost for application as fertilizers (Deshmukh et al., 2016). For increasing the potential and rates of these methods, the fungi and its hydrolytic enzymes could be used for the transformation of complex polymeric compounds of MSW to simple one acting as mediators for VFA and biogas production. The enzymes cellulolytic, hemicellulolytic, pectinolytic, and amylolytic of *Aspergillus niger* are used to increase the efficiency of steam, acid, and base pretreated kitchen waste residues for solid-state fermentation of enzymes (Rani et al., 2014). The using of these enzymes in pretreatment enhance the potential for hydrolysis and saccharification of certain biomasses such as willow and rice straw, by application of a fungal consortium containing two fungi *Armilleria gemina* and *Pholiota adipose*. The white-rot fungi also increase the efficiency for composition of other residual biomass by utilizing the used biomass in soil application. The accumulated copper in wood could be remediated through highly efficient wood-rotting fungi *Antrodia xanthan* and *Fomitopsis palustris* (Deshmukh et al., 2016; Voberkova et al., 2017). *Trichoderma viride* is one of the promising fungi for conversion of organic municipal solid waste to harmless substances. The main factors such as pH and temperature are good indicator for the end of the bioconversion of municipal solid waste (Gautam et al., 2012).

3. Fungi in bioremediation

Fungi could survive in different habitats that colonize in soil matrix along with freshwater and also in marine habitats. Fungi also grow in different climatic conditions including the stress condition and inseminate the spores through air that aid in balancing of ecosystem. It has been reported that the fungi can also survive in effluent treatment plants (ETPs) during treatment of wastewaters. The surviving ability in diverse habitats and potential for producing the diverse enzymes indicates fungi as potential candidates for bioremediation.

3.1 White-rot fungi

White-rot fungi are used for biodegradation of lignininous substances in atmosphere that attribute in the carbon recycling. The accumulation of endocrine disrupting chemicals (EDCs) and TrOCs such as pharmaceuticals and personal care products (PPCPs) develop acute and chronic toxicity to aquatic organisms and also harmful for human health have been remediated by white-rot fungi. Many previous studies explained the efficiency of white-rot fungi for bioremediation, namely *Phanerochaete chysosporium, Bjerkandera adjusta,* and *Pleurotus* sp., which secrete the varieties of ligninolytic enzymes, e.g., laccases and peroxidases (Voberkova et al., 2017; Yang et al., 2013). The ligninolytic enzymes of white-rot fungi have been used for conversion of different organic pollutants such as pesticides of the contaminated wastewaters through enhancing the microbial activity using a biopurification system (BPS). The deposited lignin granules in the lignocellulosic fibers are separated by applying the pressure refining that has been enhanced through using of ligninolytic enzymes of white-rot fungus *Ceriporiopsis subvermispora* exhibited greater delignification in comparison to pressure refined *Miscanthus* than milled *Miscanthus*. Extracellular ligninolytic enzymes of Agaricomycete and white-rot from Amazon forest fungi are also capable for adsorption, degradation or decolourization of dyes such as Direct Blue 14 manufactured of Pleurotus and Remazol Brilliant Blue-R. The fungal groups such as *Coriolus versicolor, Inonotus hispidus,* and *Phlebia tremellosa* have capacity to decolorize of dye effluent, although 38 species of white-rot fungi have been explained in reduction of total phenolics (>60%) and color (≤70%) from olive-mill wastewater. The remediation of cresolate in the contaminated soil has been accomplished through two strains of white-rot fungi: *T. versicolor* and *Lentinus tigrinus*. The cresolate-polluted soil is generally mixture of contaminants such as residual petroleum hydrocarbons and high molecular weight PAH fraction after a biopiling treatment. The potential degradation of the contaminants could be held through biostimulation of lignocellulosic substrate with bioaugmentation of fungi (Yadav et al., 2006; Yang et al., 2013). However, such type of treatment possibly chances to promote the growth of neighbor microbes that may influence to the augmented organism, hence before applying such treatment in filed it is required to validate the study at a small scale. Moreover, other properties of white-rot fungi secreting a laccase enzyme are also applied for degradation of substituted organic. Looking to such important features in bioremediation, it has been approached by escalating the laccase production in white-rot fungi, *T. versicolor* and *P. ostreatus*, through solid state fermentation on orange peels followed by further investigating of its efficiency for bioremediation of PAHs such as phenanthrene and pyrene. The production of laccase (3000 U/L) from *T. versicolor* cultures and *P. ostreatus* produced 2700 U/L laccase exhibited better removal of phenanthrene and pyrene. However, more study at the molecular level to understand the potential role of fungi in bioremediation is required.

3.2 Marine fungi

The marine fungi produce the efficient secondary metabolites, biosurfactants, enzymes, polysaccharides and fatty acids that are used in bioremediation of hydrocarbons and heavy metals. The adaptation of marine fungi to high saline conditions and pH makes more biological advantageous in extreme condition than the terrestrial fungi. The potential of marine microbes for remediating the metal ion indicate to the promising nature of extremophilic microbes for bioremediation and in nanotechnology. Thatoi et al. have explained the diverse potential applications of marine fungi from mangrove with emphasizing on its diversity, promising ecological role, and biotechnological application for source of drugs, enzymes, biodiesel, biopesticides, and bioremediation (Thatoi et al., 2013). Recently, it has been elucidated the potential role of the derived enzymes from marine fungi and its biotechnological application (Bonugli-Santos et al., 2015). Marine fungi are highly tolerable to high concentrations of heavy metals and its interaction to metal ions of marine ecosystems could be employed for preparation of metal nanoparticles. Fungi have capability to synthesize nanoparticles both extra- and intracellularly that are attributed for applications in different areas from textile industries, food preservations, to medicines and clinical microbiology, etc.

The various factors of fungi have been explored for increasing the efficiency of bioremediation of toxic and persistent organic pollutants. The production of laccase enzyme from marine fungi is highly tolerable to salinity and phenolics have been implied in the isolated *Trichoderma viride Pers* NFCCI-2745 of an estuary polluted with phenolics (Divya et al., 2014). The enzyme mediated bioremediation has also been explained for decolorizing Remazol Brilliant Blue-R dye applying basidiomycetes of marine sponges, and anthraquinone dye Reactive Blue 4 by *C. unicolor*, a marine white-rot basidiomycete. Gao et al. have reported that the biostimulation and bioaugmentation have adverse affect to the transformation of persistent organic pollutants (POPs) such as PCB 118 by marine fungi belonging to genus *Penicillium* containing maifanite. Another POP such as pentachlorophenol has been transformed even at high concentrations through marine-derived fungus, *Trichoderma harzianum* and other marine fungi such as *Mucor*, *Aspergillus*, and slime mold exhibited efficient bioremediation for water-soluble crude oil fractions (0.01−0.25 mg/mL) at higher level causing toxicity to the organisms (Gao et al., 2013).

3.3 Extremophilic fungi

The extremophilic fungi are very imperative for industries producing extremophilic enzymes posses several unique properties such as thermo tolerance, pH tolerance, and sustain to other stress conditions. The effluent treatment plant has strong niche that is targeted for fungi with different ways of bioremediation applications at high levels of pollutants from industrial effluents.

The extremophilic fungi with such properties are better candidates for economical level and eco-friendly along with bio-transformation of raw materials, i.e., in food industries, textiles, and animal fodder preparation, etc. The implications of metallophilic microbes in bioremediation produce problems with heavy metals from the environment that is generated by formation of nanoparticle for efficient bioremediation (Sinha et al., 2011). A psychrophilic fungus, *Cryptococcus* sp. occurring in deep-sea sediments have capability to tolerate at high amount of heavy metals (100 mg/L) $ZnSO_4$, $CuSO_4$, $Pb(CH_3COO)_2$ and $CdCl_2$. Several hydrolytic enzymes also exhibit activity in extremophilic conditions and participate in remediation methods under stress conditions such as high salinity and extra-heavy crude oil (ECHO) contamination of oil belts. The laccases activity under extreme environment are used in bioremediation processes in *Pestalotiopsis palmarum* in presence of wheat bran and lignin peroxidases are produced from extra heavy crude oil as a source for carbon and energy (Majeau et al., 2010). The enzymes such as chitinases from psychrophilic fungus, *Lecanicillium muscarium*, are used to enhance the activity of insecticides. The isolation of extremophilic fungi for bioremediation processes from extreme environments like deep biosphere habitat having fumarolic ice caves on Antartica's Mt. Erebus are also used to characterize the other unique fungi using as energy sources to remediate the human contamination in such extreme places.

3.4 Symbiotic association of fungi with plants and bacteria

Fungi have close relation with plants and bacteria to reduce the hurdle of restricted growth under different environmental conditions. Arbuscular mycorrhizal fungi (AMF) are common symbiotic association with fungi and higher plants, there fungal partner involves in removal of pollutant with larger surface area by absorbing the contaminants through its hyphae and spores. AMF colonized in root of plants are used as phytoremediation for polluted groundwater with several hazardous substances in a constructed wetland. Some plant-associated fungi, e.g., *A. nidulans*, *Funalia trogii*, *Irpex lacteus*, *P. ostreatus*, etc., are capable to survive in and decolorize textile industry effluents. The colonized AM fungus *Rhizophagus custos* in root organ are able to survive in high levels of PAHs mainly anthracene with less production of toxic from anthraquinone. The ectomycorrhizal fungi, *Suillus bovinus* and *Rhizopogon roseolus* associated with Pinus involve in removal of cadmium that is also subjected to the effect of other environmental factors such as types of nutrients and pH (Mao and Guan, 2016). Another implication of such fungi is to overcome the technical rifts of algal bio-fuels and photosynthetic biorefineries through co-cultivation of microalgae and fungi for removal of single algal cells from fermentation medium. This allows their extraction and harvest through simple filtration that increases the yielding of biomass, lipid, and bioproduct. Instead of co-culture studies for bioremediation, a keen understanding of its application associated with the interaction between multitudes of metabolic pathways of different organisms is required.

4. Technology advancement

Bioremediation of toxic substances is one of reliable methods for cleanup of polluted sites and fungi play important role in remediation of hazardous substances containing different constituents in ecosystem with multiple modes for overcoming the problems of contamination. Nonetheless, their implications are mainly associated with environmental factors and long lag phase, sludge generation, and difficult methods, which may have direct impact on application of fungal biomass in bioremediation. Many advance technologies have been developed in field of fungal bioremediation to reduce the associated problems. One such advance processes associated with enzymes of the fungal biomass reduces bioremediation time, no lag phase, minimal sludge generation, and easy process control. Though enzymes create itself other problems of high cost and less shelf life due to lower stability, although enzymes production in whole cell and its immobilization have been enhanced their stability by increasing shelf life and achieved for reusing the enzyme with affordable costs. To date, the various bioreactors; fluidized beds and rotating biological contactors have been developed for remediating of contaminants with fungi (Gautam et al., 2012). However, novel bioreactor systems are being constructed for elimination of dyes, e.g., Reactive Green 19 by white-rot fungi (Sari et al., 2016). A two-stage reactor have been succeed in remediation of azo dye Reactive Blue 222 along with Photo-Fenton's and aerobic treatment through two white-rot fungi *P. ostreatus* IBL-02 and *P. chrysosporium* IBL-03 (Kiran et al., 2013). A white-rot fungus, *T. versicolor*, revealed potential removal of TrOCs through fungal membrane bioreactor in non-sterile condition at 2 days hydraulic retention time (HRT) (Yang et al., 2013). Bhattacharya et al. have developed a novel approach for degradation of HMW-PAHs using white-rot fungus *P. chrysosporium*. The remediation of benzo[a]pyrene under ligninolytic culture medium increases the PAH oxidizing monooxygeneses with formation of P450-hydroxylated metabolite that is eliminated through non-ligninolytic phase (Bhattacharya et al., 2014). The other significant strategy associated with biopurification systems in stimulating bioremediation of pesticides having wastewaters through highly active biological mixture of particular white-rot fungi has been explained by Rodrı́guez-Rodrı́guez et al. (Rodríguez-Rodríguez et al., n.d.). The sustainability and eco-friendly nature of bioremediation has been displayed for the remediation of sewage sludge containing filamentous inoculum in a large-scale bioreactor by implying a continuous process (Rahman et al., 2014). In spite of the fungi alone, its co-cultures with bacteria playing as a synergistic degradation system containing *Fusarium* sp. PY3, *Bacillus* sp. PY1, and *Sphingomonas* sp. PY2 remove pyrene by 96.0% and volatilized arsenic by 84.1% (Liu et al., 2013). Other significant and innovative approach for remediation of PAHs have been initiated by Cobas et al. (2013) in which 90% phenanthrene has been removed within 14 days by formation of permeable novel reactive biobarriers of *Trichoderma longibrachiatum* on nylon sponge. Fungal biocatalysis is employed in all cell systems associated with textile wastewater treatment (Spina et al., 2015).

4.1 Conclusions and future prospective

The unique lifestyle and biochemical properties of fungi elucidate its importance in biotechnological implications to transfer the hazardous substances to nonhazardous. It is a highly demand, sustainable, cheap, and tailor-made technology developed using fungi to translate powerful ecosystem services into ecology-based technologies. The remediation of contaminated land is a passive process for monitoring the natural attenuation, and such measures require more time than active *ex situ* remediation, so it is relied on the autochthonous microbial communities. The low efficiency of mechanical process in spontaneous attenuation of soil facilitates toward the establishment of filamentous fungi. Most fungi produce diverse exudates, which serve as carbon sources for bacteria involving in remediation of pollutants. The intrusion of plants along with mycorrhizal or other autochthonous fungi could be promising approach for increasing the efficiency of rhizosphere bioremediation of the trapped pollutants and organic contaminants. Unfortunately, relative contributions of fungi to degradation and detoxification processes have seldom been quantified, and the same is true for the positive effects of fungi on bacterial remediation processes.

Because of lack of knowledge of methodologies and ecological approaches, it is important to sustain sufficient fungal biomass, activity, and enzyme secretion at pollutant site, thereby causing inhibition of mycoremediation. Moreover, more studies on bacteria–fungi interactions are required to understand such interactions for development of novel, ecologically stable bioremediation approaches. The strong isotope techniques analyze a better appreciation of fungal components in purification of nature, i.e., the fungal removal of air-borne organic contaminants attributing into plant litter or humus that provides a relevant reservoir and secondary source in the global cycling of these compounds. Fungi as potential driver of passive remediation technologies with low cost could be widely accepted for risk-based cleanup standards that have been recently comprehended in the legislation of countries: the United States and the United Kingdom. This chapter suggests that the fungi as potential approaches with important financial, ecological, and legal reasons could be used in bioremediation of hazardous materials.

References

Al-Maqdi, K.A., Hisaindee, S.M., Rauf, M.A., Ashraf, S.S., 2017. Comparative degradation of a thiazole pollutant by an advanced oxidation process and an enzymatic approach. Biomolecules 7.

Alneyadi, A.H., Rauf, M.A., Ashraf, S.S., 2018. Oxidoreductases for the remediation of organic pollutants in water – a critical review. Critical Reviews in Biotechnology 38, 971–988. https://doi.org/10.1080/07388551.2017.1423275.

Ayangbenro, A.S., Babalola, O.O., 2017. A new strategy for heavy metal polluted environments: a review of microbial biosorbents. International Journal of Environmental Research and Public Health. https://doi.org/10.3390/ijerph14010094.

Balabanova, L., Slepchenko, L., Son, O., Tekutyeva, L., 2018. Biotechnology potential of marine fungi degrading plant and algae polymeric substrates. Frontiers in Microbiology 9, 1527. https://doi.org/10.3389/fmicb.2018.01527.

Baldrian, P., n.d. Fungal laccases – occurrence and properties. FEMS Microbiology Reviews 30, 215–242. https://doi.org/10.1111/j.1574-4976.2005.00010.x.

Bhattacharya, S., Das, A., Prashanthi, K., Palaniswamy, M., Angayarkanni, J., 2014. Mycoremediation of benzo[a]pyrene by *Pleurotus ostreatus* in the presence of heavy metals and mediators. 3 Biotech 4, 205–211. https://doi.org/10.1007/s13205-013-0148-y.

Bhushan, B., Halasz, A., Spain, J., Thiboutot, S., Ampleman, G., Hawari, J., 2002. Biotransformation of hexahydro-1,3,5-trinitro-1,3,5-triazine catalyzed by a NAD(P)H: nitrate oxidoreductase from *Aspergillus niger*. Environmental Science and Technology 36, 3104–3108. https://doi.org/10.1021/es011460a.

Bonugli-Santos, R.C., dos Santos Vasconcelos, M.R., Passarini, M.R.Z., Vieira, G.A.L., Lopes, V.C.P., Mainardi, P.H., dos Santos, J.A., de Azevedo Duarte, L., Otero, I.V.R., da Silva Yoshida, A.M., Feitosa, V.A., Pessoa, A., Sette, L.D., 2015. Marine-derived fungi: diversity of enzymes and biotechnological applications. Frontiers in Microbiology. https://doi.org/10.3389/fmicb.2015.00269.

Cobas, M., Ferreira, L., Tavares, T., Sanroman, M.A., Pazos, M., 2013. Development of permeable reactive biobarrier for the removal of PAHs by Trichoderma longibrachiatum. Chemosphere 91, 711–716.

Crocker, F.H., Indest, K.J., Fredrickson, H.L., 2006. Biodegradation of the cyclic nitramine explosives RDX, HMX, and CL-20. Applied Microbiology and Biotechnology 73, 274–290. https://doi.org/10.1007/s00253-006-0588-y.

Deshmukh, R., Khardenavis, A.A., Purohit, H.J., 2016. Diverse metabolic capacities of fungi for bioremediation. Indian Journal of Microbiology 56, 247–264. https://doi.org/10.1007/s12088-016-0584-6.

Divya, L.M., Prasanth, G.K., Sadasivan, C., 2014. Potential of the salt-tolerant laccase-producing strain *Trichoderma viride* Pers. NFCCI-2745 from an estuary in the bioremediation of phenol-polluted environments. Journal of Basic Microbiology 54, 542–547. https://doi.org/10.1002/jobm.201200394.

Esteve-Núñez, A., Caballero, A., Ramos, J.L., 2001. Biological degradation of 2,4,6-trinitrotoluene. Microbiology and Molecular Biology Reviews 65, 335–352.

Fournier, D., Halasz, A., Thiboutot, S., Ampleman, G., Manno, D., Hawari, J., 2004. Biodegradation of octahydro-1,3,5,7- tetranitro-1,3,5,7-tetrazocine (HMX) by *Phanerochaete chrysosporium*: new insight into the degradation pathway. Environmental Science and Technology 38, 4130–4133. https://doi.org/10.1021/es049671d.

Gao, P., Li, G., Dai, X., Dai, L., Wang, H., Zhao, L., Chen, Y., Ma, T., 2013. Nutrients and oxygen alter reservoir biochemical characters and enhance oil recovery during biostimulation. World Journal of Microbiology and Biotechnology 29, 2045–2054. https://doi.org/10.1007/s11274-013-1367-4.

Gautam, S.P., Bundela, P.S., Pandey, A.K., Jamaluddin, M.K., Awasthi, Sarsaiya, S., 2012. Diversity of cellulolytic microbes and the biodegradation of municipal solid waste by a potential strain. International Journal of Medical Microbiology 2012. https://doi.org/10.1155/2012/325907.

Halaouli, S., Asther, M., Sigoillot, J.-C., Hamdi, M., Lomascolo, A., n.d. Fungal tyrosinases: new prospects in molecular characteristics, bioengineering and biotechnological applications. Journal of Applied Microbiology 100, 219–232. doi: 10.1111/j.1365-2672.2006.02866.x.

Hammel, K.E., 1995. Mechanisms for polycyclic aromatic hydrocarbon degradation by ligninolytic fungi. Environmental Health Perspectives 103 (Suppl. l), 41–43. https://doi.org/10.1289/ehp.95103s441.

Harms, H., Schlosser, D., Wick, L.Y., 2011. Untapped potential: exploiting fungi in bioremediation of hazardous chemicals. Nature Reviews Microbiology 9, 177–192. https://doi.org/10.1038/nrmicro2519.

Hofrichter, M., Bublitz, F., Fritsche, W., n.d. Unspecific degradation of halogenated phenols by the soil fungus *Penicillium frequentans* Bi 7/2. Journal of Basic Microbiology 34, 163–172. doi: 10.1002/jobm.3620340306.

Hofrichter, M., Ullrich, R., Pecyna, M.J., Liers, C., Lundell, T., 2010. New and classic families of secreted fungal heme peroxidases. Applied Microbiology and Biotechnology 87, 871–897. https://doi.org/10.1007/s00253-010-2633-0.

Ikehata, K., Buchanan, I.D., Pickard, M.A., Smith, D.W., 2005. Purification, characterization and evaluation of extracellular peroxidase from two *Coprinus* species for aqueous phenol treatment. Bioresource Technology 96, 1758–1770. https://doi.org/10.1016/j.biortech.2005.01.019.

Jensen, K.A., Houtman, C.J., Ryan, Z.C., Hammel, K.E., 2001. Pathways for extracellular Fenton chemistry in the brown rot basidiomycete *Gloeophyllum trabeum*. Applied and Environmental Microbiology 67, 2705–2711.

Karich, A., Ullrich, R., Scheibner, K., Hofrichter, M., 2017. Fungal unspecific peroxygenases oxidize the majority of organic EPA priority pollutants. Frontiers in Microbiology 8, 1463. https://doi.org/10.3389/fmicb.2017.01463.

Karigar, C.S., Rao, S.S., 2011. Role of microbial enzymes in the bioremediation of pollutants: a review. Enzyme Research 2011, 805187. https://doi.org/10.4061/2011/805187.

Kasai, N., Ikushiro, S., Shinkyo, R., Yasuda, K., Hirosue, S., Arisawa, A., Ichinose, H., Wariishi, H., Sakaki, T., 2010. Metabolism of mono- and dichloro-dibenzo-p-dioxins by *Phanerochaete chrysosporium* cytochromes P450. Applied Microbiology and Biotechnology 86, 773–780. https://doi.org/10.1007/s00253-009-2413-x.

Kiran, S., Ali, S., Asgher, M., 2013. Degradation and mineralization of azo dye reactive blue 222 by sequential Photo-Fenton's oxidation followed by aerobic biological treatment using white rot fungi. Bulletin of Environmental Contamination and Toxicology 90, 208–215. https://doi.org/10.1007/s00128-012-0888-0.

Kues, U., 2015. Fungal enzymes for environmental management. Current Opinion in Biotechnology 33, 268–278. https://doi.org/10.1016/j.copbio.2015.03.006.

Lamar, R.T., White, R.B., Farrell, R.R., Thwaites, J., Davies, I., Blair, A., 2003. Evaluation of fungal-based remediation for treatment of a PCP/dioxin/furan-contaminated soil from several former wood-treating facilities in New Zealand. In: WasteMINZ Conf, pp. 1–8.

Lipnick, R.L., Hermens, J.L.M., Jones, K.C., Muir, D.C.G., 2000. Persistent, Bioaccumulative, and Toxic Chemicals I. American Chemical Society, Washington, DC. https://doi.org/10.1021/bk-2001-0772.

Liu, S., Hou, Y., Sun, G., 2013. Synergistic degradation of pyrene and volatilization of arsenic by cocultures of bacteria and a fungus. Frontiers of Environmental Science and Engineering 7, 191–199. https://doi.org/10.1007/s11783-012-0470-3.

Majeau, J.-A., Brar, S.K., Tyagi, R.D., 2010. Laccases for removal of recalcitrant and emerging pollutants. Bioresource Technology 101, 2331–2350. https://doi.org/10.1016/j.biortech.2009.10.087.

Mao, J., Guan, W., 2016. Fungal degradation of polycyclic aromatic hydrocarbons (PAHs) by *Scopulariopsis brevicaulis* and its application in bioremediation of PAH-contaminated soil. Acta Agriculturae Scandinavica Section B Soil and Plant Science 66, 399–405. https://doi.org/10.1080/09064710.2015.1137629.

Martínez, A.T., 2002. Molecular biology and structure-function of lignin-degrading heme peroxidases. Enzyme and Microbial Technology 30, 425–444. https://doi.org/10.1016/S0141-0229(01)00521-X.

Morel, M., Meux, E., Mathieu, Y., Thuillier, A., Chibani, K., Harvengt, L., Jacquot, J.-P., Gelhaye, E., 2013. Xenomic networks variability and adaptation traits in wood decaying fungi. Microbial Biotechnology 6, 248–263. https://doi.org/10.1111/1751-7915.12015.

Nakamiya, K., Hashimoto, S., Ito, H., Edmonds, J.S., Yasuhara, A., Morita, M., 2005. Degradation of dioxins by cyclic ether degrading fungus, *Cordyceps sinensis*. FEMS Microbiology Letters 248, 17–22. https://doi.org/10.1016/j.femsle.2005.05.013.

Rahman, R.A., Molla, A.H., Fakhru'l-Razi, A., 2014. Assessment of sewage sludge bioremediation at different hydraulic retention times using mixed fungal inoculation by liquid-state bioconversion. Environmental Science and Pollution Research International 21, 1178–1187. https://doi.org/10.1007/s11356-013-1974-5.

Rani, B., Kumar, V., Singh, J., Bisht, S., Teotia, P., Sharma, S., Kela, R., 2014. Bioremediation of dyes by fungi isolated from contaminated dye effluent sites for bio-usability. Brazilian Journal of Microbiology 45, 1055–1063.

Rodríguez-Rodríguez, C.E., Castro-Gutiérrez, V., Chin-Pampillo, J.S., Ruiz-Hidalgo, K., n.d. On-farm biopurification systems: role of white rot fungi in depuration of pesticide-containing wastewaters. FEMS Microbiology Letters 345, 1–12. https://doi.org/10.1111/1574-6968.12161.

Ruiz-Dueñas, F.J., Morales, M., García, E., Miki, Y., Martínez, M.J., Martínez, A.T., 2009. Substrate oxidation sites in versatile peroxidase and other basidiomycete peroxidases. Journal of Experimental Botany 60, 441–452. https://doi.org/10.1093/jxb/ern261.

Sanyal, A., Rautaray, D., Bansal, V., Ahmad, A., Sastry, M., 2005. Heavy-metal remediation by a fungus as a means of production of lead and cadmium carbonate crystals. Langmuir 21, 7220–7224. https://doi.org/10.1021/la047132g.

Sari, A.A., Tachibana, S., Muryanto, Hadibarata, T., 2016. Development of bioreactor systems for decolorization of reactive green 19 using white rot fungus. Desalination and Water Treatment 57, 7029–7039. https://doi.org/10.1080/19443994.2015.1012121.

Scheibner, K., Hofrichter, M., Herre, A., Michels, J., Fritsche, W., 1997. Screening for fungi intensively mineralizing 2,4,6-trinitrotoluene. Applied Microbiology and Biotechnology 47, 452–457.

Sinha, A., Singh, V.N., Mehta, B.R., Khare, S.K., 2011. Synthesis and characterization of monodispersed orthorhombic manganese oxide nanoparticles produced by *Bacillus* sp. cells simultaneous to its bioremediation. Journal of Hazardous Materials 192, 620–627. https://doi.org/10.1016/j.jhazmat.2011.05.103.

Spina, F., Tigini, V., Prigione, V., Varese, G.C., 2015. Fungal biocatalysts in the textile industry. In: Fungal Biomolecules. Wiley-Blackwell, pp. 39–50. https://doi.org/10.1002/9781118958308.ch3.

Subramanian, V., Yadav, J.S., 2009. Role of P450 monooxygenases in the degradation of the endocrine-disrupting chemical nonylphenol by the white rot fungus *Phanerochaete chrysosporium*. Applied and Environmental Microbiology 75, 5570–5580.

Thatoi, H., Behera, B.C., Mishra, R.R., 2013. Ecological role and biotechnological potential of mangrove fungi: a review. Mycology 4, 54–71. https://doi.org/10.1080/21501203.2013.785448.

Ullrich, R., Hofrichter, M., 2007. Enzymatic hydroxylation of aromatic compounds. Cellular and Molecular Life Sciences 64, 271–293. https://doi.org/10.1007/s00018-007-6362-1.

Verma, A., Singh, H., Anwar, S., Chattopadhyay, A., Tiwari, K.K., Kaur, S., Dhilon, G.S., 2017. Microbial keratinases: industrial enzymes with waste management potential. Critical Reviews in Biotechnology 37, 476–491. https://doi.org/10.1080/07388551.2016.1185388.

Voberkova, S., Vaverkova, M.D., Buresova, A., Adamcova, D., Vrsanska, M., Kynicky, J., Brtnicky, M., Adam, V., 2017. Effect of inoculation with white-rot fungi and fungal consortium on the composting efficiency of municipal solid waste. Waste Management 61, 157–164. https://doi.org/10.1016/j.wasman.2016.12.039.

Yadav, J.S., Doddapaneni, H., Subramanian, V., 2006. P450ome of the white rot fungus *Phanerochaete chrysosporium*: structure, evolution and regulation of expression of genomic P450 clusters. Biochemical Society Transactions 34, 1165–1169. https://doi.org/10.1042/bst0341165.

Yang, S., Hai, F.I., Nghiem, L.D., Price, W.E., Roddick, F., Moreira, M.T., Magram, S.F., 2013. Understanding the factors controlling the removal of trace organic contaminants by white-rot fungi and their lignin modifying enzymes: a critical review. Bioresource Technology 141, 97–108. https://doi.org/10.1016/j.biortech.2013.01.173.

Cyanobacteria: potential and role for environmental remediation

Priyanka[1], Cash Kumar[1], Antra Chatterjee[2], Wang Wenjing[3], Deepanker Yadav[4], Prashant Kumar Singh[5]

[1]Cytogenetics Laboratory, Department of Zoology, Institute of Science, Banaras Hindu University, Varanasi, India; [2]Molecular Biology Section, Centre for Advanced Study in Botany, Department of Botany, Banaras Hindu University, Varanasi, India; [3]State Key Laboratory of Cotton Biology, Henan Key Laboratory of Plant Stress Biology, School of Life Science, Henan University, Kaifeng, Henan, China; [4]Department of Vegetable and Fruit Science, Institute of Plant Science, Agriculture Research Organization (ARO), The Volcani Center, Rishon LeZion, Israel; [5]Agriculture Research Organization, Ministry of Agriculture and Rural Development Volcani Centre, Rishon LeZion, Israel

1. Introduction

Increasing human population has resulted in the utilization of natural resources to fulfill their demand for food, fuel, and land, which in turn has resulted in environmental pollution and global warming. Continuous human evolution and their necessity have resulted in the industrialization. The production of various chemical compounds from different industries and their inappropriate disposal at various sites increased the pollutants and contaminants into the environment, thereby significantly increasing contaminated sites globally (Zhu and Chen, 2002). Natural equilibrium in the atmosphere is maintained for the sustenance of flora and fauna which is disrupted by human and their activities. Various substances such as halogenated aromatic compounds, polycyclic aromatic hydrocarbons, and BTEX compounds (benzene, ethylbenzene, toluene, and three isomers of xylene) are discharged from oil refineries, gas stations, agrochemicals, petrochemicals, and pharmaceutical industries. Accumulation of these undesired substances in the atmosphere is negatively affecting the productivity and fertility of air, water, and soil, thereby causing health threats to human and another living system (Pimentel et al., 2005). These chemicals are persistent in the environment for a long time (Budavari, 1996). These are mutagenic and carcinogenic to human health. Maintenance of the natural balance of the ecosystem is the urgent need to control the productivity and

Abatement of Environmental Pollutants
https://doi.org/10.1016/B978-0-12-818095-2.00010-2

fertility for the benefit of humanity. Accumulation of toxic substances resulted in detrimental effects, and hence the removal of these chemicals has become the topic of concern worldwide (Lombi et al., 2001).

For the removal of these toxic substances, several methods have been applied, viz. physicochemical and biological. Physicochemical methods are very effective, but they are costly and require many chemicals for contaminant removal. On the other hand, the second method utilized microorganisms for the same and referred to as bioremediation. Bioremediation has become the most attractive biotechnological program to degrade the toxicological compounds from the environment (Matsunaga et al., 1999). Bioremediation using photosynthetic bacteria is cost-effective and eco-friendly approach for the ecosystem maintenance. Cyanobacteria have the capability of degrading or detoxifying gaseous, solid, and liquid pollutants of natural or xenobiotic origin (Kertesz et al., 1994). Genetically engineered microorganisms (GEMs) greatly enhanced the potential of contaminant degradation capability encompassing a wide range of aromatic hydrocarbon. This chapter aimed to discuss the general strategies involved in bioremediation and cyanobacterial potential in the removal of toxic substances from the polluted sites.

1.1 General features of cyanobacteria

Cyanobacteria represent one of the largest and widest groups of microorganisms found in a variety of habitats including aquatic, terrestrial, and extreme environments (Panosyan, 2015). They are most primitive and photosynthetic prokaryotes on the earth evolved approximately 2.5—3.5 billion years ago (Hedges et al., 2001). They thrive under extreme conditions and form algal blooms. They are morphologically diverse (Klymiuk et al., 2014) with unicellular, filamentous, and colonial forms and contain different combinations of photosynthetic pigments such as chlorophyll a, carotenoids, and phycobiliproteins (Castenholz, 2001). They exhibit the capacity to fix atmospheric nitrogen and thereby registered as one of the key components of ecosystem (Bergman et al., 1997).

1.2 Role of cyanobacteria in agriculture management

Maintenance of a healthy ecosystem and the environment by several biotechnological programs aimed at addressing food security for future generations and to resolve the problems of the environment. Cyanobacteria exhibit several unique features such as oxygenic photosynthesis, growth in highly saline/alkaline environment, biomass, and biofuel production, which makes them crucial bioresource for sustainable development. These microorganisms help in improving soil quality and plant growth, thereby reducing the cost of crop production (Higa and Wididana, 1991). The sustainable agriculture involves soil, water, and pest management, crop selection, soil preservation, and processing. These sustainable agricultural practices together with biotechnological approaches increase the productivity and quality by transgenic plants, animals, and microbes (Singh, 2000). In recent years, cyanobacteria have emerged as potential candidates for sustainable agricultural practices because of their survival in the extremely diverse conditions. They require very less amount of nutrient for high areal productivity with high protein, lipid, and carbohydrate content per gram of biomass (Williams and Laurens, 2010; Milledge, 2011; Hoekman et al., 2012). Globally 25 Gt a^{-1} of carbon can be fixed into energy-dense biomass by cyanobacteria using atmospheric CO_2 and solar energy. These characteristics make cyanobacteria as potential microorganisms for their applications as feedstocks for sustainable production of food and nonfood commodities, including chemicals and bioenergy (Wase and Wright, 2008; Rajneesh et al., 2017). Production of ethanol, butanol, fatty acids, and other acids makes them a valuable source for the fulfillment of ever-increasing energy demands (Rajneesh et al., 2017). For healthy agroecosystem, they can be used as biofertilizers, and in the food industry they can be utilized as food supplements. In future, several biotechnological and transgenic approaches can be applied to exploit these microorganisms for the production of improved biomass and food stock.

1.3 The cyanobacterial potential in environmental development

Before human intervention, the interaction between the atmosphere and living organisms was in a steady state. But industrialization and urbanization for more comfort have exploited nature. They have degraded the environment by contaminating the two most important commodities of life, air and water. The treatment of these waterbodies by traditional method requires a lot of effort and space and produces waste in bulk. To overcome these limitations, microorganisms have been utilized to reduce the concentration of various chemical pollutants such as heavy metals, pesticides, etc. Cyanobacteria-mediated purification of waterbodies is cheap and eco-friendly as they are abundant in the aquatic environment and they have a high affinity for metals/pollutants. Common waste effluents discharged from the industry include Cu, Zn, Ni, Co, Pb, Cr, and Cd (Venkateswarlu et al., 1994; Kaushik et al., 1997). Many microorganisms synthesize metal-binding proteins which have an affinity for the metals known as metallothioneins. These are small molecular weight proteins rich in Cys residue and bind to the metal ion in metal thiolate cluster. A unicellular cyanobacterium *Synechococcus* sp. had been shown to have the ability of binding to metal ion, e.g., Cu (11.3 mg/g biomass), Pb (30.4 mg/g biomass), Ni (3.2 mg/g biomass), and Cd (7.2 mg/g biomass) (Yee et al., 2004). In addition to metal ion removal, dyes disposed of the textile industry have been degraded by cyanobacteria isolated from the polluted sites. Many species of *Oscillatoria* have been reported to decolor dye in wastewater (Zhu et al., 1979). Algae can utilize Eriochrome blue SE and Black T as its source of carbon and nitrogen (Jinqi and Houtian, 1992). Therefore, cyanobacteria play an important role in stabilization of ponds not only by oxygen production but also by the removal of azo compounds from the waterbodies making it suitable for the usage of several purposes by living beings.

1.4 Cyanobacteria: role in bioremediation

The term bioremediation is generally a combination of two words; bio means biological and remediation means to remedy. Bioremediation is commonly used to define the removal of toxic waste from the contaminated sites with the help of microorganisms. Bioremediation is defined as the process which involves the utilization of microorganisms and their enzymes to remove the contaminants from the affected sites and to return to the original natural environment altered by pollutants. This process is generally employed to the toxic effluents released from the fertilizer industry to the fields and accumulates there leading to the contamination of soil. Therefore, this process is generally used to degrade the soil contaminants either by naturally occurring or tailor-made microbes (Ripp et al., 2000; Sayler and Ripp, 2000). The utilization of microbes for the detoxification purpose made the process cost-effective and eco-friendly, which is targeted to remove the heavy metals, radioactive waste, and phenolic compounds from the contaminated sites (Kertesz et al., 1994). Bioremediation is classified as in situ or ex situ. In situ bioremediation involves the removal of the contaminated material at the site with minimal disturbance, while in ex situ method contaminants are treated elsewhere. Ex situ techniques are employed for the treatment of soil and groundwater, which is taken at another site via excavation or pumping, respectively. It involves biostimulation where organic or inorganic components are added at the treatment site to enhance the growth of the microbes for the degradation of the contaminants.

In situ bioremediation can be undertaken by following means to remove the toxic substances by naturally occurring microorganisms.

Bioventing: This is the most common technique used for this purpose. It involves the supply of oxygen and other nutrients at the affected sites to enhance the growth of indigenous microbes. Contamination with toxic materials results in a reduction of the oxygen and other nutrients in the soil which inhibits the growth of these microbes. Supply of same at the target site enhances the growth of these naturally occurring microbes. The bioventing process enhances the degradation of the anaerobically degradable compound of soil contaminated with organic hydrocarbons, pesticides, and other chemicals (Dupont, 1993; Khan et al., 2004).

Biosparging: This method is utilized to degrade the petroleum constituents dissolved in water and absorbed in the soil to the water table (Norris et al., 1993, 1994). The efficiency to degrade the contaminants of soil and water by this method depends on properties of the soil, groundwater, pH, microbial population density, and nutrient concentration, etc. By using this method, oxygen is injected into the saturated zone to increase the biological activity of the microorganisms.

Bioaugmentation: Bioaugmentation is another technique used to remove the pollutants in the industrial freshwater by in situ method. Wastewater-activated sludge contains actively growing microorganisms that degrade a wide variety of contaminants, but some pollutants are resistant to degradation by these microorganisms. Several factors are responsible for the degradation mediated by these microorganisms which may include high toxicity, low water solubility, low bioavailability, high stability, and low biodegradability. The chemical structure complexities of these pollutants make them resistant to degradation where several microorganisms are necessary for this purpose. Bioaugmentation proves to be an essential method to overcome these challenges, as the treatment can be tailored to a specific pollutant that is dominant in the environment. Thus this approach is very promising for the treatment of both the increasing numbers of emerging pollutant and the pollutants present at high concentrations.

Ex situ Bioremediation: These techniques involved the removal of contaminated soil from the polluted sites through excavation and their treatment for bioremediation elsewhere. Ex situ bioremediation is the costlier approach. It takes less time for bioremediation, and the treatment is beneficial because of the uniformity of the treatment. Following methods can be utilized for ex situ bioremediation process:

Cyanobacteria in bioremediation: Cyanobacteria are excellent microorganisms for the biotechnological program because of their capability to remove the heavy metal from the polluted site and for the reclamation of the usar soil. They can degrade the naturally occurring hydrocarbon and xenobiotics (Meghraj et al., 1997) at the contaminated sites. Cyanobacteria have great potential for the reclamation of usar soil. Usar soil is characterized by high pH, high salinity, impermeable, and hardness because of the presence of salt over the surface of the soil. Excess salt in the soil increases pore size and hence less water retention. Earlier methods applied to modify the chemical structure of soil included the application of pyrite, gypsum, and excessive irrigation to dissolve the salt in the land, which remained very challenging. This leads to the possibility of cyanobacterial application for the same purpose. Reclamation of usar soil using cyanobacteria was first proposed by Singh (1950); later Thomas and Apte (1984) suggested the reduction of soil salinity of "Kharland" by the usage of the salt-tolerant cyanobacterium, *Anabaena torulosa*. Some saline-tolerant and alkali-tolerant cyanobacteria have been reported, viz. *Nostoc, Calothrix, Plectonema, Scytonema, Cylindrospermum, Westiellopsis*, and *Hapalosiphon*. Chatterjee et al. (2017) have extensively described the potential role of cyanobacteria in amelioration of sodic soil. Table 10.1 provides an example of cyanobacterial species involved in heavy metal removal.

Landfarming: Landfarming is an essential method of ex situ bioremediation. The contaminated soil is excavated and applied into linen beds and turned over to aerate the waste. Ex situ bioremediation by landfarming has been utilized for hydrocarbons, pyrene, and other petrochemical waste. There are several limitations of this method such as the requirement of ample space, uncontrolled rain, and temperature. Moisture content, microbial density as well as composition, and nutrients affect the biodegradation process of contaminants in landfarming technique as reported by Hezaji and Husain (2004).

TABLE 10.1   Heavy metal and organic pollutant removal from some species of Cyanobacteria.

Heavy metal removal		
Heavy metals	Cyanobacteria	References
Cd	*Nostoc linckia, Nostoc rivularis, Tolypothrix tenuis, Microcystis*	Inthorn et al. (1996); El-Enany and Issa (2000) Rai et al. (1998)
Ni	*Microcystis* sp.	Rai et al. (1998)
Co	*Nostoc muscorum, Anabaena subcylindrica*	El-Sheekh et al. (2005)
Cr	*N. calcicola, Chroococcus* sp.	Anjana et al. (2007)
Cu	*N. muscorum, A. subcylindrica*	El-Sheekh et al. (2005)
Hg	*Spirulina platensis, Aphanothece flocculosa*	Cain et al. (2008)
Mn	*N. muscorum, A. subcylindrica*	El-Sheekh et al. (2005)
Pb	*N. muscorum, A. subcylindrica, Gloeocapsa* sp.	El-Sheekh et al. (2005); Raungsomboon et al. (2006)
Zn	*N. linckia, N. rivularis*	El-Enany and Issa (2000)
Sr	*Gloeomargarita lithophora, Cyanothece* sp	Cam et al. (2016)
Ba	*G. lithophora, Cyanothece* sp	Cam et al. (2016)

Organic pollutants removal		
Organic pollutant	Cyanobacteria	References
Phenol	*Ochromonas danica* *Pseudanabaena PP16*	Semple and Cain (1996) Kirkwood et al., 2005
Tributyltin	*Chlorella vulgaris, Chlorella* sp. *Chlorella miniata*	Tsang et al. (1999) Tam et al. (2002)
Phenanthrene	*Selenastrum capricornutum* *Microcystis aeruginosa*	Chan et al. (2006) Bai et al. (2016)
Naphthalene	*Agmenellum quadruplicatum* *C. vulgaris*	Cerniglia et al. (1979) Todd et al. (2002)
Bisphenol	*Chlorella fusca*	Hirooka et al. (2005)
Azo compounds	*C. vulgaris*	Jinqi and Houtian (1992)
Diesel 99.5% (0.6% v/v)	*Phormidium* sp. *Oscillatoria* sp. *Chroococcus* sp.	Chavan and Mukherji (2008)
Total petroleum hydrocarbon 99% (diesel 0.6% v/v)	*Phormidium* sp. *Oscillatoria* sp. *Chroococcus* sp.	Chavan and Mukherji (2010)
Dimethyl phthalate	*Synechocystis* sp. PCC6803 *Synechococcus* sp. PCC7942 *Cyanothece* sp. PCC7822	Zhang et al. (2016)

Adapted and modified from Singh, J.S., Kumar A., Rai A.N., Singh, D.P., April 21, 2016. Cyanobacteria: a precious bio-resource in agriculture, ecosystem, and environmental sustainability. Frontiers in Microbiology 7 529; Subashchandrabose, S.R., Ramakrishnan, B., Megharaj, M., Venkateswarlu, K., Naidu, R., November 1, 2011. Consortia of cyanobacteria/microalgae and bacteria: biotechnological potential. Biotechnology Advances 29 (6), 896—907; Subashchandrabose, S.R., Ramakrishnan, B., Megharaj, M., Venkateswarlu, K., Naidu, R., January 1, 2013. Mixotrophic cyanobacteria and microalgae as distinctive biological agents for organic pollutant degradation. Environment International 51 59—72.

2. Conclusions and future perspectives

The utilization of natural resources for the development processes by a human being forced them to live in a threatened, dreadful, and depleted environment of the present. Continuous destruction of the natural environment has necessitated the need for healthy agroecosystem to conserve the natural resources and also to maintain the complexity and diversity of the ecosystem. A healthy ecosystem supports the production of sufficient food for the increasing population and safer living for both human being and other livestock. Physical and chemical methods utilizing chemical fertilizers and pesticides used for this purpose become a limitation for the poor farmers because of their cost ineffectiveness as well as it may also be proven to be harmful to the land. Cyanobacteria are an excellent model system because of their photosynthesis and nitrogen fixation capability, which makes soil enriched in carbon and nitrogen and enhances phosphorus bioavailability to plants. GEMs with novel genes and biomolecules having various uses in agriculture, industry, and environmental sustainability are being utilized by biotechnologists (Golden et al., 1987; Huang et al., 2010; Heidorn et al., 2011). Genetically modified cyanobacteria show potentially enhanced production of biomass, increased growth and photosynthetic efficiency, lipid and carbohydrate productivity, improved temperature tolerance, and reduced photoinhibition and photooxidation (Volkman and Gorbushina, 2006; Volkman et al., 2006). Cyanobacteria are excellent degraders of several environmental contaminants such as heavy metals, crude oil, and another aromatic compound. However, the application of genetic engineering for biofuel production is still in infancy. Metabolic engineering of cyanobacteria for the production of biofuel is likely to play a crucial role in future. Execution of these methodologies in the field is not an easy process as this requires several political and social issues that need to be sorted before any step is taken.

Acknowledgments

Antra Chatterjee is thankful to Council of Industrial and Scientific Research Fellowship (CSIR), New Delhi, India, for Senior Research Fellowship. Prashant Kumar Singh is grateful to Agriculture Research Organization (ARO), Israel, for Visiting Scientist research fellowship.

References

Anjana, K., Kaushik, A., Kiran, B., Nisha, R., September 5, 2007. Biosorption of Cr (VI) by immobilized biomass of two indigenous strains of cyanobacteria isolated from metal contaminated soil. Journal of Hazardous Materials 148 (1–2), 383–386.

Bai, L., Xu, H., Wang, C., Deng, J., Jiang, H., November 1, 2016. Extracellular polymeric substances facilitate the biosorption of phenanthrene on cyanobacteria *Microcystis aeruginosa*. Chemosphere 162, 172–180.

Bergman, B., Gallon, J.R., Rai, A.N., Stal, L.J., February 1, 1997. N2 fixation by non-heterocystous cyanobacteria. FEMS Microbiology Reviews 19 (3), 139–185.

Budavari, S., 1996. The Merk Index: An Encyclopedia of Chemicals, Drugs and Biologicals. Merk, Whitehouse Station, NJ.

Cain, A., Vannela, R., Woo, L.K., September 1, 2008. Cyanobacteria as a biosorbent for mercuric ion. Bioresource Technology 99 (14), 6578–6586.

Cam, N., Benzerara, K., Georgelin, T., Jaber, M., Lambert, J.F., Poinsot, M., Skouri-Panet, F., Cordier, L., October 18, 2016. Selective uptake of alkaline earth metals by cyanobacteria forming intracellular carbonates. Environmental Science and Technology 50 (21), 11654–11662.

Castenholz, R.W., 2001. PhylumBX.Cyanobacteria. In: Booneand, D.R., Castenholz, R.W. (Eds.), Bergey's Manual of Systematic Bacteriology, Second ed. Springer, New York, NY, pp. 473–599.

Cerniglia, C.E., Gibson, D.T., Van Baalen, C., May 14, 1979. Algal oxidation of aromatic hydrocarbons: formation of 1-naphthol from naphthalene by *Agmenellum quadruplicatum*, strain PR-6. Biochemical and Biophysical Research Communications 88 (1), 50–58.

Chan, S.M., Luan, T., Wong, M.H., Tam, N.F., July 1, 2006. Removal and biodegradation of polycyclic aromatic hydrocarbons by *Selenastrum capricornutum*. Environmental Toxicology and Chemistry 25 (7), 1772–1779.

Chatterjee, A., Singh, S., Agrawal, C., Yadav, S., Rai, R., Rai, L.C., 2017. Role of algae as biofertilizer. In: Algal Green Chemistry. Elsevier.

Chavan, A., Mukherji, S., June 2010. Effect of co-contaminant phenol on performance of a laboratory-scale RBC with algal-bacterial biofilm treating petroleum hydrocarbon-rich wastewater. Journal of Chemical Technology and Biotechnology 85 (6), 851–859.

Chavan, A., Mukherji, S., June 15, 2008. Treatment of hydrocarbon-rich wastewater using oil degrading bacteria and phototrophic microorganisms in rotating biological contactor: effect of N: P ratio. Journal of Hazardous Materials 154 (1–3), 63–72.

Dupont, R.R., February 1993. Fundamentals of bioventing applied to fuel contaminated sites. Environmental Progress 12 (1), 45—53.

El-Enany, A.E., Issa, A.A., January 1, 2000. Cyanobacteria as a biosorbent of heavy metals in sewage water. Environmental Toxicology and Pharmacology 8 (2), 95—101.

El-Sheekh, M.M., El-Shouny, W.A., Osman, M.E., El-Gammal, E.W., February 1, 2005. Growth and heavy metals removal efficiency of *Nostoc muscorum* and *Anabaena subcylindrica* in sewage and industrial wastewater effluents. Environmental Toxicology and Pharmacology 19 (2), 357—365.

Golden, S.S., Brusslan, J., Haselkorn, R., January 1, 1987. [12] Genetic engineering of the cyanobacterial chromosome. In: Methods in Enzymology, vol. 153. Academic Press, pp. 215—231.

Hedges, S.B., Chen, H., Kumar, S., Wang, D.Y., Thompson, A.S., Watanabe, H., 2001. A genomic timescale for the origin of eukaryotes. BMC Evolutionary Biology 1 (4).

Heidorn, T., Camsund, D., Huang, H.H., Lindberg, P., Oliveira, P., Stensjö, K., Lindblad, P., January 1, 2011. Synthetic biology in cyanobacteria: engineering and analyzing novel functions. In: Methods in Enzymology, vol. 497. Academic Press, pp. 539—579.

Hejazi, R.F., Husain, T., April 15, 2004. Landfarm performance under arid conditions. 2. Evaluation of parameters. Environmental Science and Technology 38 (8), 2457—2469.

Higa, T., Wididana, G.N., 1991. Changes in the soil microflora induced by effective microorganisms. In: Proceedings of the First International Conference on Kyusei Nature Farming. US Department of Agriculture, Washington, DC, USA, pp. 153—162.

Hirooka, T., Nagase, H., Uchida, K., Hiroshige, Y., Ehara, Y., Nishikawa, J.I., Nishihara, T., Miyamoto, K., Hirata, Z., August 1, 2005. Biodegradation of bisphenol A and disappearance of its estrogenic activity by the green alga *Chlorella* fusca var. vacuolata. Environmental Toxicology and Chemistry 24 (8), 1896—1901.

Hoekman, S.K., Broch, A., Robbins, C., Ceniceros, E., Natarajan, M., January 1, 2012. Review of biodiesel composition, properties, and specifications. Renewable and Sustainable Energy Reviews 16 (1), 143—169.

Huang, H.H., Camsund, D., Lindblad, P., Heidorn, T., March 17, 2010. Design and characterization of molecular tools for a synthetic biology approach towards developing cyanobacterial biotechnology. Nucleic Acids Research 38 (8), 2577—2593.

Inthorn, D., Nagase, H., Isaji, Y., Hirata, K., Miyamoto, K., January 1, 1996. Removal of cadmium from aqueous solution by the filamentous cyanobacterium *Tolypothrix tenuis*. Journal of Fermentation and Bioengineering 82 (6), 580—584.

Jinqi, L., Houtian, L., January 1, 1992. Degradation of azo dyes by algae. Environmental Pollution 75 (3), 273—278.

Kaushik, S., Sahu, B.K., Lawania, R.K., Tiwari, R.K., 1997. Occurrence of heavy metals in lentic water of Gwalior region. Journal of Postgraduate Medicine 43 (4), 137—140.

Kertesz, M.A., Cook, A.M., Leisinger, T., October 1, 1994. Microbial metabolism of sulfur and phosphorus-containing xenobiotics. FEMS Microbiology Reviews 15 (2—3), 195—215.

Khan, F.I., Husain, T., Hejazi, R., June 1, 2004. An overview and analysis of site remediation technologies. Journal of Environmental Management 71 (2), 95—122.

Kirkwood, A.E., Nalewajko, C., Fulthorpe, R.R., July 1, 2005. The impacts of cyanobacteria on pulp-and-paper wastewater toxicity and biodegradation of wastewater contaminants. Canadian Journal of Microbiology 51 (7), 531—540.

Klymiuk, V., Barinova, S., Lyalyuk, N., 2014. Diversity and ecology of algal communities from the regional landscape park" slavyansky resort", Ukraine. Research Review: Journal of Botanical Science 3 (2), 9—26.

Lombi, E., Zhao, F.J., Dunham, S.J., McGrath, S.P., November 1, 2001. Phytoremediation of heavy metal—contaminated soils. Journal of Environmental Quality 30 (6), 1919—1926.

Matsunaga, T., Takeyama, H., Nakao, T., Yamazawa, A., April 30, 1999. Screening of marine microalgae for bioremediation of cadmium-polluted seawater. Journal of Biotechnology 70 (1—3), 33—38.

Megharaj, M., Wittich, R.M., Blasco, R., Pieper, D.H., Timmis, K.N., July 1, 1997. Superior survival and degradation of dibenzo-p-dioxin and dibenzofuran in soil by soil-adapted *Sphingomonas* sp. strain RW1. Applied Microbiology and Biotechnology 48 (1), 109—114.

Milledge, J.J., March 1, 2011. Commercial application of microalgae other than as biofuels: a brief review. Reviews in Environmental Science and Biotechnology 10 (1), 31—41.

Norris, R.D., Hinchee, R.E., Brown, R.A., McCarty, P.L., Semprini, L., Wilson, J.T., Kampbell, D.H., Reinhard, M., Bower, E.J., Borden, R.C., Vogel, T.M., Thomas, J.M., Ward, C.H., 1994. Handbook of Bioremediation. CRC Press, Boca Raton, FL.

Norris, R.D., Hinchee, R.E., Brown, R.A., McCarty, P.L., Semprini, L., Wilson, J.T., Kampbell, D.H., Reinhard, M., Bower, E.J., Borden, R.C., Vogel, T.M., Thomas, J.M., Ward, C.H., 1993. In Situ Bioremediation of Groundwater and Geological Material: A Review of Technologies. EPA/5R-93/124. U.S. Environmental Protection Agency, Office of Research and Development, Ada, OK.

Panosyan, H., August 10—11, 2015. Thermophiles harbored in Armenian geothermal springs and their potential in biotechnology. Industrial Biotechnology Congress, Birmingham, The UK.

Pimentel, D., Hepperly, P., Hanson, J., Douds, D., Seidel, R., July 2005. Environmental, energetic, and economic comparisons of organic and conventional farming systems. AIBS Bulletin 55 (7), 573—582.

Rai, L.C., Singh, S., Pradhan, S., March 10, 1998. Biotechnological potential of naturally occurring and laboratory-grown *Microcystis* in biosorption of Ni^{2+} and Cd^{2+}. Current Science 461—464.

Rajneesh, Singh, S.P., Pathak, J., Sinha, R.P., 2017. Cyanobacterial factories for the production of green energy and value-added products: an integrated approach for economic viability,. Renew. Sustain. Energy Rev. 69, 578—595. https://doi.org/10.1016/j.rser.2016.11.110.

Raungsomboon, S., Chidthaisong, A., Bunnag, B., Inthorn, D., Harvey, N.W., December 1, 2006. Production, composition and Pb^{2+} adsorption characteristics of capsular polysaccharides extracted from a cyanobacterium *Gloeocapsa gelatinosa*. Water Research 40 (20), 3759—3766.

Ripp, S., Nivens, D.E., Ahn, Y., Werner, C., Jarrell, J., Easter, J.P., Cox, C.D., Burlage, R.S., Sayler, G.S., March 1, 2000. Controlled field release of a bioluminescent genetically engineered microorganism for bioremediation process monitoring and control. Environmental Science and Technology 34 (5), 846—853.

Sayler, G.S., Ripp, S., June 1, 2000. Field applications of genetically engineered microorganisms for bioremediation processes. Current Opinion in Biotechnology 11 (3), 286—289.

Semple, K.T., Cain, R.B., April 1, 1996. Biodegradation of phenols by the alga *Ochromonas danica*. Applied and Environmental Microbiology 62 (4), 1265—1273.

Singh, J.S., Kumar, A., Rai, A.N., Singh, D.P., April 21, 2016. Cyanobacteria: a precious bio-resource in agriculture, ecosystem, and environmental sustainability. Frontiers in Microbiology 7, 529.

Singh, R.N., February 1950. Reclamation of 'usar'lands in India through blue-green algae. Nature 165 (4191), 325.

Singh, R.B., June 2000. Biotechnology, biodiversity and sustainable agriculture—A contradiction. In: Regional Conference in Agricultural Biotechnology Proceedings: Biotechnology Research and Policy—Needs and Priorities in the Context of Southeast Asia's Agricultural Activities. SEARCA (SEAMEO)/FAO/APSA, Bangkok.

Subashchandrabose, S.R., Ramakrishnan, B., Megharaj, M., Venkateswarlu, K., Naidu, R., January 1, 2013. Mixotrophic cyanobacteria and microalgae as distinctive biological agents for organic pollutant degradation. Environment International 51, 59—72.

Subashchandrabose, S.R., Ramakrishnan, B., Megharaj, M., Venkateswarlu, K., Naidu, R., November 1, 2011. Consortia of cyanobacteria/microalgae and bacteria: biotechnological potential. Biotechnology Advances 29 (6), 896—907.

Tam, N.F., Chong, A.M., Wong, Y.S., September 1, 2002. Removal of tributyltin (TBT) by live and dead microalgal cells. Marine Pollution Bulletin 45 (1—12), 362—371.

Thomas, J., Apte, S.K., December 1, 1984. Sodium requirement and metabolism in nitrogen-fixing cyanobacteria. Journal of Biosciences 6 (5), 771—794.

Todd, S.J., Cain, R.B., Schmidt, S., August 1, 2002. Biotransformation of naphthalene and diaryl ethers by green microalgae. Biodegradation 13 (4), 229—238.

Tsang, C.K., Lau, P.S., Tam, N.F., Wong, Y.S., June 1, 1999. Biodegradation capacity of tributyltin by two *Chlorella* species. Environmental Pollution 105 (3), 289—297.

Venkateswarlu, V., Reddy, P.M., KUMAR, B., October 1, 1994. heavy-metal pollution in the rivers of Andhrapradesh, India. Journal of Environmental Biology 15 (4), 275—282.

Volkmann, M., Gorbushina, A.A., Kedar, L., Oren, A., March 21, 2006. Structure of *Euhalothece*-362, a novel red-shifted mycosporine-like amino acid, from a halophilic cyanobacterium (*Euhalothece* sp.). FEMS Microbiology Letters 258 (1), 50—54.

Volkmann, M., Gorbushina, A.A., February 1, 2006. A broadly applicable method for extraction and characterization of mycosporines and mycosporine-like amino acids of terrestrial, marine and freshwater origin. FEMS Microbiology Letters 255 (2), 286–295.

Wase, N.V., Wright, P.C., August 1, 2008. Systems biology of cyanobacterial secondary metabolite production and its role in drug discovery. Expert Opinion on Drug Discovery 3 (8), 903–929.

Williams, P.J., Laurens, L.M., 2010. Microalgae as biodiesel & biomass feedstocks: review & analysis of the biochemistry, energetics & economics. Energy and Environmental Science 3 (5), 554–590.

Yee, N., Benning, L.G., Phoenix, V.R., Ferris, F.G., February 1, 2004. Characterization of metal– cyanobacteria sorption reactions: a combined macroscopic and infrared spectroscopic investigation. Environmental Science and Technology 38 (3), 775–782.

Zhang, X., Liu, L., Zhang, S., Pan, Y., Li, J., Pan, H., Xu, S., Luo, F., 2016. Biodegradation of dimethyl phthalate by freshwater unicellular cyanobacteria. BioMed Research International 2016, 1–8.

Zhu, Y.K., Xie, S.Q., Dong, J.G., 1979. Primary test of testing dye waste water by rotating algal disc. Environmental Sciences 6, 37–41.

Zhu, Z.L., Chen, D.L., July 1, 2002. Nitrogen fertilizer use in China–contributions to food production, impacts on the environment and best management strategies. Nutrient Cycling in Agroecosystems 63 (2–3), 117–127.

Further reading

Pisciotta, J.M., Zou, Y., Baskakov, I.V., May 25, 2010. Light-dependent electrogenic activity of cyanobacteria. PLoS One 5 (5), e10821.

Singh, S.P., Pathak, J., Sinha, R.P., March 1, 2017. Cyanobacterial factories for the production of green energy and value-added products: an integrated approach for economic viability. Renewable and Sustainable Energy Reviews 69, 578–595.

An effective approach for the degradation of phenolic waste: phenols and cresols

Tripti Singh[1,2], A.K. Bhatiya[1], P.K. Mishra[2], Neha Srivastava[2]

[1]Department of Biotechnology, GLA University, Mathura, India; [2]Department of Chemical Engineering and Technology, Indian Institute of Technology (Banaras Hindu University), Varanasi, India

1. Introduction

Heavy industrialization around the world has resulted in several unwarranted consequences to the environment (Bhandari and Garg, 2015). It has not only led to the depletion of natural resources but also caused drastic climate changes (Chopra, 2016; Misra and Pandey, 2005). The hazardous waste released from industries is one of the major reasons of soil, air, and water pollution; thereby, its safe disposal is a major challenge (Soni et al., 2018). Industries such as paper, textile, chemical, and petrochemical industries produce various toxic organic wastes, resulting in combined hazardous impact on the environment and posing significant risks on human health—related issues, aquatic organisms, and animals (Geissen et al., 2015). The petrochemical industry generates effluents consisting of many toxic organic compounds as well as grease and oil, causing enormous concern worldwide (Abd El-Gawad, 2014). In particular, water is an environmental resource that is considerably affected by heavy industrialization (Pullanikkatil and Urama, 2011). The petrochemical industry consumes large amounts of water for refining crude oil. Consequently, a significant volume of wastewater is discharged continually into waterbodies (Diya'uddeen et al., 2011). In addition to surface and groundwater pollution, continuous wastewater discharge and inappropriate toxic organic compound disposal are the major sources of soil pollution (Ellis, 2013). Although continuous efforts have been made to treat the effluents produced by the petrochemical industry, the water pollution level has not decreased. Common pollutants from petrochemical refineries include phenols, sulfides, hydrocarbons, oil and grease, and dissolved solids and organic compounds, which can be measured by determining biochemical oxygen demand and chemical oxygen demand (COD). In particular, aromatic hydrocarbons are resistant to degradation through natural mechanisms, leading to their longer environmental persistence. Moreover, many aromatic hydrocarbons exhibit carcinogenic, teratogenic, or mutagenic properties (Ghosal et al., 2016; Abdel-Shafy and Mansour, 2016). Previously reported studies have investigated various aromatic hydrocarbons such as dihydroxybenzenes; benzene, toluene, ethyl benzene, and xylene (BTEX); and phenols as well as their derivatives [e.g., chlorophenol, bisphenol A (BPA), and methyl-substituted phenols (o-, m-, and p-cresol)] (Bretón et al., 2017; Igbinosa et al., 2013; ATSDR, 2008b). Notably, Environmental Protection Agencies (EPAs) have identified phenols as potent contaminants (Al-Hashemi et al., 2014; Glezer, 2003) because they are typically found in majority as effluents of wastewater treatment plants (Sorokhaibam and Ahmaruzzaman, 2014), discharges from crude oil—processing units and coal-converting plants, and river water in some cases (Jiang et al., 2013). The fact that they are water-soluble phenolic compounds can readily mobilize and can thus easily contaminate aquatic and terrestrial environments. Phenols and their derivatives are potential carcinogens, and at concentrations varied from 5 to 25 mg/L, they are toxic (and occasionally lethal) to fish (Abha and Singh, 2012). Even at the considerably lower concentration of 2 µg/L, phenols can impart medicinal taste and objectionable odor to drinking water (Wang et al., 2012). Because of nonfavorable health issues, the World Health Organization (WHO) has established a guideline to phenol concentration limitation in drinking water to 1.0 µg/L (Igbinosa et al., 2013). Because high volumes of phenols are used and owing to their potential toxicity, the US EPA has defined them as priority pollutants (Keith, 2014), thus, for surface water, the US EPA has fix phenol concentration of <1 µg/L as the standard of water purification (Loncar et al., 2011). Regarding the phenol concentration in drinking water, the European Council Directive has fixed limit to 0.5 µg/L (Sarma et al., 2018).

Cresols are another group of phenols that significantly contribute to surface and groundwater pollution (Sarma et al., 2018). The phenol ring of a cresol molecule is substituted with a methyl group (Michałowicz and Duda, 2007). Cresols are synthetically generated through either methylation of phenols or the hydrolysis of chlorotoluenes (Choquette-Labbé et al., 2014). Furthermore, cresols is characterized as ortho-, meta-, and para-cresols based on their position on the carbon—carbon bond. Among these cresols, p-cresol is a toxic pollutant with adverse effect on the central nervous system disease, cardiovascular system, lungs, and kidneys diseases even at very low concentration (Basheer and Farooqi, 2012; ATSDR, 2008b). Additionally, p-cresol is generally used as fumigants, disinfectants, and explosives, and it is used to manufacture synthetic resins (Balarak and Mahdavi, 2016; Berge-Lefranc et al., 2012). Furthermore, p-cresol has been classified as C group pollutant by the EPA (ATSDR, 1990), and the WHO recommends 0.001 mg/L as the acceptable p-cresol concentration in portable water (WHO, 1996). p-Cresol may be toxic; thus, its levels in wastewater must be reduced before discharge into the environment (Fang and Zhou, 2000). As not much research has been done yet on reducing the effect of p-cresol released as effluents, hence, from the perspective of environmental management, the appropriate treatment of these phenolic wastes is challenging. Henceforth, the current chapter is focused on various degradation technologies for vanishing organic pollutants from wastewater.

1.1 Cresol production

Cresols are organic compounds in which methyl and hydroxyl groups are directly attached with benzene ring. Cresols (also known as methyl phenols; formula: $CH_3C_6H_4OH$) exist in three isomers: ortho-, meta-, and p-cresols, which differ in the relative positions of their methyl and hydroxyl groups. Coal tar distillation generates a mixture of the three isomers; this mixture is applied as a germicide and antiseptic (Fardhyanti et al., 2013; Duan et al., 2018).

1.2 Adverse effects of phenols and cresols on the environment and human health

Phenolic compounds are highly toxic and stable and exhibit bioaccumulation ability; thus, they can still exist in environment for longer time period (Gami et al., 2014). Therefore, they are significant threat to the environment. Phenols are typically carcinogenic; therefore, they not only significantly damage and endanger the aquatic ecosystem but also adversely affect human health (Abdelwahab et al., 2009; Banerjee, 2011). Phenols, such as cresols, are present in air, water, and soil and may undergo rapid biological and chemical transformation. They adversely affect the ecosystem through the rapid depletion of dissolved oxygen; moreover, cresols released into waterbodies limit oxygen exchange and thus exert toxic effects on fish and birds (ATSDR, 2008a; ATSDR, 2008b; Duan et al., 2018). A few phenols, for example, alkyl phenols, BPA, 2,4-dichlorophenol, and pentachlorophenol, can disrupt the function of sexual hormones and may eventually lead to animal and human sterility (Michałowicz and Duda, 2007). Moreover, cresol poisoning causes an acute burning sensation in the mouth following throat, abdominal pain, headache, hypotension, irregular pulse, low body temperature, paralysis of the nervous system, shock, coma, and even death (ATSDR, 2008b). In addition, nontoxic compounds such as tyrosine may be transformed into p-cresol in the digestive tract of mammals, including humans (Baquerizo et al., 2015). Moreover, p-cresol is a marker indicating the exposure of organisms to toluene. A previous study reported that in the existence of hydrogen peroxide, p-cresol results in development of DNA adducts in HL-60 cells. Researchers have revealed that a p-cresol (4-methylphenol) metabolite—quinone methide—started induction in DNA damage, and this metabolite may also be used as biomarker indicating the exposure of organisms to toluene (Gaikwad and Bodell, 2003; Duan et al., 2018). For adults, a 4-methylphenol dose of 30−60 g is lethal, whereas the lethal dose for animals differs and depends on the chemical structure of cresols (ATSDR, 2008b).

2. Treatment technologies for phenolic compound removal

As it is known that wastewater contains high levels of phenolic compounds and represents a threat for living system even at lower concentrations because of their toxicity, it is imperative to remove them before its discharge into water system (Anku et al., 2017). Employing appropriate strategies for wastewater treatment is highly significant for reducing environmental issues. Several technologies have been applied for wastewater treatment (Klein and Lee, 1978; Luan et al., 2017; Kumar and Lee, 2012; Awaleh and Soubaneh, 2014). A single treatment method or a combination of treatment methods have been used to remove phenols to ensure allowable discharge limitation. The optimal method used based on the concentration and volume of the treated effluent, by-products, space availability, durable operation, and treatment cost (Igwe et al., 2013). Basic treatment methods used for degrading phenolic waste are categorized into physical, chemical, and biological methods.

2.1 Physical method

Among physical treatment methods, coagulation and adsorption are the most prominent (Preethy, 2016). Coagulation is a common method for treating phenolic compounds; however, coagulation treatment generates a substantial amount of sludge that requires further treatment, which is time-consuming (Tzoupanos and Zouboulis, 2008; Robens Institute, 1996). Furthermore, adsorption is considered as natural phenomenon, in which dissolved molecules accumulate on the surface of an adsorbent material. Molecule adsorption happens when attractive forces from the carbon surface overawed attractive forces of the concern liquid (Noonan and Curtis, 1990). Because of high surface area to volume ratio, granular activated carbon is an effective adsorbent medium (Múazu et al., 2017). The surface area of 1 g of typical commercial activated carbon is 1000 m^2. Phenolic compounds and heavy metals are easily adsorbed on the porous surface of activated carbon (Sulaymon et al., 2011). Although activated carbon is effective for the removal of trace organic compounds, it is costly (Thomas and George, 2015). Therefore, for phenolic waste removal, promising alternatives have been developed, including chemically modified activated carbon, activated carbon impregnated with nanoparticles such as TiO_2, activated carbon with different carbon sources such as rice husk, activated carbon generated through different activation methods, and low-cost biosorbents, for example, lignocellulose and chitin/chitosan (Thomas and George, 2015; Bagheri and Julkapli, 2016; Tran et al., 2015; De Gisi et al., 2016).

Membrane technologies are reliable and cost-effective for the degradation of phenolic compounds in wastewater because of their low power consumption, small carbon footprint, high-quality effluent, and smooth scale up through membrane modules (Villegas et al., 2016; NRC, 2012). The most important membrane technologies are extractive membrane bioreactors, photocatalytic membrane reactors, hollow fiber membranes, and high-pressure membrane processes such as reverse osmosis, pervaporation, nanofiltration, and membrane distillation (Ochando-Pulido and Martinez-Ferez, 2015; Villegas et al., 2016). However, membrane processes have some limitations, including their high cost because of high effluent volume and fouling and the degradation of polymeric membranes because of usage. Thus, the membranes require frequent replacement, which may increase the operating costs (Dvoraka et al., 2015; Attiogbe, 2013; Iorhemen et al., 2016).

2.2 Chemical method

Chemical oxidation is another method used for the treatment of phenolic compounds. The chemicals most commonly employed in this method are chlorine, chlorine dioxide, ozone, chloramines, permanganate, and ferrate [Fe (VI)] (Villegas et al., 2016; Guan et al., 2010). However, cost is the main disadvantage of all of these chemicals. Alternatively, phenolic compounds can be treated through electrochemical oxidation, which does not require any reagent but involves energy costs and equipment use (Hurtado et al., 2016). Advanced oxidation processes (AOPs) are techniques in which hydroxyl radicals (OH$^\bullet$) are formed in situ, and these free radicals can mineralize most organic compounds, including phenolic compounds (Sina and Mohsen, 2017; Villegas et al., 2016). AOPs include ozone treatment, photocatalytic degradation, Fenton process, wet air oxidation, and UV/H_2O_2 treatment (Ebrahiem et al., 2017; Muruganandham et al., 2014; Umar and Aziz, 2013). These processes are promising for the removal of ground, surface, and wastewaters contaminated with phenolic compounds. Among various AOPs, the Fenton reagent (H_2O_2/Fe^{2+}) is effective for phenolic compound degradation. Furthermore, no energy input is required for activating hydrogen peroxide, which represents an advantage of the Fenton reagent (Aljuboury et al., 2014; Daraei et al., 2015). Therefore, this reagent is an economical source of hydroxyl radicals. However, the drawbacks of the Fenton reagent are that substantial $Fe(OH_3)$ precipitate is produced and that the homogeneous catalyst causes additional water pollution (Awaleh and Soubaneh, 2014; Rodríguez, 2003). Wet air oxidation is an AOP that oxidizes aromatic compounds, but its operation conditions are high temperature and pressure alone or a combination of catalyst use, high temperature, and high pressure (Tungler et al., 2015; Kim and Ihm, 2011). UV/H_2O_2 treatment processes utilize large amounts of oxidants and are associated with high operating costs; thus, they considered uneconomical (Diya'uddeen et al., 2011). In ozone treatment, the gas must be generated and used on-site, limiting its application (Naidoo; Olaniran, 2014). Moreover, these processes produce carcinogenic by-products, which lead to further environmental pollution, thus limiting their applications.

2.3 Biological method

Microbial degradation is a valuable strategy for eliminating phenolic compounds and detoxifying polluted environments and wastewaters (Ghosal et al., 2016; Ayangbenro and Babalola, 2017). Phenol and its compounds, such as methyl-substituted phenol (ortho, meta, and para), are products of the petroleum industry and are degraded by various microorganisms, such as yeasts, bacteria, and fungi as shown in Tables 11.1 and 11.2. Because methyl-substituted phenols, such as p-cresol, are widespread in the environment, microorganisms, including both aerobic and anaerobic, utilize p-cresol as sole carbon and energy sources (Basha et al., 2010; Du et al., 2016). Degradation by bacteria is more efficient than that by other microbes. Extensive research has been conducted on the bacterial degradation of phenolic compounds because it is a valuable method for removing environmental toxins (Villegas et al., 2016; Lakshmi and Sridevi, 2009). According to a literature review of cresols biodegradation, most studies have used bacteria (Bakr and Mohamed, 2013; Hamitouche et al., 2012; Tallur et al., 2006; Surkatti and El-Naas, 2014; Yao et al., 2011), whereas some studies have used yeast (Claussen and Schmidt, 1998; Chaurasiaet al., 2015), and the remaining have used algae (Papazi et al., 2012; Hughes and Bayly, 1983). The p-cresol biodegradation potential of few microbial fungi has been recognized; however, studies of the capabilities of fungi for the degradation of p-cresol present in wastewater are limited (Singh et al., 2008; Du et al., 2016).

2.3.1 Bacteria

Bacteria are the most active agents for the degradation of phenolic compound containing industrial wastewaters. Although phenolic compounds, such as cresols, exhibit toxic

TABLE 11.1   Comprehensive list of bacteria employed to degrade phenolic compounds.

S.No.	Phenolic compounds	Bacteria	References
1	m-Cresol and p-cresol	*Pseudomonas putida* NCIB9869	Hopper and Taylor (1975)
2	p-Cresol	*Desulfobacterium cetonicum*	Muller et al. (2001)
3	2-chlorophenol, phenol, and p-cresol	*Rhodococcus erythropolis* M1	Goswamia et al. (2005)
4	Phenol	*Pseudomonas* sp. (IES-S) and *Bacillus subtilis* (IES-B)	Hasana and Jabeen (2015)
5	Phenol and wattle	*Pseudomonas aeruginosa* and *Bacillus subtilis*	Aravindhan et al. (2014)
6	Phenol and catechol	*Pseudomonas putida* (MTCC 1194)	Kumar et al. (2005)
7	p-Cresol	*Pseudomonas putida*	Lakshmi et al. (2011)
8	m-Cresol	*Lysinibacillus cresolivorans*	Yao et al. (2011)
9	Phenol	*Pseudomonas aeruginosa* (NCIM, 2074)	Lakshmi et al. (2011)
10	Phenol	*Ewingella americana*	Khleifat (2006)
11	Phenol	*Acinetobacter calcoaceticus* PA	Liu et al. (2016)
12	p-Cresol	*Pseudomonas putida* (ATCC 17484)	Yu and Loh (2002)
13	p-Cresol	*Bacillus* sp. strain PHN 1	Tallur et al. (2009)
14	p-Cresol	*Advenella* sp. LVX-4	Xenofontos et al. (2016)
15	m-Cresol	*Desulfotomaculum* sp.	Londry et al. (1997)
16	m-Cresol and quinoline	*Lysinibacillus* sp. SC03 and *Achromobacter* sp. DN-06	Zhao et al. (2014)
17	m-Cresol	*Pseudomonas*-like strain S100	Bonting et al. (1995)
18	o-Cresol	*Pseudomonas putida* DSM 548 (pJP4)	Kaymaz et al. (2012)
19	o-Cresol	*Pseudomonas* sp. CP4	Ahamad et al. (2001)
20	o-Cresol	*Pseudomonas* sp.	Hamitouche et al. (2015)
21	Phenol	*Azospirillum brasilense*	Barkovskii et al., 1985; Arutchelvan et al. (2005)
22	Phenol	*Bacillus brevis*	Arutchelvan et al. (2006)
23	Phenol	*Burkholderia* sp.	Salmeron- Alcocer et al. (2007)
24	Phenol	*Microbacterium phyllosphere*	Salmeron- Alcocer et al. (2007)
25	Phenol	*Acinetobacter* sp. W-17	Beshay et al. (2002); Abd-EI-Haleem et al. (2003)

TABLE 11.1 Comprehensive list of bacteria employed to degrade phenolic compounds.—cont'd

S.No.	Phenolic compounds	Bacteria	References
26	Phenol	*Pseudomonas putida* CCRC 14365	Chung et al. (2005)
27	Phenol	*Ralstonia eutropha*	Tepe and Dursun (2007), Dursun and Tepe (2005)
28	Phenol	*Halophilic bacteria* CA00, CA08, SL03, SL08, SP04	Peyton et al. (2002)
29	Phenol	*Nocardia* sp.	Tibbles and Baecker (1989), Vijaygopal and Viruthagiri (2005)
30	Phenol	*Comamonas testosteroni* P15	Yap et al. (1999)

TABLE 11.2 Comprehensive list of fungi employed to degrade phenolic compounds.

S.No.	Phenolic compounds	Fungi	References
1	Phenol	*Fusarium* sp., *Aspergillus* sp., *Penicillium* sp., and *Graphium* sp.	Santos and Linardi (2004)
2	Phenol	*Trametes*	Udayasoorian and Prabu (2005)
3	Phenol	*Aspergillus niger*	Supriya and Neehar (2014)
4	Phenolic resin	*Phanerochaete chrysosporium*	Gusse et al. (2006)
5	Phenol	*Aspergillus fumigatus*	Gerginova et al. (2013)
6	Phenol	*Paecilomyces variotii* JH6.	Wang et al., 2010
7	Gallic, protocatechuic, vanillic, syringic, Caffeic, and ferulic acids	*Fusarium flocciferum*	Mendonça et al. (2004)
8	Phenol	*Macrotermes gilvus* and a *Termitomyces* strain	Taprab et al. (2005)
9	Phenol	*Penicillium* sp.	Vanishree et al. (2014)
10	Chlorophenols	*Trametes villosa*	Bollag et al. (2003)
11	Phenol and p-cresol	*Phanerochaete chrysosporium*	Kennes and Lema (1994)
12	Phenol	*Fusarium flocciferum*	Anselmo et al. (1989)
13	Phenol and p-cresol	*Scedosporium apiospermum*	Claussen and Schmidt (1998)
14	Phenol and p-cresol 4 ethyl phenol	*Aspergillus fumigates*	Yemendzhiev et al. (2008)

properties, cresols are utilized by numerous bacteria as sole energy and carbon sources (Singh et al., 2017; Bakr and Mohamed, 2013; Hamitouche et al., 2012; Tallur et al., 2006; Surkatti and El-Naas, 2014; Yao et al., 2011). Thus, bacteria require phenols, even at high concentrations, for their growth and survival (El-Naas, 2012). In one of the recent investigations, an alkaliphilic nature of *Advenella species* (LVX-4) has been reported for the breaking of p-cresol (Xenofontos et al., 2016). They revealed p-cresol the degradation capacity of LVX-4 strain at concentration of 750 mg/L, which is considered as high concentration, at pH 9.0 (Xenofontos et al., 2016). In addition, *Pseudomonas putida* is also well-known species that has been identified and characterized for p-cresol degradation with Gram-negative aerobic nature and ability to grow at room temperature (Bouallegue et al., 2004). Surkatti and El-Naas (2014) investigated the p-cresol degradation efficiency of *P. putida* immobilized on polyvinyl alcohol gel. Furthermore, they conducted batch experiments at different concentrations of p-cresol and found that the maximum degradation occurs at a concentration of 200 mg/L (Surkatti and El-Naas, 2014). In an earlier study in 2011, Yao et al. discovered *Lysinibacillus cresolivorans*, which is a novel bacterium that degrades m-cresol. The degradation rate for m-cresol concentrations of 54.1−529.1 mg/L was analyzed. The results revealed a maximum reaction rate of 46.80 mg/h when the initial m-cresol concentration was 224.2 mg/L (Yao et al., 2011). A comprehensive list of bacteria that degrade phenolic compounds is presented in Table 11.1.

2.3.2 Biodegradation mechanism

The majority of organic pollutants can be rapidly and completely degraded under aerobic conditions (Fritsche and Hofrichter, 2000; Ghosal et al., 2016; Kristensen, 1995). Fig. 11.1 shows the main principle underlying hydrocarbon degradation under aerobic conditions (Das and Chandran, 2011). Initially, organic pollutants are activated and oxidized intracellularly through key enzymatic reactions catalyzed by peroxidases and oxygenases. Peripheral degradation pathways transformed organic pollutants into intermediates that are utilized in essential intermediary metabolic pathways, such as the tricarboxylic acid cycle (Das and Chandran, 2011). Bacteria are well-known active agents for organic pollutant degradation, and bacteria initially oxidize the ring of aromatic hydrocarbons to produce cis-dihydrodiols (Ghosal et al., 2016; Seo et al., 2009) by integrating both atoms of molecular oxygen; this reaction is catalyzed via dioxygenase, a multicomponent enzyme system. In organic pollutant biodegradation, the initial reaction involving ring oxidation denotes a rate-limiting reaction (Gupte et al., 2018). Through a NAD^+-dependent dehydrogenation reaction, cis-dihydrodiols are further oxidized to catechols (Jouanneau and Meyer, 2006). The ortho pathway and ring cleavage pathways are utilized for the oxidation of catechols. The ortho pathway includes the cleavage of the bond among carbon atoms and a hydroxyl group (Bhakta, 2017). In the ring cleavage pathway, fumaric acid, succinic acid, pyruvic acid, and acetic acid and aldehydes are produced; microorganisms use these products for the synthesis of energy and cellular components (Evans, 1963). Carbon dioxide and water are by-products of these reactions.

2.3.3 Aerobic degradation of phenolic waste

Aerobic biodegradation includes the microbial degradation of organic compounds via oxidation; therefore, the chemistry of the system, organism, or environment is characterized

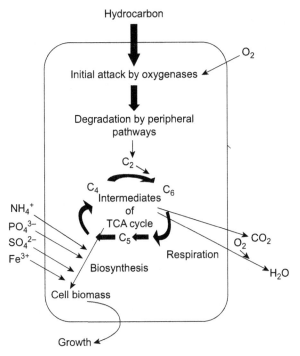

FIGURE 11.1 Principle of aerobic degradation of hydrocarbons by microorganisms. *Adopted from Das, N., Chandran, P. 2011. Microbial degradation of petroleum hydrocarbon contaminants: an overview. Biotechnology Research International 2011, 13. Article ID 941810. http://doi.org/10.4061/2011/941810.*

by aerobic conditions (Das and Chandran, 2011). In chemical reactions, oxygen is used to break down organic pollutants into water and carbon dioxide. Similar to other aromatic compounds, phenols and their derivatives in polluted environments are degraded by various aerobic microorganisms (Krastanov et al., 2013; Evans, 1963; El-Naas, 2012). Generally, oxidation is the initial step in the aerobic degradation of phenol by microorganisms (Das and Chandran, 2011; Mohanty, 2012) in the oxidation reaction, at a position ortho to the preexisting hydroxyl group, a monooxygenase phenol hydroxylase initially monohydroxylates the aromatic ring, producing catechols, which are the main intermediates obtained from the metabolism activity of phenol using different microorganisms (Sridevi et al., 2012). Depending on the strain type, catechols then undergo ring cleavage at the ortho position, initiating the ortho pathway for the formation of acetyl Co-A and succinyl Co-A, or at the meta position, initiating the meta pathway for the formation of acetaldehyde and pyruvate (Fig. 11.2) (Sridevi et al., 2012; Singh, 2017; Seo et al., 2009)

Two routes are employed for the microbial degradation of p-cresol under aerobic conditions. The first route entails the hydroxylation of the ring, producing 4-methylcatechol, which function as the substrate during the ring fission dioxygenase (Hopper and Taylor, 1975; Singh et al., 2017). In the second route, initially, the methyl group attached to the ring is oxidized to a carboxylic group (Jones et al., 1993; Hopper and Taylor, 1975). However, some bacteria use alternative routes for p-cresol degradation, such as *Bacillus* sp. strain PHN 1, which degrades

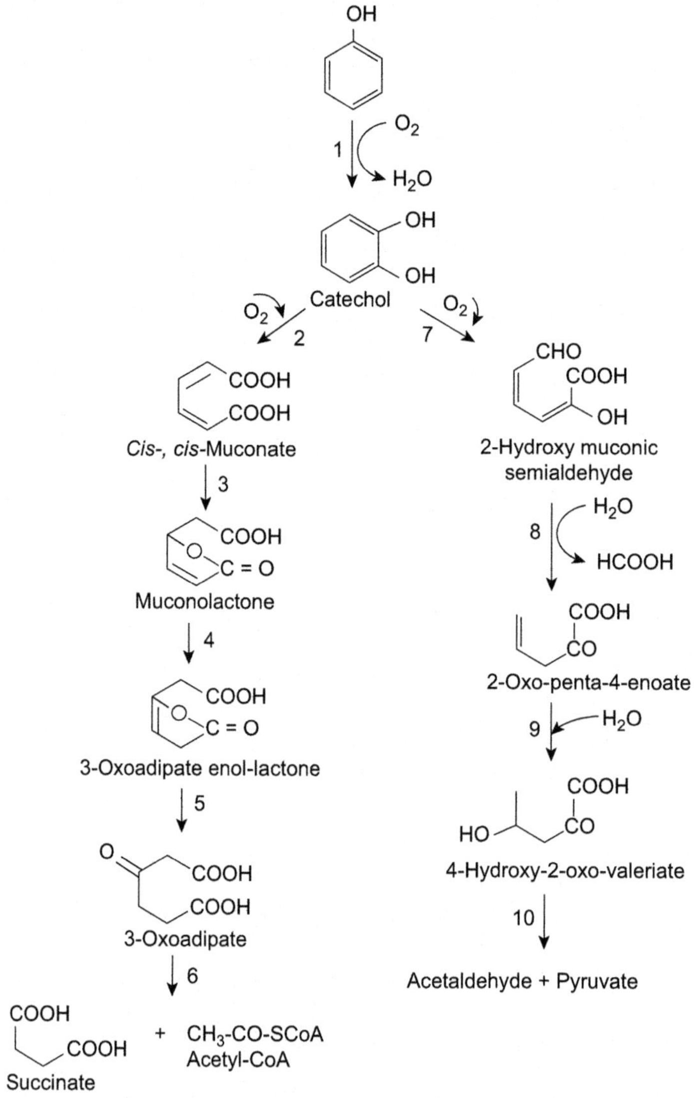

FIGURE 11.2 The two alternative pathways of aerobic degradation of phenol: *o*- and *m*-cleavage: *o*- and *m*-cleavage, (1) phenol monooxygenase, (2) catechol 1,2-dioxygenase, (3) muconate lactonizing enzyme, (4) muconolactone isomerase, (5) oxoadipate enol-lactone hydrolase, (6) oxoadipate succinyl-CoA transferase, (7) catechol 2,3-dioxygenase, (8) hydroxymuconic semialdehyde hydrolase, (9) 2-oxopent-4-enoic acid hydrolase, (10) 4-hydroxy-2-oxovalerate aldolase.

p-cresol by initially oxidizing the methyl group (Tallur et al., 2006). The initial reactions for p-cresol oxidation to 4-hydroxybenzoic acid in *Bacillus* sp. strain PHN 1 may be similar to those in *Pseudomonas* species (Hopper and Taylor, 1975). However, the subsequent reactions for 4-hydroxybenzoic acid metabolism in *Bacillus* sp. strain PHN 1 differ from those in *Pseudomonas* species. *Pseudomonas* species metabolizes 4-hydroxybenzoic acid to protocatechuic acid (Hopper, 1978; Hopper and Taylor, 1975), by contrast, in *Bacillus* sp. strain PHN1, 4-hydroxybenzoic acid is metabolized to gentisic acid (Crawford, 1976). Diverse bacterial genera can degrade low molecular weight polycyclic aromatic hydrocarbons (PAHs) such as acenaphthene, naphthalene, and phenanthrene; these genera comprise Gram-negative bacteria, most of them belonging to the *Pseudomonas* genus (Ghosal et al., 2016; Mrozik et al., 2003).

2.3.4 *Anaerobic degradation of phenolic waste*

The anaerobic removal of aromatic compounds is less progressive than aerobic degradation, and only a few anaerobic bacterial species have been extensively studied (Ghosal et al., 2016; Ghattas et al., 2017). The phototrophic proteobacterium (K-subgroup) *Rhodopseudomonas palustris* is one such known species that uses aromatic compounds as a cellular carbon source during photosynthetic growth in anoxic environments (Harwood et al., 1999). *Thauera aromatica* and *Azoarcus evansii* and related species are denitrifying bacteria belonging to the L-subgroup of *Proteobacteria*; for growth, they use aromatic compounds as cellular carbon and energy sources (Schennen et al., 1985; Blake and Hegeman, 1987; Springer et al., 1998; Bak and Widdel, 1986). Three main pathways are employed by anaerobic bacteria for degrading mononuclear aromatic compounds: resorcinol, benzoyl-CoA, and phloroglucinol pathways (Harwood et al., 1999). A general characteristic feature of the aforementioned pathways is more reductive nature than oxidative attack and is used to destabilize the aromatic nucleus. Among these pathways, the benzoyl-CoA pathway is potentially the most important one because it degrades diverse compounds, including phenol, hydroxybenzoates, some cresols, and even the aromatic hydrocarbon toluene (Harwood et al., 1999; Heider and Fuchs, 1997; Schink et al., 1992).

The denitrifying bacterium *Paracoccus* sp. has been used to study the anaerobic metabolism of methylphenols (cresols) (Rudolphi et al., 1991). This strain could grow by using o- or p-cresol as the sole organic substrate and completely oxidized it to CO_2, and nitrate was reduced to N_2. Depending on the substitution type, the anaerobic degradation of cresols occurs through three different pathways (Schink et al., 2000). As suggested in an earlier study of an aerobic *Pseudomonas* strain, hydroxylation of p-cresol occurs at the methyl group in an anaerobic reaction probably involving the quinomethide intermediate. The carboxylation of o-cresol produces 3-methyl-4-hydroxybenzoate, which can be degraded further (Bisaillon et al., 1991; Rudolphi et al., 1991). A recent study elucidated the anaerobic degradation pathway for m-cresol by using a pure culture of *Desulfobacterium cetonicum*, which is a sulfate-reducing bacterium. Noticeably, this pathway is comparable to the toluene anaerobic degradation pathway employed nitrate-reducing bacteria (Muller et al., 2001); the methyl group of m-cresol adds to fumarate, forming 3-hydroxybenzyl succinate. Through B-oxidation and activation, 3-hydroxybenzoyl-CoA and succinyl-CoA are produced.

2.3.5 *Fungi biodegradation*

Fungi are reported for significant contribution in degradation of aromatic compounds (Schink et al., 2000; Vanishree et al., 2014; Gerginova et al., 2013). Their enzymatic systems are essential for the degradation of a wide variety of structurally diverse organic compounds (Dashtban et al., 2010). Fungi are ubiquitous microorganisms that occur in diverse habitats—from deserts and tropical forests to Antarctica and the Arctic (Krastanov et al., 2013).

Fungi survive under extreme environments, for example, those with high and low temperatures, low humidity, high oxygen saturation, high UV irradiation, and drastic climatic changes (Tosi et al., 2010). Their diverse enzymatic activities enable their adaptation to unfavorable (stress) environments, including environments containing highly toxic pollutants (Santos and Linardi VR, 2004; Abrashev et al., 2008).

Many fungi oxidize PAHs in a cytochrome P-450 monooxygenase-catalyzed reaction, in which one atom of the oxygen molecule is incorporated into water and the other atom is incorporated into the PAH, forming arene oxides, most of which are unstable (Ghosal et al., 2016). They are hydrated by epoxide hydrolase, forming trans-dihydrodiols, or they undergo nonenzymatic rearrangement, forming phenols (Al-Hawash et al., 2018). The resulting phenols can be conjugated to glucose, sulfate, glucuronic acid, or xylose. The diverse group of nonligninolytic fungi oxidizes PAHs to quinones, trans-dihydrodiols, phenols, tetralones, dihydrodiol epoxides, and various conjugates of hydroxylated intermediates, and few nonligninolytic fungi are capable of degrading PAHs to CO_2 (Národní, 2002; Koul and Fulekar, 2013). Fungal genera (i.e., *Talaromyces, Amorphoteca, Graphium*, and *Neosartorya*) have been isolated from soil contaminated with petroleum and have been proven to exhibit potential for aromatic compound degradation (Chaillan et al., 2004). Laboratory studies of fungi that can degrade phenols and their derivatives have been performed (Mendonca et al., 2004). For example, the *Fusarium* sp. HJ01 strain showed high potential for phenol degradation. A previous study determined the biodegradation activity of *Fusarium flociferum* cells immobilized on different carriers for phenol concentrations up to 4 g/L (Cai et al., 2007). Similarly, in 2008, Singh et al. studied p-cresol biodegradation by using *Gliomastix indicus* MTCC 3869, which is a novel filamentous fungus. Biodegradation experiments revealed that p-cresol (700 mg/L) was 90% degraded within 108 h (Singh et al., 2008). Moreover, this technique is costly because of high demand of fungal inoculum for bioremediation process. Moreover, ensuring the homologous distribution of mycelia in the soil remains a drawback when fungal strains are used for bioremediation (Koul and Fulekar, 2013). The list of fungi involved in degradation phenolic compounds is presented in Table 11.2.

2.3.6 Enzymes participating in degradation of phenolic compounds

Similar to biological treatment, enzymatic treatment uses a biocatalyst, namely an enzyme, to transform phenolic compounds for their removal from water (Pandey et al., 2017; Agarwal et al., 2016). Microorganisms degrade phenol using their different enzymatic systems. Among all enzymes, oxidoreductases such as peroxidases as well as laccases and tyrosinases, which are copper-containing enzymes, can catalyze the removal of phenolic pollutants (Villegas et al., 2016; Xu and Yang, 2013). Monophenols are hydroxylated by tyrosinases to o-diphenols, which are subsequently oxidized to o-quinones. Molecular oxygen (four electrons) and hydrogen peroxide (two electrons) are required for the activation of laccase and heme-protein peroxidase (Mukherjee et al., 2013). These enzymes then catalyze the oxidation of reducing compounds such as phenols through successive one-electron transfers, forming the corresponding phenoxy radicals, and these radicals form dimers through nonenzymatic coupling (Ba et al., 2013). Additionally, horseradish peroxidase (HRP) is the most examined heme-protein peroxidase; researchers have paid considerable attention to soybean peroxidase (SBP) because of its many advantages over HRP. Studies have proven SBP as an effective enzyme for significant removal of phenols (Ba et al., 2013; Steevensz et al., 2014). To realize large-scale enzymatic treatment, a previous study investigated immobilization techniques such as carrier or support binding via physicochemical interaction; encapsulation, in which enzymes are trapped inside the pores of the support medium; or covalent cross-linking of functional groups of enzymes (Ba et al., 2013; Villegas et al., 2016).

2.3.7 Biosurfactants

Several studies have reported on biosurfactants, which comprise heterogeneous surface-active chemical compounds that are generated by diverse microorganisms (Santos et al., 2016; Muthusamy et al., 2008; Mahmound et al., 2008; Obayori et al., 2009). As a treatment strategy, surfactants have been extensively applied to increase the mass transfer of hydrophobic organic compounds (Zhang, 1997; Alvarez and Polti, 2014). Biosurfactants or surface-active agents produced by microorganisms are amphiphilic molecules containing hydrophobic and hydrophilic moieties which intermingle with surfaces with different polarities, reducing interfacial and surface tensions (Otzen, 2017). Thus, biosurfactants enhance the area of contact and solubility of insoluble compounds as well as their mobility, bioavailability, and subsequent biodegradation (Silva et al., 2014). Moreover, compared with chemical surfactants, they have several advantages such as better environmental compatibility, low toxicity, high biodegradability, high foaming ability, and higher selectivity and specific activity at extreme temperatures, pH values, and salinities, and they can be synthesized from renewable food stocks (Gautam and Tyagi, 2005; Silva et al., 2014). Through biosurfactant release, microorganisms influence the uptake of recalcitrant organic aromatic compounds such as phenolic compounds (BTEX), PAHs, and other halogenated compounds (Kronenberg et al., 2017; Basak et al., 2014). *Pseudomonads* are a bacterial genus widely known to produce biosurfactants and utilize hydrocarbons as carbon and energy sources (Rahman et al., 2003; Cameotra and Singh, 2008; Pornsunthorntawee et al., 2008). Among *Pseudomonads* species, *Pseudomonas aeruginosa* has been extensively studied for its production of glycolipid-type biosurfactants. Several studies have also successfully used biosurfactants to degrade organic pollutants in soil and water (Suryantia et al., 2016; Das and Chandran, 2011). Recently, Suryanti et al. (2016) applied the biosurfactant produced by *Pseudomonas fluorescens* for phenol degradation. The biosurfactant was identified as a rhamnolipid type, and in the presence of the rhamnolipid-type biosurfactant, up to 68% of phenol was degraded within 15 days, whereas in the absence of the biosurfactant, only 56% of phenol was degraded. Overall, the results demonstrated that rhamnolipids possess favorable surface-active properties and can enhance phenol degradation (Suryantia et al., 2016). Furthermore, Uysal and Turkman, in 2011, investigated the biodegradation of 4-chlorophenol (4-CP), which is a highly recalcitrant phenol, by adding biosurfactants (JBR 425 rhamnolipid) to an activated sludge bioreactor system. The sludge retention time was varied. The test reactors R2 and R3 supplemented with the biosurfactant provided 77% COD and 46% 4-CP removal and 81% COD and 63% 4-CP removal, respectively, whereas a control reactor (without biosurfactant addition; R1) resulted in 61% COD and 19.15% 4-CP removal (Uysal and Turkman, 2011).

2.3.8 Genetically modified bacteria

Research has been demonstrated that genetically engineered microorganism (GEM) application in bioremediation can improve aromatic compound degradation under the influence of laboratory conditions. There are studies on environmental pollutants degradation via different bacteria such as *Escherichia coli* AtzA, *P. fluorescens* HK44, *Burkholderia cepacia VM1468*, and *Pseudomonas putida* PaW85 and are some examples of GMOs (genetically modified microorganisms) that are used for degrading toxic organic compounds (Strong et al., 2000; Sayler and Ripp, 2000; Taghavi et al., 2005; Jussila et al., 2007). Genetically engineered bacteria and fungi showed higher degradation capacity for recalcitrant, xenobiotic organic compounds contaminating the environment. However, field tests of GEM are mainly impeded by environmental and ecological concerns and regulatory constraints (Das and Chandran, 2011).

3. Factors influencing bioremediation of phenolic waste

A bioremediation process is based on aerobic or anaerobic microbial activities, which are affected by numerous physicochemical environmental factors. Factors directly affecting bioremediation are temperature, pH, oxygen content, and substrate concentration. Organic compounds serve as carbon and energy sources as well as sources of other macronutrients, such as phosphorous, nitrogen, and sulfur. Each of these factors must be optimized to achieve the maximum degradation of the desired organic compounds, such as phenol and p-cresol.

3.1 Temperature

Temperature is a physical factor that plays a key role in hydrocarbon biodegradation; it directly alters organic pollutant chemistry, and it affects the diversity and physiology of microbial flora (Sihag et al., 2014; Das and Chandran, 2011; Okoh, 2006; Jain et al., 2011). Hydrocarbon biodegradation occurs over a wide temperature range, and the biodegradation rate generally declines with decreases in the temperature (Foght et al., 1996). The maximum rates of hydrocarbon degradation are typically detected within temperature ranges of 30°C–40°C, 20°C–30°C, and 15°C–20°C in soil environments, some freshwater environments, and marine environments, respectively (Bartha and Bossert, 1984; Cooney, 1984). However, the majority of laboratory studies of phenol degradation have used 30°C as the optimum temperature (Shah, 2014). Moreover, as the temperature increases from 30°C to 34°C, because of cell decay, no phenol degradation is observed. This suggests that the phenol degradation process is dependent on temperature (Basha et al., 2010). Microbial growth rates approximately increase twofold when the temperature increases by 10°C within the typical mesophilic operational temperature range of 10°C–30°C. No changes are observed in growth rates in the temperature range of 35°C–40°C, but higher temperatures lead to a slow growth rate for mesophiles because of protein denaturation (Kumar and Libchaber, 2013). However, microorganisms have adapted to metabolize hydrocarbons at very high temperatures. For example, *Rhodococcus* species isolated from a sample of Antarctic soil succeeded in degrading numerous alkanes at −2°C, whereas at higher temperatures, their degradation activity was considerably inhibited (de Carvalho et al., 2014). Moreover, at a temperature as low as 0°C, the PAHs of naphthalene and phenanthrene were efficiently degraded from crude oil in contaminated seawater (Sihag et al., 2014). By contrast, manganese and laccase peroxidase enzymes of ligninolytic fungi degraded hydrocarbon degradation at the optimum temperatures of >75°C and 50°C, respectively (Kadri et al., 2017). Therefore, the optimum temperature mainly falls within the mesophilic temperature range of 30°C—40°C or occasionally the thermophilic temperature 60°C (Sihag et al., 2014).

3.2 Nutrient availability

Besides readily degradable carbon sources, mineral nutrients (e.g., potassium, nitrogen, and phosphate, i.e., K, N, and P, respectively) are also required for the cellular metabolism of microorganisms for successful growth (Sihag et al., 2014). The available nutrients in contaminated sites, with high organic carbon owing to the pollutant nature, can be quickly depleted during microbial metabolism (Sutherland et al., 1995). Thus, for stimulating in situ microbial community and increasing biodegradation, the contaminated site is commonly supplemented with nutrients where N (nitrogen) and P (phosphorus) are considered as major source (Alexander, 1994; Atagana et al., 2003). The P and N amounts for achieving the optimal growth of microorganisms and thus promoting bioremediation based on estimated C:N:P ratio within the microbial biomass (between 100:15:3 (Zitrides, 1978) and 120:10:1 (Alexander, 1977)). However, biodegradation is inhibited by very high levels of nutrients (Chaillan et al., 2006). Several studies have reported that high NPK levels negatively affect hydrocarbon biodegradation (Oudot et al., 1998; Chaîneau et al., 2005), particularly aromatic compound biodegradation (Carmichael and Pfaender, 1997). The growth of bacteria-degrading hydrocarbons and the hydrocarbon degradation rate can be strongly enhanced by applying fertilizers with inorganic N and P to the contaminated environment. Pelletier et al. evaluated the efficacy of fertilizers for biodegrading crude oil in subarctic intertidal sediments (Pelletier et al., 2004). In contaminated soil, poultry manure has also been employed as organic fertilizers (Okolo et al., 2005), wherein the addition of poultry manures alone enhanced biodegradation. Degradation of phenanthrene by natural microflora present in Guayanilla Bay seawater samples increased by 10-fold after applying KNO_3 as an inorganic nitrogen source (Wilson and Jones, 1993).

3.3 Effect of pH on phenol degradation potential

It is believed that the pH of the internal cellular environment is approximately neutral. Most organisms cannot tolerate environments with pH values less than 4.0 and more than 9.0 (Kim and Maier, 1986; Krulwich et al., 2011; Kastbjerg et al., 2009). When the cellular environment has low or high (4.0 or 9.0, respectively) pH values, bases or acids can more easily enter cells because under these conditions, they exist as dissociated bases or acids, and electrostatic forces cannot prevent their entry into cells (Rajani and Reshma, 2016). Biodegradation rates of organic pollutants such as phenols were found to be the highest at neutral or near neutral pH values in most instances (Annadurai et al., 2000). Lakshmi and Sridevi, in 2009, investigated the effect of the pH range of 6.0—9.0 on the degradation of phenols by *P. aeruginosa*. The result showed the bacterium rapidly degraded phenols at pH 7.0. Phenol degradation at this pH was higher than that at other pH values. Incidentally, the degradation at pH 6.0, 8.0, and 9.0 was slow, showing that more phenol was degraded by *P. aeruginosa* at pH 7.0 than at any other pH value (Lakshmi and Sridevi, 2009). Similarly, Mohanty and Jena, in 2017, observed that the pH value of the solution highly influences the biodegradation of phenol in an aqueous solution and influences the ionization degree and surface charge of the absorbent. At a temperature of 30°C, an increase in the pH of the media from 6.0 to 8.0 increased the phenol degradation rate (Mohanty and Jena, 2017). Further increases in the pH caused a reverse effect on the phenol removal efficiency of the isolate *Pseudomonas* sp. NBM11. Both alkaline and acidic pH values markedly decreased phenol removal efficiency. Crude oil biodegradation is mainly performed by fungi, which are more tolerant to low pH (Das and Mukherjee, 2007). At the other extreme, alkaliphilic bacteria isolated from highly alkaline lake and industrial effluents effectively degraded phenol in wastewater within the optimal pH range of 7.5–10.6 (Gaokar, 2018).

3.4 Effect of additional carbon sources on phenol degradation potential

Phenol can be biologically degraded using pure and mixed microbial cultures and various studies have used pure cultures of *P. putida* for phenol degradation and have revealed that phenols are degraded through the metapathway (Allsop et al., 1993; Wang and Li, 2007). However, the phenol degradation activity of these bacteria was inhibited by high substrate concentrations (i.e., substrate inhibition); a previous study showed that high phenol concentrations inhibited growth and consequently phenol degradation (Hill and Robinson, 1975). Thus, to overcome substrate inhibitory effect, numerous methods have been developed for bioremediating concentrated phenolic wastewater and included adopted cells to higher phenol concentrations (Masque et al., 1987), immobilizing cells (Loh et al., 2000), and using GMOs (Soda et al., 1998). Moreover, the tolerance of cells to this effect can be increased by supplementing the growth medium with conventional carbon sources, such as glucose or yeast extract. A study also showed that the addition of yeast extract to the medium enhanced the affinity of *P. putida* for phenol (Armenante et al., 1995). In the study of Rozich and Colvin, it was reported that glucose attenuated the phenol removal rate of phenol-consuming cells (Rozich and Colvin, 1985). *P. aeruginosa* showed the highest rate of phenol degradation when glucose and peptone were applied as additional nutrients. The rate of phenol degradation improved when the medium was supplemented with peptone at concentrations between 0.25 and 1.0 g/L, and the optimum concentration was 0.25 g/L (Basha et al., 2010). Low concentrations of peptone influenced the degradation rate; however, higher peptone concentrations (more than 1.0 g/L) exerted inhibitory effects. Glucose is a conventional carbon source that supports growth, and its addition substantially increases cell density (Loh and Wang, 1998).

3.5 Effect of dissolved oxygen concentration on phenol degradation potential

Although it has been well-established that biodegradation of aromatic hydrocarbons, such as BTEX, dihydroxybenzenes, phenols, and their compounds (o-, p, and m-cresol), can occur under both anaerobic and aerobic conditions, most studies have focused on the dynamics of organic pollutant degradation under aerobic conditions (Kim and Maier, 1986; Rajani and Reshma, 2016). This is in part because aerobic microorganisms are easier to study and culture than anaerobic microorganisms. In the aerobic metabolism of aromatic hydrocarbons, the aromatic ring is initially oxidized, and oxygen is essential for dioxygenase and monogenase activity (Seo et al., 2009). Moreover, the aerobic biodegradation rate of aromatic hydrocarbons has been investigated to be up to an order of magnitude higher than the anaerobic biodegradation rate (Rockne and Strand, 1998). However, it has also been described that the anaerobic degradation rates of aromatic hydrocarbon under denitrifying conditions are similar to those under aerobic conditions (McNally et al., 1998). Although anaerobic bioremediation is a promising technique, anaerobic biodegradation is associated with several drawbacks. For example, under anaerobic conditions, the reduction of electron acceptors such as ferric iron, nitrate, and sulfate results to generate excess phosphorous and ferrous iron and hence toxic for the environment. In addition, greenhouse gases (CH_4, NO_2, etc.) are released and pH values increase during the anaerobic degradation of aromatic hydrocarbons (Koul and Fulekar, 2013). Air tilling or sparging increases oxygen levels in the soil. Hydrogen peroxide or magnesium peroxide can also be introduced into the environment.

3.6 Microbial growth kinetics

The biodegradation of phenols by a pure microbial strain has been extensively studied for more than two decades (Kumar et al., 2005; El-Naas, 2012; Sridevi et al., 2012; Fathepure, 2014; Gupta et al., 2016). A comprehensive biodegradation study analyzed the kinetics parameters of microbial growth and substrate removal (Krastanov et al., 2013; Mohanty, 2012). During the modeling of biodegradation processes, the specific growth rate of the biomass is related to the rate of substrate (contaminant) consumption. Various kinetics models have been developed to describe the dynamics of growth of microorganisms degrading phenols (Okpokwasili and Nweke, 2006; Krastanov et al., 2013; Gharibzahedi et al., 2013). Recently, Mohanty and Jena in 2017 studied the kinetics of biological degradation of phenol by the *Pseudomonas* sp. strain NBM11. The isolated strain completely degraded up to 1000 mg/L of phenol. As phenol concentrations were increased, the lag phase extended owing to the inhibitory nature of phenol. The growth kinetics data were fit to the Haldane model of substrate inhibition to estimate kinetics parameters such as *Ks* (half-saturation coefficient), μmax (maximum specific growth rate), and *Ki* (substrate inhibition constant), which were calculated as 7.79 mg/L, 0.184 1/h, and 319.24 mg/L, respectively (Mohanty and Jena, 2017).

Similarly, Aravindhan et al., in 2014, studied the biological degradation of phenolic compounds by employing a mixed bacterial culture comprising *P. aeruginosa* and *Bacillus subtilis*. The mixed microbial culture could degrade 250 mg/L of phenol in 36 h. The Haldane growth kinetics model sufficiently described the degradation of phenolic compounds by the mixed microbial culture. Table 11.3 presents the kinetics parameters of the Haldane model for various phenolic waste systems/microorganisms.

4. Limitations of biodegradation

Bioremediation, in which microorganisms are used for pollutant removal, is a highly promising method and is cost-effective and efficient technology (Kumar et al., 2011; Azubuike et al., 2016). However, current bioremediation approaches have numerous limitations, including the lower bioavailability of contaminants on temporal and spatial scales, low efficiency of microorganism in the area, and lack of benchmark values for testing of bioremediation for its road application in the field (Head, 1998; Megharaj et al., 2011). Other factors affecting biodegradation are suitable environmental conditions for microbial growth,

TABLE 11.3 Summary of the Haldane kinetics model parameters for various microorganisms/phenolic waste system.

S.No.	Authors	Bacterial strain	System	Concentration range (mg/L)	Types of phenolic waste	Monod's model μ_{max} (h-1)	Monod's model K_s (mg/L)	Halden's model μ_{max} (h^{-1})	Halden's model K_s (mg/L)	Halden's model K_i (mg/L)
1	Mohanty and Jena (2017)	*Pseudomonas putida* NBM11	Batch	1000 mg/L	Phenol	—	—	0.184	7.79	319.2
2	Aravindhan et al. (2014)	*Pseudomonas aeruginosa* and *Bacillus subtilis*	Batch	250 mg/L	Phenol and wattle	—	—	0.305 and 0.233	3.02 and 4.48	399 and 155
3	Kumar et al. (2005)	*P. putida* MTCC1194	Batch	10–1000 mg/L	Phenol	0.216	20.59	0.305	36.33	129.7
					Catechol	0.143	9.66	0.326	29.8	99.85
4	Mathur et al. (2010)	*P. putida*	Batch/continuous	20–200 mg/L	p-Cresol	—	—	0.339	110.95	497.6
5	Firozjaee et al. (2011)	Mixed consortia	Batch	50–500 mg/L	Phenol	—	—	0.067	25.32	200
6	Surkatti and El-Naas (2014)	*P. putida*	Continuous	25–200 mg/L	p-Cresol	357.9	21.6	357.9	21.6	1.9×10^8
7	Singh et al. (2008)	*Gliomastix indicus*	Batch	10–700 mg/L	p-Cresol	—	—	0.8009	42.3	43.2
8	Yao et al. (2011)	*Lysinibacillus cresolivorans*	Batch	54.1–529.1 mg/L	m-Cresol	—	—	0.89	426	51.2
9	Acuna-Arguelles et al. (2003)	Mixed culture	Series batch system	Multisubstrate each 100	p-Cresol	—	—	1.004	75.6	680

the presence of metabolically active microbial populations, and suitable nutrient and contaminant levels (Gkorezis et al., 2016; Das and Chandra, 2011). Monoculture strains are efficient enough to degrade organic pollutants. However, microbial consortia (Diez, 2010; Das and Chandran, 2011; Kahru et al., 1998; Zouari and Ellouz, 1996) and combining bacteria with fungi (Ghosal et al., 2016; Fulekar, 2017; Abdel-Shafy and Mansour, 2016; Kim and Lee, 2007) may efficiently degrade PAHs and volatile organic compounds. Moreover, not all compounds can be rapidly and completely degraded; thus, bioremediation is limited to biodegradable compounds.

5. Photocatalytic degradation

To overcome the limitations of biodegradation, photocatalytic degradation is effective technology for organic pollutant degradation (Ratnakar, 2016; Anjuma et al., 2016; Barakat et al., 2013; Misra and Pandey, 2005; Lazar et al., 2012; Ameta et al., 2018; Mahlambi et al., 2015). Photocatalysis is known as effective oxidation processes (AOP) and has been broadly investigated to remove phenolic waste from water system in the presence of semiconductor catalyst and UV illumination (Muruganandham et al., 2014; Ameta et al., 2018). Furthermore, AOPs give faster degradation rate along with active participation of hydroxyl radical ($HO^{\bullet}$), and phenols are mineralized to CO_2 and water instead of transformation of pollutants into intermediates (Hurtado et al., 2016; Rawindran et al., 2017; Deng and Zhao, 2015). Photocatalytic oxidation process gives high degradation rate with considerable mineralization at ambient condition (Krishnan et al., 2017; Nickheslat et al., 2013). Moreover, photocatalysts process provides way out for effective decomposition of organic waste because of their property to convert light energy to chemical energy (Singh et al., 2016).

5.1 Photo catalyst and its description

Many materials are broadly used as photocatalyst or semiconductors, for example, TiO_2, ZnO, WO_3, Fe_2O_3, CeO_2, CdS, ZnS, in which titanium dioxide (TiO_2) is the most frequently used photocatalyst for removal of ultraviolet light (Barka et al., 2013; Wang et al., 2017; Saratale et al., 2014). Moreover, TiO_2 show extensive range of application because of its low cost, nontoxic, and stable nature (Helali et al., 2011; Li et al., 2012; Vega et al., 2011; Herrmann, 1999). Apart from TiO_2, other photocatalytic materials have also been tried for phenolic degradation such as ZnO because of its fairly good optoelectronic, catalytic, and photochemical properties with nontoxic nature (Lavand and Malghe, 2015; Tao et al., 2013; Singh et al., 2016; Aslam et al., 2014). Furthermore, WO_3 is another catalyst and the valence band (VB) edge of WO_3 is situated at around 3eV (+3.1−3.2 V vs. NHE); nonetheless, the lower conduction bands (CB) edge value (+0.3−0.5 V vs. NHE) have lost reduction of O_2. The nonefficiency of O_2 to scavenge CB electron in WO_3 makes nonsuitable to separate the photogenerated electrons from the hole, and hence WO_3 is not able to function alone as photocatalyst. Similarl to WO_3, the conduction band of iron oxide (+0.3 V vs. NHE) is found below the H^+/H_2 redox potential, which means that reduction power is insufficient to reduce water (Bandara et al., 2007). Additionally, metal oxides are also regarded as potential photocatalyst because of their low band gap energy (1.3–2.40 eV) (Singh et al., 2016; Filippoa et al., 2015; Singh et al., 2016). Furthermore, TiO_2 is related to transition metal oxides and *possess* specific characteristics such as cost-effective, easy to handle, nontoxic nature, and resistance to photochemical and chemical erosion (Hengerer et al., 2000; Puma et al., 2008). In addition, it is versatile range of material with different applications (Meacock et al., 1997). Moreover, the photocatalytic behavior of TiO_2 is characteristic feature that is significantly known for its properties such as crystal structure, density, and crystalline size (Gupta and Tripathi, 2011).

TiO_2 is a very useful large band semiconductor, showing band gaps ranging from 3.2, 3.02, and 2.96 eV for anatase, rutile, and brookite phases, respectively. The valence band of TiO_2 is composed of the 2p orbitals of oxygen hybridized with the 3d orbitals of titanium, whereas the conduction band is only the 3d orbitals of titanium (Paxton and Thiên-Nga, 1998). When in presence of UV light, electrons of TiO_2, present in the valence band, are excited toward conduction band parting back of holes (h^+). The excited electrons (e^-) in the conduction band are in a purely 3d state and because of dissimilar parity, the transition probability of e− to the valence band reduces, the resultant reduction in the probability of e−/h+ recombination (Banerjee et al., 2006). Based on charge carrier dynamics, chemical properties, and the activity of photocatalytic degradation of organic compounds, anatase TiO_2 is considered to be the most active photocatalytic component. Various studies have been conducted for photodegradation of phenolic waste from water using the photocatalytic properties of TiO_2, which are outlined in Table 11.4.

5.2 Mechanism of TiO_2 in photocatalytic degradation of phenolic compounds

The initial step in the photocatalytic degradation of phenols and phenolic compound, under UV light, is achieved via the formation of electron–hole pairs within the TiO_2 photocatalyst. Generally the electron–hole pairs are recombined producing heat energy. However, hydroxyl radicals ($HO^\cdot$) are produced in the presence of electron acceptor (dissolved O_2), whereas hole (h^+) oxidizes water or TiO_2 surface-active —OH group. Meanwhile, the negative electrons react with the dissolved oxygen molecule to develop a superoxide anion ($O2^{\cdot-}$) (Singh et al., 2016). Finally, the $HO^\cdot$ reacts with either phenol or phenolic compounds until complete mineralization.

Furthermore, Devi and Rajashekhar (2011) explained the feasible breakage mechanism of phenol in presence of natural sunlight/UV light and nitrogen-doped TiO_2. It is predicted that phenol mineralization occurs via the generation of dihydroxybenzene (catechol or resorcinol), pent-2-enedioic acid, and oxalic acid. In a parallel reaction path, benzoquinone and maleic acid were formed during the mineralization.

6. Factors affecting photocatalytic degradation of TiO_2

6.1 Light intensity

Light is the most crucial factor on which rate initiation reaction for both photocatalysis and electron–hole formation in photochemical reaction is based on (Cassano and Alfano, 2000; Schneider et al., 2014; Ibhadon and Fitzpatrick, 2013). The light intensity calculates rate of

TABLE 11.4 Degradation of Different phenolic pollutants using TiO$_2$.

S.No.	Phenolic pollutant	Catalyst	Light source	Degradation efficiency	References
1	P-Cresol	TiO$_2$/ZnO/PANI composite	Ultraviolet (UV) irradiation	99.00%	Brooms et al. (2017)
2	P-Cresol	Pt/Al$_2$O$_3$−TiO$_2$	UV light	85%	Gómeza et al. (2012)
3	o-Cresol	Pt/TiO$_2$	LED irradiation	>65%	Su et al. (2011)
4	o-Cresol	H$_2$O$_2$/TiO$_2$/UV	UV irradiation	94%	Nguyen and Juang (2014)
5	Phenol	TiO$_2$/AC	Solar light	100	Alalm et al. (2014)
6	o-Cresol	TiO$_2$/C$_{active}$	UV-vis irradiation	60%−80%	Zmudziński (2010)
7	Phenol	TiO$_2$	UV radiation	50%	Nickheslat et al. (2013)
8	Phenol	UV/TIO$_2$/H$_2$O$_2$	UV irradiated	92%	Akbal and Onar (2003)
9	Phenol	TiO$_2$/Gr/PW	UV light	91%	Rafiee et al. (2016)
10	Phenol	TiO$_2$/zeolite composite	UV irradiated	100%	Sulaiman et al. (2017)
11	Phenol	N, S codoped TiO$_2$	Visible light	100%	Yunus et al. (2017)
12	Phenol	(CM)-n-TiO$_2$	UV light/ sunlight illuminations	100%	Shaban et al. (2013)
13	4-Chlorophenol	Metal ion doped TiO$_2$ (M-TiO$_2$)	Visible light irradiation	−	Nishiyama et al. (2015)
14	Phenol	Nano-TiO$_2$/ZSM-5/silica gel (SNTZS)	UV	90%	Zainudin et al. (2009)
15	Phenol	TiO$_2$ coated on perlite	UV	−	Hosseini et al. (2007)
16	Phenol	MWNT-TiO$_2$	Visible light irradiation	95%	Wang et al. (2005)
17	Phenol	CNT/TiO$_2$ composites	UV irradiation	−	Woan et al. (2009)
18	4-Nitrophenol	Graphite-supported TiO$_2$	UV irradiation	−	Palmisano et al. (2009)
19	Phenol	TiO$_2$	UV irradiation	−	Jansanthea et al. (2016)
20	Phenol	SnO$_2$−TiO$_2$	UV	99.8	Zhang et al. (2011)
21	phenol and m-nitrophenol (m-NP)	TiO$_2$	UV irradiation	72.50%	Chiou et al. (2008)

(Continued)

TABLE 11.4   Degradation of Different phenolic pollutants using TiO_2.—cont'd

S.No.	Phenolic pollutant	Catalyst	Light source	Degradation efficiency	References
22	Phenol	$N-TiO_2/SiO_2/Fe_3O_4$ $(N-TiO_2/FM)$	Visible light irradiation	64%	Vaiano et al. (2016)
23	Phenol	Carbon-doped TiO_2 nanoparticles wrapped with nanographenes	Visible light	–	Yu et al. (2014)
24	Phenol	$CNT/Ce-TiO_2$	UV	94	Shaari et al. (2012)
25	2,4-Dichlorophenol	Fe/TiO_2	UV	97	Liu et al. (2012)
26	2-sec-butyl-4, 6-dinitrophenol (dinoseb)	$TiO_2/MWCNTs$	Solar irradiation	95	Li and Wang (2010)
27	2-Chlorophenol	TiO_2	Solar radiation	100	Mogyorósi et al. (2002)
28	2,4-Dichlorophenol	TiO_2	Solar radiation	100	Takriff et al. (2011)
29	2,4,6-Trichlorophenol	TiO_2	UV	–	Tanaka and Saha (1994)
30	Phenol and m-cresol	$MW-UV-TiO_2$	UV	phenol (85) and m-cresol (92)	Karthikeyan and Gopala-krishnan (2011)

photoelectron and photo-hole production and its concentration in an illuminated semiconductor. In one of the study by Shivaraju (2011), he concluded the effect of various light sources along with different intensities such as UV light (intensity 2.3775 x 1015 quanta sec-1m-2), visible light (intensity 1.7481 x 1015 quanta sec-1m-2), and sunlight (average intensity 2.918 x 1015 quanta sec-1m-2) for the photocatalytic degradation process. It has been emphasized that photodegradation is enhanced in UV light and sunlight than visible light source. Aso, poor degradation efficiency of organic pollutants was recorded in dark because of absence of photon energy, which required for TiO_2 activation. It is also noteworthy point of this study that degradation of organic pollutant increases with increased intensity of UV light as the catalyst tries to absorb more photons which produces more electron—hole pairs on catalyst surface and hence this improves the hydroxyl radical concentration and later increases the removal rate (Shivaraju, 2011). In one of the recent study by Escudero et al. (2017), authors studied the impact of radiation on p-cresol (50 mg/L) degradation at optimum concentration of 1.0 g/L of TiO_2. In this study, it was explained that high radiation increases p-cresol degradation and mineralization because of photocatalyst excitation. However, beyond an optimum range, an increase in radiation does not affect the degradation and mineralization considerably, and this might be because of dependency of the reaction rate on the square root of the intensity of radiation at high radiation values (Li et al., 2008). Based on reports of recent researches, it is estimated that UV light emitting diodes (LEDs) give a feasible option because of nontoxic nature and low energy consumption

(Steiner, 2017; Shifu and Yunzhang, 2007; Daniel and Gutz, 2007). It was mentioned in one of the study that 800 mW LEDs is 100 times more effective over 12 and 16 W conventional UV lamps (Shie et al., 2008). To explain this concept, Jamali et al., 2013 investigated the photocatalytic degradation of phenol using TiO$_2$ illuminated via one light releasing diode (LED) in batch photocatalytic reactor (Jamalia et al., 2013). It was concluded that under optimum conditions (phenol concentration ~ 10 ppm, hydrogen peroxide–phenol molar ratio of 100 and at pH 4.8) and with LED beam width at angle of 120°, a degradation rate of 42% was attained after 4 h. Furthermore, reducing the beam width to 40°c elevated degradation up to 87% in 4 h of experiment. Therefore, the finding might be summarized as UV and visible LED in combined with TiO$_2$ is suitable for effective degradation of organic pollutants.

6.2 Reaction temperature

After light and its intensity, temperature is other important parameter that needs to be addressed for the photocatalytic behavior of any catalyst including TiO$_2$ (Schneider et al., 2014; Tolosa et al., 2011; Kumar and Pandey, 2017) because increasing temperature of reaction accelerates photocatalytic action of TiO$_2$. Nevertheless, reaction temperature >80°C endorses the charge carriers recombination and has adverse effect on adsorption process of organic compounds on TiO$_2$ surface (Gaya and Abdullah, 2008). On the other hand, reaction temperature lower than 80°C supports adsorption efficiency, and additional reduction in reaction temperature up to 0°C results in an increment in the superficial activation energy of the catalyst (Chong et al., 2010). Thus, the most suitable temperature range mentioned for the photodegradation process is between 20 and 80°C (Mozia, 2010). Additionally, lower (0°C) and higher (80°C) temperature range is responsible for rate-limiting behavior of the process because of product desorption from the catalyst surface followed by the exothermic adsorption of the pollutant on the surface of catalyst, respectively. In the study by Tolosa et al. (2011), the authors investigated the effect of temperature in the progressive range of 25°C–40°C on the photocatalytic oxidation of 2,4-DCP at 0.2 mM concentration and pH 3.0. It was reported that oxidation rate of 2,4-DCP was more at 30°C and 40°C than 25°C. Furthermore, temperature at 30°C and 40°C offered similar half-life values with first-order rate constant and applying linearized method, 30°C temperature showed highest correlation coefficient. LED lamp light is considered as the main source for activating the electron and hole pairs of the semiconductor at ambient temperature and conditions. As the band gap energy of TiO$_2$ is very high to overawe by two thermal excitations, the further increase in temperature will also increase the rate of reaction (Lu et al., 1993) because of the collision frequency of the molecules, which also showed increasing order as temperature rises.

6.3 Catalyst loading

The overall photocatalytic reaction rate is directly proportional to concentration of catalyst, for example, TiO$_2$ (Shivaraju, 2011; Chong et al., 2010; Gaya and Abdullah, 2008; Yerkinova et al., 2018). However, the reaction rate depends on the amount of catalyst used to achieve saturation level showing a linear dependency pattern. Beyond the optimum value, increasing TiO$_2$ concentration has no effect on photocatalytic reaction and it becomes independent on

TiO_2 concentration (Ahmed et al., 2011). It is noteworthy to mention here that the optimum catalyst loading for photo mineralization is also dependent on the diameter of the photo-reactor and because of this quality, TiO_2 catalysts can be mixed in uniform pattern with the targeted water before the introduction into the reactor system. In the study by Escudero et al. (2017), the authors observed that the fastest elimination of p-cresol was achieved at concentrations of 1.0 g/L and 3.0 g/L of TiO_2, and the maximum DOC removal of 80% attained after 8 h. This happens because of increase in the number of active sites with increasing TiO_2 loading that improves production of HO^- (Abdollahi et al., 2012). The reason for reaction independency beyond the optimal value might be because of increase in the number of available active sites and the increased turbidity of the solution reduced the light penetration (Ahmed et al., 2011; Dominguez et al., 2015; Singh et al., 2016).

6.4 pH of solution

The impact of pH on the photocatalytic reaction via catalyst such as TiO_2 has been extensively studied because it influences the charge of the used photocatalyst particles, the size of aggregates with the conductance position (Bahnemann et al., 2007; Jallouli et al., 2017; Delnavaz et al., 2012; Daghrir et al., 2013). In addition, TiO_2 surface can be protonated or deprotonated under the influence of acidic or alkaline conditions based on the following reactions (11.1) and (11.2):

$$TiOH + H^+ \rightarrow TiOH_2^+ \tag{11.1}$$

$$TiOH + OH^- \rightarrow TiO^- + H_2O \tag{11.2}$$

Thus, Titania surface will continue positively charged in medium poured with acidic condition (pH < 6.9) and negatively charged under basic condition (pH > 6.9). Furthermore, titanium dioxide has exposed higher oxidizing activity at lower pH with excess H^+ at very low pH, which further decreases the reaction rate (Gaya and Abdullah, 2008). Hence, pH selection is very crucial for achieving the maximum degradation efficiency and different phenolic compounds performed differently at optimum pH during photodegradation process. In a study by Venkatachalam et al. (2007), the photocatalytic behavior of Mg^{2+} and Ba^{2+}-doped TiO_2 nanoparticles for the degradation of 4-chlorophenol (4-CP) has been shown. In this study, more degradation rate is observed in acidic pH than alkaline pH. Same result has been observed in the study of Nguyen and Juang (2014) who has shown improved photodegradation of o-cresol in acidic medium using TiO_2/UV system. Additionally, Chen and Ray (1998) reported very earlier that pH plays significant role in photodegradation process.

6.5 Inorganic ions

Number of different inorganic ions such as magnesium, iron, zinc, copper, nickel, etc., found in wastewater may also participate in the photocatalytic degradation process of organic pollutant by altering rate of photocatalytic degradation of catalyst such as TiO_2 (Abdullah et al., 1990; Kudlek et al., 2016; Piscopo et al., 2001). In addition, photocatalytic deactivation has been recorded by inorganic ions on TiO_2 (Crittenden et al., 1996). Many reports are available on effect investigation of inorganic ions on photocatalytic degradation using various catalysts, most promptly TiO_2 (Kudlek et al., 2017; Habibi et al., 2005). It is reported that many cations such as copper, iron, and phosphate decrease the photodegradation efficiency at specific concentrations, whereas calcium, magnesium, and zinc have been observe to have very less effect on the photodegradation of organic anions (Chong et al., 2010). Furthermore, impact of inorganic anions on TiO_2 photocatalysis has been observed in study of Barndõk et al. (2012), and it was shown that used sulfate and chloride ion did not have any significant effect on photocatalytic oxidation of phenol, whereas nitrate improved the removal of COD ($\approx 8\%-15\%$).

A number of reports are available to confirm that TiO_2 is a very demanding used photocatalyst for remediation of wastewater polluted with toxic organic pollutant (Ollis and Al-Elkabi, 1993; Green and Rudham, 1993). Nevertheless, there are two major bottlenecks of TiO_2 as photocatalyst limits its commercial viability for wastewater treatment: (1) its unavailability to use at large-scale engineering and (2) its low rate of catalytic reaction because of the interaction with some pollutant during photodegradation process and coagulation of TiO_2 particles (Szabóa et al., 2017). For restricted limitations, many possible solutions have been proposed, including the fusion of TiO_2 with any noble metal such as Ag, Ni, Cu, Pt, Rh, and Pd (Maicu et al., 2011). While fusion with noble metal will raise its cost and it becomes nonviable for commercial application, it is supported to be replaced by more economical transition or nonmetal dopings (Halasi et al., 2011). Furthermore, the possible reaction mechanism of these types of transition and nonmetal dopings may vary based on TiO_2 and doping metal crystal lattice (Asahi et al., 2001; Irie et al., 2003; Ihara, 2003). The application and utility of nonmetal dopants (e.g., N, C, F, S, etc.) have applied well and effectively for improving the photoactivity property of TiO_2 (Fujishima et al., 2008). Additionally, one drawback associated with nonmetal doping is that the content of a doped nonmetal would decrease in annealing process, resulting in reduction in the visible light photoactivity (Chen et al., 2009). The respective shortcomings of the above-mentioned processes lead to their restricted application and focus on devising a novel process that combined sequential treatment techniques (biodegradation and photodegradation) to effectively reduce organic pollutants. Therefore, this chapter is oriented toward the combination of two degradation processes, namely biodegradation followed by photocatalytic for the effective treatment of pollutants, including p-cresol.

6.6 Conclusion

At present, several advancements toward the biodegradation have been evolved suggested by number of available research reports. Pure cultures for the microbial biodegradation of organic compounds can generate toxic intermediates; mixed cultures, which have a wider spectrum of metabolic properties, can be used to overcome this problem. Thus, special consideration should be given to cooperative processes involving a consortium of strains with complementary capacities. A universal and efficient system can be designed through the simultaneous application of all microorganisms to obtain synergistically enhanced rates. Moreover, the immobilization of biomass has for biodegradation received considerable attention. Immobilized cells have been widely used for incorporating bacterial biomass into the engineering process. In addition, exploring and developing new types of bioreactors present research challenges. It is essential to develop and design efficient reactors that reduce mass transfer limitations and enhance the degradation rate, such as the sequential batch biofilm reactor. Kinetics studies of biodegradation reactions have indicated the extent of effectiveness of the functioning of the microbial system. Knowledge of such kinetics can improve process control and hydrocarbon removal efficiency. Moreover, the implementation of advanced technologies such as genetic modification of microorganisms with biodegradation potential is promising for bioremediation. Therefore, various advanced processes are now evolving and developing to enable the efficient biodegradation of organic compounds.

A novel process combining a sequential treatment system to effectively reduce organic pollutants has also been used. Overall, the combination of biodegradation with other technologies (e.g., TiO_2-based technologies) should be explored to expand the scope of application. The combination of biodegradation with photocatalytic degradation may be the most promising technology for the remediation of environmental pollutants in the future. Therefore, in the current chapter, we focused on the combination of two degradation processes, namely biodegradation, followed by photocatalytic, for the effective treatment of phenolic pollutant, which is a major constituent of effluents from petrochemical refineries and which poses a serious threat to the environment by causing water pollution.

Acknowledgments

Authors thankfully acknowledge the Department of Chemical Engineering & Technology, IIT (BHU), Varanasi, for technical support.

References

Abd El-Gawad, H.S., 2014. Oil and grease removal from industrial wastewater using new utility approach. Advances in Environmental Chemistry 6. https://doi.org/10.1155/2014/916878. Article ID 916878.

Abd-EI-Haleem, D., Beshay, U., Abdelhamid, A.O., et al., 2003. Effects of mixed nitrogen sources on biodegradation of phenol by immobilized *Acinetobacter* sp., strain W-17. African Journal of Biotechnology 2 (1), 8–12.

Abdel-Shafy, H.I., Mansour, M.S.M., 2016. A review on polycyclic aromatic hydrocarbons: source, environmental impact, effect on human health and remediation. Egyptian Journal of Petroleum 25 (1), 107–123.

Abdelwahab, O., Amin, N.K., El-Ashtoukhy, E.-S.Z., 2009. Electrochemical removal of phenol from oil refinery wastewater. Journal of Hazardous Materials 163, 711–716.

Abdollahi, Y., Abdullah, A.H., Zainal, Z., et al., 2012. Photocatalytic degradation of p-cresol by zinc oxide under UV irradiation. International Journal of Molecular Sciences 13, 302–315. URL: https://doi.org/10.3390/ijms13010302.

Abdullah, M., Low, G.K.C., Matthews, R.W., 1990. Effects of common inorganic anions on rates of photocatalytic oxidation of organic carbon over illuminated titanium dioxide. Journal of Physical Chemistry 94, 6820–6825.

Abha, S., Singh, C.S., 2012. Hydrocarbon pollution: effects on living organisms, remediation of contaminated environments, and effects of heavy metals co-contamination on bioremediation. In: Laura Romero-Zerón, D. (Ed.), Introduction to Enhanced Oil Recovery (EOR) Processes and Bioremediation of Oil Contaminated Sites. InTech, ISBN 978-953-51-0629-6.

Abrashev, R.I., Pashova, S.B., Stefanova, L.N., et al., 2008. Heat-shock-induced oxidative stress and antioxidant response in *Aspergillus niger* 26. Canadian Journal of Microbiology 54, 977–983.

Acuna-Arguelles, M.E., Olguin-Lora, P., Razo-Flores, E., 2003. Toxicity and kinetic parameters of the aerobic biodegradation of the phenol and alkyl phenols by a mixed culture. Biotechnology Letters 25, 559–564.

Agarwal, P., Gupta, R., Agarwal, N., 2016. A review on enzymatic treatment of phenols in wastewater. Journal of Biotechnology and Biomaterials 6 (4), 1–6.

Agency for Toxic Substances and Disease Registry (ATSDR), 1990. Toxicological Profile for Cresols (Draft). US Public Health Service, US Department of Health & Human Services, Altanta.

Agency for Toxic Substances and Disease Registry (ATSDR), 2008b. Toxicological Profile for Cresols. U.S. Department of Health and Human Services, Public Health Service, Altanta.

Ahamad, P.Y.A., Kunhi, A.A.M., Divakar, S., 2001. New metabolic pathway for o-cresol degradation by *Pseudomonas* sp. CP4 as evidenced by 1H NMR spectroscopic studies. World Journal of Microbiology and Biotechnology 17 (4), 371–377.

Ahmed, S., Rasul, M.G., Brown, R., et al., 2011. Influence of parameters on the heterogeneous photocatalytic degradation of pesticides and phenolic contaminants in wastewater: a short review. Journal of Environmental Management 92, 311–330. URL: https://doi.org/10.1016/j.jenvman.2010.08.028.

Akbal, F., Onar, A.N., 2003. Photocatalytic degradation of phenol. Environmental Monitoring and Assessment 83, 295–302.

Al-Hashemi, W., Maraqa, M.A., Rao, M.V., et al., 2014. Characterization and removal of phenolic compounds from condensate-oil refinery wastewater. Desalination and Water Treatment 54 (3), 1–12. https://doi.org/10.1080/19443994.2014.884472.

Al-Hawash, A.B., Alkooranee, J.T., Zhang, X., Ma, F., 2018. Fungal degradation of polycyclic aromatic hydrocarbons. International Journal of Pure and Applied Bioscience 6 (2), 8–24.

Alalm, M.G., Tawfik, A., Ookawara, S., 2014. Solar photocatalytic degradation of phenol by TiO_2/AC prepared by temperature impregnation method. Desalination and Water Treatment 57 (2), 835–844.

Alexander, M., 1977. Introduction to Soil Microbiology. John Wiley and Sons, New York, NY.

Alexander, M., 1994. Biodegradation and Bioremediation. Academic Press, San Diego, CA.

Aljuboury, D.D.A., Palaniandy, P., Abdul Aziz, H.B., et al., 2014. A review on the Fenton process for wastewater treatment. J. Innov. Eng. 2 (3), 1–21, 4.

Allsop, P.J., Chisti, Y., Moo-Young, M., Sullivan, G.R., 1993. Dynamics of phenol degradation by *Pseudomonas putida*. Biotechnology and Bioengineering 41, 572–580.

Alvarez, A., Polti, M.A., 2014. Bioremediation in Latin America: Current Research and Perspectives. Springer, Cham, Switzerland, p. 308. https://doi.org/10.1007/978-3-319-05738-5.

Ameta, R., Solanki, M.S., Benjamin, S., et al., 2018. Chapter 6 – photocatalysis, advanced oxidation processes for waste water treatment. Emerging Green Chemical Technology 135–175. https://doi.org/10.1016/B978-0-12-810499-6.00006-1.

Anjuma, M., Miandad, R., Waqas, M., et al., 2016. Remediation of wastewater using various nano-materials. Arabian Journal of Chemistry 1, 23. https://doi.org/10.1016/j.arabjc.2016.10.004.

Anku, W.W., Mamo, M.A., Govender, P.P., 2017. Chapter 17 phenolic compounds in water: sources, reactivity, toxicity and treatment methods. In: Phenolic Compounds-Natural Sources, Importance and Applications. InTech.

Annadurai, G., Balan, S.M., Murugesan, T., 2000. Design of experiments in the biodegradation of phenol using immobilized *Pseudomonas pictorum* (NICM-2077) on activated carbon. Bioprocess Engineering 22 (2), 101–107.

Anselmo, A.M., Cabral, J.M.S., Novais, J.M., 1989. The adsorption of *Fusarium flocciferum* spores on celite particles and their use in the degradation of phenol. Applied Microbiology and Biotechnology 31 (2), 200–203.

Aravindhan, R., Naveen, N., Anand, G., et al., 2014. Kinetics of biodegradation of phenol and a polyphenolic compound by a mixed culture containing *Pseudomonas aeruginosa* and *Bacillus subtilis*. Applied Ecology and Environmental Research 12 (3), 615–625.

Armenante, P.M., Fava, F., Kafkewitz, D., 1995. Effect of yeast extract on growth kinetics during aerobic biodegradation of chlorobenzoic acids. Biotechnology and Bioengineering 47, 227–233.

Arutchelvan, V., Kanakasabai, V., Nagarajan, S., et al., 2005. Isolation and identification of novel high strength phenol degrading bacterial strains from phenol formaldehyde resin manufacturing industrial wastewater. Journal of Hazardous Materials 238–243.

Arutchelvan, V., Kanakasabai, V., Nagarajan, S., et al., 2006. Kinetics of high strength phenol degradation using *Bacillus brevis*. Journal of Hazardous Materials B 129, 216–222.

Asahi, R.Y.O.J.I., Morikawa, T.A.K.E.S.H.I., Ohwaki, T., Aoki, K., Taga, Y., 2001. Visible-light photocatalysis in nitrogen-doped titanium oxides. science 293 (5528), 269–271.

Aslam, M., Ismail, I.M., Chandrasekaran, S., et al., 2014. Morphology controlled bulk synthesis of disc-shaped WO_3 powder and evaluation of its photocatalytic activity for the degradation of phenols. Journal of Hazardous Materials 276, 120–128.

Atagana, H.I., Haynes, R.J., Wallis, F.M., 2003. Optimization of soil physical and chemical conditions for the bioremediation of creosote-contaminated soil. Biodegradation 14, 297–307.

ATSDR, 2008a. Toxicological Profile for Phenol. Agency for Toxic Substances and Disease Registry.

Attiogbe, F., 2013. Comparison of membrane bioreactor technology and conventional activated sludge system for treating bleached kraft mill effluent. African Journal of Environmental Science and Technology 7 (5), 292–306. https://doi.org/10.5897/AJEST2013.1429.

Awaleh, M.O., Soubaneh, Y.D., 2014. Waste water treatment in chemical industries: the concept and current technologies. Hydrology: Current Research 5, 164. https://doi.org/10.4172/2157-7587.1000164.

Ayangbenro, A.S., Babalola, O.O., 2017. A new strategy for heavy metal polluted environments: a review of microbial biosorbents. International Journal of Environmental Research and Public Health 14 (1), 94.

Azubuike, C.C., Chikere, C.B., Okpokwasili, G.C., 2016. Bioremediation techniques—classification based on site of application: principles, advantages, limitations and prospects. World Journal of Microbiology and Biotechnology 32 (11), 180.

Ba, S., Arsenault, A., Hassani, T., et al., 2013. Laccase immobilization and insolubilization: from fundamentals to applications for the elimination of emerging contaminants in wastewater treatment. Critical Reviews in Biotechnology 33 (4), 404–418.

Bagheri, S., Julkapli, N.M., 2016. Effect of hybridization on the value-added activated carbon materials. International Journal of Industrial Chemistry 7, 249–264. https://doi.org/10.1007/s40090-016-0089-5.

Bahnemann, W., Haque, M.M., Muneer, M., 2007. Titanium dioxide-mediated photocatalysed degradation of few selected organic pollutants in aqueous suspensions. Catalysis Today 124 (3–4), 133–148.

Bak, F., Widdel, F., 1986. Anaerobic degradation of phenol and phenol derivatives by *Desulfobacterium phenolicum* gen. nov., sp. nov. Archives of Microbiology 146, 177–180.

Bakr, S.A., Mohamed, R., 2013. Biodegradation of Cresols Using *Pseudomonas Putida* Immobilized in Poly Vinyl Alcohol (PVA) Gel. Theses, p. 510. http://scholarworks.uaeu.ac.ae/all_theses/510.

Balarak, D., Mahdavi, Y., 2016. Survey of efficiency agricultural waste as adsorbent for removal of p-Cresol from aqueous solution. International Research Journal of Pure and Applied Chemistry 10, 1–11.

Bandara, J., Klehm, U., Kiwi, J., 2007. Raschig rings Fe_2O_3 composite photocatalyst. activate in the degradation of 4–chlorophenol and Orange II under daylight irradiation. Applied Catalysis B: Environmental 34, 321–333.

Banerjee, A., 2011. Biodegradation of Phenolic and Petroleum Waste Water by Isolated bacillus Cereus (Thesis). http://gyan.iitg.ernet.in/handle/123456789/223.

Banerjee, S., Gopal, J., Muraleedharan, P., et al., 2006. Physics and chemistry of photocatalytic titanium dioxide: visualization of bactericidal activity using atomic force microscopy. Current Science 90, 1378–1383.

Baquerizo, A., Bañares, R., Saliba, F., 2015. Chapter 107 − Current Clinical Status of the Extracorporeal Liver Support Devices, Transplantation of the Liver, third ed., pp. 1463–1487

Barakat, M.A., Al-Hutailah, R.I., Qayyum, E., et al., 2013. Pt nanoparticles/TiO_2 for photocatalytic degradation of phenols in wastewater. Environmental Technology 35 (2), 137–144. https://doi.org/10.1080/09593330.2013.820796.

Barka, N., Bakas, I., Qourzal, S., et al., 2013. Degradation of phenol in water by titanium dioxide photocatalysis. Oriental Journal of Chemistry 29 (3), 1055–1060. URL: https://doi.org/10.13005/ojc/290328.

Barkovskii, A.L., Korshunova, V.E., Pozdnyacova, L.I., 1985. Catabolism of phenol and benzoate by *Azospirillium strains*. Applied Soil Ecology 2 (1), 17–24.

Barndōk, H., Hermosilla, D., Cortijo, L., et al., 2012. Assessing the effect of inorganic anions on TiO_2-photocatalysis and ozone oxidation treatment efficiencies. Advanced Oxidation Technologies 15 (1), 125–132.

Bartha, R., Bossert, I., 1984. The treatment and disposal of petroleum wastes. In: Atlas, R.M. (Ed.), Petroleum Microbiology. Macmillan, New York, NY, pp. 553–578.

Basak, B., Chakraborty, S., Bhunia, B., et al., 2014. Microbial remediation of recalcitrant aromatic compounds. Industrial and Environmental Biotechnology 171–189.

Basha, K.M., Rajendran, A., Thangavelu, V., 2010. Recent advances in the biodegradation of phenol: a review. Asian Journal of Experimental Biological Sciences 1 (2), 219–234.

Basheer, F., Farooqi, I.H., 2012. Biodegradation of p-cresol by aerobic granules in sequencing batch reactor. Journal of Environmental Sciences 24, 2012–2018. https://doi.org/10.1016/S1001-0742(11)60988-1.

Berge-Lefranc, D., Vagner, C., Calaf, R., et al., 2012. In vitro elimination of protein bound uremic toxin p-cresol by MFI-type zeolites. Microporous and Mesoporous Materials 153, 288–293.

Beshay, U., Abd-EI-Haleem, D., Moawad, H., et al., 2002. Phenol biodegradation by free and immobilized *Acinetobacter*. Biotechnology Letters 24, 1295–1297.

Bhakta, J.N., 2017. Handbook of Research on Inventive Bioremediation Techniques (Advances in Environmental Engineering and Green Technologies). IGI Global, USA, p. 652. http://doi.org/10.4018/978-1-5225-2325-3.

Bhandari, D., Garg, R.K., 2015. Effect of industrialization on environment (Indian scenario). Global Journal for Research Analysis 4 (12), 281−282.

Bisaillon, J.G., Lépine, F., Beaudet, R., et al., 1991. Carboxylation of o-cresol by an anaerobic consortium under methanogenic conditions. Applied and Environmental Microbiology 57, 2131−2134.

Blake, C.K., Hegeman, D.G., 1987. Plasmid pCBI carries genes for anaerobic benzoate catabolism in *Alcaligenes xylosoxidans* subsp. *denitriecans* PN-1. Journal of Bacteriology 169, 4878−4883.

Bollag, J.M., Chu, H.L., Rao, A., et al., 2003. Enzymatic oxidative transformation of chlorophenol mixtures. Journal of Environmental Quality 32, 63−69.

Bonting, C.F.C., Schneider, S., Schmidtberg, G., et al., 1995. Anaerobic degradation of m-cresol via methyl oxidation to 3 hydroxybenzoate by a denitrifying bacterium. Asian Journal of Experimental Biological Sciences 164, 63−69.

Bouallegue, O., Mzoughi, R., Weill, F.X., et al., 2004. Outbreak of *Pseudomonas putida* bacteraemia in a neonatal intensive care unit. Journal of Hospital Infection 57, 88−91.

Bretón, J.G.C., Bretón, R.M.C., Ucan, F.V., et al., 2017. Characterization and sources of aromatic hydrocarbons (BTEX) in the atmosphere of two urban sites located in Yucatan Peninsula in Mexico. Atmosphere 8 (6), 107. https://doi.org/10.3390/atmos8060107.

Brooms, T.J., Otieno, B., Onyango, M.S., et al., 2017. Photocatalytic degradation of p-cresol using TiO_2/ZnO hybrid surface capped with polyaniline. Journal of Environmental Science and Health. Part A, Toxic/Hazardous Substances and Environmental Engineering 53 (2), 99−107. https://doi.org/10.1080/10934529.2017.1377583.

Cai, W., Li, J., Zhang, Z., 2007. The characteristics and mechanisms of phenol biodegradation by *Fusarium* sp. Journal of Hazardous Materials 148, 38−42.

Cameotra, S.S., Singh, P., 2008. Bioremediation of oil sludge using crude biosurfactants. International Biodeterioration and Biodegradation 62 (3), 274−280.

Carmichael, L.M., Pfaender, F.K., 1997. The effect of inorganic and organic supplements on the microbial degradation of phenanthrene and pyrene in soils. Biodegradation 8 (1), 1−13.

Cassano, A.E., Alfano, O.M., 2000. Reaction engineering of suspended solid heterogeneous photocatalytic reactors. Catalysis Today 58 (2−3), 167−197.

Chaillan, F., Chaîneau, C.H., Point, V., et al., 2006. Factors inhibiting bioremediation of soil contaminated with weathered oils and drill cuttings. Environmental Pollution 144 (1), 255−265.

Chaillan, F., Le Fl`eche, A., Bury, E., et al., 2004. Identification and biodegradation potential of tropical aerobic hydrocarbon degrading microorganisms. Research in Microbiology 155, 587−595.

Chaîneau, C.H., Rougeux, G., Yéprémian, C., et al., 2005. Effects of nutrient concentration on the biodegradation of crude oil and associated microbial populations in the soil. Soil Biology and Biochemistry 37 (8), 1490−1497.

Chaurasia, A.K., Tremblay, P.-L., Holmes, D.E., et al., 2015. Genetic evidence that the degradation of para-cresol by *Geobacter metallireducens* is catalyzed by the periplasmic para-cresol methylhydroxylase. FEMS Microbiology Letters 362 (20), 1−6.

Chen, Q., Jiang, D., Shi, W., et al., 2009. Visible-light-activated Ce−Si co-doped TiO_2 photocatalyst. Applied Surface Science 255, 7918−7924.

Chen, D., Ray, A.K., 1998. Photodegradation kinetics of 4-nitrophenol in TiO_2 suspension. Water Resources 32, 3223−3234.

Chiou, C.H., Wu, C.Y., Juang, R.S., 2008. Photocatalytic degradation of phenol and m-nitrophenol using irradiated TiO_2 in aqueous solutions. Separation and Purification Technology 62, 559−564.

Chong, M.N., Chow, C.W.K., Jin, B., et al., 2010. Recent developments in photocatalytic water treatment technology: A review. Water Research 44, 2997−3027.

Chopra, R., 2016. Environmental degradation in India: causes and consequences. International Journal of Applied Environmental Sciences 11 (6), 1593−1601.

Choquette-Labbé, M., Shewa, W.A., Lalman, J.A., et al., 2014. Photocatalytic degradation of phenol and phenol derivatives using a nano-TiO_2 catalyst: integrating quantitative and qualitative factors using response surface methodology. Water 6, 1785−1806. https://doi.org/10.3390/w6061785.

Chung, T.P., Wu, P.C., Juang, R.S., 2005. Use of microporous hallow fibres for improved biodegradation of high-strength phenol solutions. Journal of Membrane Science 258, 55−63.

Claussen, M., Schmidt, S., 1998. Biodegradation of phenol and p-cresol by the hyphomycete *Scedosporium apiospermum*. Research in Microbiology 149 (6), 399–406.

Cooney, J.J., 1984. The fate of petroleum pollutants in fresh water ecosystems. In: Atlas, R.M. (Ed.), Petroleum Microbiology. Macmillan, New York, NY, pp. 399–434.

Crawford, R.L., 1976. Pathways of 4-hydroxybenzoate degradation among species of *Bacillus*. Journal of Bacteriology 127, 204–210.

Crittenden, J., Hand, D.W., Marchand, E.G., et al., 1996. Solar detoxification of fuel-contaminated groundwater using fixed-bed photocatalysts. Water Environment Researcher 68, 270–278.

Daghrir, R., Drogui, P., Robert, D., 2013. Modified TiO_2 for environmental photocatalytic applications: a review. Industrial and Engineering Chemistry Research 52 (10), 3581–3599. https://doi.org/10.1021/ie303468t.

Daniel, D., Gutz, I.G.R., 2007. Microfluidic cell with a TiO_2-modified gold electrode irradiated by an UV-LED for in situ photocatalytic decomposition of organic matter and its potentiality for voltammetric analysis of metal ions. Electrochemistry Communications 9 (3), 522–528.

Daraei, H., Davoodabadi, M., Yazdanbakhsh, A.R., 2015. Degradation of phenol with using of Fenton-like processes from water. Iranian Journal of Health, Safety and Environment 2 (3), 325–329.

Das, N., Chandran, P., 2011. Microbial degradation of petroleum hydrocarbon contaminants: an overview. Biotechnology Research International 2011, 13. Article ID 941810. http://doi.org/10.4061/2011/941810.

Das, K., Mukherjee, A.K., 2007. Crude petroleum-oil biodegradation efficiency of *Bacillus subtilis* and *Pseudomonas aeruginosa* strains isolated from a petroleum-oil contaminated soil from North-East India. Bioresource Technology 98 (7), 1339–1345.

Dashtban, M., Schraft, H., Syed, T.A., et al., 2010. Fungal biodegradation and enzymatic modification of lignin. International Journal of Biochemistry and Molecular Biology 1 (1), 36–50.

de Carvalho, C.C.C.R., Costa, S.S., Fernandes, P., Couto, I., Viveiros, M., 2014. Membrane transport systems and the biodegradation potential and pathogenicity of genus *Rhodococcus*. Frontiers in Physiology 5, 133.

De Gisi, S., Lofrano, G., Grassi, M., et al., 2016. Characteristics and adsorption capacities of low-cost sorbents for wastewater treatment: a review. Sustainable Materials and Technologies 9, 10–40.

Delnavaz, M., Ayati, B., Ganjidoust, H., et al., 2012. Kinetics study of photocatalytic process for treatment of phenolic wastewater by TiO_2 nano powder immobilized on concrete surfaces. Toxicological and Environmental Chemistry 94 (6), 1086–1098. https://doi.org/10.1080/02772248.2012.688331.

Deng, Y., Zhao, R., 2015. Advanced oxidation processes (AOPs) in wastewater treatment. Current Pollution Reports 1, 167–176.

Devi, L.G., Rajashekhar, K.E., 2011. A kinetic model based on non-linear regression analysis is proposed for the degradation of phenol under UV/solar light using nitrogen-doped TiO_2. Journal of Molecular Catalysis A: Chemical 334, 65–76.

Diez, M.C., 2010. Biological aspects involved in the organic pollutants. Journal of Soil Science and Plant Nutrition 10 (3), 244–267.

Diya'uddeen, B.H., Daud, W.M.A.W., Aziz, A.R.A., 2011. Treatment technologies for petroleum refinery effluents: a review. Process Safety and Environmental Protection 89, 95–105.

Dominguez, S., Ribao, P., Rivero, M.J., et al., 2015. Influence of radiation and TiO_2 concentration on the hydroxyl radicals generation in a photocatalytic LED reactor. Application to dodecylbenzenesulfonate degradation. Applied Catalysis B: Environmental 178, 165–169. URL: https://doi.org/10.1016/j.apcatb.2014.09.072.

Du, L., Ma, L., Qi, F., et al., 2016. Characterization of a unique pathway for 4-cresol catabolism initiated by phosphorylation in *Corynebacterium glutamicum*. Journal of Biological Chemistry 291, 6583–6594.

Duan, W., Meng, F., Cui, H., et al., 2018. Ecotoxicity of phenol and cresols to aquatic organisms: a review. Ecotoxicology and Environmental Safety 157, 441–456.

Dursun, A.Y., Tepe, O., 2005. Internal mass transfer effect on biodegradation of phenol by Ca-alginate immobilized *Ralstonia eutropha* (Online).

Dvoraka, L., Gómeza, M., Dolinaa, J., et al., 2015. Anaerobic membrane bioreactors—a mini review with emphasis on industrial wastewater treatment: applications, limitations and perspectives. Desalination and Water Treatment 57, 19062–19076.

Ebrahiem, E.E., Al-Maghrabi, M.N., Mobarki, A.R., 2017. Removal of organic pollutants from industrial wastewater by applying photo-Fenton oxidation technology. Arabian Journal of Chemistry 10 (2), 1674–1679. https://doi.org/10.1016/j.arabjc.2013.06.012.

El-Naas, M., 2012. Aerobic biodegradation of phenols: a comprehensive review. Critical Reviews in Environmental Science and Technology 42 (16), 1631.

Ellis, K.V., 2013. Surface Water Pollution and its Control. Springer Customer Service Center Gmbh, p. 388.

Escudero, C.J., Iglesias, O., Dominguez, S., et al., 2017. Performance of electrochemical oxidation and photocatalysis in terms of kinetics and energy consumption. New insights into the p-cresol degradation. Journal of Environmental Management 195, 117–124.

Evans, W.C., 1963. The microbiological degradation of aromatic compounds. Journal of General Microbiology 32, 177–184.

Fang, H.H.P., Zhou, G.M., 2000. Degradation of phenol and p-cresol in reactors. Water Science and Technology 42, 237–244.

Fardhyanti, D.S., Mulyono, P., Sediawan, W.B., et al., 2013. Extraction of phenol, o-cresol, and p-cresol from coal tar: effect of temperature and mixing. World Academy of Science, Engineering and Technology International Journal of Chemical and Molecular Engineering 7 (6), 454–458. https://doi.org/10.1999/1307-6892/13782.

Fathepure, B.Z., 2014. Recent studies in microbial degradation of petroleum hydrocarbons in hypersaline environments. Frontiers in Microbiology 5, 173.

Filippoa, E., Carluccib, C., Capodilupob, A.L., et al., 2015. Enhanced photocatalytic activity of pure anatase TiO_2 and Pt-TiO_2 nanoparticles synthesized by green microwave assisted route. Materials Research 18 (3), 473–481. URL: https://doi.org/10.1590/1516-1439.301914.

Firozjaee, T.T., Najafpour, G.D., Khavarpour, M., et al., 2011. Phenol biodegradation kinetics in an anaerobic batch reactor. Iranian Journal of Energy and Environment 2 (1), 68–73.

Foght, J.M., Westlake, D.W.S., Johnson, W.M., Ridgway, H.F., 1996. Environmental gasoline-utilizing isolates and clinical isolates of *Pseudomonas aeruginosa* are taxonomically indistinguishable by chemotaxonomic and molecular techniques. Microbiology 142 (9), 2333–2340.

Fritsche, W., Hofrichter, M., 2000. Aerobic degradation by microorganisms. In: Klein, J. (Ed.), Environmental Processes- Soil Decontamination. Wiley-VCH, Weinheim, Germany, pp. 146–155. https://doi.org/10.1002/9783527620951.ch6.

Fujishima, A., Tryk, D.A., Zhang, X., 2008. TiO_2 photocatalysis and related surface phenomena. Surface Science Reports 63, 515–582.

Fulekar, M.H., 2017. Microbial degradation of petrochemical waste-polycyclic aromatic hydrocarbons. Bioresources and Bioprocessing 4 (1), 28.

Gaikwad, N.W., Bodell, W.J., 2003. Formation of DNA adducts in HL-60 cells treated with the toluene metabolite p-cresol: a potential biomarker for toluene exposure. Chemico-Biological Interactions 145 (2), 149–158.

Gami, A.A., Shukor, M.Y., Khalil, K.A., et al., 2014. Phenol and phenolic compounds toxicity. Journal of Environmental Microbiology and Toxicology 2 (1), 11–23.

Gaokar, R.D., 2018. Bioremediation of industrial effluents by alkaliphilic bacteria isolated from marine ecosystems. Advances in Clinical Toxicology 3 (1), 1–5.

Gautam, K.K., Tyagi, V.K., 2005. Microbial surfactants: a review. Journal of Oleo Science 55 (4), 155–166.

Gaya, U.I., Abdullah, A.H., 2008. Heterogeneous photocatalytic degradation of organic contaminants over titanium dioxide: a review of fundamentals, progress and problems. Journal of Photochemistry and Photobiology C: Photochemistry Reviews 9 (1), 1–12.

Geissen, V., Mol, H., Klumpp, E., et al., 2015. Emerging pollutants in the environment: a challenge for water resource management. International Soil and Water Conservation Research 3 (1), 57–65.

Gerginova, M., Manasiev, J., Yemendzhiev, H., et al., 2013. Biodegradation of phenol by Antarctic strains of *Aspergillus fumigates*. Zeitschrift für Naturforschung C 68 (9–10), 384–393.

Gharibzahedi, S.M.T., Razavi, S.H., Mousavi, M., 2013. Kinetic analysis and mathematical modeling of cell growth and canthaxanthin biosynthesis by *Dietzia natronolimnaea* HS-1 on waste molasses hydrolysate. RSC Advances 3 (45), 23495–23502. https://doi.org/10.1039/c3ra44663h.

Ghattas, A.K., Fischer, F., Wick, A., et al., 2017. Anaerobic biodegradation of (emerging) organic contaminants in the aquatic environment. Water Research 116, 268–295.

Ghosal, D., Ghosh, S., Dutta, T.K., et al., 2016. Current state of knowledge in microbial degradation of polycyclic aromatic hydrocarbons (PAHs): a review. Frontiers in Microbiology 7, 1369. http://doi.org/10.3389/fmicb.2016.01369.

Gkorezis, P., Daghio, M., Franzetti, A., et al., 2016. The interaction between plants and bacteria in the remediation of petroleum hydrocarbons: an environmental perspective. Frontiers in Microbiology 7, 1836.

Glezer, V., 2003. Environmental Effects of Substituted Phenols. Wiley, USA, pp. 1347–1363.

Gómeza, C.M., Angela, G.D., Tzompantzia, F., et al., 2012. Photocatalytic degradation of p-cresol on Pt/γAl$_2$O$_3$–TiO$_2$ mixed oxides: effect of oxidizing and reducing pre-treatments. Journal of Photochemistry and Photobiology A: Chemistry 236, 21–25.

Goswamia, M., Shivaramanb, N., Singh, R.P., 2005. Microbial metabolism of 2-chlorophenol, phenol and q-cresol by *Rhodococcus erythropolis* M1 in coculture with *Pseudomonas fluorescens* P1. Microbiological Research 160, 101–109.

Green, K.J., Rudham, R.J., 1993. Photocatalytic oxidation of propan-2-ol by semiconductor–zeolite composites. Journal of the Chemical Society, Faraday Transactions 89 (11), 1867–1870.

Guan, X., He, D., Ma, J., et al., 2010. Application of permanganate in the oxidation of micropollutants: a mini review. Frontiers of Environmental Science and Engineering 4 (4), 405–413. https://doi.org/10.1007/s11783-010-0252-8.

Gupta, G., Kumar, V., Pal, A.K., 2016. Biodegradation of polycyclic aromatic hydrocarbons by microbial consortium: a distinctive approach for decontamination of soil. Soil and Sediment Contamination: International Journal 25 (6), 597–623. https://doi.org/10.1080/15320383.2016.1190311.

Gupta, S.M., Tripathi, M., 2011. A review of TiO$_2$ nanoparticles. Chinese Science Bulletin 56, 1639–1657. https://doi.org/10.1007/s11434-011-4476-1.

Gupte, A., Tripathi, A., Patel, H., et al., 2018. Bioremediation of polycyclic aromatic hydrocarbon (PAHs): a perspective. The Open Biotechnology Journal 12, 363–378. ISSN: 1874-0707.

Gusse, A.C., Miller, P.D., Volk, T.J., 2006. White-rot fungi demonstrate first biodegradation of phenolic resin. Environmental Science and Technology 40, 4196–4199.

Habibi, M.H., Hassanzadeh, A., Mahdavi, S., 2005. The effect of operational parameters on the photocatalytic degradation of three textile azo dyes in aqueous TiO$_2$ suspensions. Journal of Photochemistry and Photobiology A: Chemistry 172, 89–96.

Halasi, G., Solymosi, F., Ugrai, I., 2011. Photocatalytic decomposition of ethanol on TiO$_2$ modified by N and promoted by metals. Journal of Catalysis 281, 309–317.

Hamitouche, A., Bendjama, Z., Amrane, A., et al., 2012. Biodegradation of P-cresol by mixed culture in batch reactor — effect of the three nitrogen sources used. Procedia Engineering 33, 458–464.

Hamitouche, A., Bendjama, Z., Amrane, A., et al., 2015. Biodegradation of o-Cresol by a *Pseudomonas* spp. Progress Clean Energy 1, 713–724.

Harwood, C.S., Burchhardt, G., Herrmann, H., et al., 1999. Anaerobic metabolism of aromatic compounds via the benzoyl-CoA pathway. FEMS Microbiology Reviews 22, 439–458.

Hasana, S.A., Jabeen, S., 2015. Degradation kinetics and pathway of phenol by *Pseudomonas* and *Bacillus* species. Biotechnology and Biotechnological Equipment 29 (1), 45–53.

Head, I.M., 1998. Bioremediation: towards a credible technology. Microbiology 144, 599408.

Heider, J., Fuchs, G., 1997. Anaerobic metabolism of aromatic compounds. European Journal of Biochemistry 243, 577–596.

Helali, S., Guillard, C., Perol, N., et al., 2011. Methylamine and dimethylamine photocatalytic degradation-adsorption isotherms and kinetics. Applied Catalysis A General 402, 201–207.

Hengerer, R., Bolliger, B., Erbudak, M., et al., 2000. Structure and stability of the anatase TiO$_2$ (101) and (001) surfaces. Surface Science 460, 162–169.

Herrmann, J.M., 1999. Heterogeneous photocatalysis: fundamentals and applications to the removal of various types of aqueous pollutants. Catalysis Today 53, 115–129.

Hill, G.A., Robinson, C.W., 1975. Substrate inhibition kinetics: phenol degradation by *Pseudomonas putida*. Biotechnology and Bioengineering 17, 1599–1615.

Hopper, D.J., 1978. Incorporation of [^{18}O]water in the formation of p-hydroxybenzyl alcohol by the p-cresol methylhydroxylase from *Pseudomonas putida*. Biochemical Journal 175, 345–347.

Hopper, D.J., Taylor, D.G., 1975. Pathways of the degradation of m-cresol and p-cresol by *Pseudomonas putida*. Journal of Bacteriology 122, 1–6.

Hosseini, S.N., Borghei, S.M., Vossoughi, M., Taghavinia, N., 2007. Immobilization of TiO$_2$ on perlite granules for photocatalytic degradation of phenol. Applied Catalysis B: Environmental 74, 53–62.

Hughes, E.J., Bayly, R.C., 1983. Control of catechol meta-cleavage pathway in *Alcaligenes eutrophus*. Journal of Bacteriology 154, 1363–1370.

Hurtado, L., Amado-Piña, D., Roa-Morales, G., 2016. Comparison of AOPs efficiencies on phenolic compounds degradation. Journal of Chemistry 2016, 8. Article ID 4108587.

Ibhadon, A.O., Fitzpatrick, P., 2013. Heterogeneous photocatalysis: recent advances and applications. Catalysts 3, 189–218. https://doi.org/10.3390/catal3010189.

Igbinosa, E.O., Odjadjare, E.E., Chigor, V.N., et al., 2013. Toxicological profile of chlorophenols and their derivatives in the environment: the public health perspective. The Scientific World Journal 2013, 11. Article ID 460215.

Igwe, C.O., Saadi, A.A.L., Ngene, S.E., 2013. Optimal options for treatment of produced water in offshore petroleum platforms. Journal of Pollution Effects and Control 1, 102. https://doi.org/10.4172/2375-4397.1000102.

Ihara, T., Miyoshi, M., Iriyama, Y., Matsumoto, O., Sugihara, S., 2003. Visible-light-active titanium oxide photocatalyst realized by an oxygen-deficient structure and by nitrogen doping. Applied Catalysis B: Environmental 42 (4), 403–409.

Iorhemen, O.T., Hamza, R.A., Tay, J.H., 2016. Membrane bioreactor (MBR) technology for wastewater treatment and reclamation: membrane fouling. Membranes 6, 33. https://doi.org/10.3390/membranes6020033.

Irie, H., Watanabe, Y., Hashimoto, K., 2003. Nitrogen-concentration dependence on photocatalytic activity of TiO2-x N x powders. The Journal of Physical Chemistry B 107 (23), 5483–5486.

Jain, P.K., Gupta, V.K., Gaur, R.K., et al., 2011. Bioremediation of petroleum oil contaminated soil and water. Research Journal of Environmental Toxicology 5, 1–26.

Jallouli, N., Elghniji, K., Trabelsi, H., et al., 2017. Photocatalytic degradation of paracetamol on TiO2 nanoparticles and TiO2/cellulosic fiber under UV and sunlight irradiation. Arabian Journal of Chemistry 10 (2), 3640–3645.

Jamalia, A., Vanraesb, R., Hanselaerb, P., et al., 2013. A batch LED reactor for the photocatalytic degradation of phenol. Chemical Engineering and Processing: Process Intensification 71, 43–50.

Jansanthea, P., Treenattip, J., Pookmanee, P., et al., 2016. The photocatalytic degradation of phenol over titanium dioxide powder prepared by the solvothermal method. International Journal of Environmental Engineering 8 (1), 44–53.

Jiang, L., Ruan, Q., Li, R., et al., 2013. Biodegradation of phenol by using free and immobilized cells of Acinetobacter sp. BS8Y. Journal of Basic Microbiology 53, 224–230.

Jones, K.H., Trudgill, P.W., Hopper, D.J., 1993. Metabolism of p-cresol by the fungus Aspergillus fumigatus. Applied and Environmental Microbiology 59, 1125–1130.

Jouanneau, Y., Meyer, C., 2006. Purification and characterization of an arene cis-dihydrodiol dehydrogenase endowed with broad substrate specificity toward polycyclic aromatic hydrocarbon dihydrodiols. Applied and Environmental Microbiology 72, 4726–4734.

Jussila, M.M., Zhao, J., Suominen, L., et al., 2007. TOL plasmid transfer during bacterial conjugation in vitro and rhizoremediation of oil compoundsin vivo. Environmental Pollution 146 (2), 510–524.

Kadri, T., Rouissi, T., Brar, S.K., et al., 2017. Biodegradation of polycyclic aromatic hydrocarbons (PAHs) by fungal enzymes: a review. Journal of Environmental Sciences 51, 52–74.

Kahru, A., Reiman, R., Rätsep, A., 1998. The efficiency of different phenol-degrading bacteria and activated sludges in detoxification of phenolic leachates. Chemosphere 37 (2), 301–318.

Karthikeyan, S., Gopalakrishnan, A.N., 2011. Degradation of phenol and m-cresol in aqueous solutions using indigenously developed microwave-ultraviolet reactor. Journal of Scientific and Industrial Research 70 (1), 71–76.

Kastbjerg, V.G., Nielsen, D.S., Arneborg, N., et al., 2009. Response of Listeria monocytogenes to disinfection stress at the single-cell and population levels as monitored by intracellular pH measurements and viable-cell counts. Applied and Environmental Microbiology 75 (13), 4550–4556.

Kaymaz, Y., Babaoğlu, A., Pazarlioglu, N.K., 2012. Biodegradation kinetics of o-cresol by Pseudomonas putida DSM 548 (pJP4) and o-cresol removal in a batch-recirculation bioreactor system. Electronic Journal of Biotechnology 15 (1), 1–10. https://doi.org/10.2225/vol15-issue1-fulltext-4.

Keith, L.H., 2014. The source of U.S. EPA's sixteen PAH priority pollutants. Polycyclic Aromatic Compounds 35 (2–4), 147–160. https://doi.org/10.1080/10406638.2014.892886.

Kennes, C., Lema, J.M., 1994. Simultaneous biodegradation of p-cresol and phenol by the basidiomycete Phanerochaete chrysosporium. Journal of Industrial Microbiology 13 (5), 311–314.

Khleifat, K.M., 2006. Biodegradation of phenol by Ewingella americana: effect of carbon starvation and some growth conditions. Process Biochemistry 41, 2010–2016.

Kim, K.H., Ihm, S.K., 2011. Heterogeneous catalytic wet air oxidation of refractory organic pollutants in industrial wastewaters: a review. Journal of Hazardous Materials 186, 16–34.

Kim, J.D., Lee, C.G., 2007. Microbial degradation of polycyclic aromatic hydrocarbons in soil by bacterium-fungus co-cultures. Biotechnology and Bioprocess Engineering 12 (4), 410–416.

Kim, C.J., Maier, W.J., 1986. Acclimation and biodegradation of chlorinated aromatics in the presence of alternate substrates. Journal of the Water Pollution Control 58, 157–164.

Klein, J.A., Lee, D.D., 1978. Biological treatment of aqueous wastes from usual conversion processes. Biotechnology and Bioengineering 8, 379–390.

Koul, S., Fulekar, M.H., 2013. Petrochemical industrial waste: bioremediation techniques an overview. International Journal of Advanced Research and Technology 2 (7), 211.

Krastanov, A., Alexieva, Z., Yemendzhiev, H., 2013. Microbial degradation of phenol and phenolic derivatives. Engineering in Life Science 13 (1), 76–87. http://doi.org/10.1002/elsc.201100227.

Krishnan, S., Rawindran, H., Sinnathambi, C.M., et al., 2017. Comparison of various advanced oxidation processes used in remediation of industrial wastewater laden with recalcitrant pollutants. IOP Conference Series: Materials Science and Engineering 206, 012089. https://doi.org/10.1088/1757-899X/206/1/012089a, pp. 1–11.

Kristensen, E., 1995. Aerobic and anaerobic decomposition of organic matter in marine sediment: which is fastest? Limnology and Oceanography 40 (8), 1430–1437.

Kronenberg, M., Trably, E., Bernet, N., 2017. Biodegradation of polycyclic aromatic hydrocarbons: using microbial bioelectrochemical systems to overcome an impasse. Environmental Pollution 231, 509–523.

Krulwich, T.A., Sachs, G., Padan, E., 2011. Molecular aspects of bacterial pH sensing and homeostasis. Nature Reviews Microbiology 9 (5), 330–343.

Kudlek, E., Dudziak, M., Bohdziewicz, J., 2016. Influence of inorganic ions and organic substances on the degradation of pharmaceutical compound in water matrix. Water 8 (11), 532. https://doi.org/10.3390/w8110532.

Kudlek, E., Dudziak, M., Bohdziewicz, J., et al., 2017. Influence of Inorganic compounds on the process of photocatalysis of biologically active compounds. Journal of Ecological Engineering 18 (4), 123–129.

Kumar Reddy, D.H., Lee, S.M., 2012. Water pollution and treatment technologies. Journal of Environmental and Analytical Toxicology 2, 103. https://doi.org/10.4172/2161-0525.1000e103.

Kumar, A., Bisht, B.S., Joshi, V.D., et al., 2011. Review on bioremediation of polluted environment: a management tool. International Journal of Environmental Sciences 1 (6), 1079–1093.

Kumar, A., Kumar, S., Kumar, S., 2005. Biodegradation kinetics of phenol and catechol using *Pseudomonas putida* MTCC 1194. Biochemical Engineering Journal 22 (2), 151–159.

Kumar, P., Libchaber, A., 2013. Pressure and temperature dependence of growth and morphology of *Escherichia coli*: experiments and stochastic model. Biophysical Journal 105, 783–793.

Kumar, A., Pandey, G., 2017. A review on the factors affecting the photocatalytic degradation of hazardous materials. Material Science and Engineering International Journal 1 (3), 106–114.

Lakshmi, M.V.V.C., Sridevi, V., 2009. Effect of pH and inoculum size on phenol degradation by *Pseudomonas aeruginosa* (NCIM 2074). International Journal of Chemical Sciences 7 (4), 2246–2252.

Lakshmi, M.V.V.C., Sridevi, V., Rao, M.N., et al., 2011. Substrate inhibition kinetics of phenol degradation by *Pseudomonas Aeruginosa* (NCIM 2074). International Journal of Environment and Waste Management 3 (1), 103–113.

Lavand, A.B., Malghe, Y.S., 2015. Visible light photocatalytic degradation of 4-chlorophenol using C/ZnO/CdS nanocomposite. Journal of Saudi Chemical Society 19 (5), 471–478. https://doi.org/10.1016/j.jscs.2015.07.001.

Lazar, M.A., Varghese, S., Nair, S.S., 2012. Photocatalytic water treatment by titanium dioxide: recent updates. Catalysts 2, 572–601. https://doi.org/10.3390/catal2040572.

Li, Y., Leiyong, L., Chenwan, L., et al., 2012. Carbon nanotube/titania composites prepared by a micro-emulsion method exhibiting improved photocatalytic activity. Applied Catalysis A: General 427 (428), 1–7.

Li, Y., Sun, S., Ma, M., et al., 2008. Kinetic study and model of the photocatalytic degradation of rhodamine B (RhB) by a TiO_2-coated activated carbon catalyst: effects of initial RhB content, light intensity and TiO_2 content in the catalyst. The Chemical Engineering Journal 142, 147–155. URL: https://doi.org/10.1016/j.cej.2008.01.009.

Li, Z., Wang, H., 2010. Solar-Light-induced photocatalytic degradation of 2-sec-butyl-4,6-dinitrophenol (dinoseb) on TiO_2/MWCNTs composite ahotocatalyst. Second China Energy Scientist Forum (CESF 2010 E-BOOK) 282–288.

Liu, L., Chena, F., Yanga, F., et al., 2012. Photocatalytic degradation of 2,4-dichlorophenol using nanoscale Fe/TiO_2. The Chemical Engineering Journal 181 (182), 189–195.

Liu, Z., Xie, W., Li, D., et al., 2016. Biodegradation of phenol by bacteria strain *Acinetobacter calcoaceticus* PA isolated from phenolic wastewater. International Journal of Environmental Research and Public Health 13, 300. https://doi.org/10.3390/ijerph13030300.

Loh, K.C., Chung, T.S., Ang, Y.W.F., 2000. Immobilized cell membrane bioreactor for high strength phenol wastewater. Journal of Environmental Engineering 126, 75–79.

Loh, K.C., Wang, S.J., 1998. Enhancement of biodegradation of phenol and a non-growth substrate 4-chlorophenol by medium augmentation with conventional carbon sources. Biodegradation 8, 329–338.

Loncar, N., Bozic, N., Andelkovic, I., et al., 2011. Removal of aqueous phenol and phenol derivatives by immobilized potato polyphenol oxidase. Journal of the Serbian Chemical Society 76 (4), 513–522.

Londry, K.L., Fedorak, P.M., Suflita, J.M., 1997. Anaerobic degradation of m-cresol by a sulfate-reducing bacterium. Applied and Environmental Microbiology 63 (8), 3170–3175.

Lu, M.C., Roam, G.D., Chen, J.N., et al., 1993. Factors affecting the photocatalytic degradation of dichlorvos over titanium dioxide supported on glass. Journal of Photochemistry and Photobiology A 76 (1–2), 103–110.

Luan, M., Jing, G., Piao, Y., et al., 2017. Treatment of refractory organic pollutants in industrial wastewater by wet air oxidation. Arabian Journal of Chemistry 10 (1), S769–S776.

Mahlambi, M.M., Ngila, C.J., Mamba, B.B., 2015. Recent developments in environmental photocatalytic degradation of organic pollutants: the case of titanium dioxide nanoparticles—a review. Journal of Nanomaterials (790173), 29. URL: https://doi.org/10.1155/2015/790173.

Mahmound, A., Aziza, Y., Abdeltif, A., et al., 2008. Biosurfactant production by *Bacillus* strain injected in the petroleum reservoirs. Journal of Industrial Microbiology and Biotechnology 35, 1303–1306.

Maicu, M., Hidalgo, M.C., Colón, G., et al., 2011. Comparative study of the photodeposition of Pt, Au and Pd on pre-sulphated TiO_2 for the photocatalytic decomposition of phenol. Journal of Photochemistry and Photobiology A: Chemistry 217, 275–283.

Masque, C., Nolla, M., Bordons, A., 1987. Selection and adaptation of a phenol degrading strain of *Pseudomonas*. Biotechnology Letters 9, 655–660.

Mathur, A.K., Bala, S., Majumder, C.B., et al., 2010. Kinetics studies of p-cresol biodegradation by using *Pseudomonas putida* in batch reactor and in continuous bioreactor packed with calcium alginate beads. Water Science and Technology 62 (12), 2920–2929. https://doi.org/10.2166/wst.2010.736.

McNally, D.L., Mihelcic, J.R., Lueking, D.R., 1998. Biodegradation of three- and four- ring polycyclic aromatic hydrocarbons under aerobic and denitrifying conditions. Environmental Science and Technology 32, 2633–2639.

Meacock, G., Taylor, K.D.A., Knowles, M., et al., 1997. The improved whitening of minced cod flesh using dispersed titanium dioxide. Journal of the Science of Food and Agriculture 73 (2), 221–225.

Megharaj, M., Ramakrishnan, B., Venkateswarlu, K., et al., 2011. Bioremediation approaches for organic pollutants: a critical perspective. Environment International 37, 1362–1375.

Mendonca, E., Martins, A., Anselmo, A.M., 2004. Biodegradation of natural phenolic compounds as single and mixed substrates by *Fusarium flocciferum*. Electronic Journal of Biotechnology 7, 30–37.

Michałowicz, J., Duda, W., 2007. Phenols—sources and toxicity. Polish Journal of Environmental Studies 16, 347–362.

Misra, V., Pandey, S.D., 2005. Hazardous waste, impact on health and environment for development of better waste management strategies in future in India. Environment International 31 (3), 417–431.

Mogyorósi, K., Farkas, A., Dékány, I., 2002. TiO_2-Based photocatalytic degradation of 2-chlorophenol adsorbed on hydrophobic clay. Environmental Science and Technology 36 (16), 3618–3624.

Mohanty, S.S., 2012. Microbial Degradation of Phenol: A Comparative Study. Thesis URL: http://ethesis.nitrkl.ac.in/4431/1/Satyasundar_Mohanty_Final_Thesis.pdf.

Mohanty, S.S., Jena, H.M., 2017. Biodegradation of phenol by free and immobilized cells of a novel *Pseudomonas* sp. NBM11. Brazilian Journal of Chemical Engineering 34 (1), 75–84.

Mozia, S., 2010. Photocatalytic Membrane Reactors (PMRs) in water and wastewater treatment. A review. Separation and Purification Technology 73, 71–91.

Mrozik, A., Piotrowska-Seget, Z., Łabuzek, S., 2003. Bacterial degradation and bioremediation of polycyclic aromatic hydrocarbons. Polish Journal of Environmental Studies 12 (1), 15–25.

Mukherjee, S., Basak, B., Bhunia, B., et al., 2013. Potential use of polyphenol oxidases (PPO) in the bioremediation of phenolic contaminants containing industrial wastewater. Reviews in Environmental Science and Biotechnology 12 (1), 61–73.

Muller, J.A., Galushko, A.S., Kappler, A., et al., 2001. Initiation of anaerobic degradation of p-cresol by formation of 4-Hydroxybenzylsuccinate in *Desulfobacterium cetonicum*. Journal of Bacteriology 183 (2), 752–757.

Múazu, N.D., Jarrah, N., Zubair, M., et al., 2017. Removal of phenolic compounds from water using sewage sludge-based activated carbon adsorption: a review. International Journal of Environmental Research and Public Health 14 (10), 1094.

Muruganandham, M., Suri, R.P.S., Jafari, S., et al., 2014. Recent developments in homogeneous advanced oxidation processes for water and wastewater treatment. International Journal of Photoenergy 2014, 21. URL: https://doi.org/10.1155/2014/821674.

Muthusamy, K., Gopalakrishnan, S., Ravi, T.K., et al., 2008. Biosurfactants: properties, commercial production and application. Current Science 94 (6), 736−747.

Naidoo, S., Olaniran, A.O., 2014. Treated wastewater effluent as a source of microbial pollution of surface water resources. International Journal of Environmental Research and Public Health 11 (1), 249−270. https://doi.org/10.3390/ijerph110100249.

Národní, U., 2002. Bioremediation of persistent organic pollutants - a review. Technologie a biotechnologieinventura POPs v ÈR Èást VII.

Nguyen, A.H., Juang, R.S., 2014. Effect of Operating Parameters and Kinetic Study on Photocatalytic Degradation of O-Cresol in Synthetic Wastewater with Irradiated Titanium Dioxide International Conference on Advances in Engineering and Technology. ICAET'2014, Singapore, pp. 29−30.

Nickheslat, A., Amin, M.M., Izanloo, H., et al., 2013. Phenol photocatalytic degradation by advanced oxidation process under ultraviolet radiation using titanium dioxide. Journal of Environmental and Public Health (815310), 9. URL: https://doi.org/10.1155/2013/815310.

Nishiyama, N., Fujiwara, Y., Adachi, K., et al., 2015. Preparation of porous metal-ion-doped titanium dioxide and the photocatalytic degradation of 4-chlorophenol under visible light irradiation. Applied Catalysis B: Environmental 176−177, 347−353.

Noonan, D.C., Curtis, J.T., 1990. Groundwater Remediation and Petroleum: A Guide for Underground Storage Tanks. CRC Press, Technol. and Eng, p. 142.

NRC, 2012. Water Reuse:Potential for Expanding the Nation's Water Supply through Reuse of Municipal Wastewater. National research council. The national academies press, Washington, DC.

Obayori, O.S., Ilori, M.O., Adebusoye, S.A., et al., 2009. Degradation of hydrocarbons and biosurfactant production by Pseudomonas sp. strain LP1. World Journal of Microbiology and Biotechnology 25 (9), 1615−1623.

Ochando-Pulido, J.M., Martinez-Ferez, A., 2015. On the recent use of membrane technology for olive mill wastewater purification membranes, 5 (4), 513−531. https://doi.org/10.3390/membranes5040513.

Okoh, A.I., 2006. Biodegradation alternative in the cleanup of petroleum hydrocarbon pollutants. Biotechnology and Molecular Biology Reviews 1 (2), 38−50.

Okolo, J.C., Amadi, E.N., Odu, C.T.I., 2005. Effects of soil treatments containing poultry manure on crude oil degradation in a sandy loam soil. Applied Ecology and Environmental Research 3 (1), 47−53.

Okpokwasili, G.C., Nweke, C.O., 2006. Microbial growth and substrate utilization kinetics. African Journal of Biotechnology 5, 305−317.

Olajire, A.A., Essien, J.P., 2014. Aerobic degradation of petroleum components by microbial consortia. Journal of Petroleum and Environmental Biotechnology 5, 195. https://doi.org/10.4172/2157-7463.1000195.

Ollis, D., Al-Elkabi, H., 1993. Photocatalytic Purification and Treatment of Water and Air. Elsevier Sci. Ltd., pp. 481−494

Otzen, D.E., 2017. Biosurfactants and surfactants interacting with membranes and proteins: same but different? Biochimica et Biophysica Acta 1859 (4), 639−649.

Oudot, J., Merlin, F.X., Pinvidic, P., 1998. Weathering rates of oil components in a bioremediation experiment in estuarine sediments. Marine Environmental Research 45 (2), 113−125.

Palmisano, G., Loddo, V., El Nazer, H.H., et al., 2009. Graphite-supported TiO_2 for 4-nitrophenol degradation in a photoelectrocatalytic reactor. Chemical Engineering Journal 155, 339−346.

Pandey, K., Singh, B., Pandey, A.K., et al., 2017. Application of microbial enzymes in industrial waste water treatment. International Journal of Biochemistry and Molecular Biology 6 (8), 1243−1254.

Papazi, A., Assimakopoulos, K., Kotzabasis, K., 2012. Bioenergetic strategy for the biodegradation of p-Cresol by the unicellular green alga Scenedesmus obliquus. PLoS One 7 (12), e51852. https://doi.org/10.1371/journal.pone.0051852.

Paxton, A.T., Thiên-Nga, L., 1998. Electronic structure of reduced titanium dioxide. Physical Review B 57, 1579−1584.

Pelletier, E., Delille, D., Delille, B., 2004. Crude oil bioremediation in sub-Antarctic intertidal sediments: chemistry and toxicity of oiled residues. Marine Environmental Research 57 (4), 311−327.

Peyton, B.M., Wilson, T., Yonge, D.R., 2002. Kinetics of phenol degradation in high salt solutions. Water Research 36, 4811−4820.

Piscopo, A., Robert, D., Weber, J.V., 2001. Influence of pH and chloride anion on the photocatalytic degradation of organic compounds Part I. Effect on the benzamide and para-hydroxybenzoic acid in TiO_2 aqueous solution. Applied Catalysis B: Environmental 35, 117−124.

Pornsunthorntawee, O., Wongpanit, P., Chavadej, S., et al., 2008. Structural and physicochemical characterization of crude biosurfactant produced by *Pseudomonas aeruginosa* SP4 isolated from petroleum−contaminated soil. Bioresource Technology 99 (6), 1589−1595.

Preethy, P.R.V., 2016. Comparison on efficiency of various techniques in treatment of waste and sewage water − a comprehensive review. Resource-Efficient Technologies 2 (4), 175−184.

Pullanikkatil, D., Urama, K.C., 2011. The effects of industrialization on water quality and livelihoods in Lesotho. International Journal of Environmental Engineering 3 (2), 175−191.

Puma, G.L., Bono, A., Krishnaiah, D., et al., 2008. Preparation of titanium dioxide photocatalyst loaded onto activated carbon support using chemical vapor deposition: a review paper. Journal of Hazardous Materials 157, 209−219.

Rafiee, E., Noori, E., Zinatizadeh, A.A., et al., 2016. Photocatalytic degradation of phenol using a new developed TiO_2/graphene/heteropoly acid nanocomposite: synthesis, characterization and the process optimization. RSC Advances 6, 96554−96562. https://doi.org/10.1039/C6RA09897E.

Rahman, K.S.M., Rahman, T.J., Kourkoutas, Y., et al., 2003. Enhanced bioremediation of n-alkane in petroleum sludge using bacterial consortium amended with rhamnolipid and micronutrients. Bioresource Technology 90 (2), 159−168.

Rajani, V., Reshma, J.K., 2016. Factors affecting phenol degradation potential of microbes- a review. World Journal of Pharmaceutical Sciences 5 (11), 691−706.

Ratnakar, A., 2016. An Overview of biodegradation of organic pollutants. International Journal of Swarm Intelligence Research 4 (1), 73−91. URL: https://www.researchgate.net/publication/316739306.

Rawindran, S.K.H., Sinnathambi, C.M., Lim, J.W., 2017. Comparison of various advanced oxidation processes used in remediation of industrial wastewater laden with recalcitrant pollutants. Materials Science and Engineering 206, 1−11. https://doi.org/10.1088/1757-899X/206/1/012089.

Robens Institute, 1996. Fact Sheet 2.33: Turbidity Measurement. Fact Sheets on Environmental Sanitation. University of Surrey, World Health Organization.

Rockne, K.J., Strand, S.E., 1998. Biodegradation of bicyclic and polycyclic aromatic hydrocarbons in anaerobic enrichments. Environmental Science and Technology 32, 2962−2967.

Rodríguez, M., 2003. Fenton and UV-Vis Based Advanced Oxidation Processes in Wastewater Treatment: Degradation, Mineralization and Biodegradability Enhancement. Universitat de Barcelona, Spain. http://hdl.handle.net/2445/35398.

Rozich, A.F., Colvin, R.J., 1985. Effects of glucose on phenol biodegradation by heterogeneous populations. Biotechnology and Bioengineering 28, 965−971.

Rudolphi, A., Tschech, A., Fuchs, G., 1991. Anaerobic degradation of cresols by denitrifying bacteria. Archives of Microbiology 155, 238−248.

Salmeron-Alcocer, A., Ruiz-Ordaz, N., Juarez-Ramirez, C., et al., 2007. Continuous biodegradation of single and mixed chlorophenols by a mixed microbial culture constituted by *Burkholderia* sp., *Microbacterium phyllosphaerae & Candida tropicalis*. Biochemical Engineering Journal 1−11.

Santos, V.L., Linardi, V.R., 2004. Biodegradation of phenol by a filamentous fungi isolated from industrial effluents-identification and degradation potential. Process Biochemistry 39 (8), 1001−1006.

Santos, D.K.F., Rufino, R.D., Luna, J.M., et al., 2016. Biosurfactants: multifunctional biomolecules of the 21st century. International Journal of Molecular Sciences 17 (3), 401.

Saratale, R.G., Noh, H.S., Song, J.Y., et al., 2014. Influence of parameters on the photocatalytic degradation of phenolic contaminants in wastewater using TiO_2/UV system. Journal of Environmental Science and Health Part A 49 (13), 1542−1552. https://doi.org/10.1080/10934529.2014.938532.

Sarma, A.K., Singh, V.P., Bhattacharjya, R.K., et al., 2018. Urban Ecology, Water Quality and Climate Change. Water science and technology library Springer International Publishing AG, Switzerland.

Sayler, G.S., Ripp, S., 2000. Field applications of genetically engineered microorganisms for bioremediation processes. Current Opinion in Biotechnology 11 (3), 286−289.

Schennen, U., Braun, K., Knackmuss, H.-J., 1985. Anaerobic degradation of 2-£uorobenzoate by benzoate-degrading, denitrifying bacteria. Journal of Bacteriology 161, 321−325.

Schink, B., Brune, A., Schnell, S., 1992. Anaerobic degradation of aromatic compounds. In: Winkelmann, G. (Ed.), Microbial Degradation of Natural Compounds. VCH, Weinheim, pp. 219–242.

Schink, B., Philipp, B., Müller, J.A., 2000. Anaerobic degradation of phenolic compounds. Science and Nature 87 (1), 12–23.

Schneider, J., Matsuoka, M., Takeuchi, M., et al., 2014. Understanding TiO_2 photocatalysis: mechanisms materials. Chemical Reviews 114 (19), 9919–9986.

Seo, J.-S., Keum, Y.-S., Li, Q.X., 2009. Bacterial degradation of aromatic compounds. International Journal of Environmental Research and Public Health 6 (1), 278–309. http://doi.org/10.3390/ijerph6010278.

Shaari, N., Tan, S.H., Mohamed, A.R., 2012. Synthesis and characterization of CNT/Ce TiO_2 nanocomposite for phenol degradation. Journal of Rare Earths 30 (7), 651–658. URL: https://doi.org/10.1016/S1002-0721(12)60107-0.

Shaban, Y.A., El Sayed, M.A., El Maradny, A.A., et al., 2013. Photocatalytic degradation of phenol in natural seawater using visible light active carbon modified (CM)-n-TiO_2 nanoparticles under UV light and natural sunlight illuminations. Chemosphere 91, 307–313.

Shah, M.P., 2014. Environmental bioremediation: a low cost nature's natural biotechnology for environmental clean-up. Journal of Petroleum and Environmental Biotechnology 5, 191. https://doi.org/10.4172/2157-7463.1000191.

Shie, J.L., Lee, C.H., Chiou, C.S., et al., 2008. Photodegradation kinetics of formaldehyde using light sources of UVA, UVC and UVLED in the presence of composed silver titanium dioxide photocatalysis. Journal of Hazardous Materials 155, 164–172.

Shifu, C., Yunzhang, L., 2007. Study on the photocatalytic degradation of glyphosate by TiO_2 photocatalyst. Chemosphere 67, 1010–1017. URL: https://doi.org/10.1016/j.chemosphere.2006.10.054.

Shivaraju, H.P., 2011. Removal of organic pollutants in the municipal sewage water by TiO_2 based heterogeneous photocatalysis. International Journal of Environmental Sciences 1, 911–923.

Sihag, A., Pathak, H., Jaroli, D.P., 2014. Factors affecting the rate of biodegradation of polyaromatic hydrocarbons. International Journal of Pure and Applied Bioscience 2 (3), 185–202.

Silva, R.C.F.S., Almeida, D.G., Rufino, R.D., et al., 2014. Applications of biosurfactants in the petroleum industry and the remediation of oil spills. International Journal of Molecular Sciences 15 (7), 12523–12542.

Sina, M.A., Mohsen, M., 2017. Advances in Fenton and Fenton based oxidation processes for industrial effluent contaminants control-A review. International Journal of Environmental Science and Natural Resources 2 (4), 555594. https://doi.org/10.19080/IJESNR.2017.02.555594.

Singh, R.L., 2017. Principles and Applications of Environmental Biotechnology for a Sustainable Future. Springer, Singapore. URL: http://doi.org/10.1007/978-981-10-1866-4.

Singh, R.K., Kumar, S., Kumar, S., et al., 2008. Biodegradation kinetic studies for the removal of p-cresol from wastewater using *Gliomastix indicus* MTCC 3869. Biochemical Engineering Journal 40, 293–303.

Singh, T., Srivastava, N., Bhatiya, A.K., et al., 2017. Analytical study of effective biodegradation of p-cresol using *Serratia marcescens* ABHI001: application in bioremediation. 3 Biotech 7, 384, 1-8. https://doi.org/10.1007/s13205-017-1006-0.

Singh, T., Srivastava, N., Mishra, P.K., et al., 2016. Application of TiO_2 nanoparticle in photocatalytic degradation of organic pollutants. Materials Science Forum 855, 20–32.

Soda, S., Ike, M., Fujita, M., 1998. Effects of inoculation of genetically engineered bacterium on performance and indigenous bacteria of sequencing batch activated sludge process treating phenol. Journal of Fermentation and Bioengineering 86, 90–96.

Soni, N.P., Christian, R.A., Jariwala, N.D., 2018. Assessment of environment pollution potential index for re-categorization of industries in context of developing nations. International Journal of Environmental Science and Technology 5 (2), 58–63.

Sorokhaibam, L.G., Ahmaruzzaman, M., 2014. Industrial Wastewater Treatment. Recycling and Reuse Elsevier, Amsterdam, pp. 323–368.

Springer, N., Ludwig, W., Philipp, B., et al., 1998. *Azoarcus anaerobius* sp. nov., a resorcinol-degrading, strictly anaerobic, denitrifying bacterium. International Journal of Systematic Bacteriology 48, 953–956.

Sridevi, V., Chandana Lakshmi, M.V.V., Manasa, M., et al., 2012. Metabolic pathways for the biodegradation of phenol. International Journal of Engineering and Advanced Technology 2 (3), 695–705.

Steevensz, A., Madur, S., Feng, W., et al., 2014. Crude soybean hull peroxidase treatment of phenol in synthetic and real wastewater: enzyme economy enhanced by Triton X-100. Enzyme and Microbial Technology 55, 65–71.

Steiner, M.G., 2017. Photocatalytic decomposition of phenol under visible and UV light utilizing titanium dioxide based catalysts. Honors Theses and Capstones 350. URL: http://scholars.unh.edu/honors/350.

Strong, L.C., McTavish, H., Sadowsky, M.J., et al., 2000. Field-scale remediation of atrazine-contaminated soil using recombinant *Escherichia coli* expressing atrazine chlorohydrolase. Environmental Microbiology 2 (1), 91–98.

Su, T.L., Ku, Y., Hong, G.B., et al., 2011. Photodegradation of o-cresol using light emitting diodes with various wavelengths in the presence of photocatalysts. Environmental Engineering Science 28 (7), 535–542.

Sulaiman, F., Sari, D.K., Kustiningsih, I., 2017. The influence of ozone on the photocatalytic degradation of phenol using TiO_2 photocatalyst supported by Bayah natural zeolite. AIP Conference Proceedings 1840 (1), 110014. https://doi.org/10.1063/1.4982344.

Sulaymon, A.H., Abood, D.W., Ali, A.H., 2011. Removal of phenol and lead from synthetic wastewater by adsorption onto granular activated carbon in fixed bed adsorbers: predication of breakthrough curves. Hydrology: Current Research 2, 120. https://doi.org/10.4172/2157-7587.1000120.

Supriya, C., Neehar, D., 2014. Biodegradation of phenol by *Aspergillus niger*. IOSR Journal of Pharmacy 4 (7), 11–17.

Surkatti, R., El-Naas, M., 2014. Biological treatment of wastewater contaminated with p-cresol using *Pseudomonas putida* immobilized in polyvinyl alcohol (PVA) gel. Journal of Water Process Engineering 1, 84–90. https://doi.org/10.1016/j.jwpe.2014.03.008.

Suryantia, V., Marliyana, S.D., Wulandari, A., 2016. Biosurfactant production by *Pseudomonas fluorescens* growing on molasses and its application in phenol degradation. AIP Conference Proceedings 1699 (1). https://doi.org/10.1063/1.4938318.

Sutherland, J.B., Rafii, F., Khan, A.A., 1995. Mechanisms of polycyclic aromatic hydrocarbon degradation. In: Microbial Transformation and Degradation of Toxic Organic Chemicals. Wiley-Liss, New York, NY, pp. 269–306.

Szabóa, T., Veresa, A., Cho, E., et al., 2017. Photocatalyst Separation from aqueous dispersion using graphene oxide/TiO_2 nanocomposites. Colloids and Surfaces A Physicochemical and Engineering Aspects 433, 230–239.

Taghavi, S., Barac, T., Greenberg, B., et al., 2005. Horizontal gene transfer to endogenous endophytic bacteria from poplar improves phytoremediation of toluene. Applied and Environmental Microbiology 71 (12), 8500–8505.

Takriff, M.S., Ba-Abbad, M.M., Kadhum, A.A.H., et al., 2011. Solar photocatalytic degradation of 2,4-dichlorophenol by TiO_2 nanoparticle prepared by. Sol-Gel Method Advanced Materials Research 233–235, 3032–3035.

Tallur, P., Megadi, V., Kamanavalli, C., et al., 2006. Biodegradation of p-cresol by *Bacillus* sp. strain PHN 1. Current Microbiology 53, 529–533. https://doi.org/10.1007/s00284-006-0309-x.

Tallur, P.N., Megadi, V.B., Ninnekar, H.Z., 2009. Biodegradation of p-cresol by immobilized cells of *Bacillus* sp. Strain PHN 1 Biodegradation 20, 79–83. https://doi.org/10.1007/s10532-008-9201-7.

Tanaka, S., Saha, U.K., 1994. Effects of pH on photocatalysis of 2,4,6-trichlorophenol in aqueous TiO_2 suspensions. Water Science and Technology 9, 47–57.

Tao, Y., Cheng, Z.L., Ting, K.E., et al., 2013. Photocatalytic degradation of phenol using a nanocatalyst: the mechanism and kinetics. Journal of Catalysts 364275, 6. URL: https://doi.org/10.1155/2013/364275.

Taprab, Y., Johjima, T., Maeda, Y., et al., 2005. Symbiotic fungi produce laccases potentially involved in phenol degradation in fungus combs of fungus-growing termites in Thailand. Applied Environmental Microbiology 71 (12), 7696–7704.

Tepe, O., Dursun, A.Y., 2007. Combined Effects of External Mass Transfer and Biodegradation Rates on Removal of Phenol by Immobilized *Ralstonia eutropha* in a Packed Bed Reactor.

Thomas, B.N., George, S.C., 2015. Production of activated carbon from natural sources. Trends in Green Chemistry 1 (1), 7.

Tibbles, B.J., Baecker, A.A.W., 1989. Effect of pH and inoculum size on phenol degradation by bacteria isolated from landfill waste. Environmental Pollution 59 (3), 227–239.

Tolosa, N., Lu, M.C., Rollon, A., 2011. Factors affecting the photocatalytic oxidation of 2,4-dichlorophenol using modified titanium dioxide $TiO_2/KAl(SO_4)_2$ catalyst under visible light. Sustainable Environment Research 21 (6), 381–387.

Tosi, S., Kostadinova, N., Krumova, E., et al., 2010. Antioxidant enzyme activity of *Wlamentous fungi* isolated from Livingston Island, Maritime Antarctica. Polar Biology 33, 1227–1237. http://doi.org/10.1007/s00300-010-0812-1.

Tran, V.S., Ngo, H.H., Guo, W., et al., 2015. Typical low cost biosorbents for adsorptive removal of specific organic pollutants from water. Bioresource Technology 182, 353–363. https://doi.org/10.1016/j.biortech.2015.02.003. Epub 2015 Feb 8.

Tungler, A., Szabados, E., Hosseini, A.M., 2015. Wet air oxidation of aqueous wastes. In: Wastewater Treatment Engineering Mohamed Samer. IntechOpen. https://doi.org/10.5772/60935.

Tzoupanos, N.D., Zouboulis, A.I., 2008. Coagulation flocculation processes in water wastewater treatment: the application of new generation of chemical reagents. In: 6th IASME/WSEAS International Conference on Heat Transfer, Thermal Engineering and Environment Rhodes, Greece.

Udayasoorian, C., Prabu, P.C., 2005. Biodegradation of phenols by ligninolytic fungus *trametes versicolor*. Journal of Biological Sciences 5 (6), 824–827.

Umar, M., Aziz, H.A., 2013. Photocatalytic Degradation of Organic Pollutants in Water (chapter 8).

Uysal, A., Turkman, A., 2011. Biodegradation of 4-chlorophenol in biosurfactant supplemented activated sludge. Survival and Sustainability. http://doi.org/10.1007/978-3-540-95991-5_79.

Vaiano, V., Sacco, O., Sannino, D., et al., 2016. Photocatalytic removal of phenol by ferromagnetic n-TiO_2/SiO_2/Fe_3O_4 nanoparticles in presence of visible light irradiation. Chemical Engineering Transactions 47, 235–240. https://doi.org/10.3303/CET1647040.

Vanishree, M., Thatheyus, A.J., Ramya, D., 2014. Biodegradation of petrol using the fungus *Penicillium sp*. Science International 2 (1), 26–31.

Vega, A.A., Imoberdorf, G.E., Mohseni, M., 2011. Photocatalytic degradation of 2,4-dichlorophenoxyacetic acid in a fluidised bed photoreactor with composite template-free TiO2. Photocatalyst. Applied Catalysis A General 405, 120–128.

Venkatachalam, N., Palanichamy, M., Murugesan, V., 2007. Sol–gel preparation and characterization of alkaline earth metal doped nano TiO_2: efficient photocatalytic degradation of 4-chlorophenol. Molecular Catalysis 273, 177–185.

Vijaygopal, V., Viruthagiri, T., 2005. Batch kinetic studies in phenol biodegradation and comparison. Indian Journal of Biotechnology 4, 565–567.

Villegas, L.G.C., Mashhadi, N., Chen, M., et al., 2016. A short review of techniques for phenol removal from wastewater. Current Pollution Reports 2, 157–167. http://doi.org/10.1007/s40726-016-0035-3.

Wang, P., Bian, X.F., Li, Y.X., 2012. Catalytic oxidation of phenol in wastewater—a new application of the amorphous Fe78Si9B13 alloy. Chinese Science Bulletin 57 (1), 33–40.

Wang, C., Li, Y., 2007. Incorporation of granular activated carbon in an immobilized membrane bioreactor for the biodegradation of phenol by *Pseudomonas putida*. Biotechnology Letters 29, 1353–1356.

Wang, L., Li, Y., Yu, P., et al., 2010. Biodegradation of phenol at high concentration by a novel fungal strain *Paecilomyces variotii* JH6. Journal of Hazardous Materials 183 (1–3), 366–371. https://doi.org/10.1016/j.jhazmat.2010.07.033. Epub 2010 Jul 15.

Wang, W., Serp, P., Kalck, P., et al., 2005. Visible light photodegradation of phenol on MWNT-TiO_2 composite catalysts prepared by a modified sol–gel method. Journal of Molecular Catalysis A: Chemical 235, 194–199.

Wang, G., Xu, D., Guo, W., et al., 2017. Preparation of TiO_2 nanoparticle and photocatalytic properties on the degradation of phenol. IOP Conference Series: Earth and Environmental Science 59, 012046. https://doi.org/10.1088/1755-1315/59/1/012046.

WHO, 1996. Cresols Health and Safety Guide 100. Geneva Switzerland. http://apps.who.int/iris/bitstream/handle/10665/38142/9241511001_eng.pdf?sequence=1.

Wilson, S.C., Jones, K.C., 1993. Bioremediation of soil contaminated with polynuclear aromatic hydrocarbons (PAHs) a review. Environmental Pollution 81, 229–249.

Woan, K., Pyrgiotakis, G., Sigmund, W., 2009. Photocatalytic carbon-nanotube–TiO_2 composites. Advanced Materials 21, 2233–2239.

Xenofontos, E., Tanase, A.M., Stoica, I., et al., 2016. Newly isolated alkalophilic *Advenella* species bioaugmented in activated sludge for high p-cresol removal. New Biotechnology 33 (2).

Xu, D.Y., Yang, Z., 2013. Cross-linked tyrosinase aggregates for elimination of phenolic compounds from wastewater. Chemosphere 92 (4), 391–398.

Yao, H., Ren, Y., Wei, C., et al., 2011. Biodegradation characterisation and kinetics of m-cresol by *Lysinibacillus cresolivorans*. Water SA 37 (1), 15–20.

Yap, L.F., Lee, Y.K., Poh, C.L., 1999. Mechanism for phenol tolerance in phenol degrading *Comamonas testosteroni* strain. Applied Microbiology and Biotechnology 5 (6), 833–840.

Yemendzhiev, H., Gerginova, M., Zlateva, P., et al., 2008. Phenol and cresol mixture degradation by *Aspergillus awamori* strain: biochemical and kinetic substrate interactions. Proceedings of ECOpole 2 (1).

Yerkinova, A., Balbayeva, G., Inglezakis, V.J., et al., 2018. Photocatalytic treatment of a synthetic wastewater. IOP Conference Series: Materials Science and Engineering 301 (1). URL: https://doi.org/10.1088/1757-899X/301/1/012143.

Yu, Y.G., Loh, K.C., 2002. Inhibition of p-cresol on aerobic biodegradation of carbazole and sodium salicylate by *Pseudomonas putida*. Water Research 36, 1794–1802.

Yu, S., Yun, H.J., Kim, Y.H., et al., 2014. Carbon-doped TiO_2 nanoparticles wrapped with nanographene as a high performance photocatalyst for phenol degradation under visible light irradiation. Applied Catalysis B: Environmental 144, 893–899.

Yunus, N.N., Hamzah, F., Sóaib, M.S., et al., 2017. Effect of catalyst loading on photocatalytic degradation of phenol by using N, S Co-doped TiO_2. IOP Conference Series: Materials Science and Engineering 206, 012092.

Zainudin, N.F., Abdullah, A.Z., Mohamed, A.R., 2009. Characteristics of supported nano-TiO_2/ZSM-5/silica gel (SNTZS): photocatalytic degradation of phenol. Journal of Hazardous Materials 174 (1–3), 299–306.

Zhang, C., 1997. Removal of hydrophobic organic compounds and surfactants from wastewater: polyaphron-enhanced solvent extraction and biodegradation. In: LSU Historical Dissertations and Theses. UMI, USA, p. 159. https://digitalcommons.lsu.edu/gradschool_disstheses/6455.

Zhang, L., Liu, J., Tang, C., et al., 2011. Palygorskite and SnO_2–TiO_2 for the photodegradation of phenol. Applied Clay Science 51 (1–2), 68–73.

Zhao, G., Chen, S., Ren, Y., et al., 2014. Interaction and biodegradation evaluate of m-cresol and quinoline in co-exist system. International Biodeterioration and Biodegradation 86 (C), 252–257.

Zitrides, T.G., 1978. Mutant bacteria overcome growth inhibition in industrial waste facility. Industrial Waste 24, 42–44.

Zmudziński, W., 2010. Removal of o-cresol from water by adsorption/photocatalysis. Polish Journal of Environmental Studies 19 (6), 1353–1359.

Zouari, N., Ellouz, R., 1996. Microbial consortia for the aerobic degradation of aromatic compounds in olive oil mill effluent. Journal of Industrial Microbiology 16, 155–162. https://doi.org/10.1007/BF01569998.

Environmental fate of organic pollutants and effect on human health

Manita Thakur[1], *Deepak Pathania*[2]

[1]Department of Chemistry, Maharishi Markandeshwar University, Solan, India; [2]Department of Environmental Sciences, Central University of Jammu, District Samba, India

1. Introduction

The development of agrochemical and chemical industries causes environmental pollution because of the continuous release of toxic pollutants in water. The quality and quantity of water are very important for human survival. The wastewater contains a different type of contaminants such as heavy metals, dyes, phenols, persistent organic pollutants (POPs), etc. All these pollutants have adverse effects to human, animals, and plants (Kemp et al., 2006; El-Naggar et al., 2009, 2010; Khan and Alam, 2005). But among all these, POPs are of major concern because they persist for a long time in soils, sediments, and air. POPs threaten the environment and human health globally and have long durability in surroundings. POPs are pesticides, industrial compounds such as polychlorinated biphenyls (PCBs), polybrominated diphenyl ethers (PBDEs), perfluorooctanesulfonate (PFOS), etc., and by-products of different industrial routes such as dioxins and furans. POPs can transport from one place to another and bioaccumulate into food chain, causing several health hazards and environmental effects (Nabi et al., 2008; Bouju et al., 2008). These organic pollutants have been found even at poles on earth where they never been used.

1.1 Persistent organic pollutants

POPs are carbon-based chemicals with different physical and chemical properties. These are resistant to environmental degradation through chemical, biological, and photolytic processes because of their persistence for a long time in the environment before breaking down.

POPs bioaccumulate with vigorous effects on human health and surroundings (Yang et al., 2015; Katsoyiannis and Samara, 2004; Shrestha et al., 2009).

1.2 General characteristics of persistent organic pollutants

➤ POPs are synthetic chemicals released from industrial combustions either intentionally or nonintentionally such as pesticides, industrial products, or unintended by-products.
➤ Their persistence in the environment is significant and takes centuries to be degraded. They are lipophilic and have a tendency to endure in fat-rich tissues. POPs like to accumulate, persist, and bioconcentrate in an environment, which make them highly toxic (Xinying et al., 2012; Yang et al., 2013).
➤ Long-range transport leads to global pollution and can enter into a cycle in nature, accumulating in the bigger animals.
➤ Highest levels of POPs found in marine mammals increase microbial infections and reproductive disorders (Liu et al., 2009).

1.3 Sources of persistent organic pollutants

POPs are the major toxic pollutants released from both natural and man-made processes as shown in Fig. 12.1. POPs such as dibenzofurans and dioxins have been discharged into environment through natural sources such as volcanic activities or forest fires (Bergqvist et al., 2006; Pi et al., 2018; Li et al., 2013). Different manufacturing routes such as agricultural sprays, power stations, incinerating plants, evaporation from water surfaces, soil, and heating stations are intentional sources responsible for the production of POPs. Unintentional sources of POPs are wastes containing PCBs from obsolete or cooking oil, chemical facilities, sewage sludge, percolation of dumps, maintenance of tools, cement manufacture, incineration-municipal, medical surplus, incineration of fossil fuels, animal carcass incinerator, and fly ash storage.

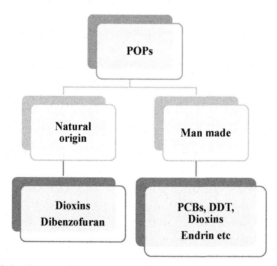

FIGURE 12.1   Sources of persistent organic pollutants on the basis of their origin.

Marine and freshwater ecosystems are the major reservoirs of POPs and deposited in the bottom of these sources through atmospheric deposition, runoff, and effluent releases. POPs have a little water solubility and bonded intensely to particulate matter in marine drugs. These pollutants have been used in circulation for long periods and can be introduced again into the network and nutrients (Katsoyiannis and Samara, 2005; Katsoyiannis et al., 2006; Li et al., 2011; Breivik et al., 2004; Jones and De Voogt, 1999). Hence, these pollutants are potentially becoming a source of global contamination.

Main anthropogenic POP pollution sources are as given below:

➢ By-products of industrial processes are resulting from incineration such as dioxins and furans.
➢ Chemicals used in industry such as polycholorobiphenyl, hexachlorobenzene (HCB), brominated compounds, etc.
➢ Chemicals used in agriculture, i.e., pesticides such as dichlorodiphenyltrichloroethane (DDT), aldrin, endrin, chlordane, toxaphene, etc.
➢ Water and wastewater treatment plants.
➢ Electronic waste recycling facilities and electric power plants.

POPs are highly toxic and their transport and bioaccumulation at high concentrations in the tissues of organisms is termed as bio-magnification. Because of the bioaccumulation and biomagnification, the concentration of POPs increases in the tissues of the organisms found at the high end of the trophic chains than in the environment (Henríquez-Hernández et al., 2016; Hafeez et al., 2016; Ali et al., 2016; Myrmet et al., 2016). POPs are nonpolar organic compounds and hydrophobic with extremely low water solubility.

Hydrophobic POPs shows high-fat solubility, which is accountable for their bioaccumulation and sustainability in the surroundings. These are lipophilic, which causes them to accumulate in the organism's fat tissue. The bioaccumulation and biomagnification processes of POPs determine chronic and acute toxic effects on organisms including humans. They have very low or high lipid solubilities so that they are bioaccumulated in body tissues, high noxiousness, and semivolatile, which helps them to travel long distances in the atmosphere before deposition (Pozo et al., 2016; Chiesa et al., 2016; Carlsson et al., 2016; Meng et al., 2016). Humans store POPs such as PCBs, HCB, etc., in their fat tissues. POPs such as pesticides cannot permanently be removed from the environment but continually circulate through soil, air, and water. Fig. 12.2 shows the cycling of POPs such as pesticides.

2. Types of persistent organic pollutants

POPs are classified on the basis of their production as shown in Fig. 12.3.

2.1 Pesticides

The pesticide is a chemical or biological agent used for destroying, preventing, or repelling the damage of any pest such as insects, plant pathogens, birds, mammals, weeds, roundworms, and microbes, etc. Most commonly, pesticides (herbicides, insecticides,

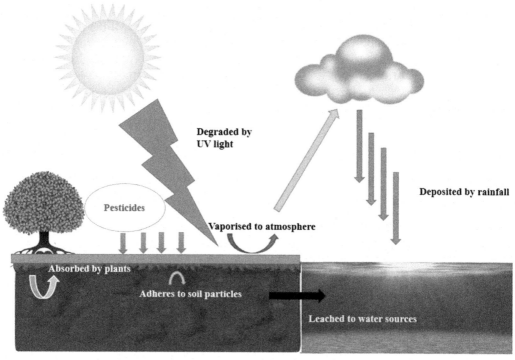

FIGURE 12.2   Pesticides cycling.

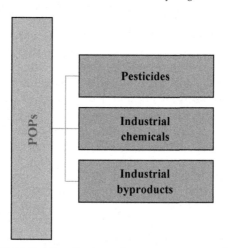

FIGURE 12.3   Classifications of persistent organic pollutants.

rodenticides, and fungicides) are used in the health sector and agricultural crops (Tette et al., 2016; Canccapa et al., 2016; Kiljanek et al., 2016). They are useful for killing vectors of disease such as mosquitoes and pests are killed by pesticides. Pesticides are potentially toxic to other surrounding nontarget organisms (Nuruzzaman et al., 2016; Masia et al., 2016;

Pang et al., 2016; Songa and Okonkwo, 2016; Zheng et al., 2016; Hladik et al., 2016). Therefore, it is necessary to use them carefully and dispose of appropriately. Some most commonly used pesticides are DDT, aldrin, chlordane, endrin, heptachlor, mirex, HCB, dieldrin, toxaphene, etc., as shown in Table 12.1.

TABLE 12.1 Different pesticides with their chemical formula and structure.

S. No.	Pesticides	Chemical formula	Structure
1	Dichlorodiphenyltrichloroethane	$C_{14}H_9Cl_5$	
2	Aldrin	$C_{12}H_8Cl_6$	
3	Chlordane	$C_{10}H_6Cl_8$	
4	Endrin	$C_{12}H_8Cl_6O$	
5	Heptachlor	$C_{10}H_5Cl_7$	
6	Mirex	$C_{10}Cl_{12}$	
7	Hexachlorobenzene	C_6Cl_6	
8	Dieldrin	$C_{12}H_8Cl_6O$	
9	Toxaphene	$C_{10}H_8Cl_8$	

2.1.1 Dichlorodiphenyltrichloroethane

DDT is a crystalline chemical compound, which is a colorless, tasteless, hydrophobic, odorless, and low water-soluble compound but shows good solubility in organic solvents, fats, and oils. It has been widely used in World War Second to protect soldiers and civilians from typhus and malaria spread by insects (Alvarez et al., 2003; Santacruz et al., 2005; Qu et al., 2015; Tian et al., 2009a,b; Pang et al., 2010). After this, DDT was continued to be used against mosquitoes to control malaria and sprayed on agricultural crops, mostly cotton. Owing to its high persistence, it can remain 50% in the soil up to 10–15 years after application. DDT residues have been found all over the place, even its residual has been found in the Arctic because of its wide applicability.

DDT is readily adsorbed to soils, sediments, and can act as sinks and long-term sources of exposure. It is hydrophobic in nature; therefore in aquatic ecosystems, DDT and its constituents are absorbed by aquatic organisms. However, a little adsorbed on suspended particles send off little DDT dissolved in the water (Kang et al., 2016; Tian et al., 2009a,b; Han et al., 2016; Neerja et al., 2016). Its breakdown metabolites such as dichlorodiphenyldichloroethylene (DDE) and dichlorodiphenyldichloroethane (DDD) are highly persistent and have similar chemical and physical properties. DDT and its breakdown products have been transported from warmer areas to the Arctic because of global distillation and accumulate in the food web.

DDT can bioaccumulate in predatory birds because of its lipophilic properties. DDT is very toxic to living organisms including marine animals such as sea shrimp, crayfish, daphnids, and many species of fish. DDT, DDE, and DDD are stored mainly in body fat and resistant to metabolism in humans. It also causes eggshell thinning in birds and a major reason for the decline of the peregrine falcon, bald eagle, brown pelican, and osprey (Gebresemati and Soha, 2016; Neitsch et al., 2016; Balawejder et al., 2016). DDT is an endocrine disruptor and carcinogen to humans. DDE acts as a weak androgen receptor antagonist but not as an estrogen.

DDT is highly carcinogenic, toxic, and hazardous but indirect exposure is relatively nontoxic for humans. Chronic exposure of DDT can affect reproductive capabilities, embryo or fetus, and breast cancer.

2.1.2 Aldrin

Aldrin is a colorless organochlorine insecticide spread to soils to kill termites, grasshoppers, corn rootworm, and other insect pests. Humans exposed to aldrin through dairy products and animal meats can also kill birds, fish, and humans. It is highly lipophilic and its solubility in water is only 0.027 mg/L, which intensifies its persistence in the environment (Leshniowsky et al., 1970).

2.1.3 Chlordane

Chlordane is a pesticide that has been obtained from the synthesis of heptachlor, chlordane, and nonachlor. It has low vapor pressure and volatilizes leisurely into the atmosphere. It is used broadly to control termites and insect repellent on a wide range of agricultural crops. It has remained in the soil for so many years with half-lives of 1 year (Moradas et al., 2008; Cuozzo et al., 2012). The lethal effects of chlordane on fish and birds vary and it can kill mallard ducks, bobwhite quail, and pink shrimp. It also affects the human immune system and carcinogenic. Human exposure of chlordane occurs through air and exposure to these compounds can cause cancers and many other diseases. These are resistant to degradation in the environment and accumulate in lipids of human and animals.

Chlordane is highly tenacious in the surroundings because it does not disintegrate easily. Because of its hydrophobic properties, it sticks to soil particles and moves in groundwater gradually. It is highly toxic, bioaccumulates in marine bodies and animals, and requires many years to degrade.

Chlordane and heptachlor are categorized as dirty dozen and proscribed by Stockholm convention on POPs in 2001 (Fuentes et al., 2016). It may cause diabetes, lymphoma, prostate cancer, obesity, testicular cancer, and breast cancer.

2.1.4 Heptachlor

Heptachlor is an organochlorine compound used as an insecticide. It has a highly stable structure so that it can persist in the environment for decades. It is carcinogenic, noncombustible, white waxy solid, and insoluble in water (Bandala et al., 2006). It is used to kill soil insects, termites, cotton insects, grasshoppers, and malaria-carrying mosquitoes. It is highly toxic to animals and humans and causes illness by inhalation, skin absorption, and ingestion. Drinking water and food are major sources through which heptachlor is exposed to humans. Long-term inhalation and oral exposure by humans may impart neurological effects including irritability, dizziness, salivation, and effects on the blood (Snyder and Mulder, 2001; Ratola et al., 2003), while acute inhalation exposure to heptachlor affects the nervous system and gastrointestinal problems.

2.1.5 Endrin

This insecticide is sprayed on the leaves of crops such as cotton and grains. It is also used to control rodents such as mice and voles. Animals can metabolize endrin, so it does not accumulate in their fatty tissue to the extent as other chemicals do. It has a long half-life, however, persisting in the soil for up to 12 years (Jinxian et al., 2005). Moreover, endrin is highly toxic to fish when exposed to high levels in the water. The primary mode of its exposure to humans is food, although current dietary intake estimates are below the limits supposed safe by world health authorities (Peng et al., 2009).

2.1.6 Mirex

Mirex is a white crystalline odorless solid insecticide commonly used against ants and termites.

On direct exposure, it does not cause injury to humans but studies showed its carcinogenic nature. It is also used in plastics, rubber, and electrical goods and most stable pesticide with a half-life of up to 10 years (Makarewicz et al., 2003). It can enter the human body through inhalation, ingestion, and skin but the main route of human contact is through food, meat, fish, and wild game. It is very toxic to several plant species, fish, and crustaceans and may induce persistent chronic physiological and biochemical disorders in various vertebrates (Yu et al., 2013). It is an endocrine disruptor and affects ovulation, pregnancy, endometrial growth, and may induce liver cancer in female rodents.

2.2 Industrial chemicals

Industrial chemicals play an important role in improving our health and standard of living. It also shows the adversarial effect on the environment, human health, and living organism. In a few decades, industrial chemicals are widely used in the development of industry and society. These are extensively used in all kind of industries and becoming more serious pollution problems because of universal industrial chemicals in the environment (Mekonnen et al., 2016). Some industrial chemicals are used in industrial production processes and some act as ingredients in the commercial products. Industrial chemicals are widely used as solvents, reactants, dyes, lubricants, coatings, colorants, inks, plasticizers, fragrances, mastics, stabilizers, flame retardants, conductors, and insulators (Veith et al., 1983; Moghadam et al., 2016). Large exposure of these chemicals can cause harmful effects to human health or environment. Some industrial chemicals are termed as POPs and cancer-causing agents. It imparts serious health effects including mild skin irritation, dizziness, and headaches, chronic effects on immune, reproduction, nervous, and endocrine systems (Grosse et al., 2016). Some industrial pollutants are very toxic including PCBs, HCB, hexachlorobutadiene, and hexabromocyclododecane, etc.

2.2.1 Polychlorinated biphenyls

PCBs are pale-yellow viscous liquids and used in industry as dielectric and coolant fluids in electric transformers and capacitors and additives in paint, carbonless copy paper, and plastics. PCBs are toxic to fish causing spawning failures at lower doses. It also causes reproductive failure and suppression of the immune system in wild animals such as seals and mink (Jepson et al., 2016; Smith et al., 2007). It causes health effects such as pigmentation of nails, mucous membranes, swelling of the eyelids, fatigue, nausea, and vomiting. Persistence of PCBs in mothers' bodies can cause developmental delays and behavioral changes in children.

These are hydrophobic, low vapor pressure, less water-soluble and highly soluble in organic solvents, fats, and oils. They have dielectric constants with very high thermal conductivity and high flash points. PCBs are resistant to acids, bases, oxidation, hydrolysis, and temperature change and are widely used in industries. They can produce enormously poisonous chemicals such as dibenzodioxins and dibenzofurans through partial oxidation. PCBs can penetrate the skin, polyvinyl chloride (PVC), and latex and show toxic effects (Liu et al., 2016; Kawano et al., 2014).

2.2.2 Hexachlorobenzene

HCB or perchlorobenzene is a fungicide used as a seed treatment to control the fungal disease bunt. It was first introduced in 1945 to treat seeds and it kills fungi that affect food crops. It is a by-product of certain industrial chemicals and exists as an impurity in many pesticide formulations. In 1954 and 1959, it developed a variety of symptoms such as photosensitive skin lesions, colic, debilitation, porphyria turcica, etc. HCB can also be passed to infants through their mother's breast milk (Chen et al., 2016; Oonnittan et al., 2010).

It is a white crystalline solid, less soluble in water but sparingly soluble in organic solvents. It is most soluble in halogenated solvents such as chloroform, esters, short chain alcohols, and hydrocarbons. It is toxic and carcinogenic to humans, aquatic organisms, and animals. Chronic exposure in humans can cause liver disease, skin lesions with discoloration, ulceration, photosensitivity, thyroid effects, bone effects, loss of hair, embryo lethality, and teratogenic effects (Jiang et al., 2018).

2.2.3 Hexachlorobutadiene

Hexachlorobutadiene is a colorless, odorless chlorinated aliphatic diene most commonly used as a solvent for other chlorine-containing compounds. It is produced in chlorinolysis plants as a by-product in the production of carbon tetrachloride and tetrachloroethene. It acts as an algicide in industrial cooling systems and a potent herbicide. But in recent years, its application has been reduced because of its high toxicity at low concentrations. It produces universal toxicity through oral, inhalation, and dermal routes (Rodrigues et al., 2017a,b). It can cause various diseases such as fatty liver degeneration, epithelial necrotizing nephritis, central nervous system depression, and cyanosis.

2.2.4 Short-chain chlorinated paraffins

Chlorinated paraffins are nonflammable complex mixtures of certain organic compounds containing chloride, i.e., polychlorinated n-alkanes. Short-chain chlorinated paraffins (SCCPs) dissolve in organic solvents but insoluble in water. These are persistent, bioaccumulative, carcinogenic, and highly toxic to aquatic organisms at low concentrations (Parera et al., 2004). Primary sources of SCCPs are metalworking fluids and PVC processing. SCCPs have been used as lubricants, coolants, chlorinated paraffin's, chlorinated paraffin plasticizers, flame retardants, in paints, adhesives and sealants, leather fat liquors, plastics, rubber, textiles, drawing, tube bending, and cold heading and polymeric materials (Štejnarová et al., 2005; Zeng et al., 2013).

SCCPs have acute toxicity at low concentration and may cause skin and eye irritation but do not induce skin sensitization. SCCPs ingestion, both directly and through contaminated food, skin contact, in paints, adhesives, and sealants may affect liver, kidneys, adipose tissue, and breast milk.

2.3 Industrial by-products

Disposal of industrial by-products (IBPs) is a serious concern for many industries because of the increasing volume of waste by-product generated and their potentially hazardous effects on the environment. IBPs are the residual material from industrial, commercial, mining, or agricultural processes that are not produced separately in the process. IBPs contain paper mill sludge, ash from energy recovery, ferrous and steel foundry sand and slag, coal ash and slag, lime kiln dust, and harmless solid waste (Reddy and Yang, 2005; Agrawal et al., 2011). These can also be regulated from food, beverage, and agricultural operations through industrial water quality including vegetable, dairy, meat processing wastes, solids from the pretreatment of wastewater, ethanol production wastes, and co-products, etc.

IBPs from food, beverage, and agricultural operations usually contain nitrogen, potassium, phosphorus, and other nutrients which are good candidates for land application. To reuse these nutrients in the soil, using IBP reduces the use of water and commercial fertilizer and fills up the soil. IBPs become harmful if applied too much and excess nutrients run off into lakes and streams and affect groundwater (Lopes et al., 2013; Baqueiro-Pena and

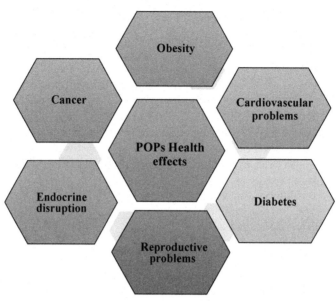

FIGURE 12.4   Health effects due to persistent organic pollutants.

Guerrero-Beltran, 2017). People applying IBP to land must know proper application rates and monitoring soil and runoff to protect water from contamination.

2.4 Health and environmental effects of persistent organic pollutants

A large number of industrial and agricultural activities continuously discharge harmful chemicals such as POPs in the environment through soil, air, and water. The existence of POPs in the surroundings is a serious matter. These are continuously used up by human beings, animals, and marine bodies. These toxic POPs impart serious health problems such as endocrine disruption, reproductive problems, cancer, diabetes, obesity, cardiovascular problems, and many other health problems as shown in Fig. 12.4.

2.4.1 Endocrine disorder

The endocrine system is an assortment of glands such as pituitary, pancreas, adrenals, and testes, which secrete different hormones to regulate metabolism, maturation, growth and development, reproduction, tissue function, sexual function, sleep, and mood. The disruption in the endocrine system because of interfering chemicals results in several health problems such as adversarial effects on the neurological, reproductive, and immune system. POPs are endocrine disruptors disturbing the hormones functioning and fetus development. It has been observed that the main endocrine disturbing pesticides are dieldrin, toxaphene, chlordane, DDT, mirex, etc. These may behave as estrogens and reorient the sex organs (Geyer et al., 2000; Gregoraszczuk and Ptak, 2013; Bergman et al., 2013).

Some toxic neurological effects, neurobehavioral deficits, low motor reflex, neuromuscular functions, and depressed responsiveness were found in kids. The insufficiencies such as depressed responsiveness, weight gain, and reduced visual recognition were observed at 4 years age (Hung et al., 2016a,b). Various diseases such as thyroid, urinary iodine, metabolic behavioral, psychoneuromotor disorders, and changes of the reproductive systems were observed because of PCBs and some pesticides.

2.4.2 Reproductive problems

Reproductive systems have been affected by POPs contamination and impart notorious effects. The reproductive problems include birth defects, preterm birth, developmental disorders, low birth weight, impotence, reduced fertility, and menstrual disorders. POP exposure affected sperm quality and quantity, altered sex ratio, early puberty in females, pregnancy, and endometriosis (Hung et al., 2016a,b; Odabasi et al., 2016; Jepson and Law, 2016). It has been found that POP concentrations were higher in mother serum as compared to umbilical cord and placentas. Poor growth of fetus has been observed and newborn babies have low length, weight, head circumference, and chest circumference.

2.4.3 Cancer

In spite of advanced medical technologies, cancer is not curable in late and chronic stages because of lack of appropriate cure and its erratic nature. It is a serious disease in which cells in human body grow suddenly in an uncontrolled way, producing abnormal growths and spread of abnormal cells in the body. Normal cells work differently in cancer because of voluminous internal and external changes. Cancer is the second most dangerous disease accountable for about 21% of deaths per year all over the world (Khairy et al., 2016; Li et al., 2016; Pegoraro et al., 2016).

Higher concentrations of POPs in low-density lipoproteins are accountable for numerous cancers. Cancers causing POPs include organochlorine pesticides, PBDEs, polychlorinated dibenzo-p-dioxins, dibenzofurans, PCBs, etc. High-level POPs cause breast cancer and also affect polymorphisms in genes during xenobiotic metabolism and estrogen biosynthesis. It has been observed that there is high risk of breast cancer in women having PFOS and perfluorooctanoic acid (Anttila et al., 2016; Tsai et al., 2016).

2.4.4 Diabetes

Diabetes mellitus is a metabolic disease which causes high blood sugar. It causes kidney failure, stroke, heart disease, foot ulcers, erectile dysfunction, poor muscle strength, urinary tract infections, yeast infections, eyes damage, and dry and itchy skin. The chlorinated pesticides and PCBs affect the development of metabolic syndrome, i.e., insulin resistance. POPs such as polychlorinated dibenzo-p-dioxins and polychlorinated dibenzofurans show diabetic problems in children (Muñoz-Arnanz et al., 2016; Lammel et al., 2016).

2.4.5 Obesity

Accumulation of extra fat into human body causes adversarial health effects. Obesity causes a number of diseases such as cancer, osteoarthritis, cardiovascular, and apnea problems. Obesity is a serious health problem of the 21st century and increasing in adults and children continuously. Bioaccumulation of POPs at high concentration increases the chances of obesity and dyslipidemia (Ugranli et al., 2016; Fujii et al., 2007).

2.4.6 Cardiovascular diseases

Cardiovascular diseases include hypertension, angina pectoris, cardiac arrhythmias, etc.

It has been reported that one billion populations is suffering from hypertension globally and the fraction of individuals suffering from hypertension diverges frequently. Hypertension is higher in low- and middle-income countries as compared with high-income countries. POP-contaminated food is responsible for developing health problems such as diabetes, obesity, and others. POPs are lipophilic, bioaccumulated into high concentration lipoproteins, and cause a number of cardiovascular problems and diseases. Dioxins and PCBs at high concentration cause high blood pressure, metabolic syndrome, elevated triglycerides, and glucose intolerance (Ren et al., 2018; Mwakalapa et al., 2018).

2.5 Environmental effects

POPs continuously affected our surroundings through biotic, social, cultural, and technological intrusions. POP pollutions destruct natural balance, menacing the survival and health of all living organisms. Direct or indirect exposure of POPs infected all living organisms such as insects, birds, and animals. The environmental effects cause wildlife cancer, a shift in sex ratios, impaired fertility, and other physical defects. Accretion of POPs in aquatic bodies leading to their death causes disparity in sea ecology. In the 19th and 20th centuries, the industrial revolution increases the consumption and disposal of POPs greatly. POPs are volatile, reach to our surroundings with dust particles, and enter in the human body through breathing (Gavrilescu et al., 2015; Weber et al., 2011; Appenzeller and Tsatsakis, 2012).

Dust and air are primary sources of POPs and enter in the human body through inhalation and ingestion. POPs also affect environmental factors including temperature, precipitation, salinity, food webs, lipid dynamics, ice melt, and organic carbon cycling. POPs also show grasshopper effect involving cyclic volatilizations and condensations. These can travel from hot weather to cool but sometimes deposited in extremely cool environments and evaporate when the temperature increases. This cyclic movement of POPs causes global warming problem. The presence of DDE, PCBs, HCB, and oxychlordane leads to ecological disruption because of hormonal disorder in animals and seabirds (Persson et al., 2005). These also affect cortisol, thyroid, and sex steroid hormones, resulting in developmental and morphological changes.

2.6 Remediation of persistent organic pollutants

It should be essential to investigate appropriate methods for the exclusion of POPs. There are different ways for the remediation of POPs (Ren et al., 2018; Mwakalapa et al., 2018; Agrawal et al., 2011).

➢ POPs production and use should be controlled. There should be preventions on the manufacturing and applications of POPs. Some countries such as the United States and US Congress banned the production and use of PCBs.
➢ Development of bioremediation methods of POPs includes genetically amended microorganisms. The microorganisms are rehabilitated so as to make them proficient for the removal of POPs.

➢ The enzymes employed for the exclusion of POPs are adapted at enzymatic expression levels, activities, and specificities. These techniques are cost-effective and viable and limit the applicability of bioremediation technologies.

➢ Development of amputation techniques includes electrochemical, membrane, adsorption, photocatalysis, etc. These methods can remove POPs from the atmosphere using specific effective adsorbents. More advanced techniques should be established in electrochemical and membrane technologies to remove POPs from water resources.

➢ The popular methods for the exclusion of POPs are sodium reduction, solvent extraction, bioremediation, verification, solidification, stabilization, gas-phase chemical reduction, incineration, pyrolysis, thermal desorption, soil washing chemical dehalogenation, mechanochemical dehalogenation, supercritical extraction, and ball milling.

3. Conclusion

POPs are extremely dangerous and impart numerous health and environmental problems. Bioaccumulation and long persistence of POPs made them life-threatening hazardous materials. They cause various diseases such as diabetes, obesity, endocrine disturbance, cancer, cardiovascular, reproductive, etc. POPs will change our environment slowly in upcoming years and constant climatic change processes become life-threatening era. POPs are our tacit foes and have to combat for these considerately. Lack of knowledge about these chemicals is one of the major drawbacks. Maximum agronomists are not conscious of poisonous properties of POPs. Hence, it is necessary to develop fast, effective, and economic remediation approaches for the exclusion of POPs from the environment. We should also organize seminars, conferences at village, and managements should severe interdict the manufacturing and consumption of POPs worldwide. There is no standardized concord model for predicting risk profiles of POPs in future. Therefore, academicians and researchers should investigate diverse simulations to determine different characteristics of POPs. Nowadays, the world is under great threat of pollution because of POPs; therefore, we need to fight against these mutually, scientifically, and precisely.

References

Agrawal, S.G., King, K.W., Fischer, E.N., Woner, D.N., 2011. PO_4^{3-} removal by and permeability of industrial byproducts and minerals: granulated blast furnace slag, cement kiln dust, coconut shell activated carbon, silica sand, and zeolite. Water, Air, and Soil Pollution 219 (1–4), 91–101.

Ali, U., Li, J., Zhang, G., Mahmood, A., Jones, K.C., Malik, R.N., 2016. Presence, deposition flux and mass burden of persistent organic pollutants (POPs) from Mehmood Booti Drain sediments, Lahore. Ecotoxicology and Environmental Safety 125, 9–15.

Alvarez, M., Calle, A., Tamayo, J., Lechuga, L.M., Abad, A., Montoya, A., 2003. Development of nanomechanical biosensors for detection of the pesticide DDT. Biosensors and Bioelectronics 18 (5–6), 649–653.

Anttila, P., Brorström-Lundén, E., Hansson, K., Hakola, H., Vestenius, M., 2016. Assessment of the spatial and temporal distribution of persistent organic pollutants (POPs) in the Nordic atmosphere. Atmospheric Environment 140, 22–33.

Appenzeller, B.M., Tsatsakis, A.M., 2012. Hair analysis for biomonitoring of environmental and occupational exposure to organic pollutants: state of the art, critical review and future needs. Toxicology Letters 210 (2), 119–140.

Balawejder, M., Józefczyk, R., Antos, P., Pieniażek, M., 2016. Pilot-scale installation for remediation of DDT-contaminated soil. Ozone Science and Engineering 38 (4), 272–278.

Bandala, E.R., Andres-Octaviano, J., Pastrana, P., Torres, L.G., 2006. Removal of aldrin, dieldrin, heptachlor, and heptachlor epoxide using activated carbon and/or Pseudomonas fluorescens free cell cultures. Journal of Environmental Science and Health 41 (5), 553–569.

Baqueiro-Peña, I., Guerrero-Beltrán, J.Á., 2017. Vanilla (*Vanilla planifolia* Andr.), its residues and other industrial by-products for recovering high value flavor molecules: a review. Journal of Applied Research on Medicinal and Aromatic Plants 6, 1–9.

Bergman, Å., Heindel, J.J., Kasten, T., Kidd, K.A., Jobling, S., Neira, M., Brandt, I., 2013. The impact of endocrine disruption: a consensus statement on the state of the science. Environmental Health Perspectives 121 (4), a104.

Bergqvist, P.A., Augulytė, L., Jurjonienė, V., 2006. PAH and PCB removal efficiencies in Umeå (Sweden) and Šiauliai (Lithuania) municipal wastewater treatment plants. Water, Air, and Soil Pollution 175 (1–4), 291–303.

Bouju, H., Buttiglieri, G., Malpei, F., 2008. Perspectives of persistent organic pollutants (POPS) removal in an MBR pilot plant. Desalination 224 (1–3), 1–6.

Breivik, K., Alcock, R., Li, Y.F., Bailey, R.E., Fiedler, H., Pacyna, J.M., 2004. Primary sources of selected POPs: regional and global scale emission inventories. Environmental Pollution 128 (1–2), 3–16.

Carlsson, P., Crosse, J.D., Halsall, C., Evenset, A., Heimstad, E.S., Harju, M., 2016. Perfluoroalkylated substances (PFASs) and legacy persistent organic pollutants (POPs) in halibut and shrimp from coastal areas in the far north of Norway: small survey of important dietary foodstuffs for coastal communities. Marine Pollution Bulletin 105 (1), 81–87.

Canccapa, A., Masiá, A., Navarro-Ortega, A., Picó, Y., Barceló, D., 2016. Pesticides in the Ebro River basin: occurrence and risk assessment. Environmental Pollution 211, 414–424.

Chen, J., Zhong, D., Hou, H., Li, C., Yang, J., Zhou, H., et al., 2016. Ferrite as an effective catalyst for HCB removal in soil: characterization and catalytic performance. Chemical Engineering Journal 294, 246–253.

Chiesa, L.M., Labella, G.F., Panseri, S., Pavlovic, R., Bonacci, S., Arioli, F., 2016. Distribution of persistent organic pollutants (POPs) in wild Bluefin tuna (*Thunnus thynnus*) from different FAO capture zones. Chemosphere 153, 162–169.

Cuozzo, S.A., Fuentes, M.S., Bourguignon, N., Benimeli, C.S., Amoroso, M.J., 2012. Chlordane biodegradation under aerobic conditions by indigenous Streptomyces strains. International Biodeterioration and Biodegradation 66 (1), 19–24.

El-Naggar, I.M., Ibrahim, G.M., El-Kady, E.A., Hegazy, E.A., 2009. Sorption mechanism of Cs^+, Co^{2+} and Eu^{3+} ions onto EGIB sorbent. Desalination 237 (1–3), 147–154.

El-Naggar, I.M., Mowafy, E.A., Abdel-Galil, E.A., El-Shahat, M.F., 2010. Synthesis, characterization and ion-exchange properties of a novel 'organic–inorganic'hybrid cation-exchanger: polyacrylamide Sn (IV) molybdophosphate. Global Journal of Physical Chemistry 1 (1), 91–106.

Fuentes, M.S., Colin, V.L., Amoroso, M.J., Benimeli, C.S., 2016. Selection of an actinobacteria mixed culture for chlordane remediation. Pesticide effects on microbial morphology and bioemulsifier production. Journal of Basic Microbiology 56 (2), 127–137.

Fujii, S., Polprasert, C., Tanaka, S., Lien, H., Pham, N., Qiu, Y., 2007. New POPs in the water environment: distribution, bioaccumulation and treatment of perfluorinated compounds—a review paper. Journal of Water Supply: Research and Technolog 56 (5), 313–326.

Gavrilescu, M., Demnerová, K., Aamand, J., Agathos, S., Fava, F., 2015. Emerging pollutants in the environment: present and future challenges in biomonitoring, ecological risks and bioremediation. New Biotechnology 32 (1), 147–156.

Gebresemati, M., Sahu, O., 2016. Sorption of DDT from synthetic aqueous solution by eucalyptus barkusing response surface methodology. Surface and Interface Analysis 1, 35–43.

Geyer, H.J., Rimkus, G.G., Scheunert, I., Kaune, A., Schramm, K.W., Kettrup, A., et al., 2000. Bioaccumulation and occurrence of endocrine-disrupting chemicals (EDCs), persistent organic pollutants (POPs) and other organic compounds in fish and other organisms including humans. In: Bioaccumulation—New Aspects and Developments. Springer, Berlin, Heidelberg, pp. 1–166.

Gregoraszczuk, E.L., Ptak, A., 2013. Endocrine-disrupting chemicals: some actions of POPs on female reproduction. International Journal of Endocrinology 2013.

Grosse, Y., Loomis, D., Guyton, K.Z., El Ghissassi, F., Bouvard, V., Benbrahim-Tallaa, L., et al., 2016. Carcinogenicity of some industrial chemicals. The Lancet Oncology 17 (4), 419–420.

Hafeez, S., Mahmood, A., Syed, J.H., Li, J., Ali, U., Malik, R.N., et al., 2016. Waste dumping sites as a potential source of POPs and associated health risks in perspective of current waste management practices in Lahore city, Pakistan. The Science of the Total Environment 562, 953–961.

Han, Y., Shi, N., Wang, H., Pan, X., Fang, H., Yu, Y., 2016. Nanoscale zerovalent iron-mediated degradation of DDT in soil. Environmental Science & Pollution Research 23 (7), 6253–6263.

Henríquez-Hernández, L.A., Luzardo, O.P., Arellano, J.L.P., Carranza, C., Sánchez, N.J., Almeida-González, M., et al., 2016. Different pattern of contamination by legacy POPs in two populations from the same geographical area but with completely different lifestyles: Canary Islands (Spain) vs. Morocco. The Science of the Total Environment 541, 51–57.

Hladik, M.L., Vandever, M., Smalling, K.L., 2016. Exposure of native bees foraging in an agricultural landscape to current-use pesticides. The Science of the Total Environment 542, 469–477.

Hung, H., Katsoyiannis, A.A., Brorström-Lundén, E., Olafsdottir, K., Aas, W., Breivik, K., Skov, H., 2016a. Temporal trends of persistent organic pollutants (POPs) in arctic air: 20 years of monitoring under the Arctic monitoring and assessment programme (AMAP). Environmental Pollution 217, 52–61.

Hung, H., Katsoyiannis, A.A., Guardans, R., 2016b. Ten years of global monitoring under the Stockholm convention on persistent organic pollutants (POPs): trends, sources and transport modelling. Environmental Pollution 1–3.

Jepson, P.D., Deaville, R., Barber, J.L., Aguilar, À., Borrell, A., Murphy, S., et al., 2016. PCB pollution continues to impact populations of orcas and other dolphins in European waters. Scientific Reports 6, 18573.

Jepson, P.D., Law, R.J., 2016. Persistent pollutants, persistent threats. Science 352 (6292), 1388–1389.

Jiang, Y., Shang, Y., Yu, S., Liu, J., 2018. Dechlorination of hexachlorobenzene in contaminated soils using a nanometallic Al/CaO dispersion mixture: optimization through response surface methodology. International Journal of Environmental Research and Public Health 15 (5).

Jinxian, H., Huijuan, L., Jiuhui, Q., Jia, R., Haining, L., Guoting, L., 2005. Dieldrin and endrin removal from water by triolein-embedded adsorbent. Science Bulletin 50 (23), 2696–2700.

Jones, K.C., De Voogt, P., 1999. Persistent organic pollutants (POPs): state of the science. Environmental Pollution 100 (1–3), 209–221.

Kang, S., Liu, S., Wang, H., Cai, W., 2016. Enhanced degradation performances of plate-like micro/nanostructured zero valent iron to DDT. Journal of Hazardous Materials 307, 145–153.

Katsoyiannis, A., Samara, C., 2004. Persistent organic pollutants (POPs) in the sewage treatment plant of Thessaloniki, northern Greece: occurrence and removal. Water Research 38 (11), 2685–2698.

Katsoyiannis, A., Samara, C., 2005. Persistent organic pollutants (POPs) in the conventional activated sludge treatment process: fate and mass balance. Environmental Research 97 (3), 245–257.

Katsoyiannis, A., Zouboulis, A., Samara, C., 2006. Persistent organic pollutants (POPs) in the conventional activated sludge treatment process: model predictions against experimental values. Chemosphere 65 (9), 1634–1641.

Kawano, S., Kida, T., Miyawaki, K., Noguchi, Y., Kato, E., Nakano, T., et al., 2014. Cyclodextrin polymers as highly effective adsorbents for removal and recovery of polychlorobiphenyl (PCB) contaminants in insulating oil. Environmental Science and Technology 48 (14), 8094–8100.

Kemp, R.V., Bennett, D.G., White, M.J., 2006. Recent trends and developments in dialogue on radioactive waste management: experience from the UK. Environment International 32 (8), 1021–1032.

Khairy, M.A., Luek, J.L., Dickhut, R., Lohmann, R., 2016. Levels, sources and chemical fate of persistent organic pollutants in the atmosphere and snow along the western Antarctic Peninsula. Environmental Pollution 216, 304–313.

Khan, A.A., Alam, M.M., 2005. Preparation, characterization and analytical applications of a new and novel electrically conducting fibrous type polymeric–inorganic composite material: polypyrrole Th (IV) phosphate used as a cation-exchanger and Pb (II) ion-selective membrane electrode. Materials Research Bulletin 40 (2), 289–305.

Kiljanek, T., Niewiadowska, A., Semeniuk, S., Gaweł, M., Borzecka, M., Posyniak, A., 2016. Multi-residue method for the determination of pesticides and pesticide metabolites in honeybees by liquid and gas chromatography coupled with tandem mass spectrometry—Honeybee poisoning incidents. Journal of Chromatography A 1435, 100–114.

Lammel, G., Meixner, F.X., Vrana, B., Efstathiou, C.I., Kohoutek, J., Kukučka, P., et al., 2016. Bidirectional air–sea exchange and accumulation of POPs (PAHs, PCBs, OCPs and PBDEs) in the nocturnal marine boundary layer. Atmospheric Chemistry and Physics 16 (10), 6381–6393.

Leshniowsky, W.O., Dugan, P.R., Pfister, R.M., Frea, J.I., Randles, C.I., 1970. Aldrin: removal from lake water by flocculent bacteria. Science 169 (3949), 993–995.

Li, A., Tanabe, S., Jiang, G., Giesy, J.P., Lam, P.S., 2011. Persistent Organic Pollutants in Asia: Sources, Distributions, Transport and Fate, vol. 7. Elsevier.

Li, J., Chen, C., Li, F., 2016. Status of POPs accumulation in the Yellow River Delta: from distribution to risk assessment. Marine Pollution Bulletin 107 (1), 370–378.

Li, K., Xiong, J., Chen, T., Yan, L., Dai, Y., Song, D., et al., 2013. Preparation of graphene/TiO$_2$ composites by nonionic surfactant strategy and their simulated sunlight and visible light photocatalytic activity towards representative aqueous POPs degradation. Journal of Hazardous Materials 250, 19–28.

Liu, H., Ru, J., Qu, J., Dai, R., Wang, Z., Hu, C., 2009. Removal of persistent organic pollutants from micro-polluted drinking water by triolein embedded absorbent. Bioresource Technology 100 (12), 2995–3002.

Liu, Y., Wang, X., Wang, J., Nie, Y., Du, H., Dai, H., et al., 2016. Graphene oxide attenuates the cytotoxicity and mutagenicity of PCB 52 via activation of genuine autophagy. Environmental Science and Technology 50 (6), 3154–3164.

Lopes, F.C., Tichota, D.M., Pereira, J.Q., Segalin, J., de Oliveira Rios, A., Brandelli, A., 2013. Pigment production by filamentous fungi on agro-industrial byproducts: an eco-friendly alternative. Applied Biochemistry and Biotechnology 171 (3), 616–625.

Makarewicz, J.C., Damaske, E., Lewis, T.W., Merner, M., 2003. Trend analysis reveals a recent reduction in mirex concentrations in coho (*Oncorhynchus kisutch*) and chinook (*O. tshawytscha*) salmon from Lake Ontario. Environmental Science and Technology 37 (8), 1521–1527.

Masia, A., Suarez-Varela, M.M., Llopis-Gonzalez, A., Pico, Y., 2016. Determination of pesticides and veterinary drug residues in food by liquid chromatography-mass spectrometry: a review. Analytica Chimica Acta 936, 40–61.

Mekonnen, T., Mussone, P., Bressler, D., 2016. Valorization of rendering industry wastes and co-products for industrial chemicals, materials and energy. Critical Reviews in Biotechnology 36 (1), 120–131.

Meng, G., Feng, Y., Nie, Z., Wu, X., Wei, H., Wu, S., et al., 2016. Internal exposure levels of typical POPs and their associations with childhood asthma in Shanghai, China. Environmental Research 146, 125–135.

Moghadam, P.Z., Fairen-Jimenez, D., Snurr, R.Q., 2016. Efficient identification of hydrophobic MOFs: application in the capture of toxic industrial chemicals. Journal of Materials Chemistry 4 (2), 529–536.

Moradas, G., Auresenia, J., Gallardo, S., Guieysse, B., 2008. Biodegradability and toxicity assessment of trans- photochemical treatment. Chem 73 (9), 1512–1517.

Muñoz-Arnanz, J., Roscales, J.L., Ros, M., Vicente, A., Jiménez, B., 2016. Towards the implementation of the Stockholm Convention in Spain: five-year monitoring (2008–2013) of POPs in air based on passive sampling. Environmental Pollution 217, 107–113.

Mwakalapa, E.B., Mmochi, A.J., Müller, M.H.B., Mdegela, R.H., Lyche, J.L., Polder, A., 2018. Occurrence and levels of persistent organic pollutants (POPs) in farmed and wild marine fish from Tanzania. A pilot study. Chemosphere 191, 438–449.

Myrmel, L.S., Fjære, E., Midtbø, L.K., Bernhard, A., Petersen, R.K., Sonne, S.B., et al., 2016. Macronutrient composition determines accumulation of persistent organic pollutants from dietary exposure in adipose tissue of mice. The Journal of Nutritional Biochemistry 27, 307–316.

Nabi, S.A., Ganai, S.A., Khan, A.M., 2008. Effect of surfactants and temperature on adsorption behavior of metal ions on organic–inorganic hybrid exchanger, acrylamide aluminum tungstate. Journal of Surfactants and Detergents 11 (3), 207–213.

Neerja, Grewal, J., Bhattacharya, A., Kumar, S., Singh, D.K., Khare, S.K., 2016. Biodegradation of 1, 1, 1-trichloro-2, 2-bis (4-chlorophenyl) ethane (DDT) by using *Serratia marcescens* NCIM 2919. Journal of Environmental Science and Health, Part B 51 (12), 809–816.

Neitsch, J., Schwack, W., Weller, P., 2016. How do modern pesticide treatments influence the mobility of old incurred DDT contaminations in agricultural soils. Journal of Agricultural and Food Chemistry 64 (40), 7445–7451.

Nuruzzaman, M., Rahman, M.M., Liu, Y., Naidu, R., 2016. Nanoencapsulation, nano-guard for pesticides: a new window for safe application. Journal of Agricultural and Food Chemistry 64 (7), 1447–1483.

Odabasi, M., Dumanoglu, Y., Falay, E.O., Tuna, G., Altiok, H., Kara, M., et al., 2016. Investigation of spatial distributions and sources of persistent organic pollutants (POPs) in a heavily polluted industrial region using tree components. Chemosphere 160, 114–125.

Oonnittan, A., Isosaari, P., Sillanpää, M., 2010. Oxidant availability in soil and its effect on HCB removal during electrokinetic Fenton process. Separation and Purification Technology 76 (2), 146–150.

Pang, S., Yang, T., He, L., 2016. Review of surface enhanced Raman spectroscopic (SERS) detection of synthetic chemical pesticides. Trends in Analytical Chemistry 85, 73–82.

Pang, W., Gao, N., Xia, S., 2010. Removal of DDT in drinking water using nanofiltration process. Desalination 250 (2), 553–556.

Parera, J., Santos, F.J., Galceran, M.T., 2004. Microwave-assisted extraction versus Soxhlet extraction for the analysis of short-chain chlorinated alkanes in sediments. Journal of Chromatography A 1046 (1–2), 19–26.

Pegoraro, C.N., Harner, T., Su, K., Chiappero, M.S., 2016. Assessing levels of POPs in air over the South Atlantic ocean off the coast of South America. The Science of the Total Environment 571, 172–177.

Peng, X., Jia, J., Luan, Z., 2009. Oxidized carbon nanotubes for simultaneous removal of endrin and Cd (II) from water and their separation from water. Journal of Chemical Technology and Biotechnology 84 (2), 275–278.

Persson, N.J., Gustafsson, Ö., Bucheli, T.D., Ishaq, R., Næs, K., Broman, D., 2005. Distribution of PCNs, PCBs, and other POPs together with soot and other organic matter in the marine environment of the Grenlandsfjords, Norway. Chemosphere 60 (2), 274–283.

Pi, Y., Li, X., Xia, Q., Wu, J., Li, Y., Xiao, J., et al., 2018. Adsorptive and photocatalytic removal of Persistent Organic Pollutants (POPs) in water by metal-organic frameworks (MOFs). Chemical Engineering Journal 337, 351–371.

Pozo, K., Palmeri, M., Palmeri, V., Estellano, V.H., Mulder, M.D., Efstathiou, C.I., et al., 2016. Assessing persistent organic pollutants (POPs) in the Sicily Island atmosphere, Mediterranean, using PUF disk passive air samplers. Environmental Science & Pollution Research 23 (20), 20796–20804.

Qu, J., Xu, Y., Ai, G.M., Liu, Y., Liu, Z.P., 2015. Novel *Chryseobacterium* sp. PYR2 degrades various organochlorine pesticides (OCPs) and achieves enhancing removal and complete degradation of DDT in highly contaminated soil. Journal of Environmental Economics and Management 161, 350–357.

Ratola, N., Botelho, C., Alves, A., 2003. The use of pine bark as a natural adsorbent for persistent organic pollutants—study of lindane and heptachlor adsorption. Journal of Chemical Technology and Biotechnology 78 (2-3), 347–351.

Reddy, N., Yang, Y., 2005. Biofibers from agricultural byproducts for industrial applications. Trends in Biotechnology 23 (1), 22–27.

Ren, X., Zeng, G., Tang, L., Wang, J., Wan, J., Liu, Y., et al., 2018. Sorption, transport and biodegradation—an insight into bioavailability of persistent organic pollutants in soil. The Science of the Total Environment 610, 1154–1163.

Rodrigues, R., Betelu, S., Colombano, S., Masselot, G., Tzedakis, T., Ignatiadis, I., 2017a. Reductive dechlorination of hexachlorobutadiene by a Pd/Fe microparticle suspension in dissolved lactic acid polymers: degradation mechanism and kinetics. Industrial and Engineering Chemistry Research 56 (42), 12092–12100.

Rodrigues, R., Betelu, S., Colombano, S., Masselot, G., Tzedakis, T., Ignatiadis, I., 2017b. Influence of temperature and surfactants on the solubilization of hexachlorobutadiene and hexachloroethane. Journal of Chemical & Engineering Data 62 (10), 3252–3260.

Santacruz, G., Bandala, E.R., Torres, L.G., 2005. Chlorinated pesticides (2, 4-D and DDT) biodegradation at high concentrations using immobilized Pseudomonas fluorescens. Journal of Environmental Science and Health, Part B 40 (4), 571–583.

Shrestha, R.A., Pham, T.D., Sillanpää, M., 2009. Effect of ultrasound on removal of persistent organic pollutants (POPs) from different types of soils. Journal of Hazardous Materials 170 (2–3), 871–875.

Smith, K.E., Schwab, A.P., Banks, M.K., 2007. Phytoremediation of polychlorinated biphenyl (PCB)-contaminated sediment. Journal of Environmental Quality 36 (1), 239–244.

Snyder, M.J., Mulder, E.P., 2001. Environmental endocrine disruption in decapod crustacean larvae: hormone titers, cytochrome P450 and stress protein responses to heptachlor exposure. Aquatic Toxicology 55 (3–4), 177–190.

Songa, E.A., Okonkwo, J.O., 2016. Recent approaches to improving selectivity and sensitivity of enzyme-based biosensors for organophosphorus pesticides: a review. Talanta 155, 289–304.

Štejnarová, P., Coelhan, M., Kostrhounová, R., Parlar, H., Holoubek, I., 2005. Analysis of short chain chlorinated paraffins in sediment samples from the Czech Republic by short-column GC/ECNI-MS. Chemosphere 58 (3), 253–262.

Tette, P.A.S., Guidi, L.R., de Abreu Glória, M.B., Fernandes, C., 2016. Pesticides in honey: a review on chromato-graphic analytical methods. Talanta 149, 124—141.

Tian, H., Li, J., Shen, Q., Wang, H., Hao, Z., Zou, L., Hu, Q., 2009a. Using shell-tunable mesoporous Fe3O4@ HMS and magnetic separation to remove DDT from aqueous media. Journal of Hazardous Materials 171 (1—3), 459—464.

Tian, H., Li, J., Zou, L., Mu, Z., Hao, Z., 2009b. Removal of DDT from aqueous solutions using mesoporous silica materials. Journal of Chemical Technology and Biotechnology 84 (4), 490—496.

Tsai, J.H., Chen, S.J., Huang, K.L., Chang-Chien, G.P., Lin, W.Y., Feng, C.W., et al., 2016. Characteristics of persistent organic pollutant emissions from a diesel-engine generator fueled using blends of waste cooking oil-based biodiesel and fossil diesel. Aerosol and Air Quality Research 16, 2048—2058.

Ugranli, T., Gungormus, E., Kavcar, P., Demircioglu, E., Odabasi, M., Sofuoglu, S.C., et al., 2016. POPs in a major conurbation in Turkey: ambient air concentrations, seasonal variation, inhalation and dermal exposure, and associated carcinogenic risks. Environmental Science and Pollution Research 23 (22), 22500—22512.

Veith, G.D., Call, D.J., Brooke, L.T., 1983. Structure—toxicity relationships for the fathead minnow, *Pimephales promelas*: narcotic industrial chemicals. Canadian Journal of Fisheries and Aquatic Sciences 40 (6), 743—748.

Weber, R., Watson, A., Forter, M., Oliaei, F., 2011. Persistent organic pollutants and landfills-a review of past experiences and future challenges. Waste Management and Research 29 (1), 107—121.

Xinying, Z., Fasheng, L.I., Duanping, X.U., 2012. Removal of POPs pesticides from soil by thermal desorption and its effect on physicochemical properties of the soil. Chinese Journal of Envionmental Engineering 4, 061.

Yang, Y., Huang, J., Wang, S., Deng, S., Wang, B., Yu, G., 2013. Catalytic removal of gaseous unintentional POPs on manganese oxide octahedral molecular sieves. Applied Catalysis B 142, 568—578.

Yang, Y., Zhang, S., Wang, S., Zhang, K., Wang, H., Huang, J., et al., 2015. Ball milling synthesized MnO x as highly active catalyst for gaseous POPs removal: significance of mechanochemically induced oxygen vacancies. Environmental Science and Technology 49 (7), 4473—4480.

Yu, Y., Huang, J., Zhang, W., Zhang, K., Deng, S., Yu, G., 2013. Mechanochemical destruction of mirex co-ground with iron and quartz in a planetary ball mill. Chemosphere 90 (5), 1729—1735.

Zeng, L., Li, H., Wang, T., Gao, Y., Xiao, K., Du, Y., et al., 2013. Behavior, fate, and mass loading of short chain chlorinated paraffins in an advanced municipal sewage treatment plant. Environmental Science and Technology 47 (2), 732—740.

Zheng, S., Chen, B., Qiu, X., Chen, M., Ma, Z., Yu, X., 2016. Distribution and risk assessment of 82 pesticides in Jiu long River and estuary in South China. Chemosphere 144, 1177—1192.

Further reading

Pan, X., Lin, D., Zheng, Y., Zhang, Q., Yin, Y., Cai, L., et al., 2016. Biodegradation of DDT by Stenotrophomonas sp. DDT-1: characterization and genome functional analysis. Scientific Reports 6, 21332.

Qin, W., Fang, G., Wang, Y., Wu, T., Zhu, C., Zhou, D., 2016. Efficient transformation of DDT by peroxymonosulfate activated with cobalt in aqueous systems: kinetics, products, and reactive species identification. Chemosphere 148, 68—76.

Rhizospheric remediation of organic pollutants from the soil; a green and sustainable technology for soil clean up

Akanksha Gupta, Amit Kumar Patel, Deepak Gupta, Gurudatta Singh, Virendra Kumar Mishra

Institute of Environment & Sustainable Development, Banaras Hindu University, Varanasi, India

1. Introduction

The present era of development is associated with increased rate of urbanization and ever-increasing population, which we have to be fed and needs all the essential stuffs. These processes lead to emergence of a new problem, soil pollution, which is because of pollutants released into environment by our actions. Any kind of undesirable elements, whether it is in the form of energy (e.g., radiation and sound) or other materials (e.g., heavy metals, particulate matter), can be classified as a pollutant. After getting released into the environment at one place, pollutants spread into the environment and become global from the state of local (Megharaj et al., 2011). Among different categories of pollutants, organic pollutants are pulling attention because of their water repellant with high toxic nature and high persistence. Because of these specific properties, these compounds can affect our environment for longer period of time, which also increase the risk of them being entered into food chain and causing more harm (Havelcová et al., 2014).

2. Organic contaminants in soil and their sources

Organic contaminants have enormously large list, which include hydrocarbons, different solvents, PAHs (polycyclic aromatic hydrocarbons), petroleum products, pesticides, herbicides, PCBs (polychlorinated biphenyls), phthalate esters, phenols, and their derivatives (Oleszczuk, 2006; Moore et al., 2006; MacKinnon and Duncan, 2013). These pollutants may affect the properties of soil, with the presence of very little amount by altering the microbial population and their metabolic activities as well. Unluckily, the classical technologies for elimination of these contaminants from the soil system are very costly, which further discourage industries to take care of their effluent discharge (Alkorta and Garbisu, 2001).

To remove organic pollutants from the contaminated soil, bioremediation not only serves as a very good option as it is eco-friendly and cost-effective but also has some boundaries (Glazer and Nikaido, 1995). To deal with the limitations of bioremediation, a branch of it is called rhizoremediation and can be opted in its place. Rhizoremediation is a process in which plants are used to remediate pollutants from soil with the help of microbes present in their rhizospheric region (Oberai and Khanna, 2018). Here plants serve homely environment to microbes by providing them carbon source in the form of root exudates, and the microorganisms do the degradation work and the residues can be taken by the plants with the water transport (Cannon and Bowles, 1962). Soil acts as bed not only for seeds and microbes but also for the environmental pollutants (Fig. 13.1). There are many organic pollutants of the soil which are categorized in three classes by UNEP in 2001; these are industrial chemicals,

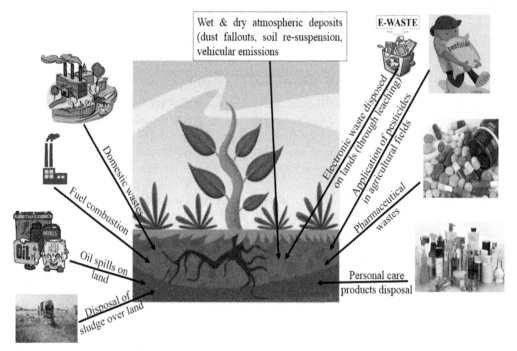

FIGURE 13.1   Major sources of organic pollutants in soil.

pesticides, and by-products. They have placed PCBs in the category of industrial chemicals, polychlorinated dibenzo-p-dioxins, and polychlorinated dibenzofurans in by-product class. The compounds which are being placed in the by-products group are completely anthropogenic in nature and are produced because of incomplete combustion of fuel.

Among the various soil organic pollutants, PAHs and PCBs are the two that are gaining more attention because of their partially volatile nature. They are water repelling and chemically quite stable compounds with very low rate of degradation (Fabietti et al., 2010; Gupta et al., 2018; Ma et al., 2009; Zhang et al., 2007). PAHs can be introduced into the environment when someone ignites organic materials, which results in partially burned residue; the fire which gives birth to the PAHs can be caused naturally or by human activities, whereas polychlorinated biphenyls are purely anthropogenic in their origin (Cachada et al., 2012).

To meet the demands of growing population, humans started using chemical manure and pesticides. Initially the application of synthetic fertilizers and pesticides were useful, and because of this fact, the application of these substances had enhanced. However, in later years, the harmful consequences on soil began to appear (Merrington et al., 2002). Currently, 2 million ton yr^{-1} of chemical pesticides are being used worldwide; in which the United States and Europe have 69% of shares in uses and the rest 31% consumption is by the remaining countries of the world. According to Atapattu and Kodituwakku (2009), South Asia is the largest region with 14% land in agricultural area, with active utilization of chemically synthesized pesticides. Around 37%−79% of agricultural production is lost because of overgrowth of weeds; to stop this enormous loss, farmers apply herbicides and weedicides (Behera and Singh, 1999). Not only weeds but also insects are a problem for agricultural sector as well as for human health; to decrease their harmful effects, most of the people rely on chemical insecticides (Ross, 2005). According to many studies, these synthesized chemicals are being less utilized in developed countries but are still continuously consumed by the developing countries like India (Eqani et al., 2011; Syed and Malik, 2011; Tariq et al., 2007).

The application of pesticides and insecticides is quite prevalent in Indian scenario where these compounds are regularly used to eliminate the harmful pests and insects. To accomplish this aim, India has started the production of BHC and DDT in 1952; since then the production of these chemicals has been enhanced uninterruptedly. According to a study by Gupta (2004) only in 1958, India alone has produced more than 5000 metric tons of pesticides, which mainly included BHC and DDT. Around 145 pesticides were registered during the middle of 90s, which produced nearly 85,000 metric tons of products. Nowadays, people are more aware of what they are using and consuming which leads to decrease in use of chemically synthesized pesticides; farmers are now switching to its green options like plant-derived products and other option (i.e., biopesticides). Currently, India is producing huge amount of pesticide products that stand this country on the first position in Asia and 12th across the globe. According to Khan (2010), India is manufacturing 90,000 tons of pesticides per year, which is a huge amount. India not only produces pesticides but also exports these to other nations (Pozo et al., 2011).

There is a category of pollutants which include chemicals that have the nature to stick to the environment after being used and have the tendency of bioaccumulation and scattered around for very long distance, so they are called persistent organic pollutants abbreviated as POPs (Buccini, 2003). Some researchers have reported POPs at different places of world where these had not even used once, which is justified by their scattering property with the rain and wind (Barra et al., 2005; Zhang et al., 2008). Vos et al. (2000) stated that some of the compounds in POP category are responsible of altering the function of hormones by disturbing their molecular structure; this kind of studies make people more aware of these chemicals and their effects on them (Table 13.1). In 2001, UNEP recognized 17 chemicals from the POP category and classified them into pesticides, industrial chemicals, and by-products. In the fourth meeting of UNEP in 2009, some more chemicals were added in each category; in the next meeting in 2011, they have added endosulfan and its related isomer as a new POP. Because of their harmful effects on environment and humans, these chemicals have been banned very long ago but are still being used from some part of the world, i.e., South Asian nations. Many studies have illustrated that India and Pakistan are regularly using DDT (Alamdar et al., 2014; Chakraborty et al., 2010; Eqani et al., 2011; Syed and Malik, 2011; Syed et al., 2013, 2014; Zhang et al., 2008). POPs are able of wide spread scattering; because of this, they may be able to contaminate untouched areas like South Pole and North Pole.

TABLE 13.1 Organic pollutants, their sources, and health risk to humans.

Pollutants	Example	Properties	Source of emission	Health risk	References
PAH	Hexabromodiphenyl ether	Water-soluble solid	Natural, anthropogenic (traffic, industries, domestic heating)	Carcinogenic, mutagenic	Barraclough (2007)
	Pentabromodiphenyl ether	Water-insoluble, white crystalline solid			
PCB	2,2′,4,4′,5,5′-Hexachlorobiphenyl	Chemically inert	Anthropogenic (transformers, capacitors, hydraulic fluids, cutting oil)	Carcinogenic	Barraclough (2007)
Organochlorine pesticides	Chlordane	Water insoluble, brown-amber colored, very viscous liquid, chlorine-like odor, highly poisonous	Agricultural runoff, pesticide storage, domestic uses	Carcinogenic, immunological disorders, reproductive disorders	Sanpera et al. (2003)
	Aldrin	Water insoluble, brown-white crystalline solid, mild chemical odor, used as an insecticide			
	Dieldrin	Water-insoluble, light-tan flaked solid, mild chemical odor, used as an insecticide			
	Endrin	Water-emulsifiable, white crystalline solid, odorless, used as pesticide			
	Hexachlorobenzene	Water-insoluble, white crystalline solid, used as an agricultural fungicide			

TABLE 13.1 Organic pollutants, their sources, and health risk to humans.—cont'd

Pollutants	Example	Properties	Source of emission	Health risk	References
	Mirex	White crystalline solid, odorless, used as an insecticide			
	Heptachlor	Water-insoluble, white to light-tan waxy looking solid, camphor-like odor, used as an insecticide			
	Toxaphene	Yellow waxy solid, pleasant-piney odor, used as an insecticide			
	Lindane	Very poorly soluble in water, white-yellow power or flakes, musty odor, used as an insecticide			
	Pentachlorobenzene	White crystalline solid, characteristic odor			
	Chlordecone	Colorless, odorless, crystalline solid, used as an insecticide			
	Endosulfan	Water insoluble, cream to brown color crystalline or flaky solid, turpentine-like odor, used as an insecticide			

According to Minh et al. (2006), South Asian countries lack the facilities to process the municipal wastes; to get rid of waste material, they use open dumping sites that have been poorly managed. Syed and Malik (2011) studied some regions of Pakistan which have shut down factories and store houses contained pesticides such as DDT; this study reveals that runoff from the surface of these places can turn out as a source of pollution for soil and waterbodies. Similar kind of studies by other scientists educated people to think about the possible health threats of inhabitants of affected areas (Agusa et al., 2003).

3. Fate of organic pollutants in soil

Once an organic pollutant enters into the ecosystem, its movement within the ecosystem will depend on its chemical nature; broadly, if it is biodegradable, then it will be mineralized in its components with time and if it is nonbiodegradable then it will persist in the environment and may impose a series of events on the surrounding environment. The concentration of pollutants which is deposited during a long course of time can be known by measuring the concentration of the same in sediments, which is a well-accepted compartment for accumulation of these pollutants (Bhattacharya et al., 2003; Guzzella et al., 2005). Sediments are made up of several components (biological, organic, and inorganic) which are originated from different sources and contain various features. Bioaccumulative and POPs get incorporated into the sediments through different mechanisms, i.e., vaporization, surface runoff, and leaching (Sarkar et al., 2008), and one can easily study POPs because of their low degradation property by studying sediments (El Nemr et al., 2013).

In Indian scenario, out of different organic chemicals in soil, pesticides are the most common one which is used in agricultural fields to protect the crops from pests, while they have been used in municipal areas to avoid diseases by pest control. Their elevated levels can be seen at several spots which are situated near the discharge points of domestic and industrial garbage (Pozo et al., 2011) (Table 13.2).

Ali et al. (2014) has studied the sediments from river Gomti in India, where he found hexachlorocyclohexane (HCH) and DDTs in high concentrations (BDL-81.2 $\mu g\, g^{-1}$); he also studied east/west coast of India where he has monitored the concentrations of HCHs and DDTs (BDL 109 $\mu g\, g^{-1}$). Similarly, Pandey et al. (2011) keep an eye on the changes in concentrations of POPs according to various seasons by studying the sediments of river Ganga and branches of other rivers which are connected to this. They have recorded concentrations of OCPs during monsoon (196−578 ng/g), before monsoon (158−308 ng/g), and after monsoon (307−844 ng/g) seasons and emphasized that weathered and used soils from agriculture lands were the major reasons of DDT pollution in the sediments. Malik et al. (2009) have advised that illegitimate use of DDT can be the other reason for its high contamination in riverine areas. Elevated levels of DDTs, HCHs, and endosulfan sulfate were recorded before monsoon season by Bhattacharya et al. (2003). The concentration of the above were found in order of HCHs > N-endosulfan sulfate > DDTs> α-endosulfan. In all forms of DDTs, the concentrations of p-p′-DDT and p′-p′-DDE were found higher than other forms and constitute around 70%−90% of the all DDTs in the sediments. This shows that DDT is being degraded via dehydrochlorination reaction, which may be produced by abiotic or biotic decay. High levels of pesticides in the command area are the result of input from agricultural fields, household releases, and industrial drop off in that area.

Khwaja et al. in 2006, have studied some samples of soil, spare materials, plaster samples of the factory walls, and damaged bags from a closed DDT factory in Pakistan; among these samples, discarded bags possess highest concentration of DDT (2822−2841 $\mu g/g$). By then the levels of DDT were not exceeding the 10 km radius of the banned manufacturing unit. One year later, Khwaja et al. showed that almost 91% of the collected samples were highly polluted with DDTs, in which half of the samples possess more than allowed limit of DDT residues in soil (0.05 $\mu g/g$). Similar kind of study was carried out by Jan et al. (2009) in same area, where they revealed that 500 m of area around the factory was still polluted with the residues of DDT. They have found that the concentration of DDT metabolites such as p′,p′-DDT were reduced with respect to depth and distance from the sealed manufacturing unit. Yadav et al. (1981) have published their findings on contamination of DDT levels in soil and earthworms in Delhi, the capital of India, which showed that the area around a DDT-synthesizing industry had an ascendant pattern of DDT concentration.

TABLE 13.2 Concentration of pollutants in soil at different places.

Pollutant	Place where contamination is found	Concentration of the pollutant (ng/g)	References
DDT	Amman garh, Pakistan in 1974	340	Yadav et al. (1981)
	Amman garh, Pakistan in 1978	1430	
	Amman garh, Pakistan in 1983	1670	
	Bay of Bengal, India	20,000–790,000	Sarkar and Sen gupta (1988)
	Bay of Bengal, India	0.04–4.79	Rajendran et al. (2005)
	Kalashah kaku, Pakistan	441–1812	Syed and Malik (2011)
	National park of sagarmatha, Nepal	0.19 ± 0.27	Guzzella et al. (2011)
	Paddy fields from Dibrugarh	873 ± 504	Mishra et al. (2012)
	Tea gardens from Dibrugarh	732 ± 446	
	Paddy fields from Nagaon	1005 ± 579	
	Tea gardens from Nagaon	872 ± 440	
HCBs	National park of sagarmatha, Nepal	0.08 ± 0.54	Guzzella et al. (2011)
Aldrin	Bay of Bengal, India	20,000–530,000	Sarkar and Sen gupta (1988)
Total HCH	Paddy fields from Dibrugarh	861 ± 416	Mishra et al. (2012)
	Tea gardens from Dibrugarh	701 ± 355	
	Paddy fields from Nagaon	1056 ± 392	
	Tea gardens from Nagaon	756 ± 424	
γ-HCH	Bay of Bengal, India	10,000–210,000	Sarkar and Sen gupta (1988)
Dieldrin	Bay of Bengal, India	50,000–510,000	
OCPs	Eastern indo-Gangetic alluvial plains	0.36–104	Singh et al. (2007)
PAHs	Borholla oil field of upper Brahmaputra Valley (pre-monsoon)	2840 ± 1770	Deka et al. (2016)
	Borholla oil field of upper Brahmaputra Valley (post monsoon)	3920 ± 2810	
	Guwahati (monsoon)	6877.9 ± 8016.3	Hussain and Hoque (2015)
	Guwahati (Pre-monsoon)	10,904.3 ± 11,807.2	
	Guwahati (Post-monsoon)	21,290.2 ± 20,390.6	
	Dhanbad, India	3488	Suman et al. (2016)

4. Rhizoremediation: a conventional approach

Rhizoremediation is a method where soil contaminants are degraded by the rhizospheric microorganisms. It is also known as rhizodegradation, microbe-assisted phytoremediation technology, and rhizosphere bioremediation. The rhizosphere is the very active region around roots (1−2 mm in length) affected by plant activity (Brink, 2016; Dzantor, 2007; Liu et al., 2014). The rhizospheric microbes in most of the cases are the main contributors, while plant provides a positive niche and carbon source to rhizospheric microorganism for degradation of pollutants. In turn microorganisms provide plant nutrients which enhance the plant growth, protect the plants against the pathogens, degrade the contaminants (Macek et al., 2000; Qixing et al., 2011), and decreases the level of stress hormones (Dams et al., 2007).

The initial studies on rhizospheric microbial degradation were mainly focused toward the protection of plants against the herbicides and the pesticides (Hoagland et al., 1994; Jacobsen, 1997; Zablotowicz et al., 1994). Now, at present several cases are reported on the degradation of harmful organic compounds such as PAHs (Radwan et al., 1995), trichloroethylene (Walton and Anderson, 1990), and PCBs (Brazil et al., 1995). However, the compositions of microbial population were remained to be analyzed in detailed in these studies. In addition to this, there is still lack of information about the proliferation, survival, and the activity of microbial population in the rhizosphere. Various studies showed that leguminous plants such as alfalfa and different grass varieties were the most appropriate plant species for rhizoremediation; this is because of the fact that these plants have the ability to maintain large number of bacteria on their well-branched root systems (Kuiper et al., 2001; Qiu et al., 1994; Shann and Boyle, 1994). Ecological interactions with other organisms, primary and secondary metabolism, survival, and establishment are the factors that influence the success of rhizoremediation.

Microbial growth in rhizosphere is significantly supported by root exudation, which is considered as the most important potential driving force for rhizoremediation (Martin et al., 2014; Phillips et al., 2012). The compounds such as sugars, amino acids, and organic acids are the major components of plant root exudates (Vancura and Hovadik, 1965). In addition to these compounds; last root cap cells, mucilage secretion via root cells, the decay of complete roots, or the starvation of root cells also provides nutrients (Lynch and Whipps, 1990; Lugtenberg and de Weger, 1992; Lynch and Whipps 1990; Rovira, 1956). On the basis of mode of production, the root exudates can be divided into four different groups, and it includes secondary plant metabolites, lysates (from senescent tissues and roots), passive exudates, and mucilages (from root tip and epidermal cell). Low molecular weight (LMW) carbohydrates, secondary metabolites, amino acids, and organic anions are water-soluble fractions of root exudates, whereas organic polymers are considered as insoluble fractions (Martin et al., 2014; Walker et al., 2003a,b). These compounds are used by soil microbes as a carbon and energy source at the time of their metabolic process (Chaudhry et al., 2005). There are two different ways by which the root exudates are supposed to trigger a shift in microbial community of the contaminated soils: first by alteration in expression of catabolic gene of microbes and second by choosing the specific microbial strains (Baudoin et al., 2003; Benizri et al., 2002; Butler et al., 2003; Hartmann et al., 2009; Siciliano et al., 2003). Plant roots provide the sites for attachment of microbes and soil aeration for biodegradation of pollutants such as TPH (Martin et al., 2014).

Usually, it is well known that root exudates play a unique role in determining rhizospheric microbiome in soil contaminated with nitro aromatic compounds and petroleum (Siciliano et al., 2003; Thijs et al., 2016). The root exudates from the slender oat collected by Miya and Firestone (2001) found that it maintains the maximum populations of PHC-degrading microbes and enhances the biodegradation of phenanthrene. Further study revealed that the addition of synthetic root exudates and mineral nutrients mixture enhanced the biodegradation speed of three to five ring PAHs (Joner et al., 2002). Yoshitomi and Shann (2001) confirmed that sterile exudates from corn (*Zea mays* L.) roots caused the increased microbial mineralization of pyrene. On the other hand, for these studies, there were no clear researches done to recognize the liability of individual compounds such as carboxylates, carbohydrates, secondary metabolites, and amino acids in root exudates. However, in some cases, it has also been found that root exudates cause suppressive effect on PHC degradation. As discovered by Phillips et al. (2012), in case of alfalfa (*Medicago sativa* L.) and wild rye (*Elymus angustus* Trin.), root exudates decreased the mineralization speed of n-hexadecane, naphthalene, and phenanthrene (Table 13.3).

Different types of root exudates compounds secreted by individual plants are very diverse, which depend on growth stages of plants, nutrient accessibility, light, soil pH, and temperature of environment (Gransee and Wittenmayer, 2000; Hodge et al., 1997; Hütsch et al., 2002; Leigh et al., 2002; Neumann, 2007; Rovira, 1959). Furthermore, other biological parameters such as mycorrhizal fungi also affect the exudation; in some plants, it has been observed that these fungi decrease the speed of root exudation (Marschner et al., 1997; Ryan et al., 2012). Concentration of exudates varies at different part of the root, highest exudates released at the site of root tips, and lateral branching (Marschner et al., 2011; McDougall and Rovira, 1970; Neumann, 2007). In ryegrass (*Lolium perenne* L.) nearby to root surface (within 3 mm), maximum amount of PHC degrader microbes are found, therefore the maximum degradation of PHC occurred at this site (Corgié et al., 2004).

4.1 Mechanism of Rhizoremediation of organic contaminants

Bacterial degradations of the organic contaminants in rhizospheric zone take place by both aerobic and anaerobic processes in the natural soils (Boopathy, 2004). Root exudates increase the degradation of organic contaminants such as TPH and PHC by four different possible ways, which include direct degradation in rhizosphere by the enzymes derived from the plant, increasing the bioavailability of contaminants, cometabolic process, and stimulation by energy/nutrient flow (Martin et al., 2014). A brief summarized sketch of the rhizoremediation process is given in Fig. 13.2. As per the researches on the rhizoremediation of organic contaminants, the following important mechanisms are supposed be involved in this process.

4.1.1 Direct degradation

Enzymes of the microbial origin are the primary causes of organic contaminants degradation pathway; however, plant exudates also contain various extracellular enzymes such as peroxidases and laccases which increase the direct degradation of various polyaromatic hydrocarbons (Martin et al., 2014). Moreover, mycorrhizal fungi may also take part in the treatment of PHC molecules through releasing the extracellular enzymes (Gramss et al., 1999; Harms et al., 2011).

TABLE 13.3   Interaction of plants with microbes in rhizoremediation of pollutants.

Organic pollutants	Microbes	Plants	References
3-Methylbenzoate	*Pseudomonas putida*	*Zea mays*	Ronchel and Ramos (2001)
2,4-D	*Burkholderia cepacia*	Barley	Jacobsen (1997)
Polychlorinated biphenyl	*Pseudomonas fluorescens*	Sugar beet	Brazil et al. (1995)
Polycyclic aromatic hydrocarbons	Not identified	Alfalfa	Nichols et al. (1997)
2,4-D	*Pseudomonas putida*	Wheat	Kingsley et al. (1994)
1,4- dioxane	*Actinomycete amycolata* sp.	Poplar root	Kelley et al. (2001)
Trichloroethylene.	*Pseudomonas fluorescens*	Wheat	Yee et al. (1998)
Polycyclic aromatic hydrocarbons	*Rhizophagus irregularis*	Wheat	Ingrid et al. (2016)
Total petroleum hydrocarbons	*Rhodococcus* ITRH43	Italian ryegrass	Ingrid et al. (2016)
Pyrene	*Pseudomonas putida*	*Salix purpurea*	Khan et al. (2014)
Fixed nitrogen	*Nitrospira* sp. and *Nitrosomonas* sp.	Reed (*Phragmitis australies*)	El Haleem et al. (2000)
Napthalene	*Pseudomonas putida*	Grasses	Kuiper et al. (2001)
Polycyclic aromatic hydrocarbons	*Micrococcus* species *Bacillus* species *Kurthia* species	*Populus deltoides*	Bisht et al. (2010) Bisht et al. (2014)
Mefenoxam	*Pseudomonas fluorescens* and *Chrysobacterium indologenes*	*Zinnia anguistifolia*	Pai et al. (2001)

4.1.2 Increased pollutant bioavailability

The uptake of xenobiotics from the environments is directly linked to the logarithm of the compound's octanol water partition coefficient (K_{ow}) for a given plant. Briggs et al. (1982) assured the first structural activity relationship, which was later termed as transpiration stream concentration factor (TSCF) for the uptake of a compound as a relation to the compound's log K_{ow} value in a soilless systems. For the organic compounds, TSCF values go to a maximum of 0.8 at a corresponding log K_{ow} of 2.1. Compounds with log K_{ow} value greater than 2.1 are more hydrophobic and bind to lipid membrane of the root while for less hydrophobic compounds with log K_{ow} values < 2.1 do not move through lipid

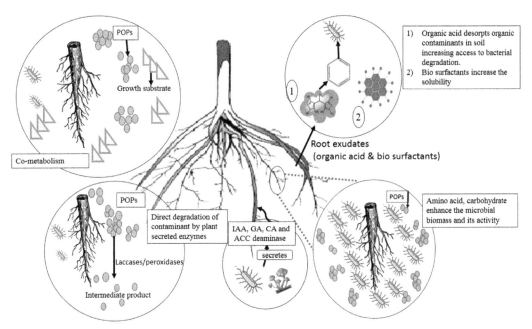

FIGURE 13.2   Plant—microbe interaction during biodegradation of organic contaminants.

membranes of the root epidermal layers. Most of the organic pollutants are hydrophobic, so they are insoluble in soil and not easily available to microbes for degradation. Root exudates increased the bioavailability of pollutants by enhancing their solubility, hence making them more accessible for microbial attack. The compounds that increase the bioavailability of pollutants by biosurfactants production, biofilms formation, and organic acids are described briefly below:

4.1.2.1 Production of biosurfactants

The bioavailability of the pollutant is a major dilemma for soil bioremediation. Because of hydrophobic nature of organic contaminants, they are poorly dissolved in polar molecules such as water, and many of these contaminants can form complex with soil particles. This deficit of bioavailability usually decreases their removal efficiency (Johnsen et al., 2005). With the help of different strategies such as excretion of biosurfactants, formation of biofilms, and production of extracellular polymeric substances (EPS), bacteria increase the bioavailability of hydrophobic compounds (i.e., PAHs). Biosurfactants are amphiphilic molecules (both hydrophobic and hydrophilic in nature) which form micelles with organic contaminants; as a result of this, bioavailability increases and enhances the rhizoremediation process. Because of hydrophobic nature of core region of micelles, it solubilized the hydrophobic contaminants in it, which results in an increase in transfer of compound from solid phase to water phase and thus makes them easily availability for bacterial degradation. Glycolipids are an important biosurfactants produced by bacteria. It has been observed that rhamnolipids, the most important representative of glycolipids, increase the biodegradation speed of contaminants (Cui et al., 2008; Mulligan, 2005; Providenti et al., 1995; Shreve et al., 1995; Zhang and Miller, 1994). Two lipopeptide biosurfactants named as putisolvins produced by *Pseudomonas putida* strain isolated from plant roots by Kuiper et al. (2004) at a PAH-polluted site increased the emulsion formation with toluene. From the bioremediation perspective, the search of rhizobacteria, which helps to enhance the bioavailability of contaminants, is thus of great interest. The positive chemotaxis nature of a large number of microbes toward the pollutants makes them a subject of more curio for research (Parales and Haddock, 2004). Therefore, the mutual action of the chemotaxis and biosurfactants may help in microbial spread and bacterial propagation in polluted soils leads to more plentiful zones clearing.

4.1.2.2 Biofilms formation

Biofilms are the cluster of microorganisms in which the cells are often surrounded by self-producing EPS. The microflora in biofilm is protected by matrix, so their chances of survival increase and thus help in degradation and immobilization of the contaminants. Contaminants such as chlorinated aromatic compounds are present in many chemical industry effluents and can be transferred rapidly into the soils. They are carcinogenic at very low concentrations, extensively present in groundwater and soils (Kargi and Eker, 2005; Puhakka et al., 1995). With the help of mixed microbial biomass of activated sludge culture and *P. putida* in a rotating perforated tube biofilm reactor, Kargi and Eker (2005) eliminated nearly 100% of the 2,4-dichlorophenol from artificial wastewater. Likewise, degradation of PAHs was assisted by bacteria specific in adhesion to PAH (Bastiaens et al., 2000; Johnsen et al., 2005). It was observed that in case of diclofop-methyl, a chlorinated herbicide gets adsorbed to microbial exopolymers and accumulated in biofilms, which was later metabolized by biofilms community during starvation (Wolfaardt et al., 1995). Nitroaromatic compounds used in various fields, such as pharmaceuticals, synthesis of foams, and in explosives, etc., contain nitro group which makes these compounds resistant to biodegradation, and microbial conversion frequently produces harmful metabolites. With the help of mixed culture, dinitrotoluene was degraded in a fluidized-bed biofilm reactor (Gisi et al., 1997; Lendenmann et al., 1998). Dehalococcoides (a group of bacteria) are the only acknowledged microorganisms as described by Löffler et al. (2013), which reduce the chlorinated ethenes via specializing in reductive dechlorination.

4.1.2.3 Organic acid production

Various rhizospheric microbes secrete a variety of organic acids such as gluconic acid, phytic acid, etc., which lowers the soil pH, thus enhancing the solubility of contaminants and increasing their degradation (Oberai and Khanna, 2018). Recently, it has been shown that LMW aliphatic carboxylates, for example, oxalate and citrate, increase the PHCs bioavailability by encouraging their desorption from the matrix of soil (An et al., 2010, 2011; Gao et al., 2010). However, their mechanism of desorption is not well known, and it is hypothesized that the released soil organic matter may causes the liberation of sorbed organic contaminants or due to competition for available adsorption sites between carboxylates and certain PHCs may decrease the adsorption of contaminants (Reilley et al., 1996; An et al., 2010).

4.1.3 Structural analogy and cometabolism

Various secondary metabolites secreted by the plant roots such as flavonoids show similar structural resemblance to organic pollutants, in particular to aromatic hydrocarbons (Bais et al., 2008; Singer, 2006). Because of this structural analogy, degradation of pollutants is enhanced by influencing suitable enzymatic pathways in addition to increase in cometabolic processes (Fletcher and Hegde, 1995). In cometabolism, oxidation of nongrowth substrates takes place in the presence of other growth-causing substrates and it is assumed that this is the major pathway to degrade the recalcitrant hydrocarbons such as PAHs (Cunningham and Berti, 1993; Kuiper et al., 2004). According to Kanaly and Bartha (1999), the mineralization of benzo[a] pyrene occurred in the presence of other primary substrates by cometabolic processes.

4.1.4 Energy and nutrient flow

The root exudates of plants deliver a constant amount of nutrients and energy to microorganism to enhance the degradation of organic pollutants (Anderson et al., 1993; Kuiper et al., 2004). Up to 10%–40% photosynthetic products of plants may go to the soil, which corresponds to around $15-60 \, kg \, N \, ha^{-1} \, yr^{-1}$ and $800-4500 \, kg \, C \, ha^{-1} \, yr^{-1}$ (Gerhardt et al., 2009; Grayston et al., 1997; Hütsch et al., 2002; Lynch and Whipps, 1990; Singh et al., 2004). Nitrogen and carbon are delivered by plant root exudates to rhizospheric microorganisms either in the form of water-insoluble, high molecular weight (HMW) organic polymers or as soluble, LMW organic molecules. The HMW organic polymers take part in formation of mucilage, sloughing cells, and other forms of root debris (Bais et al., 2008). Because HMW organic polymers are mostly insoluble and complex in nature, it therefore has somewhat long biodegradation times (Kalbitz et al., 2003; Kuzyakov, 2002). It was found that when root debris was added in phenanthrene-contaminated soils, it took more time for biodegradation because of insoluble and complex nature as compared with water-soluble root exudates (Miya and Firestone, 2001).

LMW exudates have higher solubility in water, and because of this reason, microbes quickly uptake these compounds. The main components of LMW root exudates are amino acids, organic acid anions, and carbohydrates, and these components are considered as the basic source of nutrients and energy to the rhizosphere because they are easily degradable (Van Hees et al., 2005). Rapid response is shown in terms of their respiration (within 1 h) by soil microbial communities when LMW root exudates are added in soil; however, observation of gene expression changes may take several days (Darrah, 1991; Jones and Murphy, 2007). Da Silva et al. (2006) discovered that exudates released by mulberry roots (*Morus rubra* L.) influenced the growth of PAH degraders in parallel manner to that of total bacteria. However, catabolic repression may occur because of presence of some other more labile carbon sources as compared with root exudates in PHC-polluted soils (Cases and de Lorenzo, 2005; Singer, 2006). For example, research of Yuste et al. (1998) revealed that when *Pseudomonas* species was grown in rich carbon sources, then alkane degradation inhibited because of catabolic repression in the alk B operon of bacterium. Further studied found that when root extracts and exudates are added, then *P. putida* showed decrease in phenanthrene-degrading action in contaminated soils (Rentz et al., 2004).

5. Factors affecting rhizoremediation

The process of rhizoremediation is affected by many factors such as environmental factors, which include temperature, pH, and organic matter present in soil. Besides this, the other factors such as plant species and microbial population involved in the method also affect the process; bioavailability of the pollutants is another and important aspect in the procedure. These elements ultimately influence the pollutants appearance toward microbial diversity.

5.1 Environmental factors

5.1.1 Temperature

Bandowe et al. (2014) stated that temperature serves as a crucial role in the remediation of PAH by incrementing its bioavailability as its solubility tends to rise with the increase of temperature. The enzymatic activities mediated by bacterial population get speed up with the rise of temperature to optimal level, i.e., Liang et al. (2003) have conducted an experiment in which they used to measure the amount of oxygen as a catalog for activity of microbes during the process of composting. During this process, the concentration of increasing oxygen was much higher at 43°C than at 22°C, 29°C, or 36°C. According to Chung et al. (2007), temperature also affects the efficiency of bacterial cell and soil particles for their adsorption and desorption of pollutants.

5.1.2 pH

Most of the enzymes secreted by microorganism for biodegradation of compounds are basically dependent on pH which ranges from 6.5 to 7.5, almost equal to their intracellular pH. The degradation of HCH isomers by *Pandoraea* sp., which was isolated by Okeke et al. (2002) from an enrichment culture, occurred over a pH range of 4–9 (Siddique et al., 2002). According to Singh et al. (2006), the biodegradation of organophosphate pesticides in the soil was slower at low pH as compared with neutral and alkaline soils. In the remediation of PAH, pH acts as a principal factor; different strains of microbes have different optimum pH for their growth and so do for their enzymatic activity. Presence of pollutants such as PAHs can change the pH of surrounding soil of the microbial community, which affects the reduction—oxidation reaction and solubility of pollutants in rhizosphere (Brito et al., 2015). According to Bamforth and Singleton (2005), some of the microbial strains have shown the potential to degrade PAHs in situ, but they get rigorously inhibited if there was extreme change in pH; so, adjustment in pH is advised at the polluted sites.

5.1.3 Soil organic matter

Organic matter in soil acts as a nutrient source for the rhizospheric microbiota and hence helps them in proliferation and controls the movement of pollutants by adsorption and desorption mechanism. For example, the bioremediation process of herbicides was improved when sewage sludge was added into the soil.

5.1.4 Low molecular weight organic acids

Organic acids that are broadly disseminated in the soil system and low in their molecular weight play a key role in the rhizoremediation mechanism of organic pollutants. These organic acids and humic acids remediate the pollutants by adsorbing pollutants on their surfaces; they also reduce the binding of pollutants on surface of soil particles or enhance the discharge of pollutant molecules from the surface of soil particles; as a result, they increase the mobilization and availability of organic pollutants in soil, which ultimately augments the rate of deprivation via microbial activity (Van de Kreeke et al., 2010). Gao et al. (2015) have conducted an experiment in which they have supplemented 10−100 mmol/kg LMW organic acids to the experimental sets and found that after 40 days of incubation period the supplemented sets have more amount of all kinds of PAH in soil than control groups that were not supplemented; there was 55%−75% increase in the availability of PAH in soil when LMW organic acids were there in rhizosphere.

5.2 Plant species

Some plants have the capacity to remove or degrade pollutants such as polychlorinated biphenyls in soil, and microbes present in association with these plants add extra effort to their properties. Some microbes promote the growth of plants by their PGP activity; in return, plants secrete some metabolites such as limonene and carvone, which helps microbes in their propagation and removal efficiency of contaminant. Besides this, plants secrete root exudates which serve as a carbon source to the microbial population and help them in achieving higher population in the rhizospheric zone; in addition to this, root exudates may contain biosurfactants which can enhance the bioavailability of the contaminants in rhizospheric soil (Vergani et al., 2017). Plants have the ability to affect organic pollutants and particulate matter level in air; McLachlan (1999) called it "forest filter effect"; but polychlorinated biphenyls can bind to soil particles tightly with less solubility in capillary water, which make the availability harder for degradation (Terzaghi et al., 2018).

5.3 Microbial activity

Activity of microbes in rhizosphere can be affected by many factors, some of which are their genes, screening conditions, and species used, but the presence of contaminants can alter the function of microbes which are naturally occurring in rhizosphere. These pollutants create a stress condition for the microbial community, which leads them to adopt and develop new characters against the contaminants. Species isolated from these polluted sites normally have greater ability to handle contaminants, and they can be used to bioremediate contaminants at the sites that have high pollution index; but they can replace the existing microbes at that site and the competition cannot be ignored (Momose et al., 2008). Moreover, these microbes may also contain resistance against the pollutant such as PAHs or can reflect the ability for their degradation, so we can also get the opportunity to isolate those genes and use in other ways of bioremediation (Mahmoudi et al., 2011). Li et al. (2015) have done the work on a strain of *Arthrobacter* SA02, which have been showing the properties of degrading intermediates of PAHs. This team had isolated the genes that are coding for RHDase (ring hydroxylating dioxygenase) and 1H2Nase (1-hydroxy-2-naphthoate dioxygenase); these enzymes play key roles in the removal of phenanthrene.

5.4 Bioavailability of pollutants

The bioavailability of the contaminants can be affected by its solubility, concentration level in the soil, adsorption, and structure. Surfactants have been widely used to enhance the availability of the pollutant. For example, when sodium dodecylbenzenesulfonate was added, D9 fatty acid desaturase concentration was boosted, which leads to amplify fluidity of cell membrane by rising unsaturated fatty acid level, thereby intensifying the transport of contaminant molecule across membrane (Li et al., 2015). Surfactants increase the degradation of PAHs in microorganisms by upregulation of some genes, i.e., 1H2Nase and RHDase and regulation of the cell membrane.

6. Rhizoremediation potential, challenges, and future perspectives

The contamination of organic substances is posing threat to the human health and harming our environment, so there is an immediate need to resolve this problem with efficient solution. With the past studies, some plants such as *M. sativa, Hordeum vulgare, Triticum aestivum,* and *Phragmites australis* are reported for successful removal of organic contaminants from the soil (Kuiper et al., 2004). For rhizoremediation process, a rhizospheric microbial strain should be inoculated with the plant seeds by coating them with microbes, so that microbes can grow in association with the roots of that plant easily and increase the speed of remediation of pollutant. This type of inoculation of bacteria will help the microbes to grow along the underground part and increase the chances of spreading in soil (Kuiper et al., 2004). Kuiper et al. (2001) have screened out a plant microbe relation, which showed the ability to degrade naphthalene and the capacity to provide tolerance to plant seed against high concentration of this pollutant in soil. When the seeds of grasses were coated with this bacterial strain and then grown, it was recorded that the bacterial community was able to grow and spread deeper in soil with the roots of plant. Ronchel and Ramos (2001) modified a strain of *Pseudomonas* with recombinant DNA technology; now this improved strain has the capacity to tolerate pollutants in the rhizosphere, which was not possible for the parent bacterial strain. Although there are many studies which shows that rhizoremediation is a sustainable solution for the removal of organic contaminants from soil, but for using this technique at large scale, we need more studies, newer plants, and microbial strains. Overall, we conclude that the rhizoremediation has proven a successful strategy for the removal of organic pollutants from the soil. However, much of the current studies are lab scale studies and more efforts are needed to establish this technique in the field. There is a great emphasis toward sustainable and green rhizoremediation technologies and the use of plant-associated bacteria to degrade toxic synthetic organic compounds from environmental soil. If the efforts are taken seriously, then this technology may provide an efficient, economic, and sustainable green remediation technology for the abatement of organic pollutants from the soil.

Acknowledgments

Authors are thankful to University Grant Commission and CSIR for providing the fellowship in the form of JRF.

References

Agusa, T., Kunito, T., Nakashima, E., Minh, T.B., Tanabe, S., Subramanian, A., Viet, P.H., 2003. Preliminary studies on trace element contamination in dumping sites of municipal wastes in India and Vietnam. In: Journal de Physique IV (Proceedings), vol. 107. EDP sciences, pp. 21–24.

Alamdar, A., Syed, J.H., Malik, R.N., Katsoyiannis, A., Liu, J., Li, J., Zhang, G., Jones, K.C., 2014. Organochlorine pesticides in surface soils from obsolete pesticide dumping ground in Hyderabad City, Pakistan: contamination levels and their potential for air–soil exchange. The Science of the Total Environment 470, 733–741.

Ali, U., Syed, J.H., Malik, R.N., Katsoyiannis, A., Li, J., Zhang, G., Jones, K.C., 2014. Organochlorine pesticides (OCPs) in South Asian region: a review. The Science of the Total Environment 476, 705–717.

Alkorta, I., Garbisu, C., 2001. Phytoremediation of organic contaminants in soils. Bioresource Technology 79 (3), 273–276.

An, C., Huang, G., Yu, H., Wei, J., Chen, W., Li, G., 2010. Effect of short-chain organic acids and pH on the behaviors of pyrene in soil–water system. Chemosphere 81 (11), 1423–1429.

An, C.J., Huang, G.H., Wei, J., Yu, H., 2011. Effect of short-chain organic acids on the enhanced desorption of phenanthrene by rhamnolipid biosurfactant in soil–water environment. Water Research 45 (17), 5501–5510.

Anderson, T.A., Guthrie, E.A., Walton, B.T., 1993. Bioremediation in the rhizosphere. Environment Science Technology 27 (13), 2630–2636.

Atapattu, S.S., Kodituwakku, D.C., 2009. Agriculture in South Asia and its implications on downstream health and sustainability: a review. Agricultural Water Management 96 (3), 361–373.

Bais, H.P., Broeckling, C.D., Vivanco, J.M., 2008. Root exudates modulate plant–microbe interactions in the rhizosphere. In: Secondary Metabolites in Soil Ecology. Springer, Berlin, Heidelberg, pp. 241–252.

Bamforth, S.M., Singleton, I., 2005. Bioremediation of polycyclic aromatic hydrocarbons: current knowledge and future directions. Journal of Chemical Technology and Biotechnology 80 (7), 723–736.

Bandowe, B.A.M., Bigalke, M., Boamah, L., Nyarko, E., Saalia, F.K., Wilcke, W., 2014. Polycyclic aromatic compounds (PAHs and oxygenated PAHs) and trace metals in fish species from Ghana (West Africa): bioaccumulation and health risk assessment. Environment International 65, 35–146.

Barra, R., Popp, P., Quiroz, R., Bauer, C., Cid, H., von Tümpling, W., 2005. Persistent toxic substances in soils and waters along an altitudinal gradient in the Laja River Basin, Central Southern Chile. Chemosphere 58 (7), 905–915.

Barraclough, D., 2007. UK Soil and Herbage Pollutant Survey. Introduction and Summary. Environment Agency (EA), UK.

Bastiaens, L., Springael, D., Wattiau, P., Harms, H., Verachtert, H., Diels, L., 2000. Isolation of adherent polycyclic aromatic hydrocarbon (PAH)-degrading bacteria using PAH-sorbing carriers. Applied and Environmental Microbiology 66 (5), 1834–1843.

Baudoin, E., Benizri, E., Guckert, A., 2003. Impact of artificial root exudates on the bacterial community structure in bulk soil and maize rhizosphere. Soil Biology and Biochemistry 35, 1183–1192.

Behera, B., Singh, G.S., 1999. Studies on weed management in monsoon season crop of tomato. Indian Journal of Weed Science 31 (1and2), 67–70.

Benizri, E., Dedourge, O., Dibattista-Leboeuf, C., Piutti, S., Nguyen, C., Guckert, A., 2002. Effect of maize rhizodeposits on soil microbial community structure. Applied Soil Ecology 21, 261–265.

Bhattacharya, B., Sarkar, S.K., Mukherjee, N., 2003. Organochlorine pesticide residues in sediments of a tropical mangrove estuary, India: implications for monitoring. Environment International 29 (5), 587–592.

Bisht, S., Pandey, P., Kaur, G., Aggarwal, H., Sood, A., Sharma, S., Kumar, V., Bisht, N.S., 2014. Utilization of endophytic strain Bacillus sp. SBER3 for biodegradation of polyaromatic hydrocarbons (PAH) in soil model system. European Journal of Soil Biology 60, 67–76.

Bisht, S., Pandey, P., Sood, A., Sharma, S., Bisht, N.S., 2010. Biodegradation of naphthalene and anthracene by chemotactically active rhizobacteria of Populus deltoides. Brazilian Journal of Microbiology 41 (4), 922–930.

Boopathy, R., 2004. Anaerobic biodegradation of no. 2 diesel fuel in soil: a soil column study. Bioresource technology 94 (2), 143–151.

Brazil, G.M., Kenefick, L., Callanan, M., Haro, A., De Lorenzo, V., Dowling, D.N., O'gara, F., 1995. Construction of a rhizosphere pseudomonad with potential to degrade polychlorinated biphenyls and detection of bph gene expression in the rhizosphere. Applied and Environmental Microbiology 61 (5), 1946–1952.

Briggs, G.G., Bromilow, R.H., Evans, A.A., 1982. Relationships between lipophilicity and root uptake and translocation of non-ionised chemicals by barley. Pesticide Science 13 (5), 495–504.

Brink, S.C., 2016. Unlocking the secrets of the rhizosphere. Trends in Plant Science 21 (3), 169–170.

Brito, E.M., De la Cruz Barrón, M., Caretta, C.A., Goñi-Urriza, M., Andrade, L.H., Cuevas-Rodríguez, G., Malm, O., Torres, J.P., Simon, M., Guyoneaud, R., 2015. Impact of hydrocarbons, PCBs and heavy metals on bacterial communities in Lerma River, Salamanca, Mexico: investigation of hydrocarbon degradation potential. The Science of the Total Environment 521, 1–10.

Buccini, J., 2003. The development of a global treaty on persistent organic pollutants (POPs). In: Persistent Org Pollut. Springer, Berlin, Heidelberg, pp. 13–30.

Butler, J.L., Williams, M.A., Bottomley, P.J., Myrold, D.D., 2003. Microbial community dynamics associated with rhizosphere carbon flow. Applied and Environmental Microbiology 69 (11), 6793–6800.

Cachada, A., Pato, P., Rocha-Santos, T., da Silva, E.F., Duarte, A.C., 2012. Levels, sources and potential human health risks of organic pollutants in urban soils. The Science of the Total Environment 430, 184–192.

Cannon, H.L., Bowles, J.M., 1962. Contamination of vegetation by tetraethyl lead. Science 137 (3532), 765–766.

Cases, I., de Lorenzo, V., 2005. Genetically modified organisms for the environment: stories of success and failure and what we have learned from them. International microbiology 8 (3), 213–222.

Chakraborty, P., Zhang, G., Li, J., Xu, Y., Liu, X., Tanabe, S., Jones, K.C., 2010. Selected organochlorine pesticides in the atmosphere of major Indian cities: levels, regional versus local variations, and sources. Environmental Science and Technology 44 (21), 8038–8043.

Chaudhry, Q., Blom-Zandstra, M., Gupta, S.K., Joner, E., 2005. Utilising the synergy between plants and rhizosphere microorganisms to enhance breakdown of organic pollutants in the environment (15 pp). Environmental Science and Pollution Research International 12, 34–48.

Chung, M.K., Tsui, M.T., Cheung, K.C., Tam, N.F., Wong, M.H., 2007. Removal of aqueous phenanthrene by brown seaweed *Sargassum hemiphyllum*: sorption-kinetic and equilibrium studies. Separation and Purification Technology 54 (3), 355–362.

Corgie, S.C., Beguiristain, T., Leyval, C., 2004. Spatial distribution of bacterial communities and phenanthrene degradation in the rhizosphere of *Lolium perenne* L. Applied and Environmental Microbiology 70 (6), 3552–3557.

Cui, C.Z., Zeng, C., Wan, X., Chen, D., Zhang, J.Y., Shen, P., 2008. Effect of rhamnolipids on degradation of anthracene by two newly isolated strains, *Sphingomonas* sp. 12A and *Pseudomonas* sp. 12B. Journal of microbiology and biotechnology 18 (1), 63–66.

Cunningham, S.D., Berti, W.R., 1993. Remediation of contaminated soils with green plants: an overview. In Vitro Cellular and Developmental Biology Plant 29 (4), 207–212.

Da Silva, M.L., Kamath, R., Alvarez, P.J., 2006. Effect of simulated rhizodeposition on the relative abundance of polynuclear aromatic hydrocarbon catabolic genes in a contaminated soil. Environmental Toxicology and Chemistry 25 (2), 386–391.

Dams, R.I., Paton, G.I., Killham, K., 2007. Rhizoremediation of pentachlorophenol by *Sphingobium chlorophenolicum* ATCC 39723. Chemosphere 68 (5), 864–870.

Darrah, P.R., 1991. Models of the rhizosphere. Plant and Soil 133 (2), 187–199.

Deka, J., Sarma, K.P., Hoque, R.R., 2016. Source contributions of polycyclic aromatic hydrocarbons in soils around oilfield in the Brahmaputra valley. Ecotoxicology and Environmental Safety 133, 281–289.

Dzantor, E.K., 2007. Phytoremediation: the state of rhizosphere 'engineering'for accelerated rhizodegradation of xenobiotic contaminants. Journal of Chemical Technology and Biotechnology 82 (3), 228–232.

El Haleem, D.A., Von Wintzingerode, F., Moter, A., Moawad, H., Göbel, U.B., 2000. Phylogenetic analysis of rhizosphere-associated β-subclass proteobacterial ammonia oxidizers in a municipal wastewater treatment plant based on rhizoremediation technology. Letters in Applied Microbiology 31 (1), 34–38.

El Nemr, A., Moneer, A.A., Khaled, A., El-Sikaily, A., 2013. Levels, distribution, and risk assessment of organochlorines in surficial sediments of the Red Sea coast, Egypt. Environmental Monitoring and Assessment 185 (6), 4835–4853.

Eqani, S.A.M.A.S., Malik, R.N., Mohammad, A., 2011. The level and distribution of selected organochlorine pesticides in sediments from River Chenab, Pakistan. Environmental Geochemistry and Health 33 (1), 33–47.

Fabietti, G., Biasioli, M., Barberis, R., Ajmone-Marsan, F., 2010. Soil contamination by organic and inorganic pollutants at the regional scale: the case of Piedmont, Italy. Journal of Soils and Sediments 10 (2), 290–300.

Fletcher, J.S., Hegde, R.S., 1995. Release of phenols by perennial plant roots and their potential importance in bioremediation. Chemosphere 31 (4), 3009−3016.

Gao, Y., Ren, L., Ling, W., Kang, F., Zhu, X., Sun, B., 2010. Effects of low-molecular-weight organic acids on sorption−desorption of phenanthrene in soils. Soil Science Society of America Journal 74 (1), 51−59.

Gao, Y., Yuan, X., Lin, X., Sun, B., Zhao, Z., 2015. Low-molecular-weight organic acids enhance the release of bound PAH residues in soils. Soil and Tillage Research 145, 103−110.

Gerhardt, K.E., Huang, X.D., Glick, B.R., Greenberg, B.M., 2009. Phytoremediation and rhizoremediation of organic soil contaminants: potential and challenges. Plant Science 176 (1), 20−30.

Gisi, D., Stucki, G., Hanselmann, K.W., 1997. Biodegradation of the pesticide 4, 6-dinitro-ortho-cresol by microorganisms in batch cultures and in fixed-bed column reactors. Applied Microbiology and Biotechnology 48 (4), 441−448.

Glazer, A.N., Nikaido, H., 1995. Biomass. In: Microbial Biotechnology: Fundamentals of Applied Microbiology. Freeman and Co, New York, p. 662.

Gramss, G., Kirsche, B., Voigt, K., Gunther, T., Fritsche, W., 1999. Conversion rates of five polycyclic aromatic hydrocarbons in liquid cultures of fifty-eight fungi and the concomitant production of oxidative enzymes. Mycological Research 103, 1009−1018.

Gransee, A., Wittenmayer, L., 2000. Qualitative and quantitative analysis of water-soluble root exudates in relation to plant species and development. Journal of Plant Nutrition and Soil Science 163, 381−385.

Grayston, S.J., Vaughan, D., Jones, D., 1997. Rhizosphere carbon flow in trees, in comparison with annual plants: the importance of root exudation and its impact on microbial activity and nutrient availability. Applied Soil Ecology 5 (1), 29−56.

Gupta, P.K., 2004. Pesticide exposure—Indian scene. Toxicology 198 (1−3), 83−90.

Gupta, A., Singh, M., Pandey, K.D., Kumar, A., 2018. Biodegradation of xenobiotics using cyanobacteria. Journal of Scientific Research ISSN 62, 35−40.

Guzzella, L., Poma, G., De Paolis, A., Roscioli, C., Viviano, G., 2011. Organic persistent toxic substances in soils, waters and sediments along an altitudinal gradient at Mt. Sagarmatha, Himalayas, Nepal. Environmental Pollution 159 (10), 2552−2564.

Guzzella, L., Roscioli, C., Vigano, L., Saha, M., Sarkar, S.K., Bhattacharya, A., 2005. Evaluation of the concentration of HCH, DDT, HCB, PCB and PAH in the sediments along the lower stretch of Hugli estuary, West Bengal, northeast India. Environment International 31 (4), 523−534.

Harms, H., Schlosser, D., Wick, L.Y., 2011. Untapped potential: exploiting fungi in bioremediation of hazardous chemicals. Nature Reviews Microbiology 9, 177−192.

Hartmann, A., Schmid, M., Tuinen, D., Berg, G., 2009. Plant-driven selection of microbes. Plant and Soil 321, 235−257.

Havelcová, M., Melegy, A., Rapant, S., 2014. Geochemical distribution of polycyclic aromatic hydrocarbons in soils and sediments of El-Tabbin, Egypt. Chemosphere 95, 63−74.

Hoagland, R.E., Zablotowicz, R.M., Locke, M.A., 1994. Propanil metabolism by rhizosphere microflora. In: Anderson, T.A., Coats, J.R. (Eds.), Bioremediation through Rhizosphere Technology. American Chemical Society, Washington, D.C, pp. 160−183.

Hodge, A., Paterson, E., Thornton, B., Millard, P., Killham, K., 1997. Effects of photon flux density on carbon partitioning and rhizosphere carbon flow of *Lolium perenne*. Journal of Experimental Botany 48, 1797−1805.

Hussain, K., Hoque, R.R., 2015. Seasonal attributes of urban soil PAHs of the Brahmaputra Valley. Chemosphere 119, 794−802.

Hütsch, B.W., Augustin, J., Merbach, W., 2002. Plant rhizodeposition—an important source for carbon turnover in soils. Journal of Soil Science and Plant Nutrition 165 (4), 397−407.

Ingrid, L., Sahraoui, A.L.H., Frédéric, L., Yolande, D., Joël, F., 2016. Arbuscular mycorrhizal wheat inoculation promotes alkane and polycyclic aromatic hydrocarbon biodegradation: microcosm experiment on aged-contaminated soil. Environmental Pollution 213, 549−560.

Jacobsen, C.S., 1997. Plant protection and rhizosphere colonization of barley by seed inoculated herbicide degrading *Burkholderia* (*Pseudomonas*) *cepacia* DBO1 (pRO101) in 2, 4-D contaminated soil. Plant and Soil 189 (1), 139−144.

Jan, M.R., Shah, J., Khawaja, M.A., Gul, K., 2009. DDT residue in soil and water in and around abandoned DDT manufacturing factory. Environmental Monitoring and Assessment 155 (1−4), 31−38.

Johnsen, A.R., Wick, L.Y., Harms, H., 2005. Principles of microbial PAH-degradation in soil. Environmental Pollution 133 (1), 71−84.

Joner, E.J., Corgié, S.C., Amellal, N., Leyval, C., 2002. Nutritional constraints to degradation of polycyclic aromatic hydrocarbons in a simulated rhizosphere. Soil Biology and Biochemistry 34, 859—864.

Jones, D.L., Murphy, D.V., 2007. Microbial response time to sugar and amino acid additions to soil. Soil Biology and Biochemistry 39 (8), 2178—2182.

Kalbitz, K., Schmerwitz, J., Schwesig, D., Matzner, E., 2003. Biodegradation of soil-derived dissolved organic matter as related to its properties. Geoderma 113 (3—4), 273—291.

Kanaly, R.A., Bartha, R., 1999. Cometabolic mineralization of benzo [a] pyrene caused by hydrocarbon additions to soil. Environmental Toxicology and Chemistry 18 (10), 2186—2190.

Kargi, F., Eker, S., 2005. Removal of 2, 4-dichlorophenol and toxicity from synthetic wastewater in a rotating perforated tube biofilm reactor. Process Biochemistry 40 (6), 2105—2111.

Kelley, S.L., Aitchison, E.W., Deshpande, M., Schnoor, J.L., Alvarez, P.J., 2001. Biodegradation of 1, 4-dioxane in planted and unplanted soil: effect of bioaugmentation with *Amycolata* sp. CB1190. Water Research 35 (16), 3791—3800.

Khan, M.J., Zia, M.S., Qasim, M., 2010. Use of pesticides and their role in environmental pollution. World Academy of Science Engineering and Technology 72, 122—128.

Khan, Z., Roman, D., Kintz, T., delas Alas, M., Yap, R., Doty, S., 2014. Degradation, phytoprotection and phytoremediation of phenanthrene by endophyte *Pseudomonas putida*, PD1. Environmental Science and Technology 48 (20), 12221—12228.

Khwaja, M.A., Jan, M.R., Gul, K., 2006. Physical verification and study of contamination in and around an abandoned DDT factory in north west frontier province (NWFP) Pakistan. IPEP 1—48.

Kingsley, M.T., Fredrickson, J.K., Metting, F.B., Seidler, R.J., 1994. Environmental Restoration Using Plant-Microbe Bioaugmentation. Bioremediation of Chlorinated and Polyaromatic Hydrocarbon Compounds. Lewis Publishers, Boca Raton, Fla, pp. 287—292.

Kuiper, I., Bloemberg, G.V., Lugtenberg, B.J., 2001. Selection of a plant-bacterium pair as a novel tool for rhizostimulation of polycyclic aromatic hydrocarbon-degrading bacteria. Molecular Plant-Microbe Interactions 14 (10), 1197—1205.

Kuiper, I., Lagendijk, E.L., Bloemberg, G.V., Lugtenberg, B.J., 2004. Rhizoremediation: a beneficial plant-microbe interaction. Molecular Plant-Microbe Interactions 17 (1), 6—15.

Kuzyakov, Y., 2002. Factors affecting rhizosphere priming effects. Journal of Plant Nutrition and Soil Science 165 (4), 382—396.

Leigh, M.B., Fletcher, J.S., Fu, X., Schmitz, F.J., 2002. Root turnover: an important source of microbial substrates in rhizosphere remediation of recalcitrant contaminants. Environmental Science and Technology 36, 1579—1583.

Lendenmann, U., Spain, J.C., Smets, B.F., 1998. Simultaneous biodegradation of 2, 4-dinitrotoluene and 2, 6-dinitrotoluene in an aerobic fluidized-bed biofilm reactor. Environmental Science and Technology 32 (1), 82—87.

Li, F., Zhu, L., Wang, L., Zhan, Y., 2015. Gene expression of an *Arthrobacter* in surfactant-enhanced biodegradation of a hydrophobic organic compound. Environmental Science and Technology 49 (6), 3698—3704.

Liang, C., Das, K.C., McClendon, R.W., 2003. The influence of temperature and moisture contents regimes on the aerobic microbial activity of a biosolids composting blend. Bioresource Technology 86 (2), 131—137.

Liu, R., Xiao, N., Wei, S., Zhao, L., An, J., 2014. Rhizosphere effects of PAH-contaminated soil phytoremediation using a special plant named Fire Phoenix. The Science of the Total Environment 473, 350—358.

Löffler, F.E., Ritalahti, K.M., Zinder, S.H., 2013. Dehalococcoides and reductive dechlorination of chlorinated solvents. In: Bioaugmentation for Groundwater Remediation. Springer, New York, NY, pp. 39—88.

Lugtenberg, B.J.J., de Weger, L.A., 1992. Plant root colonization by *Pseudomonas* spp. In: Galli, E., Silver, S., Witholt, B. (Eds.), Pseudomonas: Molecular Biology and Biotechnology. Am. Soc. Microbiol, Washington, D.C, pp. 13—19.

Lynch, J.M., Whipps, J.M., 1990. Substrate flow in the rhizosphere. Plant and Soil 129 (1), 1—10.

Ma, W.L., Li, Y.F., Sun, D.Z., Qi, H., 2009. Polycyclic aromatic hydrocarbons and polychlorinated biphenyls in topsoils of Harbin, China. Archives of Environmental Contamination and Toxicology 57 (4), 670.

Macek, T., Mackova, M., Káš, J., 2000. Exploitation of plants for the removal of organics in environmental remediation. Biotechnology Advances 18 (1), 23—34.

MacKinnon, G., Duncan, H.J., 2013. Phytotoxicity of branched cyclohexanes found in the volatile fraction of diesel fuel on germination of selected grass species. Chemosphere 90 (3), 952—957.

Mahmoudi, N., Slater, G.F., Fulthorpe, R.R., 2011. Comparison of commercial DNA extraction kits for isolation and purification of bacterial and eukaryotic DNA from PAH-contaminated soils. Canadian Journal of Microbiology 57 (8), 623—628.

Malik, A., Ojha, P., Singh, K.P., 2009. Levels and distribution of persistent organochlorine pesticide residues in water and sediments of Gomti River (India)—a tributary of the Ganges River. Environmental Monitoring and Assessment 148 (1–4), 421–435.

Marschner, P., Crowley, D., Higashi, R., 1997. Root exudation and physiological status of a root-colonizing fluorescent pseudomonad in mycorrhizal and non-mycorrhizal pepper (Capsicum annuum L.). Plant and Soil 189, 11–20.

Marschner, P., Crowley, D., Rengel, Z., 2011. Rhizosphere interactions between microorganisms and plants govern iron and phosphorus acquisition along the root axis—model and research methods. Soil Biology and Biochemistry 43 (5), 883–894.

Martin, B.C., George, S.J., Price, C.A., Ryan, M.H., Tibbett, M., 2014. The role of root exuded low molecular weight organic anions in facilitating petroleum hydrocarbon degradation: current knowledge and future directions. The Science of the Total Environment 472, 642–653.

McDougall, B.M., Rovira, A.D., 1970. Sites of exudation of 14C-labelled compounds from wheat roots. New Phytologist 69, 999–1003.

McLachlan, M.S., 1999. Framework for the interpretation of measurements of SOCs in plants. Environmental Science and Technology 33 (11), 1799–1804.

Megharaj, M., Ramakrishnan, B., Venkateswarlu, K., Sethunathan, N., Naidu, R., 2011. Bioremediation approaches for organic pollutants: a critical perspective. Environment International 37 (8), 1362–1375.

Merrington, G., Nfa, L.W., Parkinson, R., Redman, M., Winder, L., 2002. Agricultural Pollution: Environmental Problems and Practical Solutions. CRC Press.

Minh, N.H., Minh, T.B., Kajiwara, N., Kunisue, T., Subramanian, A., Iwata, H., Tana, T.S., Baburajendran, R., Karuppiah, S., Viet, P.H., Tuyen, B.C., 2006. Contamination by persistent organic pollutants in dumping sites of Asian developing countries: implication of emerging pollution sources. Archives of Environmental Contamination and Toxicology 50 (4), 474–481.

Mishra, K., Sharma, R.C., Kumar, S., 2012. Contamination levels and spatial distribution of organochlorine pesticides in soils from India. Ecotoxicology and Environmental Safety 76, 215–225.

Miya, R.K., Firestone, M.K., 2001. Enhanced phenanthrene biodegradation in soil by slender oat root exudates and root debris. Journal of Environmental Quality 30 (6), 1911–1918.

Momose, Y., Hirayama, K., Itoh, K., 2008. Competition for proline between indigenous Escherichia coli and E. coli O157: H7 in gnotobiotic mice associated with infant intestinal microbiota and its contribution to the colonization resistance against E. coli O157: H7. Antonie van Leeuwenhoek 94 (2), 165–171.

Moore, F.P., Barac, T., Borremans, B., Oeyen, L., Vangronsveld, J., Van Der Lelie, D., Campbell, C.D., Moore, E.R., 2006. Endophytic bacterial diversity in poplar trees growing on a BTEX-contaminated site: the characterisation of isolates with potential to enhance phytoremediation. Systematic and Applied Microbiology 29 (7), 539–556.

Mulligan, C.N., 2005. Environmental applications for biosurfactants. Environmental pollution 133 (2), 183–198.

Neumann, G., 2007. Root exudates and nutrient cycling. In: Marschner, P., Rengel, Z. (Eds.), Nutrient Cycling in Terrestrial Ecosystems. Springer, Berlin, pp. 123–157.

Nichols, T.D., Wolf, D.C., Rogers, H.B., Beyrouty, C.A., Reynolds, C.M., 1997. Rhizosphere microbial populations in contaminated soils. Water, Air, and Soil Pollution 95 (1–4), 165–178.

Oberai, M., Khanna, V., 2018. Rhizoremediation—plant microbe interactions in the removal of pollutants. International Journal of Current Microbiology and Applied Science 7 (1), 2280–2287.

Okeke, B.C., Siddique, T., Arbestain, M.C., Frankenberger, W.T., 2002. Biodegradation of γ -hexachlorocyclohexane and α-hexachlorocyclohexane in water and soil slurry by Pandoraea sp. Journal of Agricultural and Food Chemistry 50, 2548–2555.

Oleszczuk, P., 2006. Persistence of polycyclic aromatic hydrocarbons (PAHs) in sewage sludge-amended soil. Chemosphere 65 (9), 1616–1626.

Pai, S.G., Riley, M.B., Camper, N.D., 2001. Microbial degradation of mefenoxam in rhizosphere of Zinnia angustifolia. Chemosphere 44 (4), 577–582.

Pandey, P., Khillare, P.S., Kumar, K., 2011. Assessment of organochlorine pesticide residues in the surface sediments of River Yamuna in Delhi, India. Journal of Environmental Protection 2 (05), 511.

Parales, R.E., Haddock, J.D., 2004. Biocatalytic degradation of pollutants. Current opinion in biotechnology 15 (4), 374–379.

Phillips, L.A., Greer, C.W., Farrell, R.E., Germida, J.J., 2012. Plant root exudates impact the hydrocarbon degradation potential of a weathered-hydrocarbon contaminated soil. Applied Soil Ecology 52, 56–64.

Pozo, K., Harner, T., Lee, S.C., Sinha, R.K., Sengupta, B., Loewen, M., Geethalakshmi, V., Kannan, K., Volpi, V., 2011. Assessing seasonal and spatial trends of persistent organic pollutants (POPs) in Indian agricultural regions using PUF disk passive air samplers. Environmental Pollution 159 (2), 646−653.

Providenti, M.A., Flemming, C.A., Lee, H., Trevors, J.T., 1995. Effect of addition of rhamnolipid biosurfactants or rhamnolipid-producing Pseudomonas aeruginosa on phenanthrene mineralization in soil slurries. FEMS microbiology ecology 17 (1), 15−26.

Puhakka, J.A., Melin, E.S., Järvinen, K.T., Koro, P.M., Rintala, J.A., Hartikainen, P., Shieh, W.K., Ferguson, J.F., 1995. Fluidized-bed biofilms for chlorophenol mineralization. Water Science and Technology 31 (1), 227−235.

Qiu, X., Shah, S.I., Kendall, E.W., Sorensen, D.L., Sims, R.C., Engelke, M.C., 1994. Grass-enhanced bioremediation for clay soils contaminated with polynuclear aromatic hydrocarbons. In: Anderson, T.A., Coats, J.R. (Eds.), Bioremediation through Rhizosphere Technology. American Chemical Society, Washington, D.C, pp. 142−157.

Qixing, Z., Zhang, C., Zhineng, Z., Weitao, L., 2011. Ecological remediation of hydrocarbon contaminated soils with weed plant. Molecular Ecology Resources 2 (2), 97−105.

Radwan, S., Sorkhoh, N., Ei-Nemr, I., 1995. Oil biodegradation around roots. Nature 376, 302.

Rajendran, R.B., Imagawa, T., Tao, H., Ramesh, R., 2005. Distribution of PCBs, HCHs and DDTs, and their ecotoxicological implications in Bay of Bengal, India. Environment International 31 (4), 503−512.

Reilley, K.A., Banks, M.K., Schwab, A.P., 1996. Dissipation of polycyclic aromatic hydrocarbons in the rhizosphere. Journal of Environmental Quality 25 (2), 212−219.

Rentz, J.A., Alvarez, P.J., Schnoor, J.L., 2004. Repression of *Pseudomonas putida* phenanthrene-degrading activity by plant root extracts and exudates. Environmental Microbiology 6 (6), 574−583.

Ronchel, M.C., Ramos, J.L., 2001. Dual system to reinforce biological containment of recombinant bacteria designed for rhizoremediation. Applied and Environmental Microbiology 67 (6), 2649−2656.

Ross, G., 2005. Risks and benefits of DDT. The Lancet 366 (9499), 1771−1772.

Rovira, A.D., 1956. Plant root excretions in relation to the rhizosphere effect. Plant and Soil 7, 209−217.

Rovira, A.D., 1959. Root excretions in relation to the rhizosphere effect. Plant and Soil 11, 53−64.

Ryan, M.H., Tibbett, M., Edmonds-Tibbett, T., Suriyagoda, L.D.B., Lambers, H., Cawthray, G.R., et al., 2012. Carbon trading for phosphorus gain: the balance between rhizosphere carboxylates and arbuscular mycorrhizal symbiosis in plant phosphorus acquisition. Plant, Cell and Environment 35, 2170−2180.

Sanpera, C., Ruiz, X., Jover, L., Llorente, G., Jabeen, R., Muhammad, A., Boncompagni, E., Fasola, M., 2003. Persistent organic pollutants in little egret eggs from selected wetlands in Pakistan. Archives of Environmental Contamination and Toxicology 44 (3), 0360−0368.

Sarkar, A., Gupta, R.S., 1988. DDT residues in sediments from the Bay of Bengal. Bulletin of Environmental Contamination and Toxicology 41 (4−6), 664−669.

Sarkar, S.K., Bhattacharya, B.D., Bhattacharya, A., Chatterjee, M., Alam, A., Satpathy, K.K., Jonathan, M.P., 2008. Occurrence, distribution and possible sources of organochlorine pesticide residues in tropical coastal environment of India: an overview. Environment International 34 (7), 1062−1071.

Shann, J.R., Boyle, J.J., 1994. Influence of plant species on in situ rhizosphere degradation. In: Anderson, T.A., Coats, J.R. (Eds.), Bioremediation Through Rhizosphere Technology. American Chemical Society, Washington, DC, pp. 70−81.

Shreve, G.S., Inguva, S., Gunnam, S., 1995. Rhamnolipid biosurfactant enhancement of hexadecane biodegradation by Pseudomonas aeruginosa. Molecular marine biology and biotechnology 4 (4), 331−337.

Siciliano, S.D., Germida, J.J., Banks, K., Greer, C.W., 2003. Changes in microbial community composition and function during a polyaromatic hydrocarbon phytoremediation field trial. Applied and Environmental Microbiology 69 (1), 483−489.

Siddique, T., Okeke, B.C., Arshad, M., Frankenberger Jr., W.T., 2002. Temperature and pH effects on biodegradation of hexachlorocyclohexane isomers in water and soil slurry. Journal of Agricultural and Food Chemistry 50, 5070−5076.

Singer, A.C., 2006. The chemical ecology of pollutant biodegradation: bioremediation and phytoremediation from mechanistic and ecological perspectives. In: Phytoremediation Rhizoremediation. Springer, Dordrecht, pp. 5−21.

Singh, B.K., Walker, A., Wright, D.J., 2006. Bioremedial potential of fenamiphos and chlorpyrifos degrading isolates: influence of different environmental conditions. Soil Biology and Biochemistry 38, 2682−2693.

Singh, B.K., Millard, P., Whiteley, A.S., Murrell, J.C., 2004. Unravelling rhizosphere−microbial interactions: opportunities and limitations. Trends in Microbiol 12 (8), 386−393.

Singh, K.P., Malik, A., Sinha, S., 2007. Persistent organochlorine pesticide residues in soil and surface water of northern Indo-Gangetic alluvial plains. Environmental Monitoring and Assessment 125 (1–3), 147–155.

Suman, S., Sinha, A., Tarafdar, A., 2016. Polycyclic aromatic hydrocarbons (PAHs) concentration levels, pattern, source identification and soil toxicity assessment in urban traffic soil of Dhanbad, India. The Science of the Total Environment 545, 353–360.

Syed, J.H., Malik, R.N., 2011. Occurrence and source identification of organochlorine pesticides in the surrounding surface soils of the Ittehad Chemical Industries Kalashah Kaku, Pakistan. Environmental Earth Sciences 62 (6), 1311–1321.

Syed, J.H., Malik, R.N., Li, J., Chaemfa, C., Zhang, G., Jones, K.C., 2014. Status, distribution and ecological risk of organochlorines (OCs) in the surface sediments from the Ravi River, Pakistan. The Science of the Total Environment 472, 204–211.

Syed, J.H., Malik, R.N., Liu, D., Xu, Y., Wang, Y., Li, J., Zhang, G., Jones, K.C., 2013. Organochlorine pesticides in air and soil and estimated air–soil exchange in Punjab, Pakistan. The Science of the Total Environment 444, 491–497.

Tariq, M.I., Afzal, S., Hussain, I., Sultana, N., 2007. Pesticides exposure in Pakistan: a review. Environment International 33 (8), 1107–1122.

Terzaghi, E., Zanardini, E., Morosini, C., Raspa, G., Borin, S., Mapelli, F., Vergani, L., Di Guardo, A., 2018. Rhizoremediation half-lives of PCBs: role of congener composition, organic carbon forms, bioavailability, microbial activity, plant species and soil conditions, on the prediction of fate and persistence in soil. The Science of the Total Environment 612, 544–560.

Thijs, S., Sillen, W., Rineau, F., Weyens, N., Vangronsveld, J., 2016. Towards an enhanced understanding of plant–microbiome interactions to improve phytoremediation: engineering the metaorganism. Frontiers in Microbiology 7, 341.

UNEP, 2001. Final Act of the Conference of Plenipotentiaries on the Stockholm Convention on Persistent Organic Pollutant. United Nations Environment Program, Geneva, Switzerland.

UNEP, 2011. Draft Revised Guidance on the Global Monitoring Plan for Persistent Organic Pollutants, UNEP/POPS/COP.5/INF/27. United Nations Environment Programme, UNEP Chemicals, Geneva, Switzerland.

UNEP, U., 2009. Report of the conference of the parties of the Stockholm convention on persistent organic pollutants on the work of its fourth meeting. In: United Nations Environment Programme: Stockholm Convention on Persistent Organic Pollutants. Geneva, p. 112.

van de Kreeke, J., de la Calle, B., Held, A., Bercaru, O., Ricci, M., Shegunova, P., Taylor, P., 2010. IMEP-23: the eight EU-WFD priority PAHs in water in the presence of humic acid. Trends in Analytical Chemistry (Reference Ed.) 29 (8), 928–937.

Van Hees, P.A., Jones, D.L., Finlay, R., Godbold, D.L., Lundström, U.S., 2005. The carbon we do not see—the impact of low molecular weight compounds on carbon dynamics and respiration in forest soils: a review. Soil Biology and Biochemistry 37 (1), 1–13.

Vančura, V., Hovadik, A., 1965. Root exudates of plants: II. Composition of root exudates of some vegetables. Plant and Soil 21–32.

Vergani, L., Mapelli, F., Zanardini, E., Terzaghi, E., Di Guardo, A., Morosini, C., Raspa, G., Borin, S., 2017. Phytorhizoremediation of polychlorinated biphenyl contaminated soils: an outlook on plant-microbe beneficial interactions. The Science of the Total Environment 575, 1395–1406.

Vos, J.G., Dybing, E., Greim, H.A., Ladefoged, O., Lambré, C., Tarazona, J.V., Brandt, I., Vethaak, A.D., 2000. Health effects of endocrine-disrupting chemicals on wildlife, with special reference to the European situation. Critical Reviews in Toxicology 30 (1), 71–133.

Walker, T.S., Bais, H.P., Grotewold, E., Vivanco, J.M., 2003a. Root exudation and rhizosphere biology. Plant Physiology 132 (1), 44–51.

Walker, T.S., Bais, H.P., Halligan, K.M., Stermitz, F.R., Vivanco, J.M., 2003b. Metabolic profiling of root exudates of *Arabidopsis thaliana*. Journal of Agricultural and Food Chemistry 51, 2548–2554.

Walton, B.T., Anderson, T.A., 1990. Microbial degradation of trichloroethylene in the rhizosphere: potential application to biological remediation of waste sites. Applied and Environmental Microbiology 56 (4), 1012–1016.

Wolfaardt, G.M., Lawrence, J.R., Robarts, R.D., Caldwell, D.E., 1995. Bioaccumulation of the herbicide diclofop in extracellular polymers and its utilization by a biofilm community during starvation. Applied and Environmental Microbiology 61 (1), 152–158.

Yadav, D.V., Mittal, P.K., Agarwal, H.C., Pillai, M.K., 1981. Organochlorine insecticide residues in soil and earthworms in the Delhi area, India, August—October, 1974. Pesticides Monitoring Journal 15 (2), 80—85.

Yee, D.C., Maynard, J.A., Wood, T.K., 1998. Rhizoremediation of trichloroethylene by a recombinant, root-colonizing *Pseudomonas fluorescensstrain* expressing toluene ortho-monooxygenase constitutively. Applied and Environmental Microbiology 64 (1), 112—118.

Yoshitomi, K.J., Shann, J.R., 2001. Corn (*Zea mays* L.) root exudates and their impact on 14C-pyrene mineralization. Soil Biology and Biochemistry 33, 1769—1776.

Yuste, L., Canosa, I., Rojo, F., 1998. Carbon-source-dependent expression of the PalkB promoter from the *Pseudomonas oleovorans* alkane degradation pathway. Journal of Bacteriology 180 (19), 5218—5226.

Zablotowicz, R.M., Hoagland, R.E., Locke, M.A., 1994. Glutathione S-transferase activity in rhizosphere bacteria and the potential for herbicide detoxification. In: Anderson, T.A., Coats, J.R. (Eds.), Bioremediation Through Rhizosphere Technology. American Chemical Society, Washington, DC, pp. 184—198.

Zhang, Y., Miller, R.M., 1994. Effect of a Pseudomonas rhamnolipid biosurfactant on cell hydrophobicity and biodegradation of octadecane. Appl. Environ. Microbiol. 60 (6), 2101—2106.

Zhang, G., Chakraborty, P., Li, J., Sampathkumar, P., Balasubramanian, T., Kathiresan, K., Takahashi, S., Subramanian, A., Tanabe, S., Jones, K.C., 2008. Passive atmospheric sampling of organochlorine pesticides, polychlorinated biphenyls, and polybrominated diphenyl ethers in urban, rural, and wetland sites along the coastal length of India. Environmental Science and Technology 42 (22), 8218—8223.

Zhang, H.B., Luo, Y.M., Wong, M.H., Zhao, Q.G., Zhang, G.L., 2007. Concentrations and possible sources of polychlorinated biphenyls in the soils of Hong Kong. Geoderma 138 (3—4), 244—251.

Further reading

Burken, J.G., Schnoor, J.L., 1996. Phytoremediation: plant uptake of atrazine and role of root exudates. Journal of Environmental Engineering 122 (11), 958—963.

Khwaja, M.A., Jan, M.R., Gul, K., 2007. Study of Contamination of Soil in Surrounding of Abandoned Persistent Organic Pollutant (DDT) Nowshera Factory in North West Frontier Province (NWFP) Pakistan. Sustainable Development Policy Institute (SDPI), Islamabad Jan, Pakistan.

The role of scanning probe microscopy in bacteria investigations and bioremediation

Igor V. Yaminsky[1], Assel I. Akhmetova[2]

[1]Lomonosov Moscow State University, Moscow, Russian Federation; [2]Advanced Technologies Center, Moscow, Russian Federation

Summary

Scanning probe microscopy (SPM) provides three-dimensional (3D) imaging of bacterial cells in air and liquid with subnanometer space resolution.

SPM appeared in 1981 firstly in the form of a scanning tunneling microscope (STM), which transformed in 1986 into atomic force microscopy (AFM). Later it developed to several dozen other types, revolutionizing in many scientific fields related to the study at scales ranging from microns to subnanometers because it often allowed receiving unique information that cannot be obtained by other methods. The main trends in the development of SPM for the three decades of its existence are the development of the methods and the wide penetration of SPM in different areas of science and technology, such as physics, chemistry, biology, ecology, medicine, materials science, geology, and others. These two trends cannot be separated, and their development is highly dependent on another important factor—on the development of investigation protocols, including the sample preparation and the substrate, the choice of microscope operation modes, and data processing, analysis, and interpretation.

SPM has proved to be a powerful method for the investigation of bacterial cell in different environmental conditions—in air and liquid. Nanometer space resolution can be achieved in the morphological studies. This method is permanently developing for the last 30 years.

1. Introduction

Bacteria were the first life forms that appear on the Earth about 3 billion years ago. At present time, about 10,000 bacteria cells are characterized, while it is believed that more than 1 million different bacteria exist in the world. The first systemization of bacteria was done and published by Bergey in 1923. Since that time, this prominent manual was several times improved and several editions were published (Garrity et al., 2001; Brenner et al., 2005; Vos et al., 2009; Krieg et al., 2010; Whitman et al., 2012).

Bacterial cells have an ability to change the genome in the process of adaptation to changing external conditions. Because of high speed of division and possible changes in genome, they can withstand antibiotics (Sagitova et al., 2016). This creates serious risks on the way to effective prevention of bacterial infectious diseases.

The present review deals with the SPM, which was applied to the study of bacteria about 20 years ago (Yaminsky et al., 1997; Yaminsky, 1997). This approach has several distinguished features, which cannot be accessed in other methods. Bacterial cells are studied with nanometer space resolution in environmental conditions: in air and in liquid. A real 3D topography of bacteria is visualized using SPM. Additional mechanical properties of bacteria—rigidity, viscoelasticity, strength against damage, surface friction, and adhesion—can also be obtained.

To date, AFM is the only method that allows microsurgery on the nanoscale, and it meets the increased requirements for accuracy, reproducibility, and automation of exposure to cells and gives new impetus to the cellular technologies development, including for ecological practice. AFM is a modern and promising method used in scientific research.

2. Bacterial biofilms

Bacterial cells surrounded by a matrix of extracellular polymeric substances produced by the cells themselves and attached to the surface are called biofilms. About 99% of bacteria on earth live in biofilms. Biofilm formation is a unique property of bacterial cells. Aggregates occur during binding of bacteria to different surfaces. Biofilm is not only an irregular cellular aggregate but also has unique biochemical characteristics that differ from isolated cells. Bacteria in biofilms are more resistant to antibiotics and disinfectants than free-floating bacteria. The biofilms formation involves several stages. First, the bacteria attach to the surface. The surfaces are usually charged, which contributes to the adsorption of various inorganic ions and charged organic molecules, proteins. The surface velocity of the fluid is minimal, which creates favorable conditions for the bacteria adhesion to the surface. Electrostatic forces, hydrophobic interaction, and van der Waals attraction force play the main role in this process. The bacterial cells begin to divide, forming microcolonies and producing an extracellular matrix consisting of proteins and polysaccharides. Through the "feeling of the quorum" (communication and coordination of the microorganism behavior because of the molecular signals secretion), the further development of microcolonies leads to the formation of macrocolonies separated by fluid channels and surrounded by an extracellular matrix. Microorganisms irreversibly bind to the surface and maturation begins. The channel's development inside the film allows water and nutrients to reach the cells located inside the biofilm.

The biofilm architecture depends on many factors, such as the nutrients availability, surface properties, the composition of microorganisms, and the presence of shear forces. The most representative structure of biofilms involves the formation of mushroom aggregates separated by channels with liquid. A mature biofilm can periodically eject individual cells into the environment, thereby enabling the new biofilms formation on other surfaces. At present, it is believed that biofilm consists of 10%−50% volume of bacterial cells. Biofilm can include polysaccharides, proteins, nucleic acids, fats, and humic substances.

The process of biofilm formation is influenced by many factors:

1. Texture, hydrophobicity, charge, surface chemical structure, the presence of adsorbates;
2. Temperature, pH, ionic strength, flow rate, nutrient content in the environment;
3. Chemical composition of the bacterial cells surface, their hydrophobicity, the presence of pili and flagella, the composition of the extracellular matrix, and the expression of "quorum sensation."

The film formation on the surface of the anodic electrode is still of a great interest.

Detailed structural studies of bacterial surface may help to understand molecular mechanisms of the biofilm formation and functioning.

In bacterial AFM investigations, a rigid cellular surface becomes a key feature. Cell wall contains specific antigenic determinants, responsible for bacteria-host binding. Rigid cellular surface also plays important role in the bacteria survival in different environmental conditions. It is much more rigid than animal cell wall.

Bacterial systems play an important role in maintaining ecological balance. Many dimensions of diverse bacterial systems are still to be harnessed for their successful application in environmental management programs. Most importantly, the bacteria may also serve as a sensitive and an early indicator of the environmental pollutants appearance before starting to eliminate them.

3. Scanning probe microscopy is a necessary tool in bioremediation investigations

AFM is the best tool to characterize and visualize the degradation process in details. AFM is applied to various investigations of bioremediation processes.

AFM is used to determine the biomass morphological modification during the process of anthracene degradation (Das et al., 2017). It was shown that biomass filamentous growth demonstrated significant transformation on the cell volume. It also leads to cell surface modification and formation of exopolymeric matrix. Filamentous growth of the *Bacillus cereus* strain JMG-01 biomass in the form of biofilm can also be reasoned to the chemotaxis behavior of the strain JMG-01 in enhanced anthracene degradation.

Di-(2-ethylhexyl) phthalate (DEHP) is one of the most extensively used plasticizers in plastic products. According to obtained AFM results, well growth of strain *Rhodococcus pyridinivorans* XB under DEHP stress was confirmed (Zhao et al., 2018). AFM pictures proved that XB could rapidly utilize DEHP (200 mg/L, 98% of removal rate within 48 h) and might produce some extracellular polymeric substances as a response to DEHP stress.

Pseudomonas putida KT2442 is a bacterium that could degrade chlorinated hydrocarbons in soils, and it is believed that its behavior in subsurfaces is controlled by surface polymers (Bell et al., 2005). AFM was applied to understand the mechanism of surface polymers interactions. Force analysis was fulfilled by means of AFM. The obtained data can be quantified as the discrete pull-off forces versus pull-off distances.

Bioremediation is a considerable method for treating Cr(VI) contamination. Bacterial surface changes of *Ochrobactrum anthropi* during Cr biosorption was investigated in study of Li et al. (2008). It was discovered that Cr adsorption capacity increased with the increase of initial Cr(VI) concentration. AFM morphologic analysis combined with surface roughness analysis indicated that the bacterial surfaces became rougher during Cr uptake process.

Surface roughness of bacteria cell wall is a very important parameter in biodegradation research. The surface roughness affects bacterial transport of substances. In study of Xiaohong et al. (2011), biosorption mechanism of Ni(II) by living *B. cereus* was investigated. AFM analysis showed that the bacterial surface roughness increased from 7.9 ± 0.5 nm to 12.6 ± 1.6 nm during this process.

The growth of *Beauveria bassiana* was observed in the absence and presence of Pb(II). AFM was performed to determine the morphological changes influenced by Pb(II) toxicity. Obtained results showed the increase in surface roughness compared with control cell (Gola et al., 2018). The increase in the roughness indicated the rupturing in the cell wall of fungus because of Pb(II) toxicity.

The bacterial degradation of graphite oxide was proven by means of AFM and other microscopic methods (Qu et al., 2018), which also helps to reveal the degradation mechanism.

The green bioremediation technology of toxic nitro aromatic pollutants by means of golden nanoparticles is described in study of Nag et al. (2018). AFM is used to characterize golden nanoparticles produced by bacterial strain *Staphylococcus warneri*.

Brevibacillus laterosporus are bacteria with insecticidal features because of the specific canoe-shaped crystals with antimosquito properties (Smirnova et al., 1996; Orlova et al., 1998). *B. laterosporus* is commonly found in soil, water, plants, and other places in the human environment and shows survival in harsh environmental conditions. *B. laterosporus* has the ability to biodegrade and discolor commercial textile dyes in wastewater until its removal (Gomare and Govindwar, 2009).The dead biomass of *B. laterosporus* (MTCC 1628) is used as a biosorbent to remove Cd (II) and Ni (II) ions from an aqueous solution (Kulkarni et al., 2014). Electron microscope does not allow us to work with native material because of probe preparation. That is why the spores of this bacilli type are poorly studied. Sample preparation for electron microscopy consists of following steps: fixation, dehydration, and coloration, which destroy the biological structure of objects. It is impossible to accurately measure the objects size and to observe native ultrastructure of samples according to electron microscopy images. AFM was used to visualize the molecular architecture of spores without chemical and physical influences.

FemtoScan multifunctional SPM (Yaminsky, 2013) and FemtoScan Online software (Yaminsky et al., 2016) were used for AFM of spores and crystals of *B. laterosporus*. It contains a canoe inclusion, for which function is not defined, and crystals, whose activity toward different biological objects is poorly studied. 3D image of spore of *B. laterosporus* on the surface of cleaved mica was obtained, and spores of *B. laterosporus* were studied using the AFM (Fig. 14.1). The observed size of the spore was about 1.5 μm. AFM allows us to obtain 3D

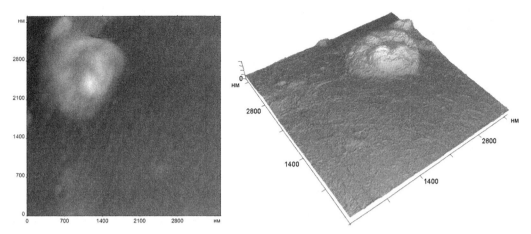

FIGURE 14.1 Atomic force microscopy image of spore of *Brevibacillus laterosporus* Lat. 006: (left) topography, (right) three-dimensional view (FemtoScan scanning probe microscope).

topography and to estimate the spore volume at its drying in the air. High biocidal effect (antibacterial, fungicidal, cyanolytic) allows us to use these strains as producers of biological plant protection from disease and to fight against toxic microscopic algae (blue-green bacteria) (Zubasheva et al., 2017).

4. Bacterial electromechanical biosensor

Biomonitoring tools (e.g., bioassays, biomarkers, microbial community analysis) have great potential when used as a tool for assessing the risk of emerging chemical pollutants (Gavrilescu et al., 2015). Biosensors can be very useful for quick and cheap first screening to detect the specific pollutants presence (Diplock et al., 2010; Xu et al., 2013). Existing designs of bacterial biosensors and bioreporters, their application in measurement of harmful chemicals present in water, air, soil, food, or biological samples are presented in this study (Van der Meer and Belkin, 2010). Biosensors can be good additional tools for the pollutants detection in situ, and they do not require heavy and expensive equipment (Kalogerakis et al., 2015). It is important to note that biosensors respond to the amount of pollutant that is bioavailable (available to cells), while chemical methods determine the total amount of compound present (bioavailable and not bioavailable), which can overestimate the real risks from a toxic point of view (Tecon and ven der Meer, 2008). Therefore, biosensors can be very useful for measuring the pollutants ecotoxicity, as well as for monitoring bioremediation processes. For example, it was found that a biosensor based on the constructed bacterial strain *Escherichia coli* is useful to detect arsenic salts in groundwater (Siegfried et al., 2012). A study of Nakada et al. (2008) in Japan showed that ibuprofen was effectively removed (~50%) from estuarine waters by microbiological degradation, and triclosan was removed by a combination of microbiological degradation, photodegradation, and absorption. Ibuprofen bacterial degradation by *Sphingomonas sp.* was shown by Kagle et al. (2009).

It is obvious that fundamental studies of bacterial cells active reactions at the biochemical and physiological levels in certain environmental conditions will allow the development of monitoring environmental pollutants methods.

The living bacterial cell is a moving microobject, and the rate of its movement is the indicator of its life. The excess of the movement power over the typical thermal motion, discovered by Robert Brown in 1827, is because of the living processes in the active cell. The SPM is one of the instruments that can detect such a movement and visualize bacteria.

The response of the individual cell because of environmental condition changes may become an important signal in the precise measurements. SPM is an instrument for the proper study, even on the level of the individual cell. SPM provides precise information of the bacteria morphology and many of its properties including mechanical rigidity, deterioration, adhesion, frictional properties.

Now we are developing a biosensor for recording the bacteria viability and the effectiveness of its application in waste biodegradation. It should be a compact and inexpensive biosensor for various biological targets detection. For example, it might check living activities of bacteria depending on environmental conditions in microbial fuel cell.

The biosensor would be applicable in bioreactor for the real-time biodegradation analysis. It can be applied to various substances that are biodegradable. The biosensor could be a simple and cheap way to test the ability to survive bacteria in a polluted environment. The biosensor also might be used for waste chemical monitoring in water. As an indicator of the bacteria state, it is suggested to register the intensity and frequency range of the membranes characteristic vibrations.

The biosensor consists of the biochip that is covered by sensory layer. To attach bacteria sensory layer contains antibodies against surface antigenic determinants, providing biospecific binding. The biochip represents a miniature of piezoceramic disk with sensory layers on the opposite sides.

Bacteria live/dead test is carried out by registration of the amplitude, phase, frequency, and the quality factor of the biochip mechanical vibrations. Measuring with SPM, you can directly calculate the number of biological agents on the biochip surface in addition to the determination of these parameters.

The biological agent attachment to biochip changes the resonant frequency of disk mechanical oscillations. The bacteria interaction with the sensor surface leads to a change in the effective mass and stiffness of biochip that can be registered by resonance frequency shift:

$$\Delta f = \frac{1}{2} f_n \left(\frac{\Delta k}{K} - \frac{\Delta m}{m} \right) \qquad (14.1)$$

where Δf, Δk, and Δm are changes of the resonance frequency f_n, stiffness k, and mass m of the biochip, and f_n, k, and m are the initial values of these parameters.

Let us consider the detection of *E. coli* bacterial cell. Length of bacteria is about 2–6 μm and diameter is about 1 μm. With an average length of 4 μm, the mass of bacteria is equal to about 3.1×10^{-12} g.

The biochip with 3 mm diameter and 0.1 mm thickness is made of piezoelectric ceramics (lead zirconate titanate 19). It has a volume of 0.7×10^{-3} cm^3 and weight of 5.3×10^{-3} g (the density of material is 7.5 g/cm^3). Single bacterial cell attachment leads to a relative change in the resonant frequency by the value:

$$\Delta f/f_n = -\frac{1}{2}\frac{\Delta m}{m} = -3,0.10^{-10} \tag{14.2}$$

According to relation (2), the relative frequency stability of the drive-pulse generator should be at the level of 10^{-10}.

Some factors limit the method sensitivity, e.g., instability of piezoceramics and the frequency measurement error Δf of the piezoelectric biochip, which electronic noise of the measuring circuit can cause.

The piezoelectric ceramic resonator with a resonant frequency of 1 MHz was used to assess the self-noise level of the biosensor measuring system.

The electronic boards of the SPM are used in the biosensor control unit, including a digital frequency synthesizer, a precision amplifier of the input signal, an interface for communication with a computer unit, a thermostat of the flowing liquid cell, a DAC–ADC, a stabilized power supply, and a digital signal processor. Such technical solutions as biochip symmetry and electronic circuit symmetry have made it possible to significantly improve the accuracy and reliability of measurements.

Thus, the response of an individual cell to changes in environmental conditions plays an important role in accurate measurements of the contamination level. This reaction allows evaluating the possibility of application of a particular bacterium for biodegradation purposes.

5. Scanning ion-conductance microscopy

"The scanning ion-conductance microscope" (SICM) is the first publication about ion-conductance microscopy that was published in 1989 (Hansma et al., 1989). The article by P.Hansma, B.Drake, and O.Marti in the Science journal gave impetus for a new microscopy area development.

SICM consists of four main components:

- an ion-sensitive glass probe (micropipette) filled with electrolyte;
- scanning piezoelectric system;
- specialized electronic measuring equipment, including a feedback system;
- digital electronics and a computer that provides a microscope user interface, control of the system, and also allows received data processing.

SICM can image the surfaces topography in the electrolyte. A micropipette filled with electrolyte contains the electrode that scans the surface. The smaller the distance between the electrode and the object, the smaller is the ion flux running through the micropipette.

Ion-conductance microscope can be used for both conductors and insulators in comparison with STM that measures only the conductors. Unlike the AFM, ion-conductance microscope does not make any impact on the sample during scanning. SICM has advanced significantly in the last 10–15 years with the modulation techniques development (Gorelik et al., 2003, 2004; Zhang et al., 2005) and especially because of hopping mode (thanks to Yuri Korchev and co-authors) (Novak et al., 2009; Shevchuk et al., 2011). The hopping mode allows to fully realize the scanning ion-conductance microscopy benefits (Yaminsky, 2016). Needle-controlled movement vertically to the sample at a distance of a few microns makes it possible to study rough and smooth objects, which include many biological systems. Advanced signal processors and FPGA in combination with high-speed operational amplifiers, analog-to-digital converters, and digital-to-analog converters enabled a delicate scan of cells, fibers, and many other biostructures without their deformation. Quartz or glass capillary probe in comparison with a conventional cantilever has a smaller convergence angle at the vertex, and hence the image broadening is visible in SICM in significantly lesser extent.

The feedback mechanism can maintain a given conductivity while simultaneously determining the distance to the surface. Ion-conductance microscopy can image not only topography but also local ion currents through the pores on the surface. Hansma and colleagues also proposed a combined atomic force and ion-conductance microscope. The design is based on the use of a curved glass pipette, which acts as a force and conductivity sensors. The deflection measurement of the pipette allowed obtaining a more stable feedback than it was possible in previous ion-conductance microscopy versions. Synthetic membranes were studied in the contact and tapping modes in liquid by means of combined microscope. A year later Prof. Yu. Korchev and colleagues fulfilled scanning ion-conductance microscopy of living cells that allowed studying topography without damaging the sample. The images resemble those obtained with scanning electron microscopy, with a significant difference in the fact that the cells remain viable and active.

The device can control the cell surfaces of small-scale dynamics, as well as the movement of whole cells (Korchev et al., 1997a). The experiments with mice melanocytes showed that SICM is more suitable for sample visualization in aqueous solutions (Korchev et al., 1997b). As the probe measures ion current without physical sample contact during scanning, no preliminary cell preparation is required (e.g., fixation on the substrate).

SICM could potentially be applicable in real-time studies of electrophysiology, micromanipulation, and drug delivery. Ion-conductance microscopy can measure changes in cell volume in the range from 10^{-19} to 10^{-9} L (Korchev et al., 2000a). A hybrid of SICM and scanning near-field optical microscope was presented by Korchev et al. (2000b). This method allows cell surface quantitative analysis with high-resolution and simultaneous topographic and optical imaging. A special method feature is a reliable controlling mechanism that manages the distance between the probe and the sample in a physiological buffer. Seifert et al. (2015) carried out comparing study of scanning ion-conductance and AFM. Living cells A6 microvilli were used as model samples for comparing the capabilities of AFM and SICM imaging. The quality of AFM images has improved significantly after fixing the cells, while on SICM images, it has not changed. In AFM, the measured height and width of whole cells depended on the force value, while in SICM they were constant within a large range of preset values.

The recently much improved SICM is another instrument for bacterium local probing as single cell chemical sensor. For example, it can detect also the concentration of active forms of oxygen in the close vicinity of the cellular membrane of *Gluconobacter oxydans* (Fig. 14.2).

The resolution of SICM is determined by the pipette tip geometry and the distance between the pipette and the sample. The typical achieved resolution is about 10 nm in the vertical direction and about 50 nm in the transverse direction (Ying et al., 2005).

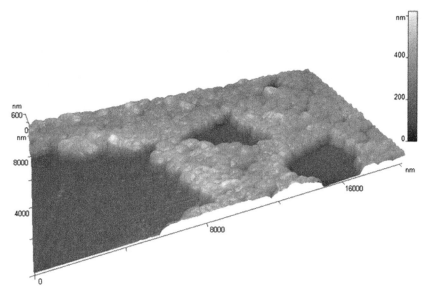

FIGURE 14.2   Three-dimensional image of a monolayer film of bacteria *Gluconobacter oxydans*. Measurement in resonance mode.

The best resolution (3–6 nm) was obtained by visualization of S-layer proteins of *Bacillus sphaericus* on the surface of mica with a 13 nm diameter nanopipette (Shevchuk et al., 2006). Hopping mode of ion-conductance microscopy allows adjusting the angle at which the nanopipette approaches to the cell (Leo-Macias et al., 2016). The angle can be adjusted in the range of 0–90 degrees to the surface. In addition to high-resolution imaging of topography, SICM can perform a multifunctional analysis of living cells, including morphological transformations caused by physiological effects, identification of intracellular signaling pathways, and characterization of mechanical responses, which demonstrates the method versatility. With the help of SICM, ventricular myocytes obtained from cardiac tissue (Lyon et al., 2009) subjected to prolonged mechanical unloading (Ibrahim et al., 2010) or dissection caused by osmotic shock (Gorelik et al., 2006) were investigated. In all these cases, scanning ion-conductance microscopy revealed obvious surface structure changes compared to healthy tissue images of myocytes. The nanopipette can be used as a pH sensor (Piper et al., 2006). Thus, we see that a capillary probe or nanopipette can act as a drug delivery device, an electrochemical sensor, a pH biosensor, a test system for detecting metal ions, and many others. Scanning ion conductance microscopy has been evolving into scanning capillary microscopy in which the probe capillary plays a lot of different functions as biosensor, electrochemical electrode, test system for metal ions, the means for delivery of biomacromolecules.

Scanning ion-conductance microscopy capabilities are much broader than just surface topography observation of rough and smooth objects with a low mechanical rigidity. The use of multichannel capillaries as a probe allows multiparametric analysis of cells. Chemical modification of one or several capillary channels can serve as electrochemical nanosensor (Bhargava et al., 2013). Capillaries with two or more channels also give the opportunity for directed substances transfer of biomacromolecules (peptides, proteins, nucleic acids, etc.) on the biological object surface or inside (Akhmetova and Yaminsky, 2017).

In our research on scanning capillary microscopy, we use a device built onto an inverted optical microscope Ti—U (Nikon, Japan). We can predict the future wide application of ion-conductance microscopy in bioremediation applications.

6. Nanolithography

AFM-based lithographic methods can be used to selectively functionalize and structure the surface in addition to creating complex quantum devices and sensory layers. Progress in nanolithography using SPM depends on speed increase and reliability of the method for mass production, as well as on the appearance of specialists advancing these methods.

AFM-based lithography makes it possible to manufacture structures with a size of only several tens of nanometers.

In 1990, Dagata and colleagues observed chemical modification of the silicon surface passivated with hydrogen by applying a voltage between the needle of a STM and the surface (Dagata et al., 1990). It was found that the composition of the modified region is silica. This phenomenon is explained by the silicon local oxidation under the electric field action in the region of the tunnel junction.

In 1993, lithography was demonstrated by local anodic oxidation (LAO) using an AFM (Day and Allee, 1993). This observation opened the way for the development and expansion of local oxidation approaches for surface modification. A sample modified by LAO must be a conductor or semiconductor, and a substrate with a conductive coating is used to install the sample to provide the voltage between the probe and the sample. For LAO, probes with a conductive coating are used. Usually for LAO of surfaces, both in the contact and dynamic modes, AFM probes with such coatings as Pt/Ir, Au, W2C coating and heavily doped Si probes are used.

LAO using an AFM is carried out in air at a sufficiently high humidity. Water vapor in the environment is necessary for this process. In a humid environment, a thin water layer is adsorbed onto the surface. When a conducting probe (AFM cantilever) approaches the sample, a water bridge is formed between the surface and the tip of the cantilever. The probe and the conductive sample form a single circuit. When a voltage of several volts is applied to the probe, a field with electric field strength of 10^8–10^{10} V/m arises at the top of the probe; under the action of this field, an electrochemical reaction occurs: water molecules dissociate into protons H^+, and O^{-2}, and OH^- anions. The ions interact with the surface of the substrate to form an oxide. As the oxide volume exceeds the volume of the initial substrate material, the oxidation reaction leads to the formation of protrusions (volume parts enlarged in volume) on the surface, which can be observed in the AFM. The height and width of the formed oxide layer can be increased by amplifying the current flowing between the probe and the surface. The size of the formed surface is also affected by the scanning speed of the probe over the surface. The higher the scanning speed, the less charge that passes through a point on the surface, and the smaller the size of the structure (Bukharaeva et al., 2012).

The main advantage of LAO is the ability to control in real time the electrical and topographic characteristics of nanoscale structures. This kind of probe nanolithography also allows the formation of dielectric barriers, resistive masks for selective etching, and patterns that can be used in the formation.

The disadvantage of the method is application only for a limited class of materials (anodically oxidized metals and heavily doped semiconductors, hydrogenated silicon).

Force lithography is based on the direct mechanical impact created by a sharp probe on the sample surface. This type of nanolithography is used in nanoelectronics, nanotechnology, materials science, etc. Power lithography allows creating electronic components on a nanometer scale, studies the mechanical properties of a material, and is mainly used to apply patterns on polymer surfaces (Xie et al., 2006). There are static and dynamic modes of power lithography. In the static mode, the mechanical action of the cantilever on the sample surface is used to obtain small grooves with a characteristic cross section determined by the shape of the tip.

The main disadvantage of this method is rapid probe destruction. Cappella et al. (2002) also pointed out the following shortcomings of the method for modifying thin films. First, only thin films (5–25 nm) on solid substrates can be modified. In addition, the study of topography after modification can be performed only with small forces and, therefore, with a reduced resolution. If the applied force is too large (the threshold value depends on the polymer and on its thickness), the polymer will be further modified. The only solution is to scan the surface after modification in another mode. The third drawback is the torsion bends of the cantilever during lithography, which causes irregularities on the edges. Because of this, the direction of lithography is limited in a certain range around the axis of the cantilever.

In the dynamic mode, the probe oscillates, and the modification is made by forming depressions in the intermittent-contact scanning mode. This method eliminates torsional distortion; the visualization of the formed pattern is made without serious impact on the substrate surface.

To obtain a qualitative modification of surfaces of different hardness, it is necessary to use hard probes with a minimum tip radius, select a surface scan area without significant artifacts and minimal elevation differences, prepare a quality template, and select the optimal settings for the SPM control program parameters.

The fabrication of nanochannels (50 and 5 nm), nanosquares (50 and 5 nm), and complex nanostructures on a gold nanowire was shown by Li et al. (2005). Such structures were created by direct impact of the AFM probe on the surface of Au. It is believed that the creation of nanostructures directly on the nanowire may have applications in the field of fluid and biotechnology because these samples can be used to perform nanoscale fluid injection, fluid separation, and chemical sensing.

6.1 Capillary stereolithography

Since its inception, AFM has found successful applications for the implementation of nanolithography of various modes using force, thermal, or electrical effects. Scanning capillary microscopy has become an effective tool for directed mass transfer of individual biomacromolecules, nanoparticles, etc. Multichannel probes can be used in 3D printing of specified molecular configurations. The supply of macromolecules through a capillary can be carried out in various ways, for example, using electrophoresis or creating an overpressure in the capillary.

Rodolfa et al. (2005) successfully performed two-dimensional lithography with fluorescent proteins on a SICM using a two-channel capillary. Thus, a color miniature copy of the "Dancers" painting by Degas was created. The size of the miniature corresponded to the diameter of the hair, about 50 microns.

Scanning capillary stereolithography has certain advantages compared with laser stereolithography, the spatial resolution of which is limited by the diffraction limit—about half the length of laser radiation. In capillary stereolithography, nanometer and subnanometer positioning accuracy is achieved. The output capillary diameter can vary from units to hundreds of nanometers, which opens up the possibility to create a molecular 3D printer.

In the manufacture of single-channel or multichannel nanocapillaries for stereolithography, glass microtubules made of quartz or borosilicate glass are widely used in electrophysiological applications.

During the 1990s, AFM-based nanolithography was used to manipulate and modify surfaces at the nanoscale in studies of various fundamental phenomena. Although early studies have focused on studying the Si surface to develop a basic understanding of the growth kinetics of oxides, further work has been devoted to the properties of nanoscale oxides acting as dielectric layers, etching masks, and creating patterns for specific similarity to functional molecules (Xie et al., 2006).

In bioremediation processes, nanolithography can be used in following ways: to create structure on the surface of graphite electrode for bacteria better adhesion. In addition, the AFM-based lithographic techniques are suitable for patterning a variety of organic and biological molecules on surfaces at ambient conditions (Liu and Zhang, 2007). It can be applied for fabrication of nanodevices and nanosensors. By means of nanolithography, one can create nanotemplates that can immobilize biomolecules (e.g., proteins) and thus form a sensory layer for biological specific binding.

7. Scanning probe microscopy measurements of bacteria—manual

The unique microscopic capability of SPM is the real-scale visualization of the surface structures down to the atomic size. This function is continuously developing, and recent instrumental improvements are helping to make such measurements routine in practically all imaging modes (contact, oscillatory resonant, and nonresonant modes). In particular, the enhanced capabilities of high-resolution imaging can be implemented for visualization of individual macromolecules on substrates when they are deposited as the isolated objects or in various self-assembled structures such as biofilms.

7.1 Substrate preparation for imaging of bacteria in air

Mica is an excellent material for the bacterial samples deposition for SPM observations. The most popular type of mica is muscovite, which is a hydrated phyllosilicate mineral of aluminum and potassium with a typical formula $(KF)_2 (Al_2O_3)_3 (SiO_2)_6(H_2O)$. Before bacteria deposition, the mica should be cleaved along the basal plane. In this case, a clean sufficiently large surface is exposed for bacteria deposition. 10×10 mm^2 sheets of mica are proper for deposition of 4 µL droplet of bacteria with concentration of about 10^9 mL.

Other substrates such as glass, quartz, and silicon also can be used for bacteria deposition. Still they are not as hydrophilic as freshly cleaved mica. Sometimes it is not easy to obtain clean surface of such substrates.

7.1.1 Muscovite mica, 10×10 mm^2 square size, 0.1—0.4 mm thickness.

7.1.2 Split mica along the basal plane into two sheets of approximately equal thickness by pressing on the end face with a sharp awl or metal needle.

7.1.3 Put the prepared mica sheets into the Petri dish so that the cleaved surface is oriented upward. Avoid touching the cleaved surface. The mica can be slightly attached to the bottom of Petri dish using double-strand adhesive tape.

7.1.4 Every sheet of mica should be labeled with numeric number or with clear alphabetic code, written on the bottom of Petri dish. Later on this numeric number of alphabetic code will be used as file name or a part of the file name for the image obtained using AFM.

7.1.5 Up to 15 freshly cleaved sheets of mica can be placed in the Petri dish 90 mm in diameter.

7.1.6 Store the prepared samples in a closed Petri dish.

7.2 Substrate preparation for imaging of bacteria in air

For AFM measurements in liquids, mica must be pretreated with polylysine.

7.2.1 Prepare muscovite mica as described earlier.

7.2.2 10^{-2} M polylysine solution: dissolve 100—300 kDa polylysine in distilled water to obtain a 10^{-2} M concentration.

7.2.3 Put 100 µL droplet of 10^{-2} M polylysine solution on the surface of freshly cleaved mica.

7.2.4 Let the droplet to dry at room temperature.

7.2.5 Store the prepared samples in a closed Petri dish.

7.3 Bacteria preparation

7.3.1 The bacteria are grown in a standard manner on a nutrient medium in a Petri dish.

7.3.2 The colony of bacteria is transferred by means of a loop to Eppendorf tube filled with distilled water.

7.3.3 The amount of transferred bacteria should be enough to the concentration of bacteria in suspension at the level about 10^9 bacteria per mL.

7.3.4 Shake quickly the tube for several or tens of seconds with bacteria suspension to achieve uniform concentration in the volume.

7.3.5 After shaking the tube, immediately put 4 µL droplet of bacteria suspension onto the freshly cleaved mica surface 10×10 mm^2 in size.

7.3.6 The preparation of bacteria with distilled water should be performed quickly within a few minutes to avoid osmotic shock and lysis of bacteria in distilled water.

7.4 Cantilever choice

Different types of cantilevers should be used for different modes of SPM. Rectangular or triangular types of cantilevers are the most proper. Cantilevers with metal coating not only give better reflected optical signal but also provide higher temperature drift.

7.4.1 Silicon or silicon nitride cantilevers with spring constant in the range 0.01−1 N/m are used for bacteria imaging in air and in liquids.

7.4.2 The specified radius of the cantilever tip curvature should be less than 10 nm for better space resolution during imaging.

7.4.3 Rectangular or triangular types of cantilevers are proper for imaging in air and in water.

7.4.4 Silicon cantilevers with spring constant in the range 1−100 N/m and mechanical resonant frequency in the range 50−500 kHz are used for imaging in air in resonant mode. Radius of the cantilever tip curvature should be less than 10 nm for better space resolution during imaging.

7.4.5 Rectangular types of cantilevers are proper for imaging in air in resonant mode.

8. Methods

8.1 Atomic force microscope installation

AFM is usually used for the study of bacterial cells. A sharp tip scans the surface of bacteria in contact or semicontact mode. The interaction in between the tip and the bacterial surface is controlled by the feedback.

8.1.1 Place the AFM in a quiet and silent place to avoid the influence of external acoustic, seismic, and electronic noise.

8.1.2 The temperature in the AFM location should be stable and comfortable for the microscope operator.

8.1.3 Place the cantilever, chosen according to the selected imaging mode, into the cantilever holder.

8.1.4 Place the mica with deposited bacteria onto the metallic tablet 16 mm in diameter and 1 mm in thickness.

8.1.5 Put the table with the sample onto the sample table of AFM.

8.1.6 Use an optical microscope to position the cantilever above the selected bacteria.

8.1.7 Adjust the laser beam on the free end of the cantilever and adjust the photodiode signal slightly below zero.

8.1.8 Choose the set point as zero, so that after switching on, the reflected beam from the cantilever laser beam will be centered at the photodiode. In this case, the dynamic range of the cantilever bending in up and down directions will be equal. In addition, this position of the reflected laser light decreases the influence of laser power fluctuations on the useful signal.

8.1.9 Approach the cantilever tip toward the sample and start scanning.

8.2 Atomic force microscopy imaging in air in contact mode

Contact mode is widely used for bacteria imaging. The bacteria shows high mechanical rigidity during observation in air and that is why the received image is not distorted by the applied normal and lateral forces during scanning.

Three different images can be obtained in contact mode: height image, deflection image, and friction image. The actual height of the object is seen in height image. In AFM, it corresponds to the position of the cantilever over the object, while the tip of the cantilever is in contact with the object's surface. In real experiments, the feedback circuit of the microscope controls the position of the cantilever or the object so that the force in between the object and cantilever remains constant. For example, the cantilever may be fixed in vertical direction, while the piezoceramic manipulator moves the object forward or backward relative to the sample. However, the imperfection of feedback leads to small displacements of the cantilever in vertical direction that is recorded in the deflection image. The deflection of the cantilever occurs mainly on sharp changes in the sample relief in those cases when the feedback is not so quick to track these changes. For the observer, the deflection image is more informative because all small details of the relief are clearly visible on it. Nevertheless, the real height of the object can be seen only in the height image. The friction image corresponds to the variation of friction properties of the bacteria surfaces. To obtain this image, it is necessary to move the sample in the direction perpendicular to the axis of the cantilever. In this case, the applied cantilever lateral force leads to torsional deflections of the cantilever along its axis.

8.2.1 Adjust the feedback parameters for better contrast and lower noise. Start with increasing the integral gain Imax until the appearance of small oscillations and access noise in the image, the proportional gain should be at lower position. Reduce the integral gain up to 30% of Imax.

8.2.2 Increase the proportional gain Pmax until the appearance of small oscillations and access in the image. Reduce the proportional gain up to 30% of Imax.

8.2.3 Check the quality of the image and start measurements if the achieved quality is sufficiently good. Otherwise repeat the previous procedure again.

8.2.4 Obtain height, deflection, and friction images.

8.3 Atomic force microscopy imaging in air in resonant mode

Resonant mode has different commercial names, for example, tapping mode (hopping), semicontact mode (semicontact). In the resonant mode, the cantilever oscillates at its mechanical resonant frequency. When the cantilever approaches the sample surface, the amplitude of its oscillations changes with changing frequency and phase. The cantilever scans the surface, and the feedback maintains the oscillation amplitude at a constant value. You can also scan the sample surface at a constant frequency or phase of the cantilever oscillation. As a rule, three types of images are obtained in the resonant mode—the image of height, the image of amplitude, and the image of phase. The phase image is very sensitive to small changes in surface properties and thereby emphasizes the relief features.

8.3.1 Adjust the amplitude and frequency of the free cantilever oscillation as described in microscope manual.

8.3.2 Choose the set point for bacteria scanning as 90% of the amplitude of free cantilever oscillations.

8.3.3 Approach the cantilever tip toward the sample and start scanning.

8.3.4 Adjust the feedback parameters according to the 3.2.7–3.2.9 procedures.

8.3.5 Obtain height, amplitude, and phase images.

8.4 Atomic force microscopy imaging in liquids

E. coli of two different strains JM109 and K12 J62 were observed in the study (Bolshakova et al., 2001). The comparative studies of bacteria with AFM in air and in liquids have shown that bacteria, imaged in liquid, show the loss of topographic features, as seen in air. In addition, the observed dimensions in liquids are larger than in air.

8.4.1 Place the sample of bacteria prepared for imaging in liquids onto the metallic tablets.

8.4.2 Place the metallic tablet with the sample into the liquid cell of the AFM.

8.4.3 Fill the liquid cell with the buffer saline solution to avoid osmotic shock of the bacteria.

8.4.4 Adjust the laser diode, photodiode, and feedback for the contact mode or the resonant mode.

8.5 Three-dimensional image processing

AFM provides bacteria imaging in 3D. Using FemtoScan Online software, it is possible to prepare movies composed of preliminary chosen flyviews.

Special attention should be paid to the image processing of the experimental data, especially when it is used for presentations and publications. Many different software packages may be used for image processing. Among them are Gwydion, VSxM, SPIP, FemtoScan Online. In present review, we use mainly FemtoScan Online software, which has many different options and permits to obtain good results in less number of steps (Yaminsky et al., 2016). Color presentation of the images is a long and fascinating story. Still a simple warm brown palette is widely used in imaging and may be regarded as standard de facto. The height of the object in the image is displayed in shades of the brown palette. Many other palettes are also used for proper presentation of bacterial morphology and its specific features. The thermal drift leads to an apparent overall slope in the image. This slope should be eliminated by algorithm of linear or nonlinear slope reduction. 3D imaging of bacterial cells is widely used for vivid presentation in different color palettes.

8.5.1 Choose the top view image for building 3D view in the FemtoScan Online software.

8.5.2 Build 3D image in the chosen palette, Z scale, and point of view.

8.5.3 The scales of all coordinate axis may be shown or removed. This is controlled by the tick against *Show legend* in the Menu View.

8.5.4 It is possible to show or to hide marks, curves, isolines, and sections that were drawn on the surface. To show these elements, items *Marks, Curves, Isoline,* and

Sections should be checked in the View/Decorations menu item. If *Labels* item is checked, than captions to visible elements will be shown. When new elements will be drawn on the image for which 3D image is built, it does not affect 3D image automatically. So new elements will not be shown there. To refresh the 3D image in this case, click it with the left mouse button and hold it for a second, and release.

8.5.5 The surface may be rotated round vertical axis. To fulfill this, press the left button of the mouse and holding it pressed move the mouse in left and right directions.

8.5.6 The scale in vertical direction may be changed. Press the left button of the mouse and move the mouse up and down without releasing it.

8.5.7 The image may be put closer or far. It can be done if the mouse is moved up and down while its left button is pressed and keyboard button shift is also pressed.

8.5.8 The surface may be rotated round horizontal axis, lying in the monitor plane. To do it, move the mouse up and down with mouse left button and button Ctrl pressed.

8.5.9 If the word Highlighted is selected in the View Menu, then the surface will be illuminated. It is possible to adjust the light position, and this can drastically change the impression of the image. Movement of the mouse with pressed its left button and keyboard buttons Shift and Ctrl leads to the movement of the light source.

8.6 Image filtering

Practice with following filters to improve the overall view of the image:

- Slope reduction
- Median filtering
- Morphological filters
- Differential filters
- Averaging
- Threshold filtering

Quantitative analysis

Options for the quantitative analysis:

- Distance measurements
- Angle measurements
- Histogram measurements
- Roughness measurements
- Calculation of the volume, restricted by the isoline

Some examples of quantitative analysis of bacterial cell images are shown below in Fig. 14.10—14.12.

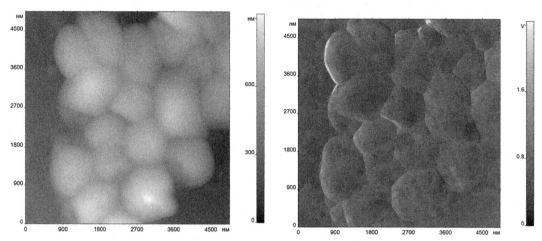

FIGURE 14.3 Height (left) and friction (right) images of *Escherichia coli*. The regions with different frictional properties are seen in the left image: Z scale in friction image is in arbitrary units.

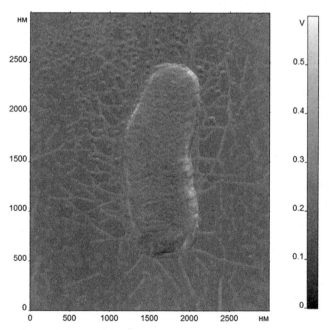

FIGURE 14.4 Atomic force microscopy friction image of *Escherichia coli* bacterial cell. Z scale is arbitrary units.

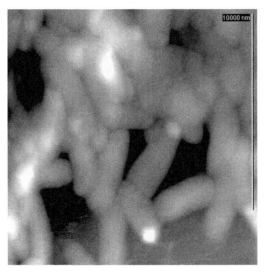

FIGURE 14.5 *Escherichia coli* JM109 is observed in liquid media. The bacteria are immobilized on mica substrate modified by polylysine.

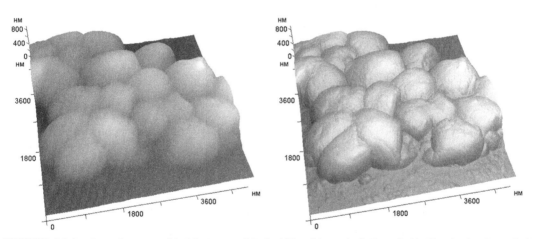

FIGURE 14.6 Three-dimensional height image of *Escherichia coli* bacteria (left), and side illumination is applied for the image (right).

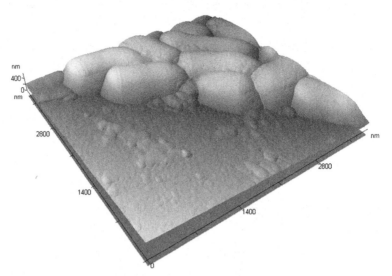

FIGURE 14.7 Three-dimensional atomic force microscopy image of *Helicobacter pylori* bacterial cells in blue palette.

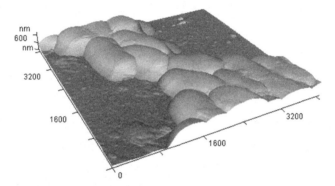

FIGURE 14.8 Three-dimensional atomic force microscopy image of *Helicobacter pylori* bacterial cells in green palette.

FIGURE 14.9 Three-dimensional atomic force microscopy image of *Escherichia coli* bacterial cell in brown palette.

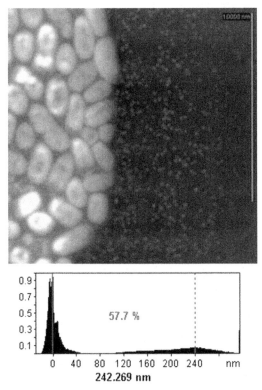

FIGURE 14.10   The height histogram (the normalized number of points for different height) is used to determine the average height of bacteria film. The measured average height is about 242 nm.

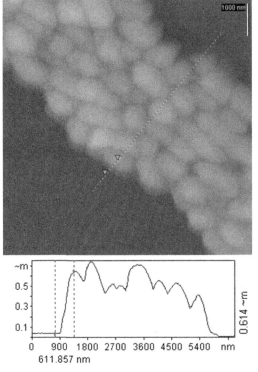

FIGURE 14.11   The size of bacteria is determined using cross section.

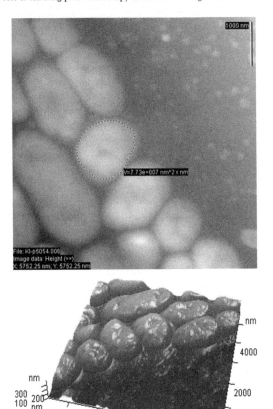

FIGURE 14.12   The isoline is used for the determination of the contour length, square, and volume of the cell. Top image with isoline (line of the constant height) (upper), and illuminated three-dimensional image of the same region (lower).

9. Conclusion

SPM is an efficient tool to be used in applied studies; among them is the involvement of bacteria in the remediation processes. It can provide valuable information on the 3D morphology of bacterial cells, its local properties such as mechanical rigidity, strength, deterioration, and adhesion. The study of bacteria using SPM may be provided in environmental conditions during, for example, the biofilm formation on the electrode surface of bacterial fuel cell. By measuring the intrinsic oscillations of the outer membrane, it becomes possible to provide life/dead test within several minutes. The indicator of the life state is the extra to the thermal (Brownian) motion oscillations, while its diminishing to the typical thermal level demonstrates the loss of its living abilities. Bacterial movement decrease also corresponds to the upsetting of its activities in the remediation processes. Scanning ion-conductance microscopy can measure the concentration of oxygen active forms near the cellular membrane of bacteria. AFM-based nanolithography can modify the substrate surface for bacteria better adhesion.

Abbreviations

AFM atomic force microscopy
LAO local anodic oxidation
SICM scanning ion-conductance microscope
SPM scanning probe microscope

Acknowledgments

The authors are thankful to Lomonosov Moscow State University and Advanced Technologies Center for the provided FemtoScan scanning probe microscopes and comfortable infrastructure facilities.
The authors are grateful for the financial support of Russian Foundation of Basic Research (projects N 15-04-07678 and N 16-29-06290), Ministry of Education and Science (project N 02.G25.31.0135), and FASIE (project N422ГРНТИС5/44715).

References

Akhmetova, A., Yaminsky, I., 2017. Scanning capillary microscopy. Nanoindustry 7 (78), 42–46.

Bell, C.H., Arora, B.S., Camesano, T.A., 2005. Adhesion of *Pseudomonas putida* KT2442 is mediated by surface polymers at the nano- and microscale. Environmental Engineering Science 22 (5), 629–641.

Bhargava, A., Lin, X., Novak, P., Mehta, K., Korchev, Y., Delmar, M., Gorelik, J., 2013. Super resolution scanning patch clamp reveals clustering of functional ion channels in adult ventricular myocyte. Circulation Research 112, 1112–1120.

Bolshakova, A.V., Kiselyova, O.I., Filonov, A.S., Frolova, O.Y., Lyubchenko, Y.L., Yaminsky, I.V., 2001. Comparative studies of bacteria with atomic force microscopy operating in different modes. Ultramicroscopy 1–2 (86), 121–128.

Brenner, D.J., Krieg, N.R., Staley, J.T., Garrity, G.M., 2005. Bergey's Manual of Systematic Bacteriology, second ed., vol. 2. Springer-Verlag, New York, NY. parts A, B and C.

Bukharaeva, A.A., Bizyaeva, D.A., Nurgazizova, N.I., Khanipov, T.F., 2012. Fabrication of magnetic micro and nanostructures by scanning probe lithography. Russian MicroElectronics 41 (2), 78–84.

Cappella, B., Sturm, H., Weidner, S.M., 2002. Breaking polymer chains by dynamic plowing lithography. Polymer 43, 4461–4466.

Dagata, J., Schneir, J., Harary, H.H., Evans, C.J., Postek, M.T., Bennett, J., 1990. Modification of hydrogen-passivated silicon by a scanning tunneling microscope operating in air. Applied Physics Letters 56, 2001–2003.

Das, M., Bhattacharya, A., Banu, S., Jibon, K., 2017. Enhanced biodegradation of anthracene by *Bacillus cereus* strain JMG-01 isolated from hydrocarbon contaminated soils. Soil and Sediment Contamination: An International Journal 26 (5), 510–525.

Day, H.C., Allee, D.R., 1993. Selective area oxidation of silicon with a scanning force microscope. Applied Physics Letters 62 (21), 269.

Diplock, E.E., Alhadrami, H.A., Paton, G.I., 2010. Application of microbial bioreporters in environmental microbiology and bioremediation. Advances in Biochemical Engineering/Biotechnology 118, 189–209.

Garrity, G.M., Boone, D.R., Castenholz, R.W. (Eds.), 2001, Bergey's Manual of Systematic Bacteriology, second ed., vol. 1. Springer-Verlag, New York, NY.

Gavrilescu, M., Demnerova, K., Aamand, J., Agathos, S., Fava, F., 2015. Emerging pollutants in the environment: present and future challenges in biomonitoring, ecological risks and bioremediation. New Biotechnology 32 (1), 147–156.

Gola, D., Malik, A., Namburath, M., Ahammad, S.Z., 2018. Removal of industrial dyes and heavy metals by *Beauveria bassiana*: FTIR, SEM, TEM and AFM investigations with Pb(II). Environmental Science and Pollution Research 25, 20486–20496.

Gomare, S.S., Govindwar, S.P., 2009. *Brevibacillus laterosporus* MTCC 2298: a potential azodye degrader. Journal of Applied Microbiology 106.

Gorelik, J., Shevchuk, A.I., Frolenkov, G.I., Diakonov, I.A., Lab, M.J., Kros, C.J., Richardson, G.P., Vodyanoy, I., Edwards, C.R.W., Klenerman, D., et al., 2003. Dynamic assembly of surface structures in living cells. Proceedings of the National Academy of Sciences of the United States of America 100, 5819–5822.

Gorelik, J., Zhang, Y., Shevchuk, A.I., Frolenkov, G.I., Sánchez, D., Lab, M.J., Vodyanoy, I., Edwards, C.R.W., Klenerman, D., Korchev, Y.E., 2004. The use of scanning ion-conductance microscopy to image A6 cells. Molecular and Cellular Endocrinology 217, 101–108.

Gorelik, J., Yang, L.Q., Zhang, Y.J., Lab, M., Korchev, Y., Harding, S.E., 2006. A novel Z-groove index characterizing myocardial surface structure. Cardiovascular Research 72, 422–429.

Hansma, P.K., Drake, B., Marti, O., Gould, S.A., Prater, C.B., 1989. The scanning ion-conductance microscope. Science 243, 641–643.

Ibrahim, M., Al Masri, A., Navaratnarajah, M., Siedlecka, U., Soppa, G.K., et al., 2010. Prolonged mechanical unloading affects cardiomyocyte excitation-contraction coupling, transversetubule structure, and the cell surface. The FASEB Journal 24, 3321–3329.

Kagle, J., Porter, A.W., Murdoch, R.W., Rivera-Cancel, G., Hay, A.G., 2009. Chapter 3 biodegradation of pharmaceutical and personal care products". Advances in Applied Microbiology 67, 65–108.

Kalogerakis, N., Arff, J., Banat, I.M., Broch, O.J., Daffonchio, D., Edvardsen, T., Eguiraun, H., Giuliano, L., Handå, A., López-de-Ipiña, K., Marigomez, I., Martinez, I., Øie, G., Rojo, F., Skjermo, J., Zanaroli, G., Fava, F., 2015. The role of environmental biotechnology in exploring, exploiting, monitoring, preserving, protecting and decontaminating the marine environment. New Biotechnology 32 (1), 157–167.

Korchev, Y.E., Bashford, C.L., Milovanovic, M., Vodyanoy, I., Lab, M.J., 1997a. Scanning ion conductance microscopy of living cells. Biophysical Journal 73, 653–658.

Korchev, Y.E., Milovanovic, M., Bashford, C.L., Bennett, D.C., Sviderskaya, E.V., Vodyanoy, I., Lab, M.J., 1997b. Specialized scanning ion-conductance microscope for imaging of living cells. Journal of Microscopy 188, 17–23.

Korchev, Y.E., Gorelik, J., Lab, M.J., Sviderskaya, E.V., Johnston, C.L., Coombes, C.R., Vodyanoy, I., Edwards, C.R., 2000a. Cell volume measurement using scanning ion conductance microscopy. Biophysical Journal 78, 451–457.

Korchev, Y.E., Raval, M., Lab, M.J., Gorelik, J., Edwards, C.R., Rayment, T., Klenerman, D., 2000b. Hybrid scanning ion conductance and scanning near-field optical microscopy for the study of living cells. Biophysical Journal 78, 2675–2679.

Krieg, N.R., Ludwig, W., Whitman, W.B., Hedlund, B.P., Paster, B.J., Staley, J.T., Ward, N., Brown, D. (Eds.), 2010, Bergey's Manual of Systematic Bacteriology, second ed., vol. 4. Springer-Verlag, New York, NY.

Kulkarni, R.M., Shetty, K.V., Srinikethan, G., 2014. Cadmium (II) and nickel (II) biosorption by *Bacillus laterosporus* (MTCC 1628). Journal of the Taiwan Institute of Chemical Engineers 45, 1628–1635.

Leo-Macias, A., Agullo-Pascual, E., Sanchez-Alonso, J.L., Keegan, S., Lin, X., Arcos, T., Feng-Xia-Liang, Korchev, Y.E., Gorelik, J., Fenyo, D., Rothenberg, E., Delmar, M., et al., 2016. Nanoscale visualization of functional adhesion/excitability nodes at the intercalated disc". Nature Communications 7.

Li, X., Nardi, P., Baek, C., Kim, J., Kim, Y., 2005. Direct nanomechanical machining of gold nanowires using a nanoindenter and an atomic force microscope. Journal of Micromechanics and Microengineering 15, 551–556.

Li, B., Pan, D., Zheng, J., et al., 2008. Microscopic investigations of the Cr(VI) uptake mechanism of living *Ochrobactrum anthropi*. Langmuir 24 (17), 9630–9635.

Liu, X.G., Zhang, H., 2007. Scanning probe microscopy-based nanofabrication for emerging applications. In: Méndez-Vilas, A., Díaz, J. (Eds.), Modern Research and Educational Topics in Microscopy, Microscopy Series, no 3. Formatex, Spain, pp. 770–778.

Lyon, A.R., MacLeod, K.T., Zhang, Y.J., Garcia, E., Kanda, G.K., et al., 2009. Loss of T-tubules and other changes to surface topography in ventricular myocytes from failing human and rat heart. Proceedings of the National Academy of Sciences of the United States of America 106, 6854–6859.

Nag, S., Pramanik, D., Chattopadhyay, D., Bhattacharyya, M., 2018. Green-fabrication of gold nanomaterials using *Staphylococcus warneri* from Sundarbans estuary: an effective recyclable nanocatalyst for degrading nitro aromatic pollutants. Environmental Science and Pollution Research 25, 2331–2349.

Nakada, N., Kiri, K., Shinohara, H., Harada, A., Kuroda, K., Takizawa, S., et al., 2008. Evaluation of pharmaceuticals and personal care products as water-soluble molecular markers of sewage. Environmental Science & Technology 42, 6347–6353.

Novak, P., Li, C., Shevchuk, A.I., Stepanyan, R., Caldwell, M., Hughes, S., Smart, T.G., Gorelik, J., Ostanin, V.P., Lab, M.J., et al., 2009. Nanoscale live-cell imaging using hopping probe ion-conductance microscopy. Nature Methods 6, 279–281.

Orlova, M.V., Smirnova, T.A., Azizbekyan, R.R., Ganushkina, L.A., Yacubovich, V.Y., 1998. Insecticidal activity of *bacillus laterosporus*". Applied and Environmental Microbiology 64 (7), 2723–2725.

Piper, J.D., Clarke, R.W., Korchev, Y.E., Ying, L., Klenerman, D., et al., 2006. A renewable nanosensor based on a glass nanopipette. Journal of the American Chemical Society 128, 16462–16463. ISSN: 0002-7863.

Qu, Y., et al., 2018. A novel environmental fate of graphene oxide: biodegradation by a bacterium *Labrys sp.* WJW to support growth. Water Research 143, 260–269.

Rodolfa, K.T., Bruckbauer, A., Zhou, D., et al., 2005. Two-component graded deposition of biomolecules with a double-barreled nanopipette. Angewandte Chemie International Edition 44, 6854–6859.

Sagitova, A., Meshkov, G., Yaminsky, I., 2016. View of the bacterial strains of *Escherichia coli* M-17 and its interaction with the nanoparticles of zinc oxide by means of atomic force microscopy. IOP Journal of Physics: Conference Series 741 (1), 012059.

Seifert, J., Rheinlaender, J., Novak, P., Korchev, Y., Schäffer, T.E., 2015. Comparison of atomic force microscopy and scanning ion conductance microscopy for live cell imaging. Langmuir 31, 6807.

Shevchuk, A.I., Frolenkov, G.I., Sanchez, D., James, P.S., Freedman, N., et al., 2006. Imaging proteins in membranes of living cells by high-resolution scanning ion conductance microscopy. Angewandte Chemie 45, 2212–2216.

Shevchuk, A.I., Novak, P., Takahashi, Y., Clark, R., Miragoli, M., Babakinedjad, B., Gorelik, J., Korchev, Y., Klenerman, D., 2011. Realizing the biological and biomedical potential of nanoscale imaging using pipette probe. Nanomedicine 6 (3), 565–575. Future Medicine Ltd.

Siegfried, K., Endes, C., Bhuiyan, A.F., Kuppardt, A., Mattusch, J., van der Meer, J.R., et al., 2012. Field testing of arsenic in groundwater samples of Bangladesh using a test kit based on lyophilized bioreporter bacteria. Environmental Science & Technology 466, 3281–3287.

Smirnova, T.D., Minenkova, I., Orlova, M.V., Azizbekyan, R.R., Lecadet, M., 1996. The crystal-forming strains of *bacillus laterosporus*. Research in Microbiology 147 (5), 343–350.

Tecon, R., ven der Meer, J.R., 2008. Bacterial biosensors for measuring availability of environmental pollutants. Sensors 8, 4062–4080.

Van der Meer, J.R., Belkin, S., 2010. Where microbiology meets microengineering: design and applications of reporter bacteria. Nature Reviews Microbiology 8, 511–522.

Vos, P., Garrity, G., Jones, D., Krieg, N.R., Ludwig, W., Rainey, F.A., Schleifer, K.-H., Whitman, W.B. (Eds.), 2009, Bergey's Manual of Systematic Bacteriology, second ed., vol. 3. Springer-Verlag, New York, NY.

Whitman, W.B., Goodfellow, M., Kämpfer, P., Busse, H.-J., Trujillo, M.E., Ludwig, W., Suzuki, K.-i., 2012. In: Bergey's Manual of Systematic Bacteriology, second ed., vol. 5. Springer-Verlag, New York, NY. parts A and B.

Xiaohong, P., Zhi, C., Yangjian, C., et al., 2011. The analysis of the immobilization mechanism of Ni(II) on *Bacillus cereus*. Journal of Nanoscience and Nanotechnology #11 (4), 3597–3603.

Xie, X.N., Chung, H.J., Sow, C.H., Wee, A.T.S., 2006. Nanoscale materials patterning and engineering by atomic force microscopy nanolithography. Materials Science and Engineering R 54, 1–48.

Xu, T., Close, D.M., Sayler, G.S., Ripp, S., 2013. Genetically modified whole-cell bioreporters for environmental assessment. Ecological Indicators 28, 125–141.

Yaminsky, I.V. (Ed.), 1997. Scanning Probe Microscopy of Biopolymers. Scientific World, Moscow.

Yaminsky, I., 2013. FemtoScan scanning probe microscope: a new tool for medicine. Nanoindustry 5 (43), 44–46.

Yaminsky, I.V., 2016. Scanning capillary microscopy. Nanoindustry 1 (63), 76–79.

Yaminsky, I.V., Demin, V.V., Bondarenko, V.M., 1997. Differences in cellular surface of hybrid bacteria Escherichia coli K12 inheriting rfb-a3,4 gene of *Shigella flexneri* as revealed by atomic force microscopy. Journal of Microbiology, Epidemiology and Immunology 6, 15.

Yaminsky, Y., Filonov, A., Sinitsyna, O., Meshkov, G., 2016. FemtoScan online software. Nanoindustry 2 (64), 42–46.

Ying, L.M., Bruckbauer, A., Zhou, D., Gorelik, J., Shevchuk, A., et al., 2005. The scanned nanopipette: a new tool for high-resolution bioimaging and controlled deposition of biomolecules. Physical Chemistry Chemical Physics 7, 2859—2866.

Zhang, Y., Gorelik, J., Sanchez, D., Shevchuk, A., Lab, M., Vodyanoy, I., Klenerman, D., Edwards, C., Korchev, Y., 2005. Scanning ion conductance microscopy reveals how a functional renal epithelial monolayer maintains its integrity. Kidney International 68, 1071—1077.

Zhao, H.-M., et al., 2018. Biodegradation pathway of di-(2-ethylhexyl) phthalate by a novel *Rhodococcus pyridinivorans* XB and its bioaugmentation for remediation of DEHP contaminated soil. The Science of the Total Environment 640—641, 1121—1131.

Zubasheva, M., Sagitova, A., Smirnov, Y., Smirnova, T., Azizbekyan, R., Zhukhovitsky, V., Yaminsky, I., 2017. Ultra-structural analysis of *Brevibacillus laterosporus* by electron and atomic force microscopy. Nanoindustry 2 (72), 74—78.

Research progress of biodegradable materials in reducing environmental pollution

Kangming Tian[1], Muhammad Bilal[2]

[1]Department of Biological Chemical Engineering, College of Chemical Engineering and Materials Science, Tianjin University of Science and Technology, Tianjin, China; [2]College of Biotechnology, Tianjin University of Science and Technology, Tianjin, China

1. Introduction

Materials are the basis for human survival and development. Human social activities depend on the constant discovery and use of new tools. Materials developed from wood and stone play an extremely important role in the early stages of human lives. Not only those materials have helped human to form the ability to defend against other biological attacks, but they also have gradually formed a more efficient and wider ability to live, laying the foundation for large-scale human reproduction. In the early 20th century, especially in 1907, Leo Beckland successfully invented the manufacturing technology of phenolic resin. Then the manufacturing technology of plastic products with coal and petroleum as raw materials was developed one after another, and the preparation of plastics was also widely used in all aspects of life and work. Since then, the efficiency of human activities has been improved qualitatively. Till now, our clothing, food, housing, and transportation have been unable to leave the plastic products. While bringing great convenience to plastic products, it destroyed human life, the earth's environment, and the ecological balance caused by its mass existence and nondegradable properties, which is increasing exponentially.

The burning of plastic products in municipal solid waste has caused serious air pollution. The plastic products after buried in soil are destroying the farming land and groundwater resources on which we depend for our survival. The plastic films used in crop cultivation are deteriorating the soil configuration because of the continuous deposition and even lead to a large number of farming because of the destruction of the residual plastic film to the water flow and nutrient transport process in the soil. Plastic products and microplastics formed by their fragmentation have become "PM2.5" particles in water and seriously endanger the aquatic organisms and the aquatic ecology. Microplastics in the ocean also cause the death of a large number of marine organisms and the destruction of marine ecology.

At present, many countries and organizations have applied legal provisions to restrict the use and spread of plastic products, but human life is totally dependable on plastic products and now it is not possible to control the harm of plastic products. Finding an ideal substitute for plastic products has become one of the most possible ways to solve the above contradictions. As early as the 1990s, the development of natural degradable plastic products to replace commonly used plastics products had become a research hotspot. However, because of cost and technology constraints, the research and development of degradable plastics is still very slow. In recent years, with the progress of raw material production and product processing technology the degradable plastics and especially biodegradable plastics have received renewed attention and become the highlight of ecological and circular economy development.

Plenty of materials like plastics were used around us, but their nondegradable property in short time and environment pollution limits their continued application toward to the substitute biodegradable materials (Fig. 15.1).

2. Biodegradable materials used for environmental protection

2.1 Main types of biodegradable materials

2.1.1 Definition and types of biodegradable materials

By definition, biodegradable materials are those which can be degraded by bacteria, fungi, or other biological means. It is usually combined with environmentally friendly products and can be decomposed into natural elements. Reasonable modification and controlled degradation can make biodegradable materials to play an extremely important role in reducing environmental pollution.

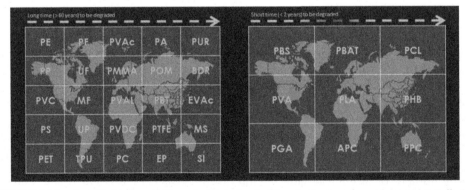

FIGURE 15.1   Possible biodegradable materials instead of the plastics to reduce environment pollution.

According to the different sources of raw materials and its processing, it can be divided into the following:

(1) Degradable natural polymer materials

Collagen (tissue, bone, cartilage, chamber, and ligament), gelatin (skin, tendon), polysaccharide (cellulose, starch, glucans, alginic acid, hyaluronic acid, heparin, chitosan), silk fibroin (Savaris et al., 2019; Younas et al., 2019).

(2) Degradable synthetic polymer materials

Mainly include fatty polyesters such as polyglycolic acid (PGA), polylactic acid (PLA), polycaprolactone (PCL), polybutylene succinate (PBS), polyvinyl alcohol (PVA), polyterephthalic acid/adipic acid/butanediol (PBAT), and their copolymers.

(3) Synthesis of polymer materials degradable by microorganisms

Polyhydroxyalkanoate (PHA) and polyhydroxybutyrate (PHB).

Owing to the different sources or different manufacturing methods of biodegradable materials, the bond composition and derivative modification methods are different and the corresponding degradation methods and degradation cycles are also very different, often from hours to years of time distribution. Table 15.1 lists the degradation times of several common biodegradable materials (including those assisted by enzymes or microbes).

To pay attention to the function and degradation controllability of biodegradable materials in the process of replacing plastic products, in recent years, a series of studies have been carried out around the monomer manufacturing technology, processing and polymerization technology, modification technology, and degradation technology of biodegradable materials, and great progress has been made.

2.1.2 Production of natural polymer raw materials and their properties

Biodegradable plastics which are made from natural macromolecule materials include thermoplastic starch, biocellulose, polysaccharides, and polyamino acids, as well as their mixtures obtained during modification and chemical modification varieties.

TABLE 15.1 Comparison of degradation of main degradable materials.

Material types	Degradation time	Typical examples
Polysaccharide	14 days	Cellulose, starch, glucans, alginic acid, hyaluronic acid, heparin, chitosan
Silk fibroin	15–17 days	Silk
Gelatin	35 days	Animal skin, bone, tendon
Microbial synthetic polymers	50 days	PHA and PHB
Fatty polyesters	6 months–2 years	PGA, PLA, PCL, PBS, PVA, PBAT, and their copolymers
Collagen	5 years	Tissue, bone, cartilage, chamber, and ligament

Starch can be extracted from corn, rice, potatoes, and other foods. It is a rich and cheap raw material. There are two types of starch, amylose formed by the binding of glucose with α-1,4 bond and amylopectin with α-1,6 bond. Starch has many branching structures and lacks crystallinity. It does not have the characteristics of plastics when used alone, but when mixed with other plastics, it can be formed into thin films and so on. Starch composites with synthetic biodegradable polymers have been extensively studied which help to improve the mechanical properties of starch films (Emadian et al., 2017; Esmaeili et al., 2019; Mangaraj et al., 2018). Concentrated research work has also been dedicated to develop blends with nonbiodegradable polymers. However, these systems are not measured to be biodegradable materials but may be partially biodegradable. Only the part of starch in the mixture which is accessible to enzymes is degradable (Firdaus et al., 2018; Mushi et al., 2018; Thomas et al., 2019). Starch particles with a diameter of 5−80 μm cannot be completely mixed with plastics. In most cases, starch particles will remain in a particulate state, resulting in a decrease in strength. So it is generally used by crushing starch and alkali together to form ester bonds with hydrophilic acrylic resin or by grafting starch obtained after heating onto plastics as a mixed solvent of starch and plastics. However, it does not matter how it is mixed with plastic, attention should be paid to the decomposition of starch when the processing temperature rises. Starch and cellulose are polysaccharides that exemplify the most representative family of these natural polymers. Some other natural polymers like proteins can be also used to produce biodegradable materials. These are the two main renewable sources of biopolymers. Another resource is lipids. Natural polymers are frequently modified chemically to improve the mechanical properties of such polymers or to modify their deprivation rate (Emadian et al., 2017; Schroeter et al., 2011; Vroman and Tighzert, 2009). The term starch modification refers to change the natural properties of starch, increase its properties, or introduce new properties. The methods of manufacturing starch derivatives include physical methods (pregelatinized starch, electro-irradiated starch, thermal-degraded starch), chemical methods (oxidized starch, esterified starch, etherified starch, cross-linked starch, grafted copolymer starch), and biological modification (enzyme-converted starch).

Cellulose is mainly produced by plants and it is among the largest organic resources on the earth. Cellulose, a well-known polysaccharide produced by plants, is a linear polymer with very long macromolecular chains of one repeating unit, cellobiose. It is crystalline, infusible, and unsolvable in all organic solvents (Emadian et al., 2017; Nair and Laurencin, 2007; Schroeter et al., 2011; Vroman and Tighzert, 2009; Zhang, 2015). Therefore, its effective utilization is an important issue for human beings. However, cellulose cannot be dissolved directly in water and ordinary solvents, and there is no thermal manipulability, so forming process is very difficult. Therefore, many chemical-substituted derivatives of hydroxyl groups on cellulose have been developed. Cellulose derivatives have many uses, such as fibers, textiles, packaging films, lighting films, emulsifying stabilizers, spray paint, separation membranes, anion exchange resins, foam gel particles, medical polymers, and so on. Fiber derivatives are difficult to be decomposed by microorganisms or enzymes because of the stereoscopic barriers of their substituents. The enzymatic oxidation helps in biodegradation of cellulose and secretes peroxidase by fungi. Bacteria are also involved in degradation of cellulose. As for starch, degradation products are nontoxic (Anderson et al., 2018; Xu et al., 2018). However, cellulose acetate derivatives, which are the most studied cellulose acetate derivatives, can be decomposed under aerobic composting or activated sludge when the degree of substitution is 2.5 (the degree of hydroxyl substitution by acetyl is 3.0, equivalent to 100%). In the future, if we can understand the relationship between the degree of substitution and biodegradability and physical properties of cellulose derivatives, cellulose derivatives will play an important role in decomposing plastics.

The sources of chitin are crustaceans such as crabs and shrimps, insects such as unicorn, clams, and other shellfish. The structure of crustaceans is similar to cellulose except that the hydroxyl group in the second position of glucose is replaced by acetylamino group. In general the protein in crab shell or shrimp shell is removed by dilute alkali first, then the minerals are removed by dilute acid, and then the chitin can be extracted. Chitin is treated to chitosan by partial alkaline N-deacetylation. In chitosan, main units are glucosamine, and the ratio of glucosamine to acetyl glucosamine is testified as the degree of deacetylation. This degree may range from 30% to 100% and it depends on the synthesis method which also affects the crystallinity, surface energy, and rate of degradation of chitosan. Because of its rigid and dense crystalline structure and strong intra- and intermolecular hydrogen bonding, chitosan is insoluble in water and alkaline media. It can only be soluble in few dilute acid solutions. So chitosan is dissolved in acidic solutions before its assimilation into biodegradable films. Enzymes such as chitosanase or lysozymes are known to degrade chitosan (Firdaus et al., 2018; Mushi et al., 2018). The applications are limited. Because of its insolubility in most solvents, the applications of both chitin and chitosan are inadequate. As chitosan has amino and hydroxyl reactive groups, so some chemical modifications can be considered. Modified chitosan has been prepared as N-carboxymethylchitosan or N-carboxyethylchitosan. Chitosan is mostly prepared for use in cosmetics and in wound treatment (Firdaus et al., 2018; Mushi et al., 2018).

Chitosan can be obtained by heating chitin in concentrated alkali solution and degreasing treatment. Usually the degree of degreasing is 0.7—0.95. The acetamide on the second position of chitosan is replaced by amino group, which is the only alkaline polymer in nature. Chitin and chitosan are widely used as water treatment agglutinators, moisturizing cosmetics, antibiotics, and functional foods to reduce blood cholesterol content. In addition, the application of cellulose or Prussian blue in agricultural materials such as films and sheets is also attempted.

2.1.3 Production of synthetic biodegradable materials and their properties

Because of the suitable degradation time and their industrial scale production, the degradable polyesters including PBS, PBAT, PCL, PVA, and PLA were regarded as the most potential materials used instead of traditional plastic for reducing environment pollution. The performance comparison of main degradable materials with polyethylene (PE) was listed in Table 15.2 and potential application of these biodegradable materials was presented especially their biodegradability.

2.1.3.1 Polybutylene succinate and PBAT

PBS is synthesized by condensation of succinic acid and butylene glycol. The products developed by PBS include foaming materials, which can be used as packaging materials for household appliances and electronic instruments. By introducing carbonate (ester) junction into PBS, water-resistant and degradable materials can be developed. By controlling the form of monomer polymerization, the degradation rate of PBS derivatives can be

TABLE 15.2 Performance comparison of main degradable materials with polyethylene (PE).

Material types	PLA	PBAT	PPC	PE
Density (g/cm^3)	1.25	1.26	1.22	0.95
Strength (μm)	>30	>15	>10	>10
Elongation at break (%)	15	800	700	800
Wvtr (g/m^2 day)	1500	2000	40	10
Processing temperature (°C)	<200	<250	<170	<220
Biodegradability	0.5–2.0 years	60 days	50 days	Undegradable

controlled and biodegradable to CO_2 and water in fresh water, seawater, or buried underground for 60 days. PBAT copolymer is synthesized from the monomers including terephthalic acid, adipic acid, and butanediol. The melting point of PBAT obtained by introducing aromatic polyesters into PBS has been greatly improved, and when the proportion of terephthalate is less than 40%, still it is biodegradable. By rational blending of monomers, the polymeric materials with good tensile properties, oxygen, and water vapor barrier properties can be obtained. The blending of PBS with starch or PLA can improve the mechanical properties, increase the elastic modulus of the material, and also reduce the cost. By adjusting the macromolecule architecture, a variety of products with PBAT as the main structure can also be developed.

2.1.3.2 Polycaprolactone

PCL is synthesized by ring-opening polymerization of ε-caprolactone with seven-membered rings. PCL has low melting point but high compatibility with various resins. It can be used as a composite of modification and other biodegradable polyesters. Usually it is used with starch, cellulose, PLA, and other polymers. It is used in plastic film, composting bags, buffer materials, fishing lines, and fishing nets.

2.1.3.3 Polyvinyl alcohol

PVA is often used as raw material for synthetic fiber nylon and is the only biodegradable polymer with C−C bond in the main chain. PVA is a water-soluble macromolecule which is formed by the decomposition of PVA nylon with water in alkali or acid. It can be widely used in adhesives, emulsifying stabilizers, and so on. The decomposition mechanism is that PVA is oxidized by oxidase, and then the main chain is cut off by hydrolase.

2.1.3.4 PLA and its copolymers

PLA is produced by the polymerization of lactic acid monomers. Lactic acid monomers have both hydroxyl and carboxyl functional groups, which have high reactivity and are easy to dehydrate and condensate to PLA under appropriate conditions. PLA can be synthesized by direct polycondensation and lactide ring-opening polymerization. Because lactic acid molecule has an asymmetric carbon atom, its monomer has two forms, L-form and D-form. The crystallinity of PLA depends largely on the ratio of L-form or D-form to the total lactic acid monomers. Various polymers are mixed with PLA to improve the toughness and properties such as optical barrier, thermal, and biodegradation (Quiles-Carrillo et al., 2018b).

As a result of polycondensation of D- or L-lactic acid or from ring-opening polymerization of lactide, that is a cyclic dimer of lactic acid, PLA is usually obtained. The two optical forms D-lactide and L-lactide are exist. Other than the natural isomer is L-lactide and the synthetic blend is DL-lactide; some synthetic methods have been studied and are reported in detail (Castro-Aguirre et al., 2016; Pretula et al., 2016).

PLA with molecular weight of 100−200,000 Da can be prepared in one step by direct dehydration and condensation. It also can be synthesized by heating polycondensation of lactic acid monomer to form lactide and ring-opening polymerization to form PLA with molecular weight of up to 300−900,000 Da. However, when PLA films were prepared by further processing, the melting point of PLA films was only 170°C, which was much lower than that of polyethylene terephthalate (PET) films at 264°C, although the tensile strength and elongation of PLA films were almost the same as that of PET films. It was found that when poly(L-lactic acid) and poly(D-lactic acid) were mixed in solution or melt at a ratio of 1:1, their molecular chains were interactively oriented in crystallization to form a stereo complex. The right helix 3-loop structure and the left helix 3-loop structure of the three-dimensional complex are interactively oriented in the crystallization. The melting point can reach to 230°C, far exceeding the melting point of homopolymer. Therefore, it overcomes the shortcoming of low heat resistance of general polyester and can be used as high strength and high heat resistance material.

Biodegradation is a phenomenon which takes place through the action of enzymes or chemical worsening link with living organisms. This is a two-step event. In the first step, the fragmentations of the polymers into lower molecular mass species occur by means of either abiotic reactions, i.e., oxidation, photodegradation, or hydrolysis, or biotic reactions such as degradations by microorganisms, and this is tailed by bioassimilation of the polymer fragments by microorganisms and their mineralization. Biodegradability depends on the origin of the polymer and on the chemical structure and the environmental-degrading conditions (Emadian et al., 2017; Thakur et al., 2018).

In addition to self-polymerization, lactic acid (LA) can also form copolymer PLGA with glycolic acid (GA). In particular, ring-opening polymerization technology is used to copolymerize GA and LA in a specific proportion. The degradation rate of copolymer is about 10 times higher than that of homopolymer. Moreover, the degradation rate of copolymer can be effectively adjusted by changing the composition of GA and LA. PLGA has been widely used in biomedical engineering by adjusting composition, molecular weight, biocompatibility, biodegradation rate, strength and modulus of materials, such as drug delivery system, absorbable suture materials, built-in isolation materials, orthopedic fixation, and tissue repair materials.

Polymer poly(lactic acid-epsilon-caprolactone) (PCLA) formed by lactic acid and ε-caprolactone has good biocompatibility, biodegradability, and drug permeability and is often used as drug delivery carrier materials and tissue engineering scaffolds. PLA offers more resistance to microbial degradation as compared to other biodegradable polymers such as PCL (Favaro et al., 2019; Raghunath, 2018).

The resistance of PLA is mainly as a result of the high molecular weight of commercial PLA ($\geq 2.0 \times 10^5$ Da). Therefore, microorganisms require more time to use PLA as their food source. To hydrolyze the Mw to a manageable limit, make it easy for the microorganisms to use it. Therefore, lower molecular weight PLA is likely a better candidate for the denitrification mechanism. Studies on PLA with Mw 1.0×10^4 Da showed a significantly greater removal rate of nitrogen than for PLAs of higher molecular weight (Bussa et al., 2018; Mokhena et al., 2018; Pacifici et al., 2018; Willberg-Keyriläinen et al., 2018).

PLA has the properties of toughness, softness, and transparency, but its rigidity makes the impact resistance of PLA a material's poor. To improve the impact resistance of PLA, it is necessary to select appropriate modifiers and plasticizers to modify PLA. However, crystalline polymers, common modifiers, and plasticizers in PLA cannot be long lasting because of the migration phenomenon, and the transparency will be affected when mixed with soft biodegradable plastics to improve the softness. After continuous research and development, the softness modifier compatible with PLA has been developed. The impact resistance and toughness of PLA have been improved.

PLA are the main components of most bio-based polymers, which are not heat-resistant and durable enough. The main reason is that the crystallization rate is too slow; which did not result in complete crystallization forming process. When layered silicate (clay mineral) was added to form clay/PLA nanocomposites, the crystallization rate was increased by about 100 times and the products with heat resistance over 100°C were obtained. Compared with PET, the easy hydrolysis of PLA limits its application in the field of long-term durable electronic products which are mostly used in products with a commercial life of only 3—5 years at room temperature. To expand the application field of PLA products, Japanese enterprises have developed a formula that can inhibit the hydrolysis rate of PLA, so it is also used in electronic, machine shell, and automobile interior materials, which require long-term durability. To examine automotive requirement properties such as seam fatigue, flammability, resistance to abrasion, and snagging, a comparing PET- and PLA-based seat fabrics based study was conducted. This study showed that PLA met most of the requirements for automotive fabrics and had equivalent performance to PET, but unsuccessful in the flammability and abrasion tests. For the processing of PLA, well-established polymer manufacturing techniques were used for other commercial polymers such as PS and PET. To produce high Mw PLA, melt processing is the main technique in which the PLA resin obtained is converted into end products such as consumer goods, packaging, and other applications.

PLA is the only transparent biodegradable plastics in current practical applications. In addition, the air permeability of PLA is higher than that of polyethylene/polypropylene (PE/PP), but the water vapor permeability (especially moisture permeability) of PLA is lower than that of polyolefin and PET because it has higher water vapor permeability in glass state.

PLA has good biodegradability. Unlike PCL, PBS, and PBAT, PLA is hydrolyzed before microbial decomposition. So in the experiment, the processed products were biodegraded smoothly in the soil and disappeared completely after 1—2 years. In addition, according to ISO 14855, composite can be rapidly biodegraded under both aerobic and anaerobic environment. Biodegradation of PLA includes hydrolysis and microbial decomposition. Its core elements include three aspects: high temperature, high humidity environment, and microbial nutrition sources. First, under the conditions of high temperature and humidity, PLA began to hydrolyze remarkably in about 15 days and then the corresponding hydrolytic enzymes were produced during the growth of microorganisms to hydrolyze PLA. Finally, PLA was decomposed into lactic acid and lactic acid oligomer and released CO_2.

PLA is widely used in biomedical engineering. PLA and its copolymers have been officially marketed as sustained-release microspheres and microcapsules such as hormones and antibiotics. In addition, research and clinical trials are being carried out on many drugs including anticancer, antibiotics, polypeptide drugs (insulin, vaccines), nervous system drugs, analgesics, contraceptives, conjugates, and so on. PLA and PGA copolymers are good scaffolds for tissue engineering. They have good biocompatibility, blood compatibility, and better mechanical strength. Tissue engineering research has involved cartilage, skin, pancreas, liver, kidney, bladder, ureter, bone marrow, nerve, skeletal muscle, tendon, heart valve, blood vessel, intestine, breast, and other organs.

Copolymers of PLA, PGA, and PCL have been approved by the FDA. They can be used as temporary medical scaffolds, fixation, suture, or loading materials in vivo. PLA replacing metal as internal fixation material in orthopedics can avoid secondary operation, simplify operation procedure, reduce patient's pain, and improve treatment effect. PLA has been used in various kinds of bone fixation and repairing operation sites in developed countries. When implanted into organisms, PGA will decompose into various harmless glucose and lactic acid after about 3−12 months and eventually metabolize into CO_2 and water and discharge out of the body. The decomposition rate can be adjusted by molecular weight and the composition of PGA and PLA copolymers. In addition, PLA is also used as release matrix in Japan, not only in medicine but also in the direction of coating matrix of pesticides and microcapsules. PGA is among one of the simplest linear aliphatic polyesters. It is prepared through ring-opening polymerization of a cyclic lactone, glycolide. PGA is highly crystalline, with a rate of crystallinity of 45%−55% due to which it is not soluble in most organic solvents. It exhibits a high melting point (220−225°C) and a glass shift temperature of 35−40°C (Savaris et al., 2019; Younas et al., 2019).

PGA also shows excellent mechanical properties. However, because of its low solubility and its high rate of degradation, yielding acidic products limited its biomedical applications. Thus copolymers of glycolide with caprolactone, lactide, or trimethylene carbonate have been prepared for medical devices (Corradini et al., 2014; Morreale et al., 2015; Wang et al., 2016). The effects of seasons and environment on biodegradation should be highly considered. In some initial trials carried out for starch/PCL composite, the sample was found to degrade in 20 weeks in Queensland waters and 30 weeks in South Australian waters and the same sample was found to fully reduced in 20−30 days in a manure environment.

PLA can also be widely used in textile, packaging, and other special areas of civil, industrial, and medical. PLA fiber has almost the same strength and elongation at break as polyester and its fabric is soft. It is a good textile material. PLA fibers can be processed into short fibers, multiple filaments, and single filaments, blended with decomposable fibers such as cotton, wool, or viscose and can be used to make silk-like fabrics, textile, and garment fabrics and to produce T-shirts, jackets, stockings, and dresses with silk-like appearance. The environmentally friendly fibers have no irritation to human skin, good moisture absorption, and form excellent fabrics through processing. PLA fibers can be used as materials in daily cosmetics with morphological stability and wrinkle resistance. Morphological stability and wrinkle resistance. PLA fiber is a kind of ecological fiber with great potential for development, which is made of cheap starch and has the characteristics of complete natural circulation and biodecomposition. PLA fiber has been recommended as "environmental recycling material in the 21st century" by many experts. At present, PLA has been widely used as hard biodegradable plastic in the production of various packaging materials and extrusion products, especially in the field of sheet materials and cosmetic ampoules. Some products are mixed with soft biodegradable plastics and used in garbage bags and general packaging. In Japan, PLA has passed the test of mildew resistance and the hygienic safety confirmation of Food Hygiene Act No. 370. Its future development in the direction of food utensils and foamed sheets for fresh food is worth looking forward to.

2.1.3.5 PHB and its copolymers

Poly(β-hydroxybutyrate) (PHB) and poly(β-hydroxybutyrate/β-hydroxybutyrate) (PHBV) copolymers are the most studied and industrialized poly(hydroxybutyrate/β-hydroxybutyrate) copolymers. When the bacterium is cultivated under strictly definite conditions, the amount of synthesized PHB may be stabilized. Like in case of *Rhizobium*, the presence of storage biopolymers enhances its survival by serving as a carbon and energy source during periods of starvation outside the nodule (Juengert et al., 2018). PHB is usually extracted from cells and its crystallinity is as high as 60%–80%. Therefore, it is very brittle, its elongation at break is very low, it is easy to crack, and its processing window is very narrow. Therefore, PHB is basically not a practical material and its properties are similar to those of polypropylene. PHBV is different from PHB. With the increase of hydroxyvalerate (HV) content in the components, the melting point of PHBV decreases and the impact strength increases, which greatly improve its processing performance. On this basis, the processing properties of PHB and PHBV were improved by copolymerization and blending. By copolymerization with 4-hydroxybutyrate and some copolymers of medium- and long-chain alkyl esters, the glass transition temperature and crystallinity of the copolymers were reduced by the internal plasticization of 4-hydroxybutyrate and medium- and long-chain alkyl esters, and the rheological properties, stiffness, and elongation at break of the copolymers were improved. Among PHAs, the main studied is a copolymer of hydroxybutyrate and HV is the major biodegradable microbial polymer among PHAs, which is studied mostly. PHA was first produced by ICI in 1983, which is also formed by adding propionic acid to nutrient feedstock supplied to bacteria. PHBV is extremely crystalline polymer with a melting point of 108°C and glass transition temperature in the range −5 to 20°C. The pure copolymer is also brittle, less than PHB (Ding et al., 2018; Quiles-Carrillo et al., 2018a). In addition, by introducing functional groups into the side chains of PHB and PHBV, changing the number of methylene in the main or side chains, polyhydroxyalkanoates (PHA) with different properties can be obtained by changing the composition of copolymers, and various polymer materials from high crystallinity rigid plastics to rubber elastomers can be prepared. In Azotobacter vinelandii strain UWD which during growth produced PHAs on a large range of raw sugar sources including sucrose, molasses, cane molasses, and corn syrup, the hyperaccumulation of PHB and poly(3-hydroxybutyrate-co-3-hydroxyvalerate) (PHBV) has been reported, which is nutrient-regulated (Anjum et al., 2016; Ghaffar et al., 2018; Sabapathy et al., 2018; Sabbagh and Muhamad, 2017).

PHB and PHBV can be merged with other biodegradable polymers, plasticizers, low molecular weight substances, and other synthetic polymers to improve the properties of materials. If mixed with starch or PCL, the toughness of PHB can be improved. Combination with PLA, polyvinyl acetate, PVA, polymethyl methacrylate, and ethylene-vinyl acetate can also improve the properties of PHB and PHBV.

PHB, which has deprived low-impact strength, is solved by assimilation of hydroxyvalerate monomers into the polymer to produce polyhydroxybutyrate-co-valerate (PHBV), which is commercially marketed under the trade name Biopol. Like PHB, PHBV completely reduces into carbon dioxide and water under aerobic conditions (Goto et al., 2018).

2.1.3.6 APC and PPC

Carbon dioxide copolymers mainly include aliphatic polycarbonate (APC) formed by ring-opening polymerization of carbon dioxide and epoxy compounds and terpolymer (PPC) of carbon dioxide, propylene oxide, and epoxides containing ester bonds. Its products are mainly used in packaging and medical tools. Polyether is a water-soluble polymer with C−O−C chain in the backbone. It has three kinds: polyethylene glycol (PEG), polypropylene glycol (PPG), and polytetramethylene ether glycol (PTMG). It is usually synthesized by ring-opening polymerization of ethylene oxide and is widely used in cosmetics, pharmaceuticals, water-soluble coatings, adhesives, thiophene, and so on. Polyaspartic acid (PAA) is a kind of polyamide produced by L-aspartic acid thermal polycondensation and then decomposed into PAA by adding water in alkali water. PAA can be used in physiological products, refrigerant, wet cloth, etc.

Because of the needed property of biodegradable materials, the methods to enhance the property of biodegradable materials were constructed based on the PLA using PBS, PBAT, PCA, and PGA as shown in Fig. 15.2.

2.2 Existing or forthcoming biodegradable materials

Up to now, many biodegradable products based on PLA and its copolymers have to be produced for their application in industry, agriculture, and food packaging. The performance comparison of main degradable materials for film manufacture was listed in Table 15.3.

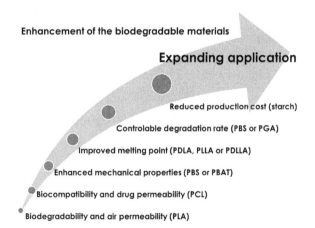

FIGURE 15.2   Enhancement of the property of biodegradable materials to expand their application area.

TABLE 15.3 Performance comparison of main degradable materials for film manufacture.

Items	Starch	PLA	PBS	PBAT	PHB
Heat resistance	Lower	Higher	High	High	High
Film forming property	Preferably	Good	Preferably	Good	Preferably
Hardness	Lower	High	Lower	Low	Low
Mechanical strength	Moderate	Higher	High	High	High
Hydrolysis resistance	Moderate	Low	High	High	High
Transparency	Low	High	Low	Low	Low
Price	Low	Lower	Higher	Higher	High

2.2.1 Biodegradable plastic films and sheets

The biodegradable plastic film made of aliphatic polyester, aromatic polyester, and modified starch can maintain the same effect as polyethylene film during its use. After harvesting, it can be planted in the soil and decomposed without special treatment.

2.2.2 Domestic waste collection bag

Domestic waste collection bags made of aliphatic and aromatic polyester, PCL, modified starch, PLA, and other biodegradable plastics are collected separately after use and then composted to reduce waste, landfill, and incineration activities.

2.2.3 Transparent window envelope

Transparent window envelopes made of PLA film can be composted with domestic waste or treated as paper. They can escape to the natural environment in time and return to nature with paper after a period of time.

2.2.4 Fresh packaging film

PLA heat-sealed bags are used to pack fruits and vegetables, manure with domestic wastes, and return to nature after a period of time.

2.2.5 Forming sheet

PLA with aliphatic and aromatic polyesters is used for food dishes and disposable tableware without transparency. In addition, foaming products are developed. At the same time, PLA series sheets have been widely used in vacuum packaging materials of household appliances and storage media.

2.2.6 Card sheets and films

PLA stretching sheets have been successfully applied to contactless IC cards.

2.2.7 Film for synthetic paper and label

PLA films are used in synthetic paper and label films. The application of inorganic filler in biaxially oriented film in various synthetic paper fields such as label, postcard, business card, POP, etc., are used

2.2.8 Shrinkage packaging—related fields

PLA film has also attracted much attention in the field of shrinkage packaging. Wine, spices, edible oil, and other cup seals are used.

2.2.9 Transparent box

PLA sheets are used in transparent boxes for cosmetics, daily necessities, and other packaging.

2.2.10 Industrial film

PLA thin films can be used for processing media tapes, and the application fields will be further expanded when conductive carbon is mixed into PLA thin films.

2.2.11 Fruit and vegetable preservation bags

PLA fruit bags have the properties of proper oxygen and water vapor permeation, which is helpful for the preservation of fruits and vegetables. PLA stretching sheets are also being gradually used in fruit and vegetable packaging bags and mesh bags.

2.2.12 Auxiliary materials in biological recovery

They include aliphatic polyester, PCL, PLA and paper powder, starch, colorant, inorganic filler made of water removal bags, domestic waste collection bags, and shopping bags.

2.3 Application of biodegradable materials

2.3.1 Application in agriculture, forestry, fisheries, and animal husbandry

Starch biodegradable plastics are used to make seedling pots and tree planting pots. Starch and fiber biodegradable plastics are used to make fumigation films in forests and film for seedling protection. PLA film was used to produce agricultural film. Modified starch binder and plant active carbon colorant were added into PLA film to make black film for paddy field paper. PLA films with clay and plant fibers are used for soil improvement and vegetation protection in wasteland and artificial mounds. Modified biodegradable plastics are used for revetment engineering and beach restoration. Aliphatic polyesters are used to make biodegradable baits, fishing lines, and aquatic ropes, fishing nets, and aquaculture nets to protect the river and sea environment.

2.3.2 Application in foamed products

PLA sheets are used in the manufacture of foaming materials. Starch and biodegradable resins (PVA, cellulose acetate, etc.) are blended to produce water-foamed bulk buffer materials. PLA is used to make qualitative foam cushioning materials (such as packaging materials for hard disk drives). PLA film is used to make bubble material.

2.3.3 Application in other daily necessities

PLA materials are used to make reusable and retrievable durable food utensils. Disposable food utensils that can be treated with food residues are made up of PLA. Isopolylactic acid biodegradable plastics used for toothbrushes protect teeth and the environment. Bamboo fibers are added to PLA materials to make clothing components such as buttons, window envelopes, short-lived desk calendars, breastplates, notepad covers, book covers, cards, etc. PCL is used to make free-form toys in hot water, etc. The PLA material is mixed with wood flour and grain husk to produce environmental protection fireworks. Fiber-protected PLA material is used to make clothes hangers, human models, large advertising flags, posters, and so on. PLA film is used to make systems for material distribution, packaging ropes, sheets and films for automatic packaging, battery packaging, etc., instead of semiconductor packaging materials, which are difficult to reuse.

2.3.4 Application in automobile industry

It is used in durable automotive products, electronic machinery shell, carpet, and spare tire cover.

2.3.5 Application in processing AIDS

PLA is used in making printing ink, resins, solvents, additives, pigments, ink adhesives, ink auxiliaries, ink pigments, adhesives, emulsions, coatings, and much more.

2.3.6 Application in medical biodegradable plastics

It is used in sustained and controlled release injection, suture, and plastic fixation.

After full modification of the property of biodegradable materials, series materials were applied in industries and daily life as shown in Fig. 15.3.

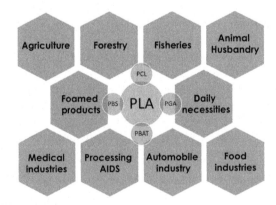

FIGURE 15.3   Application of biodegradable materials instead of plastics.

2.4 Novel biodegradable materials and their applications

Biodegradable materials are being widely used from being found and used, and their product series and manufacturing methods need to be adjusted and supplemented accordingly. To meet the needs of environmental protection, new biodegradable materials need to be studied and manufactured on a large scale. Among them, the product characteristics of new biodegradable materials should include the following: (1) the product series is complete and can replace the existing plastic products to the greatest extent; (2) the product has good application properties, especially toughness, strength, heat resistance, biocompatibility, and so on; (3) the degradation rate of the product is controllable, and it can be controlled according to different practical application requirements; (4) the cost of degradation is within the affordable range of large-scale application; and (5) the degradation products are green and environmentally friendly and will not cause secondary damage to the environment.

The manufacturing process of new degradable materials should satisfy the following characteristics: (1) There are abundant sources of raw materials, and the main body is renewable resources; (2) raw materials are cheap and can be processed on a large scale; (3) the manufacturing process is mild and environmentally friendly; (4) processing and manufacturing efficiency can meet the large-scale demand of products; and (5) processing and manufacturing process cost can be controlled, close to or even lower than the cost of traditional plastic products.

2.5 Ideal biodegradable technology for future

The ideal biodegradable material is a kind of material with excellent performance. After being discarded, it can be completely decomposed into carbon dioxide and water by environmental microorganisms and eventually be organized into a component of the carbon cycle in nature. The degradation process usually involves the digestion and absorption of filamentous fungi, yeasts, and bacteria. This process is usually a complex biophysical and biochemical process, accompanied by other physical and chemical processes, such as hydrolysis, oxidation, and so on. These effects are mutually reinforcing and have synergistic effects. The biodegradation process includes three stages: (1) the surface of biodegradable materials is adhered by microorganisms, and the way of adherence to the surface is affected by surface tension, surface structure, porosity, temperature, and humidity; (2) the microorganisms on the surface of materials break down into relative molecular weight by hydrolysis and oxidation under the action of enzymes secreted on the surface, low molecule compounds; (3) microorganisms absorb or consume small molecule compounds and metabolize them to form carbon dioxide, water, and biomass. In addition to the above biochemical effects, the degradation process also has physical effects, that is, after microorganisms erode the polymer, the cells roughly cause the mechanical damage of the polymer materials.

According to the biodegradation mechanism and environmental protection requirements, the new degradation technology should have the following characteristics: (1) establishing a quantifiable degradation evaluation system; (2) screening or breeding microorganisms that can be used for biodegradation; (3) analyzing the degradation mechanism of the above microorganisms; (4) establishing corresponding degradation conditions; (5) designing and

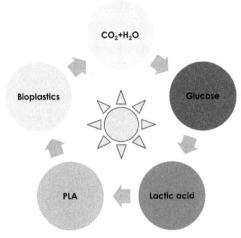

FIGURE 15.4   Ideal cycle life of biodegradable materials to reduce the environment pollution.

manufacturing equipment and sites needed for degradation process; (6) realizing controllable degradation under large-scale conditions; and (7) rational treatment and reuse of degradation products.

Up to now the industrial chain from glucose to lactic acid and then synthesis to PLA as the main material for bioplastics production was regarded as the ideal cycle for our life especially for environment protection (Fig. 15.4).

3. Conclusion

Plastics played an important role during the development of industry construction and offer really a lot to our daily life. However, the following uncontrollable environment pollution resulted by residues from plastics drives the development of biodegradable materials. Design and production of various goods based on the biodegradable materials instead of the goods produced by plastics will obviously reduce the environment pollution.

Appendix A: List of abbreviations

PBAT Poly(butylene adipate-co-terephthalate)
PBS Polybutylene succinate
PBSA Polybutylene succinate adipate
PCL Polycaprolactone
PE Polyethylene
PEF Polyethylene furanoate
PET Polyethylene terephthalate
PHA Polyhydroxyalkanoate
PHB Polyhydroxybutyrate
PHBV Poly(hydroxybutyrate-co-valerate)

PLA Polylactic acid
PP Polypropylene
PS Polystyrene
PTT Polytrimethylene terephthalate
PVC Polyvinyl chloride

References

Anderson, L.A., Islam, M.A., Prather, K.L.J., 2018. Synthetic biology strategies for improving microbial synthesis of "green" biopolymers. Journal of Biological Chemistry 293 (14), 5053–5061. https://doi.org/10.1074/jbc.tm117.000368.

Anjum, A., Zuber, M., Zia, K.M., Noreen, A., Anjum, M.N., Tabasum, S., 2016. Microbial production of polyhydroxyalkanoates (PHAs) and its copolymers: a review of recent advancements. International Journal of Biological Macromolecules 89, 161–174.

Bussa, M., Eisen, A., Zollfrank, C., Röder, H., 2018. Life cycle assessment of microalgae products: state of the art and their potential for the production of polylactid acid. Journal of Cleaner Production. https://doi.org/10.1016/j.jclepro.2018.12.048.

Castro-Aguirre, E., Iñiguez-Franco, F., Samsudin, H., Fang, X., Auras, R., 2016. Poly(lactic acid)—mass production, processing, industrial applications, and end of life. Advanced Drug Delivery Reviews 107, 333–366.

Corradini, E., Curti, P., Meniqueti, A., Martins, A., Rubira, A., Muniz, E., 2014. Recent advances in food-packing, pharmaceutical and biomedical applications of zein and zein-based materials. International Journal of Molecular Sciences 15 (12), 22438.

Ding, Y., Stoliarov, S., Kraemer, R., 2018. Development of a semiglobal reaction mechanism for the thermal decomposition of a polymer containing reactive flame retardants: application to glass-fiber-reinforced polybutylene terephthalate blended with aluminum diethyl phosphinate and melamine polyphosphate. Polymers 10 (10), 1137.

Emadian, S.M., Onay, T.T., Demirel, B., 2017. Biodegradation of bioplastics in natural environments. Waste Management 59, 526–536.

Esmaeili, M., Pircheraghi, G., Bagheri, R., Altstädt, V., 2019. Poly(lactic acid)/coplasticized thermoplastic starch blend: effect of plasticizer migration on rheological and mechanical properties. Polymers for Advanced Technologies 30 (4), 839–851. https://doi.org/10.1002/pat.4517.

Favaro, L., Basaglia, M., Caslla, S., 2019. Improving polyhydroxyalkanoate production from inexpensive carbon sources by genetic approaches: a review. Biofuels, Bioproducts and Biorefining 13 (1), 208–227. https://doi.org/10.1002/bbb.1944.

Firdaus, F.E., Purnamasari, I., Gunatama, P., 2018. Chitin and chitosan from green shell (*Perna viridis*): utilization fisheries wastes from traditional market in Jakarta. MATEC Web of Conferences 248, 04002.

Ghaffar, S.H., Madyan, O.A., Fan, M., Corker, J., 2018. The influence of additives on the interfacial bonding mechanisms between natural fibre and biopolymer composites. Macromolecular Research 26 (10), 851–863.

Goto, T., Iwata, T., Abe, H., 2018. Synthesis and characterization of biobased polyesters containing anthraquinones derived from gallic acid. Biomacromolecules. https://doi.org/10.1021/acs.biomac.8b01361.

Juengert, J.R., Patterson, C., Jendrossek, D., 2018. Poly(3-Hydroxybutyrate) (PHB) polymerase PhaC1 and PHB depolymerase PhaZa1 of *Ralstonia eutropha* are phosphorylated *in vivo*. Applied and Environmental Microbiology 84 (13), e00604–e00618. https://doi.org/10.1128/aem.00604-18.

Mangaraj, S., Yadav, A., Bal, L.M., Dash, S.K., Mahanti, N.K., 2018. Application of biodegradable polymers in food packaging industry: a comprehensive review. Journal of Packaging Technology and Research. https://doi.org/10.1007/s41783-018-0049-y.

Mokhena, T., Sefadi, J., Sadiku, E., John, M., Mochane, M., Mtibe, A., 2018. Thermoplastic processing of PLA/cellulose nanomaterials composites. Polymers 10 (12), 1363.

Morreale, M., Liga, A., Mistretta, M., Ascione, L., Mantia, F., 2015. Mechanical, thermomechanical and reprocessing behavior of green composites from biodegradable polymer and wood flour. Materials 8 (11), 5406.

Mushi, N.E., Nishino, T., Berglund, L.A., Zhou, Q., 2018. Strong and tough chitin film from α-chitin nanofibers prepared by high pressure homogenization and chitosan addition. ACS Sustainable Chemistry & Engineering. https://doi.org/10.1021/acssuschemeng.8b05452.

Nair, L.S., Laurencin, C.T., 2007. Biodegradable polymers as biomaterials. Progress in Polymer Science 32 (8), 762–798.

Pacifici, A., Polimeni, A., Pacifici, L., 2018. Additive manufacturing and biomimetic materials in oral and maxillofacial surgery: a topical overview. Journal of Biological Regulators & Homeostatic Agents 32 (6), 1579–1582.

Pretula, J., Slomkowski, S., Penczek, S., 2016. Polylactides—methods of synthesis and characterization. Advanced Drug Delivery Reviews 107, 3–16.

Quiles-Carrillo, L., Montanes, N., Lagaron, J.M., Balart, R., Torres-Giner, S., 2018a. In situ compatibilization of biopolymer ternary blends by reactive extrusion with low-functionality epoxy-based styrene–acrylic oligomer. Journal of Polymers and the Environment. ISSN: 1566-2543. https://doi.org/10.1007/s10924-018-1324-2.

Quiles-Carrillo, L., Montanes, N., Pineiro, F., Jorda-Vilaplana, A., Torres-Giner, S., 2018b. Ductility and toughness improvement of injection-molded compostable pieces of polylactide by melt blending with poly(ε-caprolactone) and thermoplastic starch. Materials 11 (11), 2138.

Raghunath, S., 2018. Microbial bioreactor systems for dehalogenation of organic pollutants. In: Vinay Mohan, P., Navneet (Eds.), Handbook of Research on Microbial Tools for Environmental Waste Management. IGI Global, Hershey, PA, USA, pp. 319–340.

Sabapathy, P.C., Devaraj, S., Parthiban, A., Kathirvel, P., 2018. Bioprocess optimization of PHB homopolymer and copolymer P3 (HB-co-HV) by *Acinetobacter junii* BP25 utilizing rice mill effluent as sustainable substrate. Environmental Technology 39 (11), 1430–1441.

Sabbagh, F., Muhamad, I.I., 2017. Production of poly-hydroxyalkanoate as secondary metabolite with main focus on sustainable energy. Renewable and Sustainable Energy Reviews 72, 95–104.

Savaris, M., Garcia, C.S.C., Roesch-Ely, M., Henriques, J.A.P., dos Santos, V., Brandalise, R.N., 2019. Polyurethane/poly(d,l-lactic acid) scaffolds based on supercritical fluid technology for biomedical applications: studies with L929 cells. Materials Science and Engineering: C 96, 539–551.

Schroeter, M., Wildemann, B., Lendlein, A., 2011. Biodegradable materials. In: Steinhoff, G. (Ed.), Regenerative Medicine. Springer Netherlands, Dordrecht, pp. 469–492.

Thakur, S., Chaudhary, J., Sharma, B., Verma, A., Tamulevicius, S., Thakur, V.K., 2018. Sustainability of bioplastics: opportunities and challenges. Current Opinion in Green and Sustainable Chemistry 13, 68–75.

Thomas, M.S., Koshy, R.R., Mary, S.K., Thomas, S., Pothan, L.A., 2019. Applications of polysaccharide based composites. In: Starch, Chitin and Chitosan Based Composites and Nanocomposites. Springer Briefs in Molecular Science, Springer, Cham, pp. 43–55. ISBN: 978-3-030-03158-9. https://doi.org/10.1007/978-3-030-03158-9_4.

Vroman, I., Tighzert, L., 2009. Biodegradable polymers. Materials 2 (2), 307.

Wang, J., Wang, L., Zhou, Z., Lai, H., Xu, P., Liao, L., Wei, J., 2016. Biodegradable polymer membranes applied in guided bone/tissue regeneration: a review. Polymers 8 (4), 115.

Willberg-Keyriläinen, P., Orelma, H., Ropponen, J., 2018. Injection molding of thermoplastic cellulose esters and their compatibility with poly(lactic acid) and polyethylene. Materials 11 (12), 2358.

Xu, W., Pranovich, A., Uppstu, P., Wang, X., Kronlund, D., Hemming, J., Öblom, H., Moritz, N., Preis, M., Sandler, N., Willför, S., Xu, C., 2018. Novel biorenewable composite of wood polysaccharide and polylactic acid for three dimensional printing. Carbohydrate Polymers 187, 51–58.

Younas, M., Noreen, A., Sharif, A., Majeed, A., Hassan, A., Tabasum, S., Mohammadi, A., Zia, K.M., 2019. A review on versatile applications of blends and composites of CNC with natural and synthetic polymers with mathematical modeling. International Journal of Biological Macromolecules 124, 591–626.

Zhang, C., 2015. Biodegradable polyesters: synthesis, properties, applications. In: C. Zhang (Ed.), Biodegradable Polyesters.

Further reading

Yutaka Tokiwa "Biodegradation of Polycarbonates" 2005. https://doi.org/10.1002/3527600035.bpol9019.

Genetically engineered bacteria for the degradation of dye and other organic compounds

Arvind Kumar[1], Ajay Kumar[2], Rishikesh Singh[3],
Raghwendra Singh[4], Shilpi Pandey[5], Archana Rai[6],
Vipin Kumar Singh[7], Bhadouria Rahul[8]

[1]State Key Laboratory of Cotton Biology, Key Laboratory of Plant Stress Biology, School of Life Science, Henan University, Kaifeng, Henan, PR China; [2]Agriculture Research Organization (ARO), Volcani Center, Rishon LeZion, Israel; [3]Institute of Environment & Sustainable Development, Banaras Hindu University, Varanasi, India; [4]Crop Production Division, ICAR-Indian Institute of Vegetable Research, Varanasi, India; [5]Department of Botany, Institute of Science, Banaras Hindu University, Varanasi, India; [6]Department of Molecular and Cellular Biology, Sam Higginbotom Institute of Agriculture, Technology and Sciences (SHIATS), Allahabad, India; [7]Center of Advanced Study in Botany, Institute of Science, Banaras Hindu University, Varanasi, India; [8]Department of Botany, University of Delhi, New Delhi, India

1. Introduction

Rapid increase in human population and agricultural activities has aggravated the problem of environmental contamination globally. Rapid increase in textile and petroleum-based industries has added large amount of unwanted dyes (Chanwala et al., 2019) and organic contaminants (Yang et al., 2019) into the environment affecting human health and environment. The environmental contaminants being toxic in nature pose negative consequences on human health and natural environment. The environmental contaminants may be of synthetic or natural in origin. The synthetic contaminants, e.g., herbicides, pesticides, weedicides, and azo dyes, are not naturally available to an environment and hence are new to any environmental system. The natural contaminants include toxic compounds present in fossil fuels and are released into natural environment under certain conditions. The rate of nature's self-cleaning process under a given condition is generally slow and in many cases insufficient to eliminate the contaminants from a system. Deployment of microbe-based processes for remediation of toxic environmental contaminants gives us not only an economically viable opportunity for management of environmental pollutants but also a suitable alternative to costly physicochemical processes (Chanwala et al., 2019). To date, numbers of microbes capable of degrading xenobiotics have been isolated in vitro. Hence, one of the important strategies to encounter the pollution problem is exploitation and desired manipulation in genetic resources of diverse microbial population (Watanabe, 2001). Although every organism is provided with capability of contaminant detoxification through degradation, conversion, or immobilization onto cell surfaces, bacterial system has been demonstrated as most suitable candidate for bioremediation purposes. Bacteria have come into origin from more than 3 billion years ago and have evolved the mechanisms to derive energy from almost every compound naturally present in an ecosystem. Because of efficient substrate utilization, these prokaryotes have contributed significantly in the cycling of various important nutrients and elements of vital importance. The large pool of microbes and high duplication rate, along with their suitability for movement of genes, have permitted them to adjust very rapidly in a given environmental system of extreme nature that are generally too harsh to support the life of other living beings on earth. The easy adaptation of bacteria under such harsh conditions has been possible only because of presence of diverse genetic system responsible for evolution of great array of metabolic activities (Timmis and Pieper, 1999; Lovley, 2003).

Rise of the recent molecular biology techniques and advancement of genetic engineering have given a new direction to combat the ever-rising pollutant problem in natural ecosystem through significant changes in metabolic activities of a host system. The developments of genetic engineering have made a considerable revolution in the field of bioremediation. The objective of the genetic engineering was thus focused on development of host system equipped with altered enzymatic activities responsible for resistance and on-site degradation of intended pollutant. Development of such genetically engineered microorganisms (GEMS) possessing the enormous ability to degrade the variety of hazardous pollutants into nontoxic forms has emerged as a nature friendly, less-expensive, and a suitable alternative to that of generally employed physicochemical techniques (Ramos et al., 2011). The GEMs have been defined as an organism possessing the desired gene/foreign gene generally not present in the host system for a particular purpose. The endings of 1980s and initial discoveries of 1990s may be considered as the revolutionary era in the field of biodegradation studies especially the research work conducted by Chakrabarty and his research team (Kellogg et al., 1981). Their group has designed the genetically engineered *Pseudomonas putida* by integrating suitable genes into plasmid and made them capable to mineralize the different compounds. The availability of several man-made foreign substances (xenobiotics) into the contaminated environment was observed to be due to absence of responsible enzymatic systems into a single host system causing degradation of targeted compound (Brenner et al., 1994; Reineke and Knackmuss, 1979). Insertion of different naturally existing genes in a suitable host related with number of enzymatic activities resulting into expression of designed pathways leading to degradation of persistent organic contaminants including polychlorinated biphenyls (PCBs) may be regarded as a useful tool to mitigate the pollution problems (Lehrbach et al., 1984; Ramos et al., 1987; Rojo et al., 1987).

Several anthropogenically constructed chemical substances are known to possess components generally not existing in natural environment. This may be considered as a possible reason for the nonavailability of degradation pathways in most of the microbes for these synthetic contaminants. Most importantly, for few chemical substances, the degradation pathways are either completely unknown or microbes are able to degrade only a fraction of contaminant. In some cases, however, the complex nature of contaminant itself may restrict the biological degradation activities of microbes. The presence of hazardous contaminants in an ecosystem, however, may be managed through microbial degradation efficiencies. Many of the factors such as concentration, nature, and availability of contaminant should also be taken into consideration at a greater extent while trying for remediation of contaminants. Before degradation by microbial actions, the desired contaminants intended for degradation must have access to cellular system. For efficient mineralization of targeted compounds, the suitable microbes must be present in sufficient number and able to perform metabolic activities responsible for contaminant degradation under existing natural environmental conditions. Therefore, a critical investigation regarding the limitations and finding the solution through genetic engineering techniques is inevitable to explore the huge microbial diversity to decontaminate the polluted sites. In this chapter, we would shed lights on application of GEMs in abatement of pollution caused by dyes and organic contaminants.

2. Constructing genetically engineered microorganisms

To adapt in an existing environmental condition, microorganisms expresses their genetic system differently against different toxic substances. The adaptation can be improved by integrating suitable genes. Rapid development in the area of recombinant DNA and advanced molecular biological have allowed for (1) increase in copy number, deletion, and/or changes in desired genes synthesizing particular enzymes of a metabolic routes, (2) reduction in limitations associated with metabolic routes, (3) improvement in energy generation processes, and (4) incorporation of desired genes with novel features (Liu et al., 2006; Shimizu, 2002; Timmis and Piper, 1999). Different genetic tools and techniques have come into light to reinforce the optimum expression of genes governing a particular enzyme and metabolic routes for efficient degradation of target compounds (Pieper and Reineke, 2000). Revealing novel facts dealing with the metabolic pathways and limiting factors pertaining to biodegradation are coming into the knowledge of current researchers, necessitating the improvement in existing molecular biology techniques (Stegmann, 2011). However, the full expression of desired genes inevitably requires complex regulatory and metabolic system of suitable host system (Cases and Lorenzo, 2005). Four basic ways relying on (1) changes in nature of enzyme, (2) control of metabolic processes causing degradation, (3) generation, assessment, and regulation of a biological pathway, and (4) development of bioaffinity bioreporter sensor have been considered for generation of GEMs to degrade the environmental contaminants (Menn et al., 2008).

3. Detection of genetically engineered microbes

Because the genetically engineered microbes demonstrate the better degradation of targeted compounds under given environmental conditions, assessment of their presence in natural system should be monitored. Few techniques of molecular biology such as denaturing gradient gel electrophoresis (DGGE), ribosomal intergenic spacer analysis (RISA), and amplified ribosomal DNA restriction analysis (ARDRA) have been applied extensively to get information on existence of engineered bacteria at contaminated sites and their impact on native microbial diversity (Zanaroli et al., 2002; Heinaru et al., 2005; Quan et al., 2004; Mufiel et al., 1999). In addition, wide application of monoclonal antibodies (Ramos-Gonzalez, 1992), bioluminescence (Ripp et al., 2000a,b), and green fluorescent protein (GFP; Wu et al., 2000) to identify and monitor the presence of genetically modified organisms (GMOs) is also presented. Most importantly, deployment of GFP can give a direct insight on the availability and unavailability of GEMs without the use of specific tools. Better understanding and improvement in available molecular biology techniques would further enhance our knowledge in this direction.

4. Need of genetically engineered microbes

The significant increase in availability of hazardous and persistent contaminants in natural ecosystem reveals the information that naturally existing diversity of microbial metabolic process in an organism is not able to degrade synthetic contaminants completely. During the last phases of 1970s and initial 1980s, genes from different bacteria responsible for degradation of xenobiotics were identified and cloned. Thereafter, numerous scientists working in the area of microbiology and molecular biology recognized the feasibility of recombinant DNA technology for bioremediation of contaminated sites (Cases et al., 2005). In general, the term GEM or modified microorganism refers to creation of a microorganism with transformed genetic information facilitated by gene transfer by the so-called recombinant DNA technology. GEMs possessing the ability to mineralize a wide array of synthetic chemical have demonstrated great potential for remediation of contaminated soil, water (surface and groundwater), and sewage sludge (Sayler and Ripp, 2000; Sarma and Prasad, 2019). Once the probability of transferring GEMs into contaminated sites for field-scale remediation purposes was considered, the need of biosafety and risk assessment was realized (Cases et al., 2005).

5. Dye degradation by engineered microbes

Currently, dye degradation is a challenging problem worldwide. Synthetic dyes have been commercialized in fabric coloration, paper industries, food system, drugs, personal care product, and other industries. As dyes are constructed in way to resist the loss of color under the different physical, chemical, and biological conditions, they are not susceptible to degradation by traditional treatment systems and hence get their path into natural environment. Very low availability of dye in an aqueous system affects its transparency, water quality characteristics, and life of aquatic organisms. It has been demonstrated that many azo dyes and their degradation products such as aromatic amine are potentially toxic and carcinogenic in nature (Khalid et al., 2008; Dafale et al., 2008). The strict regulation pertaining to water quality characteristics and raised public knowledge and awareness has compelled the industries to employ microbiological process to mitigate the pollution problem resulting from excess release of dye effluents into the natural environment. Degradation of dyes is achieved by either a single treatment method or integration of physical, chemical, and biological methods. Interestingly, microbiological treatment methodology has got momentum because of environment friendliness, improved efficacy, and less expensive nature (Moharikar et al., 2005). In present condition, the treatment of generated waste is relied on the employment of traditional process utilizing biological activities of indigenous microorganisms. Nevertheless, the dye degradation efficiency could be improved through enrichment of native microbes and GEMs harboring the genes responsible for dye degradation. Therefore, to maintain the optimum efficiency of dye degradation process, the treatment plants are designed to support the maximum population density of responsible microorganism.

Bacterial systems under anaerobic environment are known to convert dyes into colorless form with the help of so-called enzyme azoreductase and electron donors (Blümel et al., 2002; Blümel, and Stolz, 2003; Yan et al., 2004; Chen et al., 2004). As there is scanty information available on identification and sequences of genes directing the synthesis of azoreductase enzyme and other proteins participating in dye mineralization, constructing engineered bacterial system for dye degradation is a major challenge. Hence, detailed investigations on genes are responsible for dye degradation for improvement in bacterial dye degradation abilities. During preceding decades, identification of azoreductase-mediated dye degradation has been evidenced in bacteria including *Xenophilus azovorans* KF46 (Blümel et al., 2002), *Pseudomonas luteola* (Hu, 1998), *Klebsiella pneumonia* RS-13 (Wong and Yuen, 1996), and *Clostridium perfringens* (Rafii et al., 1997). Experimental investigations on aforementioned bacterial species have demonstrated that few of them may be dependent on flavin compounds for azoreductase-based dye degradation while others may not (Van der Zee and Villaverde, 2005).

The high dye decolorization efficiency of GEMs has been demonstrated (Sandhya et al., 2008). The suitable technique for remediation of textile dye effluents is thus based on the enrichment of suitable microbes and exploitation of GEMs. For the effective remediation, the regular assessment of introduced microbial strains is an inevitable step for better performance of the treatment system. The monitoring could be made possible through advanced molecular biology tools and techniques such as 16S rDNA sequencing, DGGE, temperature gradient gel electrophoresis, RISA, single-stranded conformation polymorphism, randomly amplified polymorphic DNA, and ARDRA (Quan et al., 2004; Khelifi et al., 2009; Heinaru et al., 2005).

Jin et al. (2008) have reported the successful exploitation of GEM in a treatment system for removal of Acid Red—laden wastewater. The engineered bacterial strain *Escherichia coli* JM109 (pGEX-AZR) exhibited increased azoreductase activity. Dye degradation by selected engineered bacteria followed the Andrews model. To facilitate the higher degradation of tested dye, the treatment was performed in anaerobic sequencing batch reactors. The study revealed that decolorization was influenced by the inoculum density, and the best performance was recorded at density equivalent to 10%. The degradation was also influenced by the carrier matrix used for the immobilization of engineered bacteria and that the macroporous foam was the most effective for bioaugmentation. The foam immobilized engineered bacteria were more tolerant to higher dye concentration and had high rate of dye degradation as compared with those immobilized into sodium alginate. Analysis of surviving microbial community through RISA and ARDRA resulted into high-density availability of inoculated *E. coli* JM109 (pGEX-AZR) together with raised metabolism.

The genetically engineered *E. coli* JM109 (pGEX-AZR) was also investigated for its activity to degrade the azo dye Direct Blue 71 (DB 71) (Jin et al., 2009). The engineered bacteria exhibited high rate of dye decolorization in treatment system. The control and engineered bacteria added treatment system could not degrade the dye at pH 5.0. Nevertheless, the strain *E. coli* JM109 (pGEX-AZR) was able to operate the dye degradation process at pH 9.0. The seeded biotreatment system was able to reduce the dye concentration from 150 mg L^{-1} to 27.4 mg L^{-1} within 12 hours. The bioaugmented system was not affected by the presence of salt concentration ranging from 1% to 3%. Introduction of GEM into reactor system enhanced the dye decolorization efficiency. It is noteworthy here that community structure of control and *E. coli* JM109 (pGEX-AZR) bioaugmented reactor system was nearly same.

6. Organic contaminants degradation by genetically engineered microorganisms

Rapid rise in petroleum-based industries has aggravated the problem of organic pollutant contamination. The organic contaminant treatment methods by conventional physico-chemical methods have several limitations such as large amount of sludge generation and production of secondary by-products of more toxic nature as compared with parent substrate. The application of biological processes for remediation of soil and water contaminated with organic substances is more attractive because of less expensive and environment-friendly nature. However, the application of native microorganisms has several limitations. These limitations can be overcome through development of genetically engineered microbes. One of the major events in the scientific era was the degradation of oil by different strains of *Pseudomonas*. The gene governing the degradation of oil was found to present on plasmid. The isolation and insertion of desired gene into single host system demonstrated 10–100 times higher oil degradation abilities as compared with wild strains (Time Magazine, 1975). The radiation-resistant bacterium *Deinococcus radiodurans* has also been engineered genetically to degrade the toluene. However, the field application has not been performed because of expected risks and regulatory acts (Ezezika and Singer, 2010).

The techniques of recombinant DNA technology has enabled the present-day scientists to accelerate the biodegradation potential of native microbes and synthesis of novel degradation pathways through insertion of different genes participating in degradation of a particular compound from various microbes (Ramos et al., 1994). Different genes involved in degradation of hazardous contaminants such as toluene, benzene, xylene, chlorobenzene acids, and persistent organic contaminants are now identified. As a single plasmid is not associated with the degradation of every compound, one plasmid is needed for degradation of each compound. The responsible plasmids may be classified into four types: (1) OCT plasmid with the potential to mineralize octane, hexane, and decane; (2) XYL plasmid conferring degradation abilities for xylene and toluenes, (3) CAM plasmid causing degradation of camphor, and (4) NAH plasmid involved in naphthalene degradation (Ramos et al., 1994). The development of genetically engineered strains harboring different plasmids into a single host system responsible for degradation of multiple hydrocarbons is reported (Markandey and Rajvaidya, 2004). They have reported the development of engineered *Pseudomonas* strains containing mutiplasmid with the ability to degrade different types of contaminants.

Genetically engineered *P. putida* strains containing XYL, NAH, CAM, and OCT plasmid have been prepared and shown to degrade camphor, octane, salicylate, and naphthalene (Sayler and Ripp, 2000). The designed strains containing four different genes grew rapidly on oily substrate because of efficient utilization as compared to the wild (Markandey and Rajvaidya, 2004).

7. Agent orange degradation by genetically engineered microorganisms

The hazardous chemical Agent Orange is known for its defoliating action and is reported to be employed during the Vietnam War. The carcinogenic compound is equimolar amalgamation of two different herbicides, i.e., 2,4-dichlorophenoxyacetic acid (2,4-D) and 2,4,5-trichlorophenoxyacetic acid (2,4,5-T) (Ezezika and Singer, 2010). The genetically engineered *Burkholderia cepacia* strain was developed and assessed for remediation of Agent Orange at contaminated sites in US Air Force site in Pensacola, northwestern Florida (Marwick, 2003; Chauhan et al., 2008).

8. Organophosphate and carbamate degradation by genetically engineered microorganisms

Mineralization of organophosphate and carbamate by genetically engineered strain of *Sphingomonas* sp. CDS-1 is reported by Liu et al. (2006). They cloned so-called "mpd" gene directing the synthesis of enzyme methyl parathion (MP) hydrolase in conjunction with gene associated with MP present in *P. putida* DLL-1 through shotgun approach. The vector pBBR1MCS-2 was explored to generate pBBR-mpd plasmid to transform the *Sphingomonas* sp. CDS-1. The process resulted into the development of a CDS-pBBR-mpd recombinant. Thus generated recombinant strain was observed to have seven times higher MP degradation ability as compared to native microbes (Liu et al., 2006).

Some of the organophosphorus compounds including paraoxon, parathion, chlorpyrifos, disulfoton, ruelene, carbophenothion, and dimeton have been reported to exhibit neurotoxicity (Das and Adholeya, 2012). The organophosphates are known to inhibit the enzymatic activities of acetylcholine esterase enzyme and induction of alternation in chromosome integrity and carcinogenic action on bladder cells (Webster et al., 2002). Phosphotriesterase and organophosphorus hydrolase (OPH) enzymes isolated from bacterial systems have been evidenced with the efficiency of organophosphate degradation through action on phosphorus-linked bonds such as P−O, P−F, and P−S. The enzymes act more favorably on Sp-enantiomeric forms of organophosphate as compared with Rp-enantiomeric forms. Chen-Goodspeed et al. (2001a,b) were successful in modifying the active sites by site-directed mutagenesis and hence enabled the constituted enzyme suitable for the degradation of mixtures of different enantiomeric forms (Chen-Goodspeed et al., 2001a).

The containment system for degradation of organophosphorus compound is reported by Qin and Yi-Jun (2009). The synthesized GEM has the capability to degrade the targeted compound and produce green fluorescent light with a tendency to kill itself when needed through integration of genes under the regulation of different promoters. The genes responsible for emission of green fluorescence and OPH were cloned in a vector pBV220. The genes were able to perform their associated functions only when present in an environment containing the targeted contaminant as the repressor genes had been eliminated. The suicide in designed microbial system could be possible after integration of nuclease genes devoid of leader sequence from *Serratia marcescens*. The nuclease gene was kept under the strict regulation of a promoter to generate a suicide cassette inducible in the presence of arabinose. To enhance the consistency of synthesized containment system, the suicide cassette was doubled in a provisional suicide plasmid. Hence, the plasmid harboring genes responsible for fluorescence emission and organophosphate degradation together with the suicide cassette were suitable and simultaneously present in a single host system (Qin and Yi-Jun 2009).

9. Polychlorinated biphenyls degradation by genetically engineered microorganisms

PCBs are well-recognized environmental contaminants. The enzyme biphenyl dioxygenase has been described to participate in the aerobic degradation through introduction of hydroxyl group into the aromatic ring structure (Ang et al., 2005). Biodegradation of PCBs leads to formation of intermediates of important tricarboxylic acid. The enzyme system is consisted of small and large subunits of dioxygenase, ferredoxin, and ferredoxin reductase synthesized by the bph operon system (Erickson and Mondello, 1992). By utilizing the shuffling techniques, different strains possessing a portion of bph gene recovered from *B. cepacia* strain LB400 and *Rhodococcus globerulus* P6 have been designed exhibiting the improved PCBs degradation efficiencies (Barriault et al., 2002). Similarly, improved degradation of 2,6-dichlorobiphenyl, a persistent PCB, is also reported through incorporation of hybrid BphA, II-9 genes causing approximately 58% degradation in contrast controls where only 10% (Suenaga et al., 2002) degradation was noticed.

Development of recombinant pathway by plasmid-assisted integration of genes leading to dehalogenation of PCBs aromatic rings by *Comamonas testosteroni* VP44 has been demonstrated (Hrywna et al., 1999). Thus designed recombinant strain was able to utilize the ortho- and para-chlorinated biphenyls as carbon source. Two different plasmids pE43 and pPC3, harboring the genes ohb and fcb and directing the synthesis of proteins responsible for dechlorination, were integrated into the host system to produce the strain VP44(pE43) and VP44(pPC3), respectively, and were able to completely dechlorinate the higher concentrations (up to 10 mM) of tested chlorinated biphenyls. The developed transgenic strains thus possessed the ability to detoxify the polychlorinated contaminated sites and can be extensively exploited for management of hydrocarbon-contaminated sites (Hrywna et al., 1999).

10. Degradation of polycyclic aromatic hydrocarbons

The enzyme cytochrome P450 monooxygenases (CYPs) have been described to actively engage in degradation of hazardous contaminants such as naphthalene, acenaphthalene, acenaphthylene, and 9-methylanthracene. The changes in overall activity of enzyme could be improved through modification in active site based on rational designing for enhanced degradation of aromatic hydrocarbons. Studies have revealed that alternation in CYPs in *P. putida* at location F87 and Y96 resulted into the enhancement in enzymatic activity by three times (Harford-Cross et al., 2000). In another study, the increase in CYP102 activity of *Bacillus megaterium* was observed after mutational changes executed at location R47L and Y51F. The modification was able to improve the enzymatic degradation up to 40 and 10 times for phenanthrene and fluoranthene, respectively (Carmichael and Wong, 2001). Similarly, mutational changes at three different positions corresponding to A74G/F87V/L188Q sites in CYP caused a 30-fold rise in polyaromatic hydrocarbon degradation and utilization of NADPH as compared to control sets (Liu et al., 2006).

11. Degradation of herbicide

Rapid increase in agriculture practices has surmounted the application of variety of herbicides. Many of the pesticides have long residence in soil system after application by farmers. Heavy use of toxic pesticides has caused considerable changes not only in microbial community of natural soil environment but also adversely affected the human health. The detrimental impact on human health has raised the concern for remediation of contaminated soil sites by sustainable biological approaches. The bacterial system house numbers of enzyme causing degradation of hazardous synthetic pesticides, but the efficiency is very low under natural environmental conditions. However, the efficiency could be enhanced through manipulation of enzymes causing degradation of targeted herbicides. The enzyme complex cytochrome P450cam and toluene dioxygenase from bacterial systems holds the great opportunity to detoxify the polyhalogenated C2 compounds. The integration and simultaneous expression of seven different genes responsible for the synthesis of enzyme complex in *Pseudomonas* offered their utilization as a microbe-based absolute degradation of pentachloroethane into carbon dioxide. Similarly, the enzyme responsible for degradation of atrazine was modified to enhance its degradation efficiency through DNA shuffling. The modified enzyme atrazine chlorohydrolase represented improved dechlorination of herbicide atrazine. Analysis of altered enzyme revealed the changes in a total of 11 amino acids. This enzyme has gained much attraction in the context of remediation of atrazine and is being evaluated for its better performance in model water treatment systems. Hence, the metabolic pathways that do not exist naturally can be developed under laboratory conditions to facilitate the degradation of hazardous contaminants. *Pseudomonas* sp. has been reported to synthesize two variables enzyme Atz (A, B, and C) and TriA causing degradation of atrazine. The mechanism of atrazine degradation by different enzymatic forms is different, but the final product is same, i.e., cyanuric acid. These enzymes for degradation of atrazine have been supposed to emerge very recently (Wackett, 1998).

12. Genetically modified endophytic bacteria and phytoremediation

Application of recombinant DNA technology as a promising tool in the field of contaminant removal has allowed present-day researchers with the possibility of modifying the genetic constitution of endophytes and rhizospheric microbes to facilitate the phytore-mediation of soil contaminated with toxic chemical substances (Divya et al., 2011). Selection of a suitable strain for transgenic development and transfer into rhizospheric region is relied on three important points: (1) the generated transgenic microbes should express stability along with enhanced expression of candidate gene, (2) chosen microbe should be able to withstand the harsh environment and able to survive at higher contaminant concentration, and (3) designed microbes should have easy adaptation in plant-specific rhizospheric region (Huang et al., 2004). Numerous bacterial species has been observed to have low degradation abilities for a particular contaminant in rhizosphere. Recent advances in molecular biology techniques have given the fascinating opportunity to develop genetically engineered bacteria equipped with contaminant degrading capability for removal of noxious contaminants present in rhizosphere, i.e., rhizoremediation (Glick, 2010). Experimental investigations pertaining to the molecular details of degradation of contaminants such as trichloroethylene and PCBs have also been conducted. Interestingly, the inoculation of GEMs at the time of sowing into the rhizosphere would be helpful to minimize the competition for resource utilization by microbial consortia. As the transfer of GEMs into natural ecosystem could have a potential threat on the natural microbial communities, detailed risk assessment should be conducted as an inevitable step in this context (Wackett, 2004).

A list of different bacterial species deployed for the remediation of hazardous contaminants is provided in Table 16.1.

13. Approaches to minimize the risks of genetically engineered microbes

Safety is one of the important issues regarding the release of GEMs into natural ecosystem. The problem can be resolved to a greater extent through biological containment (Lee et al., 2018). The phenomena of biological containment refer to the multiplication and availability of transgenic microbes exclusively at contaminated sites and also for the duration necessary for contaminant detoxification. To limit the dissemination of recombinant characteristics from transgenic microorganism to native surviving microbial species, numerous biocontainment systems integrated with a specific toxic compound and its antidote have been synthesized so far. The gene coding for a particular toxin is transferred under the genetic control of inducible promoter. Hence, designing of active containment systems would considerably minimize the possible dangers of release of recombinant bacteria that otherwise might happen in a natural environment (Ramos et al., 1995; Torres et al., 2004).

14. Challenges associated with the use of genetically engineered microorganism in bioremediation applications

Although lots of possibilities in the field of bioremediation have been expected after the discovery of recombinant DNA technology, the field-scale application could not have come into existence so far. The actual in situ demonstration of such programs for bioremediation purposes is under the infant stage (Sayler and Ripp, 2000). The important limitation of this approach is related with the fact that many of the in vitro isolated bacterial strains or other

TABLE 16.1 List of some genetically engineered bacterial species used for bioremediation.

S. N.	Engineered bacterial species	Contaminant	Remarks	References
1.	*Pseudomonas fluorescens* HK44	Naphthalene	The engineered microbe was luminescent; the first engineered bacterium approved for field trials; high survival under natural condition	Ripp et al. (2000a,b)
2.	*Pseudomonas fluorescens*	Naphthalene	Light induction was rapid and was highly responsive to alteration in naphthalene	King et al. (1990)
3.	*Comamonas testosteroni* VP44(pPC3) and VP44(pE43)	4-Chlorobenzoate	Useful for remediation of polychlorinated biphenyls; complete degradation of contaminants; degradation of high conc. 10 mM; transgenic microbe possessed two different genes "ohb" and "fcb"	Hrywna et al. (1999)
4.	*Burkholderia xenovorans* strain LB400 (ohb), *Rhodococcus* sp. strain RHA1(fcb)	Aroclor 1242	Contaminant favored the growth of transgenic bacteria; degradation was not significantly affected by inoculum density; the population dynamics was monitored by PCR and plate assay	Rodrigues et al. (2006)
5.	*Rhodococcus* sp. strain RHA1	4-Chlorobenzoates (CBA) and 4-chlorobiphenyl (4-CB)	Recombinant strain stored lesser content of chlorinated by-products; recombinant strains were able to grow only in presence of contaminant; *fcb* operon stable even after 60 days	Rodrigues et al. (2001)
6.	*Cupriavidus necator* RW112	Monochlorobenzoates and 3,5-dichlorobenzoate	First description of aerobic utilization of Aroclor mixtures; expression of transferred gene was measured by investigation of corresponding enzymatic activities	Wittich and Wolff (2007)
7.	*Sinorhizobium meliloti* strain USDA 1936	2′,3,4 PCB congener	Engineered microbe enhanced the degradation of contaminant; gene transfer was confirmed through molecular assays; plants associated with engineered microbes had more than two time degradation capabilities as compared to plants harboring wild strains	Chen et al. (2005)
8.	*Sinorhizobium meliloti* strain DHK1	2,4-Dinitrotoluene	In situ remediation of 2,4-dinitrotoluene; approximately 95% degradation of tested contaminant (0.55 mM)	Dutta et al. (2003)
9.	*Pseudomonas fluorescens* RE	2,4-Dinitrotoluene	The modified organism completely degraded the contaminant; the degradation was also feasible at low temperature (10°C); the inoculation of GEM with plants did not affect the growth	Monti et al. (2005)
10.	*Pseudomonas fluorescens*	Hexahydro-1,3,5-trinitro-1,3,5-triazine (RDX)	Pollutant degradation was high in presence of α-aminolevulinic acid; application of rhizosphere bacteria for degradation of explosives	Lorenz et al. (2013)

(Continued)

TABLE 16.1 List of some genetically engineered bacterial species used for bioremediation.—cont'd

S. N.	Engineered bacterial species	Contaminant	Remarks	References
11.	*Pseudomonas* sp. strain CB15	3-Chlorobiphenyl (3CB)	Emulsification raised the degradation by engineered *Pseudomonas*; degraded product was accumulated by transgenic bacteria	Adams et al. (1992)
12.	*Pseudomonas* hybrids	Trichloroethylene and cis-1,2-dichloroethylene	Engineered cells having todC1 gene were able to efficiently degrade a wide range of organic contaminants; possessed the genes	Suyama et al. (1996)
13.	*Streptomyces lividans*	Phenanthrene and 1-methoxynaphthalene	The rate of biotransformation was high; 200 μM and 2 mM of phenanthrene were converted within 6 and 32 h, respectively; genes responsible for contaminant conversion was recovered from *Nocardioides* sp. strain KP7	Chun et al. (2001)
14.	*Ralstonia* sp. KN1-10A	Trichloroethylene (TCE)	Most of the chlorine present in TCE was released as chloride ion; supply of external donors did not affect the degradation; contaminant did not show toxicity on engineered bacterium	Ishida and Nakamura (2000)
15.	*Escherichia coli* and *Pseudomonas putida* strains	TCE	Nearly absolute degradation was recorded for the contaminants; recombinant *E. coli* bacteria did not require the presence of isopropyl-β-D-thiogalactopyranoside	Fujita et al. (1995)
16.	*Escherichia coli* JM109	C.I. Direct Blue 71	The bioaugmented system was not functional at pH 5; at pH 9 there was dye content removal from 150 to 27.4 ppm within 12h	Jin et al. (2009)
17.	*Escherichia coli* JM109	Acid red GR	The dye degradation was affected by inoculums density and best was recorded at 10%; immobilization on foam carrier enhanced their degradation; the engineered cells maintained higher metabolic rate in batch reactors	Jin et al. (2008)
18.	*Rhodococcus erythropolis* strains	Phenol	The transformed cells represented enhanced phenol hydroxylase activity; engineered cells were 50% much effective in phenol degradation; the plasmids harboring the genes responsible for contaminant degradation were recovered even after 288 h indicating stability	ZÝdkovß et al. (2013)
19.	*Pseudomonas putida* BH (pS10-45)	Phenol	Engineered bacteria enhanced the degradation; higher sludge settling was observed for GEM inoculated systems; presence of engineered *Pseudomonas* cells affected the native microbial community participating in phenol degradation	Soda et al. (1998)

20.	*Sphingomonas paucimobilis* 551 (pS10-45)	Phenol	The designed floc forming phenol degrading bacteria was effective in phenol removal; engineered cells maintained population density up to four times	Soda et al. (1999)
21.	*Cupriavidus necator* JMP134 -ONP	Nitrophenol	The transgenic bacteria was able to degrade different nitrophenols simultaneously; engineered organism was unable to utilize the metanitrophenol as the sole nitrogen source	Hu et al. (2014)
22.	*Escherichia coli* BL21AI-GOS	Organophosphates	The bacteria could survive only in the presence of contaminant and commended suicide in the absence of contaminant; the bacteria was able to emit fluorescent light; the engineered microbe is safe for environmental applications	Li and Wu (2009)
23.	*Pseudomonas putida* KTUe	Organophosphates, pyrethroids, and carbamates	The engineered bacteria was able to simultaneously degrade three different pesticides; strain was able to survive at low oxygen availability; can be applied for in situ remediation of pesticide polluted soil; the modified bacteria was able to degrade 50 ppm of selected pesticides within 30 h in minimal medium containing glucose	Gong et al. (2018)

microbes for conducting remediation of contaminated sites may not be those performing contaminant degradation under natural environmental conditions and hence could not be considered as the best microbes for contaminant degradation. Because many of the microbes still could have not been cultured under laboratory conditions, development of their enrichment and identification procedures would add an important breakthrough in the area of contaminant remediation. Deployment of stable isotope probing has demonstrated that *Pseudomonas*, *Rhodococcus*, and other aerobic microorganisms are rapidly multiplying microbes, and most compatible host for transfer of genes participating in the contaminant degradation, however, are less important as wild strain under natural condition (Wackett, 2004). Moreover, exploitation of rapidly multiplying microbes for remediation purposes could essentially add unwanted biomass of the microorganism. Noteworthy, the genetically engineered microbes may suffer from survivability under natural environmental conditions limiting its field application. In addition, most of the transgenics have been developed for selected bacterial systems such as *E. coli*, *P. putida*, and *Bacillus subtilis*. Furthermore, the engineering of microorganisms should also be attempted for other organisms of environmental and agricultural relevance. Release of GEMs into natural environment may require the additional supply of energy, and they may have altered adaptability in a natural condition because of introduction of foreign genes (Saylor and Ripp, 2000; Singh et al., 2011). Most notably, there are also the considerable risks of acquisition of foreign genes by naturally existing unwanted microorganism that may affect the natural genetic diversity present in an ecosystem and human health in few cases. Although multiple technologies have been developed to circumvent the problem of environmental contaminants through GEMs, the development of suicidal GEMS, popularly recognized as "biocontainment system," relied on toxin/antitoxin gene system is considered as the most promising, efficient, and environment-friendly approach (Pandey et al., 2005).

15. Factors influencing genetically engineered microorganisms

The environmental application of GEMs is determined by following important factors:

(i) *Survival efficiency and stability under natural ecosystem*

Survival of GEMs is greatly determined by rate of multiplication, efficiency of resource utilization, presence of nutrient materials, physicochemical characteristics of environment, density of microorganisms, and availability of competing microbes. The growth rate of plasmid-bearing cells is relatively low as compared with plasmid-less cells because of presence of additional energy requirement of new metabolic pathways. Because of low multiplication rate and completion with native microorganisms, GEMs are less efficient in resource utilization and hence may be eliminated from the environment, limiting its application in bioaugmentation (Dechesne et al., 2005). The stability of engineered plasmid harboring the gene of interest linked with bioremediation is governed by the type of vector used, number of plasmids, plasmid duplication rate, presence of oxygen, and nutrient composition of the environment where they are applied along with the climatic conditions.

(ii) *Acquisition of foreign genes by natural microbial community*

The existence of GEMs at contaminated sites is influenced by horizontal transfer of gene of interest. It was demonstrated that introduction of GEMs on native microbes is not regularized and hence a detailed investigation is necessary to resolve this limitation. The phenomenon of horizontal gene transfer from transgenic microbe to indigenous microbes has been described by many workers. Transfer of genes from transgenic microbes *P. putida* UWC3 to native microbial population and enhancement of 2,4-D degradation is well mentioned (Dejonghe et al., 2000).

Contrary to the above facts, investigations have also shown the lack of transfer of plasmid-integrated genes into native microbial community after mixing and the fast degradation of DNA from GEMs; however, in few instances, the stability of genome was also documented. In an experiment conducted by Min et al. (1998), the working group has demonstrated the rapid degeneration of genetically engineered *P. putida* strains after transfer into sediment enriched with PCBs; however, very low transfer of genes rendering PCBs degradation into indigenous microbial population was observed. There are also the evidences regarding the impact of GEMs on indigenous microbial population.

Reports have also described the rapid horizontal gene transfer from recombinant microbes to wild strains under laboratory conditions as compared with natural conditions. Therefore, GEMs can be applied successfully at contaminated sites under field conditions to remove desired pollutants. Furthermore, the transfer of genes from transgenic microbes to native microbial communities would enhance the long-term stability and efficiency of bioremediation practices for intended purposes (Newby et al., 2000).

16. Regulation of genetically engineered microorganisms

The role of USEPA in controlling the release of GEMs for pesticide degradation is recognized. Toxic Substances Control Act (TSCA), 1977, facilitates the review and recommendation of experimental investigation regarding the field-scale application of GEMs within the regulations 40 of Biotechnology (Wozniak et al., 2012). Canadian government set up the Federal Regulatory Framework for Biotechnology in the year 1993 for controlling the application of GEMs for contaminant removal. The Indian government has established the Regulatory Environment Protection Act of 1986 for managing and controlling the application and utilization of GMOs and its derived products. Most notably, United Nations Environmental Program (UNEP) has also designed the regulations in Convention of Biological Diversity (CBD) to control such genetically engineered organism along with products of modern biotechnology. The European Union (EU) has made European Economic Community (EEC) directives 90/220 and EEC directives 90/219 for discharging engineered organisms into natural ecosystem.

17. Future perspective

Development of GEMs able to proliferate and degrade the contaminants under natural condition is one of the major challenges in the field of bioremediation. The release into natural ecosystem is restricted because of certain ethical and technical issues. Detailed insight into the factors governing the proper functioning of GEMS participating in bioremediation is of utmost importance. Measures should be developed to monitor the spread of engineered microbes into places that are not targeted for contaminant degradation. Development of effective biomonitoring system would be much helpful in this connection. Advances should also be made to restrict the acquisition of contaminant degrading genes into native microbial populations. More microbial host systems, apart from commonly employed ones, should be searched out for housing the contaminant degrading genes.

18. Conclusion

Rapid industrialization has generated very large amount of dyes and organic contaminant waste into the natural environment. Excess presence of these hazardous substances into the ecosystem has affected the human health and environment severely. Application of conventional treatment processes has several limitations that can be managed only through biological approaches. The application of useful microorganisms under these scenario is very much promising form environmental perspectives. The naturally occurring microbes may harbor the genes related with contaminant degradation; however, their application in wild state is limitations. This calls for the need of generation of genetically engineered microbes for the management of hazardous chemicals present into the environment. Because the release of engineered microbes in the natural conditions has long been a matter of controversy, more research work under natural conditions is sought for getting optimum results for the degradation of contaminants. The safety issues and concerned risks must be taken into account while considering their release into natural environment. Overall, there are lots of opportunities in application of engineered microbes for the management of environmental contaminants.

References

Adams, R.H., Huang, C.M., Higson, F.K., Brenner, V., Focht, D.D., 1992. Construction of a 3-chlorobiphenyl-utilizing recombinant from an intergeneric mating. Applied and Environmental Microbiology 58, 647–654.

Ang, E.L., Zhao, H., Obbard, J.P., 2005. Recent advances in the bioremediation of persistent organic pollutants via biomolecular engineering. Enzyme and Microbial Technology 37, 487–496.

Barriault, D., Plante, M.M., Sylvestre, M., 2002. Family shuffling of a targeted *bphA* region to engineer biphenyl dioxygenase. Journal of Bacteriology 184, 3794–3800.

Blümel, S., Knackmuss, H.J., Stolz, A., 2002. Molecular cloning and characterization of the gene coding for the aerobic azoreductase from *Xenophilus azovorans* KF46F. Applied and Environmental Microbiology 68, 3948–3955.

Blümel, S., Stolz, A., 2003. Cloning and characterization of the gene coding for the aerobic azoreductase from *Pigmentiphaga kullae* K24. Applied Microbiology and Biotechnology 62, 186–190.

Brenner, V., Arensdorf, J.J., Focht, D.D., 1994. Genetic construction of PCB degraders. Biodegradation 5, 359–377.

Carmichael, A.B., Wong, L.L., 2001. Protein engineering of *Bacillus megaterium* CYP102—the oxidation of polycyclic aromatic hydrocarbons. European Journal of Biochemistry 268, 3117–3125.

Cases, I., de Lorenzo, V., 2005a. Genetically modified organisms for the environment: stories of success and failure and what we have learned from them. International Microbiology 8, 213–222.

Cases, I., de Lorenzo, V., 2005b. Promoters in the environment: transcriptional regulation in its natural context. Nature Reviews Microbiology 3, 105–118.

Chanwala, J., Kaushik, G., Dar, M.A., Upadhyay, S., Agrawal, A., 2019. Process optimization and enhanced decolorization of textile effluent by *Planococcus* sp. isolated from textile sludge. Environmental Technology and Innovation 13, 122–129.

Chauhan, A., Fazlurrahman, Oakeshott, O.G., Jain, R.K., 2008. Bacterial metabolism of polycyclic aromatic hydrocarbons: strategies for bioremediation. Indian Journal of Microbiology 48, 95–113.

Chen, H.Z., Wang, R.F., Cerniglia, C.E., 2004. Molecular cloning, overexpression, purification, and characterization of an aerobic FMN-dependent azoreductase from *Enterococcus faecalis*. Protein Expression and Purification 34, 302–310.

Chen, Y., Adam, A., Toure, O., Dutta, S.K., 2005. Molecular evidence of genetic modification of *Sinorhizobium meliloti*: enhanced PCB bioremediation. Journal of Industrial Microbiology and Biotechnology 32, 561–566.

Chen-Goodspeed, M., Sogorb, M.A., Wu, F.Y., Hong, S.B., Raushel, F.M., 2001b. Structural determinants of the substrate and stereochemical specificity of phosphotriesterase. Biochemistry 40, 1325–1331.

Chen-Goodspeed, M., Sogorb, M.A., Wu, F.Y., Raushel, F.M., 2001a. Enhancement, relaxation, and reversal of the stereo selectivity for phosphotriesterase by rational evolution of active site residues. Biochemistry 40, 1332–1339.

Chun, H.K., Ohnishi, Y., Misawa, N., Shindo, K., Hayashi, M., Harayama, S., Horinouchi, S., 2001. Biotransformation of phenanthrene and 1-methoxynaphthalene with *Streptomyces lividans* cells expressing a marine bacterial phenanthrene dioxygenase gene cluster. Bioscience, Biotechnology, and Biochemistry 65, 1774–1781.

Dafale, N., Rao, N., Meshram, S., Wate, S., 2008. Decolorization of azo dyes and simulated dye bath wastewater using acclimatized microbial consortium-Biostimulation and halotolerance. Bioresource Technology 99, 2552–2558.

Das, M., Adholeya, A., 2012. Role of microorganisms in remediation of contaminated soil. In: Microorganisms in Environmental Management. Springer, Dordrecht, pp. 81–111.

Dechesne, A., Pallud, C., Bertolla, F., Grundmann, G.L., 2005. Impact of the microscale distribution of Pseudomonas strain introduced into soil on potential contacts with indigenous bacteria. Applied and Environmental Microbiology 71, 8123–8131.

Dejonghe, W., Goris, J., Fantroussi, S.E., Höfte, M., Paul De Vos, Verstraete, W., Top, E.M., 2000. Effect of dissemination of 2,4- dichlorophenoxyacetic acid (2,4-D) degradation plasmids on 2,4-D degradation and on bacterial community structure in two different soil horizons. Applied and Environmental Microbiology 66, 3297–3304.

Dutta, S.K., Hollowell, G.P., Hashem, F.M., Kuykendall, L.D., 2003. Enhanced bioremediation of soil containing 2, 4-dinitrotoluene by a genetically modified *Sinorhizobium meliloti*. Soil Biology and Biochemistry 35, 667–675.

Erickson, B.D., Mondello, F.J., 1992. Nucleotide sequencing transcriptional mapping of the genes encoding biphenyl dioxygenase, a multicomponent polychlorinatedbiphenyl-degrading enzyme in *Pseudomonas* strain-LB400. Journal of Bacteriology 174, 2903–2912.

Ezezika, O.C., Singer, P.A., 2010. Genetically engineered oil-eating microbes for bioremediation: prospects and regulatory challenges. Technology in Society 32, 331–335.

Fujita, M., Ike, M., Hioki, J.I., Kataoka, K., Takeo, M., 1995. Trichloroethylene degradation by genetically engineered bacteria carrying cloned phenol catabolic genes. Journal of Fermentation and Bioengineering 79, 100–106.

Glick, B.R., 2010. Using soil bacteria to facilitate phytoremediation. Biotechnology Advances 28, 367–374.

Gong, T., Xu, X., Dang, Y., Kong, A., Wu, Y., Liang, P., Wang, S., Yu, H., Xu, P., Yang, C., 2018. An engineered Pseudomonas putida can simultaneously degrade organophosphates, pyrethroids and carbamates. The Science of the Total Environment 628, 1258–1265.

Harford-Cross, C.F., Carmichael, A.B., Allan, F.K., England, P.A., Rouch, D.A., Wong, L.L., 2000. Protein engineering of cytochrome P450(cam) (CYP101) for the oxidation of polycyclic aromatic hydrocarbons. Protein Engineering 13, 121–128.

Heinaru, E., Merimaa, M., Viggor, S., Lehiste, M., Leito, I., Truu, J., Heinaru, A., 2005. Biodegradation efficiency of functionally important populations selected for bioaugmentation in phenol- and oil-polluted area. FEMS Microbiology Ecology 51, 363–373.

Hrywna, Y., Tsoi, T.V., Maltseva, O.V., Quensen, J.F., Tiedje, J.M., 1999. Construction and characterization of two recombinant bacteria that grow on *ortho*- and *para*-substituted chlorobiphenyls. Applied and Environmental Microbiology 65, 2163–2169.

Hu, F., Jiang, X., Zhang, J.J., Zhou, N.Y., 2014. Construction of an engineered strain capable of degrading two isomeric nitrophenols via a sacB-and gfp-based markerless integration system. Applied Microbiology and Biotechnology 98, 4749–4756.

Hu, T.L., 1998. Degradation of azo dye RP2B by *Pseudomonas luteola*. Water Science and Technology 38, 299–306.

Huang, X.D., El-Alawi, Y., Penrose, D.M., Glick, B.R., Greenberg, B.M., 2004. Responses of three grass species to creosote during phytoremediation. Environmental Pollution 130, 453–463.

Ishida, H., Nakamura, K., 2000. Trichloroethylene degradation by *Ralstonia* sp. KN1-10A constitutively expressing phenol hydroxylase: transformation products, NADH limitation, and product toxicity. Journal of Bioscience and Bioengineering 89, 438–445.

Jin, R., Yang, H., Zhang, A., Wang, J., Liu, G., 2009. Bioaugmentation on decolorization of CI Direct Blue 71 by using genetically engineered strain *Escherichia coli* JM109 (pGEX-AZR). Journal of Hazardous Materials 163, 1123–1128.

Jin, R.F., Zhou, J.T., Zhang, A.L., Wang, J., 2008. Bioaugmentation of the decolorization rate of acid red GR by genetically engineered microorganism *Escherichia coli* JM109 (pGEX-AZR). World Journal of Microbiology and Biotechnology 24, 23–29.

Kellogg, S.T., Chatterjee, D.K., Chakrabarty, A.M., 1981. Plasmid-assisted molecular breeding: new technique for enhanced biodegradation of persistent toxic chemicals. Science 214, 1133–1135.

Khalid, A., Arshad, M., Crowley, D.E., 2008. Accelerated decolorization of structurally different azo dyes by newly isolated bacterial strains. Applied Microbiology and Biotechnology 78, 361–369.

Khelifi, E., Bouallagui, H., Touhami, Y., Godon, J., Hamdi, M., 2009. Bacterial monitoring by molecular tools of a continuous stirred tank reactor treating textile wastewater. Bioresource Technologyl 100, 629–633.

King, J.M.H., DiGrazia, P.M., Applegate, B., Burlage, R., Sanseverino, J., Dunbar, P., Larimer, F., Sayler, G.A., 1990. Rapid, sensitive bioluminescent reporter technology for naphthalene exposure and biodegradation. Science 249, 778–781.

Lee, J.W., Chan, C.T., Slomovic, S., Collins, J.J., 2018. Next-generation biocontainment systems for engineered organisms. Nature Chemical Biology 1.

Lehrbach, P.R., Zeyer, J., Reineke, W., Knackmuss, H.J., Timmis, K.N., 1984. Enzyme recruitment in vitro: use of cloned genes to extend the range of haloaromatics degraded by *Pseudomonas* sp. strain B13. Journal of Bacteriology 158, 1025–1032.

Li, Q., Wu, Y.J., 2009. A fluorescent, genetically engineered microorganism that degrades organophosphates and commits suicide when required. Applied Microbiology and Biotechnology 82, 749–756.

Liu, Z., Hong, Q., Xu, J.H., Jun, W., Li, S.P., 2006. Construction of a genetically engineered microorganism for degrading organophosphate and carbamate pesticides. International Biodeterioration anf Biodegradation 58, 65–69.

Lorenz, A., Rylott, E.L., Strand, S.E., Bruce, N.C., 2013. Towards engineering degradation of the explosive pollutant hexahydro-1,3,5-trinitro-1,3,5-triazine in the rhizosphere. FEMS Microbiology Letters 340, 49–54.

Lovley, D.R., 2003. Cleaning up with genomics: applying molecular biology to bioremediation. Nature Reviews Microbiology 1, 35–44.

Markandey, D.K., Rajvaidya, N., 2004. Environmental Biotechnology, first ed. APH Publishing corporation, p. 79.

Marwick, C., 2003. Link found between Agent Orange and chronic lymphocytic leukaemia. BMJ 326, 242.

Menn, F.M., Easter, J.P., Sayler, G.S., 2008. Genetically engineered microorganisms and bioremediation. In: Rehm, H.J., Reed, G. (Eds.), Biotechnology: Environmental Processes II 2008; 11b, second ed. Wiley-VCH Verlag GmbH, Weinheim, Germany. Chapter, 21.

Min, M.G., Kawabata, Z., Ishii, N., Takata, R., Furukawa, K., 1998. Fate of a PCBs degrading recombinant Pseudomonas putida AC30 (PMFB2) and its effect on the densities of microbes in marine microcosms contaminated with PCBs. International Journal of Environmental Studies 55, 271–285.

Moharikar, A., Purohit, H.J., Kumar, R., 2005. Microbial population dynamics at effluent treatment plants. Journal of Environmental Monitoring 7, 552–558.

Monti, M.R., Smania, A.M., Fabro, G., Alvarez, M.E., Argaraña, C.E., 2005. Engineering *Pseudomonas fluorescens* for biodegradation of 2,4-dinitrotoluene. Applied and Environmental Microbiology 71, 8864–8872.

Newby, D.T., Josephson, K.L., Pepper, I.L., 2000. Detection and characterization of plasmid pJP4 transfer to indigenous soil bacteria. Applied and Environmental Microbiology 66, 290–296.

Pandey, G., Paul, D., Jain, R.K., 2005. Conceptualizing "suicidal genetically engineered microorganisms" for bioremediation applications. Biochemical and Biophysical Research Communications 327, 637–639.

Pieper, D.H., Reineke, W., 2000. Engineering bacteria for bioremediation. Current Opinion in Biotechnology 11, 262–270.

Quan, X., Shi, H., Liu, H., Lv, P., Qian, Y., 2004. Enhancement of 2,4-dichlorophenol degradation in conventional activated sludge systems bioaugmented with mixed special culture. Water Research 38, 245–253.

Rafii, F., Hall, J.D., Cerniglia, C.E., 1997. Mutagenicity of azo dyes used in foods, drugs and cosmetics before and after reduction by *Clostridium* species from the human intestinal tract. Food and Chemical Toxicology 35, 897–901.

Ramos, J.L., Díaz, E., Dowling, D., de Lorenzo, V., Molin, S., O'Gara, F., Ramos, C., Timmis, K.N., 1994. The behavior of bacteria designed for biodegradation. Biotechnology 12, 1349–1356.

Ramos, J.L., Duque, E., Huertas, M.J., HaÏDour, A.L.I., 1995. Isolation and expansion of the catabolic potential of a Pseudomonas putida strain able to grow in the presence of high concentrations of aromatic hydrocarbons. Journal of Bacteriology 177, 3911–3916.

Ramos, J.L., Marqués, S., van Dillewijn, P., Espinosa-Urgel, M., Segura, A., Duque, E., Krell, T., Ramos-González, M.I., Bursakov, S., Roca, A., Solano, J., Fernádez, M., Niqui, J.L., Pizarro-Tobias, P., Wittich, R.M., 2011. Laboratory research aimed at closing the gaps in microbial bioremediation. Trends in Biotechnology 29, 641–647.

Ramos, J.L., Wasserfallen, A., Rose, K., Timmis, K.N., 1987. Redesigning metabolic routes: manipulation of TOL plasmid pathway for catabolism of alkylbenzoates. Science 235, 593–596.

Ramos-González, M.I., Ruiz-Cabello, F., Brettar, I., Garrido, F., Ramos, J.L., 1992. Tracking genetically engineered bacteria: monoclonal antibodies against surface determinants of the soil bacterium Pseudomonas putida 2440. Journal of Bacteriology 174, 2978–2985.

Reineke, W., Knackmuss, H.J., 1979. Construction of haloaromatics utilising bacteria. Nature 277, 385–386.

Ripp, S., Nivens, D.E., Ahn, Y., Werner, C., Jarrell, J., Easter, J.P., Cox, C.D., Burlage, R.S., Sayler, G.S., 2000a. Controlled field release of a bioluminescent genetically engineered microorganism for bioremediation process monitoring and control. Environmental Science and Technology 34, 846–853.

Ripp, S., Nivens, D.E., Werner, C., Sayler, G.S., 2000b. Bioluminescent most-probable-number monitoring of a genetically engineered bacterium during a long-term contained field release. Applied Microbiology and Biotechnology 53, 736–741.

Rodrigues, J.L., Kachel, C.A., Aiello, M.R., Quensen, J.F., Maltseva, O.V., Tsoi, T.V., Tiedje, J.M., 2006. Degradation of Aroclor 1242 dechlorination products in sediments by *Burkholderia xenovorans* LB400 (ohb) and *Rhodococcus* sp. strain RHA1 (fcb). Applied and Environmental Microbiology 72, 2476–2482.

Rodrigues, J.L., Maltseva, O.V., Tsoi, T.V., Helton, R.R., Quensen, J.F., Fukuda, M., Tiedje, J.M., 2001. Development of a Rhodococcus recombinant strain for degradation of products from anaerobic dechlorination of PCBs. Environmental Science and Technology 35, 663–668.

Rojo, F., Pieper, D.H., Engesser, K.H., Knackmuss, H.J., Timmis, K.N., 1987. Assemblage of ortho cleavage route for simultaneous degradation of chloro- and methylaromatics. Science 238, 1395–1398.

Sandhya, S., Sarayu, K., Uma, B., Swaminathan, K., 2008. Decolorizing kinetics of a recombinant *Escherichia coli* SS125 strain harboring azoreductase gene from *Bacillus latrosporus* RRK1. Bioresource Technology 99, 2187–2191.

Sarma, H., Prasad, M.N.V., 2019. Metabolic engineering of rhizobacteria associated with plants for remediation of toxic metals and metalloids. In: Prasad, M.N.V. (Ed.), Transgenic Plant Technology for Remediation of Toxic Metals and Metalloids. Academic Press, pp. 299–318.

Sayler, G.S., Ripp, S., 2000. Field applications of genetically engineered microorganisms for bioremediation processes. Current Opinion in Biotechnology 11, 286–289.

Shimizu, H., 2002. Metabolic engineering–integrating methodologies of molecular breeding and bioprocess systems engineering. Journal of Bioscience and Bioengineering 94, 563–573.

Singh, J.S., Abhilash, P.C., Singh, H.B., Singh, R.P., Singh, D.P., 2011. Genetically engineered bacteria: an emerging tool for environmental remediation and future research perspectives. Gene 480, 1–9.

Soda, S., Ike, M., Fujita, M., 1998. Effects of inoculation of a genetically engineered bacterium on performance and indigenous bacteria of a sequencing batch activated sludge process treating phenol. Journal of Fermentation and Bioengineering 86, 90–96.

Soda, S., Uesugi, K., Ike, M., Fujita, M., 1999. Application of a floc-forming genetically engineered microorganism to a sequencing batch reactor for phenolic wastewater treatment. Journal of Bioscience and Bioengineering 88, 85–91.

Stegmann, R., 2011. Treatment of Contaminated Soil: Fundamentals, Analysis, Applications. Springer Verlag, Berlin.

Suenaga, H., Watanabe, T., Sato, M., Ngadiman, Furukawa, K., 2002. Alteration of regiospecificity in biphenyl dioxygenase by active-site engineering. Journal of Bacteriology 184, 3682–3688.

Suyama, A., Iwakiri, R., Kimura, N., Nishi, A., Nakamura, K., Furukawa, K., 1996. Engineering hybrid pseudomonads capable of utilizing a wide range of aromatic hydrocarbons and of efficient degradation of trichloroethylene. Journal of Bacteriology 178, 4039–4046.

Time Magazine, September 22, 1975. Environment: Oil-Eating Bug. Time.

Timmis, K.N., Pieper, D.H., 1999. Bacteria designed for bioremediation. Trends in Biotechnology 17, 200–204.

Torres, B., García, J.L., Díaz, E., 2004. Plasmids as tools for containment. In: Phillips, G., Funnell, B. (Eds.), Plasmid Biology. ASM Press, Washington DC, pp. 589–601.

Van der Zee, F.P., Villaverde, S., 2005. Combined anaerobic–aerobic treatment of azo dyes–a short review of bioreactor studies. Water Research 39, 1425–1440.

Wackett, L.P., 1998. Directed evolution of new enzymes and pathways for environmental biocatalysis. Annals of the New York Academy of Sciences 864, 142–152.

Wackett, L.P., 2004. Stable isotope probing in biodegradation research. Trends in Biotechnology 22, 153–154.

Watanabe, M.E., 2001. Can bioremediation bounce back? Nature Biotechnology 19, 1111–1115.

Webster, L.R., McKenzie, G.H., Moriarty, H.T., 2002. Organophosphate-based pesticides and genetic damage implicated in bladder cancer. Cancer Genetics and Cytogenetics 133, 112–117.

Wittich, R.M., Wolff, P., 2007. Growth of the genetically engineered strain *Cupriavidus necator* RW112 with chlorobenzoates and technical chlorobiphenyls. Microbiology 153, 186–195.

Wong, P.K., Yuen, P.Y., 1996. Decolorization and biodegradation of methyl red by *Klebsiella pnumoniae* RS-13. Water Research 30, 1736–1744.

Wozniak, C.A., McClung, G., Gagliardi, J., Segal, M., Matthews, K., 2012. Regulation of genetically engineered microorganisms under FIFRA, FFDCA and TSCA. In: Regulation of Agricultural Biotechnology: The United States and Canada. Springer, Netherlands, pp. 57–94.

Wu, C.F., Cha, H.J., Rao, G., Valdes, J.J., Bentley, W.E., 2000. A green fluorescent protein fusion strategy for monitoring the expression, cellular location, and separation of biologically active organophosphorus hydrolase. Applied Microbiology and Biotechnology 54, 78–83.

Yang, C.F., Liu, S.H., Su, Y.M., Chen, Y.R., Lin, C.W., Lin, K.L., 2019. Bioremediation capability evaluation of benzene and sulfolane contaminated groundwater: determination of bioremediation parameters. The Science of the Total Environment 648, 811–818.

Zanaroli, G., Fedi, S., Carnevali, M., Fava, F., Zannoni, D., 2002. Use of potassium tellurite for testing the survival and viability of *Pseudomonas pseudoalcaligenes* KF707 in soil microcosms contaminated with polychlorinated biphenyls. Research in Microbiology 153, 353–360.

ZÝdkovß, L., Szők, J., Ruckß, L., Pßtek, M., Nešvera, J., 2013. Biodegradation of phenol using recombinant plasmid-carrying *Rhodococcus erythropolis* strains. International Biodeterioration and Biodegradation 84, 179–184.

Further reading

Bin, Y., Jiti, Z., Jing, W., Cuihong, D., Hongman, H., Zhiyong, S., Yongming, B., 2004. Expression and characteristics of the gene encoding azoreductase from *Rhodobacter sphaeroides* AS1. 1737. FEMS Microbiology Letters 236, 129–136.

Bourrain, M., Achouak, W., Urbain, V., Heulin, T., 1999. DNA extraction from activated sludges. Current Microbiology 38, 315–319.

Divya, B., Deepak Kumar, M., 2011. Plant–microbe interaction with enhanced bioremediation. Research Journal of BioTechnology 6, 72–79.

Index

Note: 'Page numbers followed by "f" indicate figures and "t" indicate tables'.

A

A. nidulans, 186
Abiotic stresses, 170
    factors, 151
ACC deaminase activity. *See* 1-Aminocyclopropane-
    1-carboxylate deaminase activity
    (ACC deaminase activity)
Accelerated solvent extraction (ASE), 71
Acclimatization, 15
Acenaphthene, 211–213
Acephate ($C_4H_{10}NO_3PS$), 41f, 48t–53t
Acetobacter, 57–58
Acetylcholine (Ach), 33–34
Acetylcholinesterase enzyme (AChE), 33–34
    AChE-diethyl complex, 33–34
    AChE-OP complex, 33–34
    reactivation and aging, 35f
Ach. *See* Acetylcholine (Ach)
AChE. *See* Acetylcholinesterase enzyme (AChE)
Achromobacter, 87–88
    A. denitrificans, 10–13
    A. piechaudii, 156
    A. xylosoxidans Ax10, 167t–168t
Acid Blue 113, 118–119
Acid Blue 29, 109–110
Acid Red 183 dye, 111–112
Acinetobacter sp., 55
    A. johnsonii, 57–58
    A. rhizosphaerae, 57
Activated sludge process, 13
Active ingredients, 25–26
Acute toxicity, 29–31
Adsorption, 13, 206
Advanced oxidation processes (AOPs), 71, 75–76,
    206–207, 221
Advenella species, 207–210
Aeration, 141
Aerobic
    cathode chamber, 117
    degradation of phenolic waste, 210–213
Aeromonas hydrophila, 9
AFM. *See* Atomic force microscopy (AFM)

Age-related disorders, 164
Agent orange degradation, 337
Aging, 33–34
Agriculture, 325
Agriculture management, cyanobacteria role in,
    194–195
Agrobacterium sp., 58
AIDS, 326
Air, 256
5 L Airlift biological reactor, 114
Alcaligenes xylosoxidans Y234, 15
Aldicarb, 69f
Aldrin, 247, 249t, 250
Alfalfa. *See* Medicago sativa (Alfalfa)
Algae, 112
Aliphatic polycarbonate (APC), 323
Aliphatic polyesters, 325
Alkali-tolerant cyanobacteria, 196–197
Alternaria alternata, 140
Alyssum murale, 166–169
1-Aminocyclopropane-1-carboxylate deaminase
    activity (ACC deaminase activity), 156
Amorphoteca, 214
Amplified ribosomal DNA restriction analysis
    (ARDRA), 333–334
Anabaena torulosa, 196–197
Anaerobic
    anode chamber, 117
    bioremediation, 218–219
    degradation of phenolic waste, 213
    treatment, 108
Animal husbandry, 325
Anionic functional groups, 166–169
Anthracene, 133–135
    degradation pathway of, 134f
Anthropogenic emissions, 128
Antrodia vaillantii, 131–133
Antrodia xanthan, 183–184
AOPs. *See* Advanced oxidation processes (AOPs)
APC. *See* Aliphatic polycarbonate (APC)
Arabidopsis, 84–85, 94–95
    A. thaliana, 84

ARDRA. *See* Amplified ribosomal DNA restriction
    analysis (ARDRA)
Armilleria gemina, 183—184
Aromatic
    amines, 108
    compounds, 169
    hydrocarbons, 203—204, 218—219
Arthrobacter sp., 47—57, 153
    A. atrocyaneus, 57—58
Ascomycetes, 179
Ascomycota, 182
ASE. *See* Accelerated solvent extraction (ASE)
Aspergillus, 128, 131, 137, 185
    A. terreus, 47—54
Aspergillus fumigatus, 47—54
Aspergillus niger, 56, 111—112, 133—135, 137, 140,
    183—184
Atmosphere, 68
Atomic force microscopy (AFM), 287—289
    friction image of *E. coli*, 304f
    image of spore of *B. laterosporus*, 291f
    imaging in air
        in contact mode, 301
        in resonant mod, 301—302
    imaging in liquids, 302
    installation, 300
ATP-binding cassette transporters, 84
Automobile industry, 326
Auxiliary materials in biological recovery, 325
Azo-dye reduction, 9
Azoarcus evansii, 213
Azomonas sp. RJ4, 167t—168t
Azospirillum, 153, 154t, 155
    A. lipoferum, 166—169
Azotobacter, 153, 170
    A. chroococcum HKN-5, 167t—168t

B

Bacillus, 54—55, 57—58, 117, 153, 155, 167t—168t, 170
    PY1, 187
    RJ31, 167t—168t
Bacillus amyloliquefaciens, 155
    IN937a, 166
Bacillus cereus, 54—55, 87—88, 167t—168t
Bacillus licheniformis NCCP-59, 167t—168t
Bacillus megaterium, 57—58, 166—169, 338—339
    HKP-1, 167t—168t
Bacillus pumilus, 155
    SE34, 166
Bacillus safensis, 57
Bacillus subtilis, 9, 54—55, 219, 344
    FZB24, 166
    GB03, 166

SJ-101, 167t—168t
Bacillus thermoleovorans, 16
Bacillus weihenstephanensis SM3, 167t—168t
Bacteria(l), 110—111, 207—210, 288
    bacteria-mediated PAH degradation, 130
    biofilms, 288—289
    cells, 288—289
    degradation, 74
    electromechanical biosensor, 291—293
    preparation, 299
    species, 153
    symbiotic association of fungi with, 186
Basic Blue 9, 109—110
Basidiomycetes, 179
Basidiomycota, 182
Bayberry. *See Myrica rubra* (Bayberry)
Beauveria, 140
    B. bassiana, 290
Benz[a]anthracene, 69f
Benzene, toluene, ethyl benzene, and xylene (BTEX),
    3—5, 15, 193—194, 203—204
Benzo[a]pyrene, 69f, 131—133
Berknolderia sp., 167t—168t
BHC, 265
Bioaccumulation process, 29
Bioaugmentation, 73—74, 169—170, 196
Bioavailability, 15—16
    of pollutants, 278
Biocatalysts, 112—113, 214—215
Biocellulose, 315
Biocontainment systems, 340
Bioconversion, 73
Biodegradable materials, 314—315, 314f
    application
        in agriculture, forestry, fisheries, and animal
            husbandry, 325
        in automobile industry, 326
        in foamed products, 325
        in medical biodegradable plastics, 326
        in other daily necessities, 326
        instead of plastics, 326f
        in processing AIDS, 326
    comparison of degradation, 315t
    enhancement of property, 323f
    existing or forthcoming, 323—325
    ideal biodegradable technology for future,
        327—328
    natural polymer raw materials production, 315—317
    novel biodegradable materials and applications, 327
    performance comparison of degradable materials
        with PE, 318t
    synthetic biodegradable materials production,
        317—323

Biodegradable plastics, 315
  films and sheets, 324
Biodegradation, 13, 210, 319, 327
Biofilm, 288—289
  formation, 274, 289
Biological
  approach, 73—75
    microbial degradation, 73—75
  factors affecting biological removal of textile dyes,
    109—110, 109f
  immobilization of biological catalysts, 113
  reactor, 114
  remediation, 115, 161—162
  treatment strategies, 108
Biological methods, 2—3
  for phenolic compound removal, 207—216
    aerobic degradation of phenolic waste,
      210—213, 211f
    anaerobic degradation of phenolic waste, 213
    bacteria, 207—210
    biodegradation mechanism, 210
    biosurfactants, 215
    enzymes participating in phenolic compounds
      degradation, 214—215
    fungi biodegradation, 213—214
    genetically modified bacteria, 215—216
Biological oxygen demand (BOD), 107
Biomonitoring tools, 291
Biopurification system (BPS), 184—185
Bioreactor systems for dye remediation,
  advancements in, 114—115
Bioremediation, 2—3, 127—128, 130, 161—162,
  177—178, 194, 216, 219—221, 290. See also
  Phytoremediation
  bioavailability, 15—16
  biological agents, 7t
  concentration of pollutant, 15
  cyanobacteria role in, 195—197
  of dye contaminated water, 6—9
  with emphasis on petrochemical and organic
    pollutants, 14—15
  effect of environmental conditions, 16—17
  for environmental pollutants cleanup, 3—17
  factors influencing bioremediation of phenolic waste,
    216—219
    carbon sources effect on phenol degradation
      potential, 218
    effect of dissolved oxygen concentration, 218—219
    microbial growth kinetics, 219
    nutrient availability, 217
    pH effect on phenol degradation potential,
      217—218
    temperature, 216—217

fungal enzymes for, 178—179
fungi in, 184—186
GEM—based bioremediation, 14
of heavy metal contaminated water, 5—6
microbial adaptation, 15
microorganism involved in organic pollutant
  degradation, 5t
nutrients availability, 15
of pesticide contamination, 9—10
PGPB in, 165—166
PPCPs removal by biological degradation
  processes, 10—13
strategies
  for hydrocarbon contaminated water and
    soil, 3—5
  and methods, 4f
of toxic substances, 187
vermi-biofiltration of wastewater, 9
vermicomposting of solid wastes, 14
Biosensors, 291—292
Biosparging, 196
Biostimulation, 73—74, 169—170
Biosurfactants, 138—139, 215
  production, 273—274
    by fungi and application in bioremediation,
      138—139, 139t
Biotechnological approaches, 94—95
Biotic factors, 151
Biotransformation of organic contaminants,
  87—88
Bioventing, 73—74, 196
Biphenyl dioxygenase (BphA), 95
Birds, 250
Bisphenol A (BPA), 203—204
Bjerkandera adjusta, 184—185
BOD. See Biological oxygen demand (BOD)
BPA. See Bisphenol A (BPA)
BPS. See Biopurification system (BPS)
Brassica campestris L., 43—44
Brassica juncea, 166—169
Brassica oleracea (Cauliflower), 43—44
Brevibacillus sp., 167t—168t
  B. laterosporus, 290—291
Brevibacterium sp., 57
Brevundimonas sp., 54—55
Bromophenol Blue, 115
Brucella melitensis, 54—55
BTEX. See Benzene, toluene, ethyl benzene, and
  xylene (BTEX)
Burkholderia, 57, 130, 153
Burkholderia cepacia, 54—55
  strain LB400, 338
  VM1468, 215—216

C

C. unicolor, 185
Cadmium (Cd) toxicity, 165
Cadusafos ($C_{10}H_{23}O_2PS_2$), 41f, 48t–53t
Calcium, 141
Calothrix, 196–197
CAM plasmid, 336–337
Camellia sinensis, 40–43
Cancer, 255
Candida antarctica, 137
Candida rugosa, 137
Cantilever choice, 300
Capillary stereolithography, 297–298
Capra hircus (Goat), 45–46
Carbamate
    degradation, 337–338
    pesticides, 27
N-Carboxyethylchitosan, 317
N-Carboxymethylchitosan, 317
Card sheets and films, 324
Cardiovascular diseases, 256
Catalyst loading, 225–226
Catla catla, 45–46
Cauliflower. *See* Brassica oleracea (Cauliflower)
CBD. *See* Convention of Biological Diversity (CBD)
Cell-bound enzymes, 179
Cellular compartmentalization, 83–84
Cellulases, 137
Cellulose, 316–317
Ceriporiopsis subvermispora, 184–185
CETPs. *See* Common effluent treatment plants (CETPs)
Chaetomium, 137
Channa striata, 45–46
Chemical approaches, 75
Chemical method for phenolic compound removal, 206–207
Chemical oxidation method, 206–207
Chemical oxygen demand (COD), 107, 203–204
Chemical substances, 332–333
Chick. *See* Gallus gallus (Chick)
Chitin, 317
Chitosan, 317
Chlordane, 247, 249t, 250–251
Chlorfenvinphos ($C_{12}H_{14}C_{13}O_4P$), 41f, 48t–53t
Chlorinated aromatic compounds, 274
4-Chlorobiphenyl, 95
Chlorophenol (CP), 203–204, 215, 226
Chlorpyrifos ($C_9H_{11}C_{13}NO_3PS$), 34–40, 41f, 45–46, 48t–53t
    chlorpyrifos-ethyl concentration, 36–40

Chlorpyrifos-methyl ($C_7H_7C_{13}NO_3PS$), 41f, 48t–53t, 54–55
Chronic toxicity, 29–31
Chrysene, 69f
Chrysosporium pannorum, 133–135
Cladosporium, 137
Clarias gariepinus, 45–46
Clostridium perfringens, 335
Coagulation, 206
COD. *See* Chemical oxygen demand (COD)
Collagen, 315
Colocasia, 119
Comamonas testosteroni VP44, 338
Cometabolic biotransformation, 165–166
Cometabolism, 275
Common effluent treatment plants (CETPs), 108–109
    current status of bioreactor application of industrial areas, 115–116
Constructed wetland, 108
    potential for treatment of dye-contaminated effluents, 118–119
Contact mode, 298, 301
Contaminants, 245
    contaminant-glutathione complex, 84
    detoxification, 331–332
Controllable degradation, 327–328
Convention of Biological Diversity (CBD), 345
Conventional *ex situ* methods of phytoremediation, 95–96
Conventional technologies, 25–26, 72
Copper (Cu) toxicity, 165
Copper-containing laccases, 179
Cordyceps sinensis, 182
Coriolus versicolor, 109–110
Corn. *See* Zea mays L. (Corn)
Corynebacterium, 57
    C. glutamicum, 110
Cost-effectiveness of bioremediation, 162
Coumaphos ($C_{14}H_{16}ClO_5PS$), 41f, 48t–53t
Coumarins, 169
CP. *See* Chlorophenol (CP)
Cresols, 204
    production, 205
Crude oil, 127–128
    biodegradation, 217–218
Crustaceans, 317
Cryptococcus sp., 186
Cunninghamella, 128, 131
    C. elegans, 133–135
Cupriavidus, 54–55, 57

Cyanobacteria, 194
  features, 194
  heavy metal and organic pollutant removal from
      species, 197t–198t
  potential in environmental development, 195
  role
    in agriculture management, 194–195
    in bioremediation, 195–197
Cyclodextrins, 86
Cylindrospermum, 196–197
Cytochrome P450 monooxygenases (CYPs), 84,
    338–339
Cytochrome P450s, 179

D
D1101 strain, 55–56
DDD. *See* Dichlorodiphenyldichloroethane (DDD)
DDE. *See* Dichlorodiphenyldichloroethylene (DDE)
DDT. *See* Dichlorodiphenyltrichloroethane (DDT)
Dechlorination reduction approach, 74
Degradation
  bacterial/fungal strains, 48t–53t
  degradable natural polymer materials, 315
  degradable synthetic polymer materials, 315
  of organophosphate pesticides, 46–58
DEHP. *See* Di-(2-ethylhexyl) phthalate (DEHP)
Deinococcus radiodurans, 336
Delftia tsuruhatensis, 10–13
Denaturing gradient gel electrophoresis (DGGE),
    333–334
Desulfobacterium cetonicum, 213
DGGE. *See* Denaturing gradient gel electrophoresis
    (DGGE)
Di-(2-ethylhexyl) phthalate (DEHP), 289
Diabetes, 255
Diaphorobacter sp., 57–58
Diazinon ($C_{12}H_{21}N_2O_3PS$), 34–40, 41f,
    48t–53t, 55–56
Dibenzofurans, 246
Dichlorodiphenyldichloroethane (DDD), 250
Dichlorodiphenyldichloroethylene (DDE), 250
Dichlorodiphenyltrichloroethane (DDT), 25–26, 247,
    249t, 250, 265, 267–268
Dichloromethane (CH_2Cl_2), 75
Dichlorvos ($C_4H_7C_{12}O_4P$), 41f, 48t–53t, 55
Diclofenac, 10–13
Dicrotophos ($C_8H_{16}NO_5P$), 41f
Dieldrin, 249t
Dietzia natronolimnaea, 155
Dihydroxybenzenes, 203–204
Dimethoate ($C_5H_{12}NO_3PS_2$), 41f, 48t–53t, 56
Dioxins, 164–165, 177–178, 246
Dioxygenases, 169

Direct Blue 15 dye, 111–112
Direct degradation, 271
Direct phytoremediation, 85–86
Direct Red 75 dye, 111–112
Direct uptake, 85–86
Dissolved oxygen concentration effect, 218–219
Domestic waste collection bag, 324
Double-chambered MFC, 117
Dust, 256
Dye contaminated water, 6–9
  bioremediation approaches for dye
      degradation, 8–9
    aerobic treatment, 8
    anaerobic treatment, 9
    anoxic treatment, 9
    potential organisms, 8t
    sequential degradation of dyes, 9
Dye(s), 245
  bioremediation process
    advancements in bioreactor systems, 114–115
    application of enzymes as biocatalyst, 112–114
    microorganisms and mechanism, 110–112
  decolorization, 8
  degradation
    bioremediation approaches for, 8–9
    by engineered microbes, 334–336
  dye-containing industrial effluent treatment, 115
  dye-containing wastewater, 108–109
  dye-contaminated effluents, 118–119

E
EC. *See* Electrical conductivity (EC)
EC of saturation extract (ECe), 152
ECHO. *See* Extra-heavy crude oil (ECHO)
EDCs. *See* Endocrine disrupting chemicals (EDCs);
    1-Ethyl-3-(3-dimethylaminopropyl)
    carbodiimide hydrochloride (EDCs)
EEA. *See* European Environment Agency (EEA)
EEC. *See* European Economic Community (EEC)
Effective oxidation processes, 221
Effluent treatment plants (ETPs), 184
Electrical conductivity (EC), 152
Electrochemistry, 75–76
Electrostatic forces, 288–289
Elymus angustus (Wild rye), 271
Endocrine disorder, 254–255
Endocrine disrupting chemicals (EDCs),
    184–185
Endoglucanase, 137
Endophytes, 88
Endosulfan degradation, 74–75
Endrin, 247, 249t, 251
Energy flow, 275

Enterobacter sp., 54–55, 57–58, 117, 153
   E. aerogenes, 154t
   E. cloacae, 57–58, 156
Entomophthora sp., 140
   E. aphidis, 140
   E. thaxteriana, 140
Environmental conditions, 16–17
   oxygen availability, 16–17
   pH, 16
   temperature, 16
Environmental contaminants, 82
Environmental management
   future prospective, 170
   PGPB, 162–163
   xenobiotics
      compounds and classification, 163–164
      effect on health of human beings, 164–165
      effects on plant growth, 165–170
Environmental pollutants cleanup, bioremediation
     for, 3–17
Environmental Protection Agency (EPA), 3–5,
     203–204
Environmentally friendly fibers, 321–322
Environmentally friendly products, 314
Enzymes, 87–88, 135–138, 136t, 177–178
   encapsulation/immobilization, 113
   enzyme-based remediation technique, 108
   participating in phenolic compounds degradation,
     214–215
EPA. *See* Environmental Protection Agency (EPA)
EPS. *See* Extracellular polymeric substances (EPS)
Escherichia coli, 215–216, 291, 304f, 344
   AFM friction image, 304f
   JM109, 305f
   3D AFM image, 306f
   3D height image, 305f
Ethion ($C_9H_{22}O_4P_2S_4$), 34–40, 41f,
     48t–53t, 58
Ethoprophos ($C_8H_{19}O_2PS_2$), 41f, 48t–53t, 56
1-Ethyl-3-(3-dimethylaminopropyl) carbodiimide
     hydrochloride (EDCs), 182
Ethylene bromide (EtBr), 84–85
Ethylene hormone, 156
ETPs. *See* Effluent treatment plants (ETPs)
European Economic Community (EEC), 345
European Environment Agency (EEA), 68
Ex situ bioremediation, 196
Exiguobacterium sp., 47–54
Exoglucanase, 137
Explanta phytoremediation, 86–87
Extra-heavy crude oil (ECHO), 186
Extracellular cellulases, 137
Extracellular oxidoreductases, 179

Extracellular polymeric substances (EPS), 273–274
Extracellular polymers, 166–169
Extremophilic fungi, 186

F
FAO. *See* Food and Agricultural Organization (FAO)
Fatty polyesters, 315
FemtoScan multifunctional SPM, 290–291
Fenamiphos ($C_{13}H_{22}NO_3PS$), 41f, 48t–53t, 57
Fenitrothion ($C_9H_{12}NO_5PS$), 41f, 48t–53t, 57
Fenthion ($C_{10}H_{15}O_3PS_2$), 41f, 48t–53t
Fenton
   method, 75–76
   process, 75
   reagent, 206–207
Film
   card sheets and, 324
   fresh packaging, 324
   industrial, 325
   plastic, 313–314
   for synthetic paper and label, 325
Fish, 250
Fisheries, 325
Flavobacterium sp., 46–54, 56, 87–88
Flavonoids, 169
Fluoranthene degradation, 133–135
Fluorene, 133–135
Foamed products, 325
Fomitopsis palustris, 183–184
Food and Agricultural Organization (FAO),
     151–152
Forest filter effect, 277
Forestry, 325
Forming sheet, 324
Fresh packaging film, 324
Fruit and vegetable preservation bags, 325
Funalia trogii, 186
Fungal/fungi, 74–75, 111–112, 138–139, 177–178
   biodegradation, 213–214
   bioremediation, 127–128, 184–186
      extremophilic fungi, 186
      heavy metal, 183
      marine fungi, 185
      MSW, 183–184
      symbiotic association with plants and
        bacteria, 186
      toxic recalcitrant compound, 182
      white-rot fungi, 184–185
   cell-bound enzymes, 179
   degradation, 74–75
   enzymes for bioremediation,
     178–179, 178f
      extracellular oxidoreductases, 179

factors affecting growth, 139–141
  aeration, 141
  humidity, 140
  light, 140
  pH, 140
  temperature, 140
  trace elements, 141
fungal bioremediation, 182–184
fungal enzymes for bioremediation, 178–179
fungal-mediated PAH metabolism, 131
microbes, 177–178
technology advancement, 187–188
transferases, 179
Fusarium sp., 137, 187, 214
  F. aqueducturn, 140
  F. flociferum, 214
  F. lycopersici, 140
  F. oxysporum, 74–75, 140

G

GA. *See* Glycolic acid (GA)
Gallus gallus (Chick), 45–46
Ganoderma lucidum, 115, 141
Gas chromatography (GC), 43–44, 70
Gelatin, 315
GEMs. *See* Genetically engineered microbes/
    microorganisms (GEMs)
Genetic engineering, 152
Genetically engineered microbes/microorganisms
    (GEMs), 89, 194, 215–216, 332
  agent orange degradation, 337
  approaches to minimizing risks, 340
  bacterial species for bioremediation, 341t–343t
  challenges in bioremediation applications,
      340–344
  construction, 333
  degradation of herbicide, 339
  detection, 333–334
  dye degradation by, 334–336
  factors Influencing, 344–345
  future perspective, 345–346
  GEM–based bioremediation, 14
  genetically modified endophytic bacteria and
      phytoremediation, 339–340
  need, 334
  organic contaminants degradation, 336–337
  organophosphate and carbamate degradation,
      337–338
  PCBs degradation, 338
  polycyclic aromatic hydrocarbon degradation,
      338–339
  regulation, 345

Genetically engineered organisms for
    phytoremediation, 94
Genetically modified bacteria, 215–216
Genetically modified microorganisms (GMOs),
    215–216, 333–334
Geobacter, 117
Geotrichum sp., 140
GFP. *See* Green fluorescent protein (GFP)
GGTs. *See* γ-Glutamyl transpeptidases (GGTs)
Gliomastix indicus, 214
Gluconobacter oxydans, 294, 295f
Glucose, 218
β-Glucosidase, 137
γ-Glutamyl transpeptidases (GGTs), 84
Glycine max (L) CYP71A10, 84–85
Glycolic acid (GA), 319
GMOs. *See* Genetically modified microorganisms
    (GMOs)
Goat. *See Capra hircus* (Goat)
Graphium, 214
Green fluorescent protein (GFP), 333–334
Green Revolution, 25–26
Greenhouse gases, 218–219

H

Haldane kinetics model, 219, 220t
Halogenated aromatic compounds, 193–194
Hapalosiphon, 196–197
Harmful chemicals, 178–179
Hazardous waste, 203–204
HCB. *See* Hexachlorobenzene (HCB)
HCH. *See* Hexachlorocyclohexane (HCH)
Heavy industrialization, 203–204
Heavy metals, 163, 165–169, 245
  bioremediation of heavy metal contaminated
      water, 5–6
Helianthus tuberosus CYP76B1, 84–85
Helicobacter pylori, 306f
Heme peroxidases, 138
Heptachlor, 249t, 251
Herbicide degradation, 339
Hetrotis niloticus, 45–46
Hexabromocyclododecane, 69f
Hexachlorobenzene (HCB), 247, 249t, 252–253
Hexachlorobutadiene, 253
Hexachlorocyclohexane (HCH), 163, 268
Hexahydro-1,3,5-trinitro-1,3,5-triazine (RDX), 93
High molecular weight (HMW), 275
High-fat solubility, 247
High-performance liquid chromatography (HPLC),
    43–44
Highly colored wastewater, 107

HMW. *See* High molecular weight (HMW)
Homo sapiens (Human), 45—46, 177, 250
  pesticide effects on human health, 29—32
  social activities, 313
  xenobiotic effect on health, 164—165
Hordeum vulgare, 278
Horizontal gene transfer, 345
Horseradish peroxidase (HRP), 214—215
HPLC. *See* High-performance liquid
      chromatography (HPLC)
HRP. *See* Horseradish peroxidase (HRP)
HRT. *See* Hydraulic retention time (HRT)
Human. *See* Homo sapiens (Human)
Humidity, 140
HV. *See* Hydroxyvalerate (HV)
Hydraulic retention time (HRT), 187
Hydrocarbon
  contaminated water and soil, 3—5
  utilization, 138—139
Hydrolases, 135—137
  cellulases, 137
  lipases, 137
  proteases, 137
Hydrolysis, 320
Hydrolytic enzymes, 135
Hydrophobic
  interaction, 288—289
  POPs, 247
1-Hydroxy-2-naphthoate dioxygenase (1H2Nase), 277
4-Hydroxybenzoic acid, 211—213
Hydroxyl radicals (•OH), 75—76, 206—207
Hydroxyvalerate (HV), 322

I

IBPs. *See* Industrial by-products (IBPs)
Ibuprofen bacterial degradation, 291
Ideal biodegradable technology for future, 327—328
Image filtering, 303
Immobilization
  of biological catalysts, 113
  techniques, 214—215
Immunotoxic xenobiotics, 164
In situ
  bioremediation, 196
  covering for stabilization, 95—96
  phytoremediation, 95—96
Increased pollutant bioavailability, 272—274
Industrial by-products (IBPs), 253—254
Industrial chemicals, 252—253
  HCB, 252—253
  hexachlorobutadiene, 253
  PCBs, 252
  SCCPs, 253

Industrial film, 325
Industrial sludge, 2
Industrialization practices, 161
Inert components, 25—26
Inorganic ions, 226—227
Inorganic pollutants, 67, 82
Insects, 317
Intracellular fungal enzymes, 179
Irpex lacteus, 74—75, 114—115, 131—133, 186
Isofenphos ($C_{15}H_{24}NO_4PS$), 41f, 48t—53t

K

Klebsiella pneumonia RS-13, 335
Klebsiella sp., 54—55, 156
Kluyvera ascorbata SUD165, 167t—168t

L

LA. *See* Lactic acid (LA)
Lab-based reliable instrumentation, 72
LAC bacteria. *See* Lactic acid bacteria (LAC bacteria)
Lac gene, 115
Laccase, 135, 138
Lactic acid (LA), 319
Lactic acid bacteria (LAC bacteria), 55—56
Lactobacillus brevis, 54—56
Lactobacillus plantarum, 54—56
Lactobacillus sakei, 54—56
Landfarming, 197
LAO. *See* Local anodic oxidation (LAO)
LC. *See* Liquid chromatography (LC)
Lecanicillium muscarium, 186
LEDs. *See* Light emitting diodes (LEDs)
Lentinus sajor-caju, 109—110
Lentinus tigrinus, 184—185
Lethal concentration (LC_{50}), 29—31
Lethal dose (LD_{50}), 29—31
Leuconostoc mesenteroides, 54—55
Light, 140
  intensity, 222—225
Light emitting diodes (LEDs), 222—225
Lignin peroxidase (LiP), 131, 135, 138
Ligninolytic enzymes, 74—75, 138
  heme peroxidases, 138
  laccase, 138
Ligninolytic fungi, 131—133
  PAH compounds degradation by, 132t
LiP. *See* Lignin peroxidase (LiP)
Lipases, 137
Liquid chromatography (LC), 70
Liquid—liquid extraction (LLE), 70—71
Living bacterial cell, 292
LLE. *See* Liquid—liquid extraction (LLE)
LMW. *See* Low molecular weight (LMW)

Local anodic oxidation (LAO), 296–297
Lolium perenne L. (Ryegrass), 271
Low molecular weight (LMW), 275
    carbohydrates, 270
    organic acids, 276–277
Lysinibacillus cresolivorans, 207–210

M

MAE. *See* Microwave-assisted extraction (MAE)
Magnetic nanoparticles (MNPs), 72
Magnusiomyces, 111–112
Malathion ($C_{10}H_{19}O_6PS_2$), 41f, 45–46, 48t–53t
Manganese peroxidase (MnP), 131, 135, 138
Mangifera indica (Mango), 43–44
Marine fungi, 185
Mass spectrometry (MS), 70
Materials, 313
Medicago sativa (Alfalfa), 166–169, 271, 278
Medical biodegradable plastics, 326
Membrane technologies, 206
Mesorhizobium huakuii, 167t–168t
Meta-cresols, 204
Metabolic coordination, 73–74
Metallothioneins, 195
Metarhizium, 140
Methyl mercury, 164
Methyl Orange, 115
Methyl phenols, 205
Methyl-substituted phenols, 203–204
Methylobacterium oryzae, 167t–168t
4-Methylphenol, 205
MFC. *See* Microbial fuel cell (MFC)
Mica, 298–299
Microbacterium arabinogalactanolyticum, 166–169
Microbacterium esteraromaticum, 57
Microbacterium liquefaciens, 166–169
Microbacterium oxydans AY509223, 167t–168t
Microbacterium sp. G16, 167t–168t
Microbe-assisted phytoremediation technology. *See* Rhizoremediation
Microbial fuel cell (MFC), 108, 116–118
    configuration and operation, 117–118
    microorganisms used in, 117
    substrates for, 117t
Microbial/microbes, 2–3, 177–178
    activity, 277
    adaptation, 15
    communities, 153
    cultures, 117t–118t
        for dye removal, 111t

decomposition, 320
degradation, 73–75, 207
    bacterial degradation, 74
    fungal degradation, 74–75
    of xenobiotic compounds, 169–170
diversity of soil, effects on, 33
growth kinetics, 219
microbial-enhanced oil recovery, 138–139
strains, 110
Micrococcus, 57–58
Microorganisms, 112–113
    microorganism-directed oxidation–reduction reactions, 5–6
    used in MFC, 117
Microwave energy (MW energy), 71
Microwave-assisted extraction (MAE), 71
Mineral nutrients, 217
Miniaturization, 71
Minimal salt media (MSM), 57
Mining, 163
Ministry of Environment and Forests (MOEF), 116
Mirex, 249t, 251
Miscanthus, 184–185
Mixed cultures, 13
MnP. *See* Manganese peroxidase (MnP)
MNPs. *See* Magnetic nanoparticles (MNPs)
Modern agricultural practices, 161
Modified biodegradable plastics, 325
MOEF. *See* Ministry of Environment and Forests (MOEF)
Molecule adsorption, 206
Monocrotophos ($C_7H_{14}NO_5P$), 40–43, 41f, 45–46, 48t–53t
Monophenols, 214–215
Mordant Yellow, 9
Morus rubra L. (Mulberry), 86–87, 275
"mpd" gene, 337
MS. *See* Mass spectrometry (MS)
MS/MS. *See* Tandem mass spectrometry (MS/MS)
MSM. *See* Minimal salt media (MSM)
MSM supplemented with nitrogen source (MSMN), 57
MSW. *See* Municipal solid waste (MSW)
Mucor, 137, 185
Mucoromycotina, 182
Mulberry. *See* Morus rubra L. (Mulberry)
Multichannel probes, 297–298
Municipal solid waste (MSW), 183–184
Muscovite, 298–299
MW energy. *See* Microwave energy (MW energy)

Mycelia, 130
Mycobacterium sp., 47–56, 130
Mycoremediation, 127–128
    bioremediation approach, 130
    biosurfactant production by fungi and application in
        bioremediation, 138–139
    enzymes, 135–138, 136t
    factors affecting growth of fungi, 139–141
    intact potential, 130–135
    PAHs
        compounds degradation by ligninolytic
            fungi, 132t
        compounds degradation by nonligninolytic
            fungi, 134t
        effect exposure on environment and human
            health, 128–129
        environmental concern, 128
        properties, 129t
Myrica rubra (Bayberry), 43–44

N

NADPH. *See* Nicotinamide adenine dinucleotide
        phosphate (NADPH)
NAH plasmid, 336–337
Nanolithography, 296–298
Nanometer space resolution, 287
Naphthalene, 211–213
    degradation pathway of, 132f
Naphthylamine sulfonic acids, 110–111
National Environmental Methods Index, 70
Natural contaminants, 331–332
Natural microbial community, 345
Natural polymer raw materials production, 315–317
Neosartorya, 214
Neutralization of free radicals, 153–155
Nickel (Ni), 165
Nicotiana tabaccum L., 85
Nicotinamide adenine dinucleotide phosphate
        (NADPH), 74–75
Nitroaromatic compounds, 274
Nitrogen (N), 217
Nitrosomonas europaea, 10–13
Nocardia, 87–88
Nonligninolytic fungi, 131, 133–135
    enzymes, 136t
    PAH compounds degradation by, 134t
Nonresonant mode, 298
Nonsystematic pesticides, 27
Nontarget organisms, impact on, 32
Nontoxic compounds, 205
Nostoc, 196–197
Novel biodegradable materials and applications, 327
Nuclear wastes, 177

Nutrient, 162–163
    availability, 15
    availability, 217
    flow, 275

O

Obesity, 255
Ochrobactrum sp., 55, 167t–168t
    O. anthropi, 290
    O. intermedium, 167t–168t
OCT plasmid, 336–337
Oil shipping, 1–2
OPH enzymes. *See* Organophosphorus hydrolase
        enzymes (OPH enzymes)
OPP. *See* Organophosphate pesticide (OPP)
OPs. *See* Organic pollutants (OPs)
Oreochromis niloticus, 45–46
Organic acid production, 274
Organic compounds, 179
Organic contaminants, 82
    degradation, 336–337
    in soil, 264–267
        plant–microbe interaction during
            biodegradation, 273f
        rhizoremediation mechanism, 271–275
Organic matter, 14
Organic pollutants (OPs), 67, 70, 169, 184–185, 210
    in soil
        concentration, 269t
        fate, 267–268
        sources, 264f, 266t–267t
    strategies of phytoremediation, 85–87
Organic solvents, 70–71
Organochlorinated chemicals, 164–165
Organochlorine pesticides, 27
Organophosphate pesticide (OPP), 27–29
Organophosphates, 27–29, 41f
    degradation, 337–338
        of organophosphate pesticides, 46–58
    general formula and chemical structures, 30f
    samples, 37t–40t
    status of organophosphate pesticide pollution,
        34–46
    toxicological mechanism, 33–34
Organophosphorus hydrolase enzymes
        (OPH enzymes), 337
Ortho-cresols, 204
Oryza sativa L., 84–85
Oscillatoria, 195
Oscillatory resonant mode, 298
Oxidoreductases, 214–215
Oximes, 33–34
Oxygen availability, 16–17

P

PAA. *See* Polyaspartic acid (PAA)

Paecilomyces sp., 140

PAHs. *See* Polycyclic aromatic hydrocarbons (PAHs)

Para-cresols, 204, 207

Paracoccus sp., 56–58, 213

Parathion ($C_{10}H_{14}NO_5PS$), 41f, 48t–53t

Parathion methyl, 36–40

Parathion-methyl ($C_8H_{10}NO_5PS$), 41f, 48t–53t, 58

PBAT. *See* Polyterephthalic acid/adipic acid/butanediol (PBAT)

PBBs. *See* Polybrominated diphenys (PBBs)

PBDEs. *See* Polybrominated diphenyl ethers (PBDEs)

PBS. *See* Polybutylene succinate (PBS)

PCBs. *See* Polychlorinated biphenyls (PCBs)

PCDDs. *See* Polychlorinated dibenzo-p-dioxins (PCDDs)

PCDFs. *See* Polychlorinated dibenzofurans (PCDFs)

PCL. *See* Polycaprolactone (PCL)

PCLA. *See* Poly(lactic acid-epsilon-caprolactone) (PCLA)

PCP. *See* Pentachlorophenol (PCP)

PEG. *See* Polyethylene glycol (PEG)

Penicillium, 128, 131, 137, 185
   P. citrinum, 47–54
   P. oxalicum strain, 111–112

Pentachlorophenol (PCP), 163

Perchlorobenzene. *See* Hexachlorobenzene (HCB)

Perflorooctane sulfonate (PFOS), 69f, 245

Perflorooctanoic acid (PFOA), 69f

Performance-based methods, 70

Peroxidases, 179, 214–215

Persistent organic pollutants (POPs), 68, 84–85, 164–165, 185, 245–246
   analytical techniques for quantification, 72
      SERS, 72
      UV-Vis spectroscopy, 72
   characteristics, 246
   classifications, 248f
   environmental effects, 256
   health effects, 254–256, 254f
   and environmental chemistry, 68
   IBPs, 253–254
   industrial chemicals, 252–253
   method of analysis, 70–72
      conventional techniques, 72
      samples collection, extraction, storage, and preparation, 70–72
   methods for degradation, 73–76
      advanced oxidation approaches, 75–76
      biological approach, 73–75
      chemical approaches, 75

pesticides, 247–251, 249t
   and related compounds, 69f
   remediation, 256–257
   sources, 246–247, 246f

Persistent organic pollutants (POPs), 265–266

Pestalotiopsis palmarum, 186

Pesticides, 25–26, 28t, 163–165, 247–251, 249t
   aldrin, 249t, 250
   bioremediation of pesticide contamination, 9–10
   chlordane, 250–251
   cycling, 248f
   DDT, 250
   effect, 29–33
      effects on microbial diversity of soil, 33
      environmental impact, 32
      on human health, 29–32
      impact on nontarget organisms, 32
   endrin, 251
   environmental fate and degradation, 47f
   exposure modes and metabolic routes, 31f
   heptachlor, 251
   mirex, 251
   resistance, 33

PET. *See* Polyethylene terephthalate (PET)

Petrochemical refineries pollutants, 203–204

PFE. *See* Pressurized fluid extraction (PFE)

PFOA. *See* Perflorooctanoic acid (PFOA)

PFOS. *See* Perflorooctane sulfonate (PFOS)

PGA. *See* Polyglycolic acid (PGA)

PGPB. *See* Plant growth–promoting bacteria (PGPB)

PGPR. *See* Plant growth–promoting rhizobacteria (PGPR)

pH, 16, 140
   effect on rhizoremediation, 276

PHA. *See* Polyhydroxyalkanoate (PHA)

Phanerochaete chrysosporium, 74–75, 109–112, 131–133, 141, 178–179, 184–185

Phanerochaete sordida, 182

Pharmaceutical and personal care products (PPCPs), 10, 184–185
   removal by biological degradation processes, 10–13
      activated sludge process, 13
      microbes, 11t–12t
      mixed cultures, 13
      pure cultures, 10–13

PHB. *See* Polyhydroxybutyrate (PHB)

PHBV. *See* Poly(3-hydroxybutyrate-co-3-hydroxyvalerate) (PHBV)

Phenanthrene, 211–213

Phenol, 207

Phenolic compounds, 207—210
  comprehensive list of bacteria employed to
      degrading, 208t—209t
  comprehensive list of fungi employed to
      degrading, 209t
Phenolic resin, 313
Phenolic waste degradation
  adverse effects of phenols and cresols on
      environment and human health, 205
  aerobic degradation, 210—213, 211f—212f
  cresol production, 205
  factors affecting photocatalytic degradation of
      TiO_2, 222—228
  factors influencing bioremediation of phenolic waste,
      216—219
  limitations of biodegradation, 219—221
  photocatalytic degradation, 221—222
  treatment technologies for phenolic compound
      removal, 205—216
    biological method, 207—216
    chemical method, 206—207
    physical method, 206
Phenols, 203—204, 245
Pholiota adiposa, 183—184
Phorate ($C_7H_{17}O_2PS_3$), 36—40, 41f, 48t—53t
Phosphamidon ($C_{10}H_{19}ClNO_5P$), 41f, 45—46
Phosphate (P), 217
Phosphoramidates, 27—29
Phosphorothioates, 27—29
Phosphorus, carbon, and oxygen bonds
      (P—O—C bonds), 27—29
Photo catalyst, 221—222
Photocatalysis, 221
Photocatalytic degradation, 221—222
  photo catalyst and description, 221—222
  TiO_2 mechanism in photocatalytic degradation of
      phenolic compounds, 222
Photorespiration, 153
Phragmites, 118—119
Phragmites australis, 118—119, 278
Physicochemical methods, 25—26
Physostigma venenosum, 27
Phytodegradation/rhizodegradation, 82
Phytodesalination, 82
Phytoextraction, 82, 170
Phytophthora nicotianae, 140
Phytoremediation, 2—3, 81—82, 169—170, 339—340.
      See also Bioremediation
  advantages and limitations, 95—96
  challenges to, 96—97
  enzymes, 87—88
  fate and transport of organic contaminants in, 93—94
  genetically engineered organisms for, 94

physiological and biochemical aspects, 83—85
plant-associated microflora, 88—93, 90t—92t
process, 82—83
research and development in, 94—95
  biotechnological approaches, 94—95
  current status, 94
  protein engineering, 95
strategies of phytoremediation of organic
      pollutants, 85—87
  direct uptake, 85—86
  explanta phytoremediation, 86—87
Phytostabilization, 82, 170
Phytotransformation, 170
Phytovolatilization, 82, 170
Pichia pastoris, 115
Piezoelectric ceramic resonator, 293
Pisum sativum, 93
PLA. See Polylactic acid (PLA)
Plant growth—promoting bacteria (PGPB), 152,
      162—163, 162f
  in bioremediation, 165—166, 167t—168t
  mechanism of xenobiotics degradation, 166—169
Plant growth—promoting microbes/microorganism,
      153, 169
Plant growth—promoting rhizobacteria
      (PGPR), 153, 165—166
  and ACC deaminase activity, 156
  in salinity stress, 153—155
Plants
  plant P450, 84—85
  plant-associated microflora, 88—93, 90t—92t
  species, 277
  symbiotic association of fungi with, 186
Plastic films, 313—314
Plastic products burning, 313—314
Plectonema, 196—197
Plesiomonas, 58
Pleurotus ostreatus, 74—75, 115, 131—133, 184—186
Pleurotus sp., 184—185
PLGA, 319
P—O—C bonds. See Phosphorus, carbon, and oxygen
      bonds (P—O—C bonds)
Pollutants, 67, 161—162, 177
  concentration of, 15
  bioavailability of, 278
Poly(3-hydroxybutyrate-co-3-hydroxyvalerate)
      (PHBV), 322—323
Poly(lactic acid-epsilon-caprolactone) (PCLA), 319
Polyamino acids, 315
Polyaspartic acid (PAA), 323
Polybrominated diphenyl ethers (PBDEs), 245
Polybrominated diphenys (PBBs), 69f
Polybutylene succinate (PBS), 315, 317—318

Polycaprolactone (PCL), 315, 318
Polychlorinated biphenyls (PCBs), 85, 163–165, 245, 252, 264–265, 338
Polychlorinated dibenzo-p-dioxins (PCDDs), 182
Polychlorinated dibenzofurans (PCDFs), 165, 182
Polycholorobiphenyl, 247
Polycyclic aromatic hydrocarbons (PAHs), 68, 69f, 85, 127–128, 163, 178–179, 182, 193–194, 211–214, 264–265
  degradation, 338–339
  mycoremediation, 128
Polyether, 323
Polyethylene glycol (PEG), 323
Polyethylene terephthalate (PET), 319
Polyglycolic acid (PGA), 315, 321
Polyhydroxyalkanoate (PHA), 315, 322
Polyhydroxybutyrate (PHB), 315, 322–323
Polyhydroxybutyrate-co-valerate. *See* Poly (3-hydroxybutyrate-co-3-hydroxyvalerate) (PHBV)
Polylactic acid (PLA), 315, 318–322
  sheets, 325
Polymer materials synthesis degradable by microorganisms, 315
Polypropylene glycol (PPG), 323
Polysaccharides, 315
Polyterephthalic acid/adipic acid/butanediol (PBAT), 315, 317–318
Polytetramethylene ether glycol (PTMG), 323
Polyvinyl alcohol (PVA), 315, 318
Polyvinyl chloride (PVC), 252
POPs. *See* Persistent organic pollutants (POPs)
Populus spp., 83–84
Potassium (K), 217
Power lithography, 297
PPC, 323
PPCPs. *See* Pharmaceutical and personal care products (PPCPs)
PPG. *See* Polypropylene glycol (PPG)
Pressurized fluid extraction (PFE), 71
Profenofos (C₁₁H₁₅BrClO₃PS), 36–40, 41f, 48t–53t
Proteases, 137
Protein engineering, 95
Proteobacteria, 213
Proteus sp., 57–58
  P. mirabilis, 9
  P. vulgaris, 55, 57–58
Providencia rettgeri, 110–111
Pseudomonads, 215
Pseudomonas, 54–58, 87–88, 153, 154t, 166–170, 167t–168t, 217–218, 275, 340–344
  culture, 110–111
  I-24, 10–13

P. aeruginosa, 8, 10–13, 47–55, 156, 215, 218–219
  MKRh3, 167t–168t
P. entomophila, 110–111
P. frederiksbergensis, 155
P. luteola, 9, 335
P. mendocina, 154t
P. migulae, 154t
P. nitroreducens, 54–55
P. pseudoalcaligenes, 155
P. rhizophila S211, 169
P. stutzeri, 54–55, 156
P. syringae, 154t
  RJ10, 167t–168t
Pseudomonas fluorescens, 54–55, 154t, 156, 166–169, 167t–168t, 215
  G10, 167t–168t
  HK44, 215–216
Pseudomonas putida, 15, 47–55, 57, 154t, 156, 207–210, 218, 332, 344
  KNP9, 167t–168t
  KT2442, 290
  PaW85, 215–216
  PML2, 166
PTMG. *See* Polytetramethylene ether glycol (PTMG)
Putisolvins, 273–274
PVA. *See* Polyvinyl alcohol (PVA)
PVC. *See* Polyvinyl chloride (PVC)
Pycnoporus sanguineus, 115
Pyrene, 131–133
  degradation pathway of, 133f
Pyrethroid pesticides, 27

Q
Quantitative analysis, 303
Quick easy cheap effective rugged safe method (QuEChERS method), 71
Quinalphos (C₁₂H₁₅N₂O₃PS), 41f, 48t–53t

R
Radiation-resistant bacterium, 336
Radionuclides, 163
Ralstonia eutropha, 57–58
Raoultella sp., 56
Reaction temperature, 225
Reactive Blue 171, 118–119
Reactive oxygen species (ROS), 153–155
Reasonable modification, 314
Remediation technology, 25–26
REMI. *See* Restriction enzyme-mediated integration (REMI)
Reproductive problems, 255
Resonant mode, 301–302
Restriction enzyme-mediated integration (REMI), 55

Reusability, potential of biocatalysts for, 113–114
Rhamnolipids, 169
RHDase. *See* Ring hydroxylating dioxygenase (RHDase)
Rhizobacteria, 166–169
Rhizobium sp., 57–58, 154t
   R. leguminosarum CPMex46, 166–169
   RP5, 167t–168t
Rhizodegradation. *See* Rhizoremediation
Rhizofiltration, 170
Rhizophagus custos, 186
Rhizopogon roseolus, 186
Rhizopus, 137
   R. arrhizus, 109–112
   R. oryzae, 111–112
   R. stolonifer, 140
Rhizoremediation, 2–3, 162, 264–265, 270–275
   affecting factors, 276–278
      bioavailability of pollutants, 278
      environmental factors, 276–277
      microbial activity, 277
      plant species, 277
   interaction of plants with microbes, 272t
   of organic contaminants, 271–275
      direct degradation, 271
      energy and nutrient flow, 275
      increased pollutant bioavailability, 272–274
      plant–microbe interaction, 273f
      structural analogy and cometabolism, 275
   potential, challenges, and future perspectives, 278
Rhizosphere, 270, 278
   bioremediation. *See* Rhizoremediation
   of plant, 153
   rhizosphere-associated microbes, 162
   rhizospheric microorganisms, 162–163
Rhodococcus sp., 47–54, 216–217, 340–344
   R. globerulus P6, 338
   R. pyridinivorans XB, 289
Rhodopseudomonas palustris, 213
Ribosomal intergenic spacer analysis (RISA), 333–334
Ring hydroxylating dioxygenase (RHDase), 277
RISA. *See* Ribosomal intergenic spacer analysis (RISA)
Root exudation, 270
ROS. *See* Reactive oxygen species (ROS)
Ryegrass. *See* *Lolium perenne* L. (Ryegrass)

S

Saline soil, 152
Saline-tolerant cyanobacteria, 196–197
Salinity stress, PGPR in, 153–155
   functional aspects of PGPR under salt stress, 153–155
   PGPB mediated salt tolerance, 154t

Saprobic fungi, 130
SBP. *See* Soybean peroxidase (SBP)
Scanning capillary
   microscopy, 297–298
   stereolithography, 298
Scanning ion-conductance microscopy (SICM), 293–296
Scanning probe microscopy (SPM), 287–288
   atomic force microscopy
      friction image of *E. coli*, 304f
      imaging in air in contact mode, 301
      imaging in air in resonant mod, 301–302
      imaging in liquids, 302
      installation, 300
   in bioremediation investigations, 289–291
   image filtering, 303
   measurements of bacteria, 298–300
   quantitative analysis, 303
   3D image processing, 302–303
Scanning tunneling microscope (STM), 287
SCCPs. *See* Short-chain chlorinated paraffins (SCCPs)
Scytonema, 196–197
Sediments, 267–268
Semivolatile organic compounds (SVOCs), 67–68
Sepedonium sp., 140
Sepedonizcm sp., 141
Serratia ficaria, 57–58
Serratia liquefaciens, 55–56
Serratia marcescens, 54–56, 337–338
Serratia sp., 54–55, 57–58
SERS. *See* Surface-enhanced Raman scattering (SERS)
Sewage treatment plants (STPs), 2
SFE. *See* Supercritical fluid extraction (SFE)
Shewanella, 117
Short-chain chlorinated paraffins (SCCPs), 253
Shrinkage packaging–related fields, 325
SICM. *See* Scanning ion-conductance microscopy (SICM)
Silk fibroin, 315
Single-chambered MFC, 118
SOD. *See* Superoxide dismutase (SOD)
Soil, 67, 70–72
   amendments, 86
   matrix, 70
   organic contaminants in, 264–267
   organic matter effect on rhizoremediation, 276
   pollution, 263
   soil-inhabiting bacteria, 89
Solanum tuberosum L., 84–85
Solid wastes, vermicomposting of, 14
Solid-phase extraction, 71
Solid-phase extraction-gas chromatography-mass spectrometry (SPE-GC-MS), 36–40

Soxhlet method, 70–71
Soybean peroxidase (SBP), 214–215
SPE-GC-MS. *See* Solid-phase extraction-gas
  chromatography-mass spectrometry
  (SPE-GC-MS)
Sphingobacterium sp. JAS3 strain, 54–55
Sphingomonas sp., 47–57, 187
  S. macrogoltabidus, 166–169
SPM. *See* Scanning probe microscopy (SPM)
Stachybotrys, 137
Staphylococcus warneri, 290
Starch, 316
  biodegradable plastics, 325
Stenotrophomonas, 10–13, 54–55
  S. malthophilia, 57–58
STM. *See* Scanning tunneling microscope (STM)
STPs. *See* Sewage treatment plants (STPs)
Streptomyces, 154t
  MIUG, 10–13
  S. chattanoogensis, 54–55
  S. olivochromogene, 54–55
Stress factors, 151–152
Substrate preparation
  for imaging of bacteria in air, 298–299
  for imaging of bacteria in liquids, 299
Suillus bovinus, 186
Supercritical fluid extraction (SFE), 71
Supercritical fluids, 71
Superoxide dismutase (SOD), 153–155
Surface-enhanced Raman scattering (SERS), 72
Surfactants, 278
Survival efficiency and stability of GEMs under
  natural ecosystem, 344
Sustainability, 2
Sustainable agriculture, 194–195
SVOCs. *See* Semivolatile organic compounds
  (SVOCs)
Syncephalastrum racemosum, 133–135
Synechococcus sp., 195
Synthetic biodegradable materials production,
  317–323
Synthetic contaminants, 331–332
Systematic pesticides, 27

T
Talaromyces, 214
Tandem mass spectrometry
  (MS/MS), 43–44
TAP-1, 57–58
TCDD. *See* 2,3,7,8-Tetrachlorodibenzo-*p*-dioxin
  (TCDD)
TCE. *See* Trichloroethylene (TCE)
TCP. *See* 3,5,6-Trichloro-2-pyridinol (TCP)

Temperature
  control in bioremediation, 16
  effect on rhizoremediation, 276
  in phenol degradation process, 216–217
  role in fungal growth, 140
Terbufos ($C_9H_{21}O_2PS_3$), 41f
3,3′,4,4′-Tetrachlorobiphenyl, 69f
2,3,7,8-Tetrachlorodibenzo-p-dioxin (TCDD), 69f, 182
2,3,7,8-Tetrachlorodibenzofuran, 69f
Tetrachlorvinphos ($C_{10}H_9Cl_4O_4P$), 41f, 48t–53t
Thauera aromatica, 213
Thermoplastic starch, 315
Three-dimensional image processing, 302–303
Tilapia zilli, 45–46
Titanium dioxide (TiO_2), 221–222
  factors affecting photocatalytic degradation of,
    222–228, 223t–224t
  catalyst loading, 225–226
  inorganic ions, 226–227
  light intensity, 222–225
  pH of solution, 226
  reaction temperature, 225
  mechanism in photocatalytic degradation of phenolic
    compounds, 222
TNT. *See* Trinitrotoluene (TNT)
Toc GTPase, 156
Toxaphene, 247, 249t
Toxic
  metals, 163
  organic compound disposal, 203–204
  recalcitrant compound, 182
Toxic Substances Control Act (TSCA), 345
Toxicological mechanism of organophosphates,
  33–34
Trace elements, 141
Traditional caring approaches, 73
Trametes sp., 115
  T. versicolor, 74–75, 109–110, 114–115, 184–185
Transferases, 179
  enzymes producing fungal taxa and mechanism of
    action, 180t–181t
Transgenic plants, 82
Transparent box, 325
Transparent window envelope, 324
Transpiration stream concentration factor (TSCF),
  272–273
Triazophos ($C_{12}H_{16}N_3O_3PS$), 41f, 48t–53t
3,5,6-Trichloro-2-pyridinol (TCP), 54–55
Trichloroethylene (TCE), 84–85, 270
Trichoderma, 137, 140
  T. atroviride, 55
  T. harzianum, 47–54, 74–75, 185
  T. longibrachiatum, 187

Trichoderma viride, 183—184
  Pers NFCCI-2745, 185
  QM9414, 137
Trichosporon sp., 54—55
  T. cutaneurn, 140
Trinitrotoluene (TNT), 86—87, 131, 177—178, 182
Triphenylene, 69f
Triticum aestivum, 278
TSCA. *See* Toxic Substances Control Act (TSCA)
TSCF. *See* Transpiration stream concentration
  factor (TSCF)
Typha, 119
Tyrosinase, 179
Tyrosine, 205

U

Ultrahigh-performance liquid chromatography
  system-mass spectrometer (UPLC-MS/MS),
  45—46
Ultrasonic-assisted extraction (UAE), 70—71
Ultrasound (US), 70—71
Ultraviolet-visible spectroscopy (UV-Vis
  spectroscopy), 72
Unicellular cyanobacterium, 195
United Nation Environmental Protection
  organization, 70
United Nations Environment Program (UNEP),
  164, 345

V

van der Waals attraction force, 288—289
Vermi-biofiltration of wastewater, 9
Vermicomposting of solid wastes, 14
Versatile peroxidase (VP), 135, 137
Vertical stream pilot-scale constructed wetlands,
  118—119
Vibrio sp., 55
  V. metschinkouii, 57—58

Volatile fatty acids (VFAs), 183—184
Volatile organic compounds (VOCs), 67—68
Volatilization process, 13

W

Wastewaters, 107
  vermi-biofiltration of, 9
  wastewater-activated sludge, 196
Water, 67, 70—72, 203—204
  water-soluble phenolic compounds, 203—204
Westiellopsis, 196—197
Wet air oxidation, 206—207
White-rot fungi, 111—112, 184—185
Wild rye. *See Elymus angustus* (Wild rye)
Wildlife cancer, 256
World Health Organization (WHO), 203—204

X

Xanthomonas sp., RJ3, 167t—168t
Xenobiotic(s), 331—332, 334
  compounds, 161
    and classification, 163—164
    microbial degradation, 169—170
  effects on human beings health, 164—165
  effects on plant growth, 165—170
    PGPB in bioremediation, 165—166
    PGPB mechanism of degradation, 166—169
  organic compounds, 215—216
Xenophilus azovorans KF46, 335
XYL plasmid, 336—337
Xylem stream, 89

Y

Yersinia enterocolitica, 57—58

Z

Zea mays L. (Corn), 165, 271
Zizania aquatica, 119

Printed in the United States
By Bookmasters